Space Sciences Series of ISSI

Volume 58

More information about this series at http://www.springer.com/series/6592

Anny Cazenave · Nicolas Champollion
Frank Paul · Jérôme Benveniste
Editors

Integrative Study of the Mean Sea Level and Its Components

Previously published in *Surveys in Geophysics*, Volume 38, Issue 1, 2017

Springer

Editors
Anny Cazenave
ISSI International Space Science Institute
Bern
Switzerland

and

LEGOS-CNES
Toulouse
France

Nicolas Champollion
International Space Science Institute
Bern
Switzerland

Frank Paul
Department of Geography
University of Zurich—Irchel
Zürich
Switzerland

Jérôme Benveniste
ESA-ESRIN
Frascati
Italy

ISSN 1385-7525
Space Sciences Series of ISSI
ISBN 978-3-319-56489-0 ISBN 978-3-319-56490-6 (eBook)
DOI 10.1007/978-3-319-56490-6

Library of Congress Control Number: 2017936039

Cover illustration: Iceberg wall in Greenland. © anderspa / Getty Images / iStock.

Printed on acid-free paper

This Springer imprint is published by Springer Nature
The registered company is Springer International Publishing AG
The registered company address is: Gewerbestrasse 11, 6330 Cham, Switzerland

Contents

Surv Geophys (2017) 38:1–5
DOI 10.1007/s10712-016-9402-2

PREFACE

International Space Science Institute (ISSI) Workshop on Integrative Study of the Mean Sea Level and its Components

A. Cazenave[1,2] · N. Champollion[1] · J. Benveniste[3] · P. Lecomte[4]

Because it integrates changes in several components of the climate system in response to external forcing factors and internal climate variability, sea level is one of the best indicators of global climate change. The Earth is currently in a state of thermal imbalance because of anthropogenic greenhouse gas emissions. 93% of this heat excess is accumulated in the ocean, the remaining 7% being used to warm the atmosphere and continents and melt sea and land ice (von Schuckmann et al. 2016). Global mean seal level (GMSL) rise is a direct consequence of this process. Tide gauges' records indicate that the GMSL has been rising since the beginning of the twentieth century at a mean rate in the range 1.2–1.9 mm $year^{-1}$ (e.g., Church and White 2011; Jevrejeva et al. 2014; Hay et al. 2015). Since the early 1990s, sea level variations are routinely measured by high-precision altimeter satellites. These indicate that, in terms of global mean, sea level is rising at a rate of 3.2 ± 0.4 mm $year^{-1}$ since 1993, i.e., twice as fast as during the previous decades, suggesting an acceleration of the phenomenon (e.g., Cazenave et al. 2014; Ablain et al. 2015). Satellite altimetry has also allowed mapping of the regional variability in sea level evolution and shown that superimposed on the global mean, strong regional changes affect a number of regions like the western tropical Pacific, the northern Atlantic and the southern ocean. In these regions, the rate of sea level rise can be up to three times larger than the global mean (Stammer et al. 2013).

The physical processes causing global sea level rise and regional changes are not identical, although they are related (Church et al. 2013). Global sea level rise primarily results from ocean thermal expansion, land ice melt and land water storage change, while

✉ A. Cazenave
anny.cazenave@issibern.ch

1 ISSI, Hallerstrasse 6, 3012 Bern, Switzerland

2 LEGOS, CNES, 18 Avenue Edouard Belin, 31401 Toulouse Cedex 9, France

3 Earth Observation Science, Applications and Future Technologies Department, ESA-ESRIN, Via Galileo Galilei, Casella Postale 64, 00044 Frascati (Roma), Italy

4 ESA Climate Office, Earth Observation Science, Applications and Future Technologies Department, ECSAT, Fermi Avenue, Harwell Campus, Didcot, Oxfordshire OX110FD, UK

spatial patterns of sea level change result from the superimposition of 'fingerprints' caused by different processes: changes in sea water density—due to changes in temperature and salinity—and changes in ocean circulation (these phenomena are called 'steric' or 'dynamic' effects), solid Earth's deformation and gravitational changes in response to mass redistribution caused by land ice melt and land water storage changes. The land ice melt-related static factor has two components: one associated with the last deglaciation—called glacial isostatic adjustment (GIA), and the other due to the ongoing land ice melt. These phenomena—mostly known from modelling—give rise to complex regional patterns of sea level change (Spada et al. 2012): sea level drop in the immediate vicinity of the melting bodies but sea level rise in the far field (e.g., along the north-east coast of America and in the tropics). At present, the dominant contribution to observed regional sea level changes comes from non-uniform thermal expansion and salinity variations (Stammer et al. 2013). Other effects, in particular the static factors, today contribute little to the regional variability but will become important in the future (Tamisiea and Mitrovica 2011; Milne et al. 2009).

Whatever the future trajectory of greenhouse gas (GHG) emissions, sea level will continue to rise in the future. And, as for the present, future sea level rise will not be uniform (Church et al. 2013). While the largest regional changes come from ocean thermal expansion and salinity changes, factors linked to solid Earth's deformations and water/ice mass redistributions will become increasingly important in the future, causing amplification up to 20–30% of the global mean rate in several regions, in particular in the tropics (Church et al. 2013).

Today, about 10% of the world population lives in coastal zones. In many regions, populations are exposed to a variety of natural hazards (extreme weather such as storm surges and cyclones) as well as to consequences of global climate change (e.g., sea level rise) and human activities (e.g., urbanization). In low-lying coastal areas, these factors may combine negatively and amplify risk and vulnerability of coastal populations. For example, climate-related sea level rise increases the risk of flooding and coastal erosion during extreme events and causes salt water intrusion into rivers and coastal aquifers. Ground subsidence caused by groundwater extraction in coastal megacities is another factor of risk, as it amplifies the negative impacts of climate-related sea level rise (Nicholls and Cazenave 2010). Because of strong anthropogenic pressure, coastal zones also suffer ecological and biological stresses, for example poor water quality, pollution, destruction of marine ecosystems. Shoreline change and coastal flooding are critical concerns for many coasts worldwide. They are expected to be strongly aggravated due to future sea level rise (Cazenave and Le Cozannet 2014).

Precisely monitoring global mean and regional sea level variations as well as their components is crucial to understand processes at work under current climate change, to detect temporal changes in one or several components, and to validate the climate models used for future projections. In recent years, the global climate observing system (GCOS), a body of the United Nations Framework Convention on Climate Change (UNFCCC), has defined a set of 50 essential climate variables (ECVs) that should be accurately monitored in the long term to improve our understanding of the changing climate. Among these ECVs, 26 are observable from space (GCOS 2011). As a contribution to GCOS, the European Space Agency (ESA) initiated, as of 2010, a programme known as the climate change initiative (CCI). The objective of the CCI is to take advantage of the full potential of the long-term global Earth observation archive that ESA and other national agencies have established over the last 30 years in order to produce consistent and continuous space-based records for a first series of these ECVs. Sea level, as well as glaciers and ice sheets mass balance, was included in the set of 13 ECVs reprocessed during the first phase of the ESA CCI programme. These climate data records developed within the ESA CCI programme present a timely and unique opportunity to investigate in a coherent way the closure of the sea level budget (accounting also for the ocean thermal expansion, another important component of the GMSL), thus allowing us to assess the quality of these variables, and finally to constrain remaining unknown or poorly known components (e.g., deep ocean heat uptake, snow and permafrost melting).

The scientific papers presented in this special issue represent the outcome of a workshop entitled 'Integrative study of the mean sea level and its components' held in February 2015 at the International Space Science Institute (ISSI) in Bern, Switzerland (http://www.issibern.ch/workshops/sealevelbudget/). The objective of the workshop was to discuss in an integrative context the recent results obtained by the ESA CCI programme for sea level, glaciers and ice sheets ECVs. Improvement in these ECVs allows better understanding of the mass contribution to sea level rise, and hence to the global mean sea level budget. The workshop has also addressed the regional variability in sea level, sea surface temperature and ocean colour and discussed the relative contributions of the natural/internal climate variability and anthropogenic forcing (detection/attribution) associated with spatial trend patterns.

The main themes covered by the workshop were:

1. Observations of global mean sea level and its components.
2. Sea level processes at regional scale.
3. Closure of the global mean sea level.
4. Detection and attribution of sea level changes.
5. From observations to sea level projections.

The first two articles of section 1 deal with the use of satellite altimetry to measure sea level at global, regional and local scales. The paper by Ablain et al. describes recent efforts realized to improve the sea level record during the altimetry era (starting in 1993 with the launch of the high-precision Topex/Poseidon altimetry mission), with a focus on the new sea level products computed in the context of the ESA CCI sea level project. Satellite altimetry has been developed to precisely measure sea level in open oceans, about 20–30 km from the coast. Close to the coast, classical altimetry data are contaminated by signals from the land and inaccurate geophysical corrections. Cipollini et al. present recent progress in improving altimetry-based sea level in coastal zones and how these data compare with tide gauge-based sea level at the coast.

The next four papers of section 1 deal with the main contributors to present-day global mean sea level changes:

- Steric sea level changes due to ocean temperature and salinity variations (MacIntosh et al.)
- Greenland and Antarctica ice sheet mass balance (Forsberg et al.)
- Glaciers mass changes (Marzeion et al.)
- Land water storage changes (Wada)

These articles summarize the current state of knowledge based on different types of space-based and in situ observations (plus modelling results for glacier melting) and their contribution to present-day sea level rise.

The last paper of section 1 (Spada) provides an overview of the glacial isostatic adjustment phenomenon (the response of the visco-elastic solid Earth to the last deglaciation) and its impact on sea level measurements.

In section 2, papers by Meyssignac et al. and Han et al. discuss regional patterns in sea level revealed by satellite altimetry and their link with natural internal modes of variability of the ocean–atmosphere coupled system (e.g., El Nino Southern Oscillation/ENSO, Pacific Decadal Oscillation/PDO). Meyssignac et al. show that the main cause of regional sea level changes comes from steric effects in response to changes in ocean circulation. They also address regional changes in other ocean variables (sea surface temperature and ocean colour—an indicator of chlorophyll concentration in marine ecosystems), while Han et al. point out the possible signature of anthropogenic forcing in some regional sea level variations.

Carret et al. focus on the altimetry-based sea level, and steric and mass components in the Arctic Ocean and highlight the difficulty of measuring sea level in this region partly covered by sea ice. Finally, Racault et al. explain how the ENSO phenomenon impacts phytoplankton concentration (derived from space-based ocean colour measurements) via its response to sea surface temperature, surface winds and sea level.

The two papers of section 3 present current estimates of the global mean sea level and assess its budget during the twentieth century (Jevrejeva et al.) and the altimetry era (Chambers et al.). Remaining uncertainties preventing full closure of the sea level budget are presented and discussed.

Section 4 includes a single paper that discusses the detection and attribution issue, i.e., the detection of anthropogenic signals versus internal climate variability (Marcos et al.). It shows that the global mean sea level rise is mainly caused by anthropogenic global warming, especially since the early 1970s.

The last section deals with ocean circulation and climate modelling. Scharffenberg et al. show how assimilation of observations, in particular altimetry-based sea level data, significantly improves general ocean circulation models. Slangen et al. present the most recent developments in projecting future global and regional sea level changes accounting for all processes at work, in particular at regional scale. The remaining uncertainties are also discussed.

To summarize, in addition to the collection of papers presented above, another outcome of the workshop was the setting-up by ESA of a new project dedicated to the closure of the sea level budget using primarily the products developed within the CCI. Another important outcome is the proposal to the UNFCCC to include the sea level ECV among the key indicators of global climate change with the goal of routine monitoring with regular updates. Finally, this workshop contributed to uniting the international research community working on sea level and its impacts and to opening up new possibilities of fruitful collaborations.

References

Ablain M, Cazenave A, Larnicol G, Balmaseda et al (2015) Improved sea level record over the satellite altimetry era (1993–2010) from the Climate Change Initiative project. Ocean Sci 11:67–82

Cazenave A, Le Cozannet G (2014) Sea level rise and coastal impacts. Earth's Future 2(2):15–34. doi:10.1002/2013EF000188

Cazenave A, Dieng H, Meyssignac B, von Schuckmann K, Decharme B, Berthier E (2014) The rate of sea level rise. Nat Clim Change. doi:10.1038/NCLIMATE2159

Church JA, White NJ (2011) Sea-level rise from the late nineteenth to the early twenty-first century. Surv Geophys 32(4–5):585–602

Church JA, Clark PU, Cazenave A, Gregory JM, Jevrejeva S, Levermann A, Merrifield MA, Milne GA, Nerem RS, Nunn PD, Payne AJ, Pfeffer WT, Stammer D, Unnikrishnan AS (2013) Sea level change. In: Climate change 2013: The Physical Science Basis. Contribution of Working Group I to the Fifth Assessment Report of the Intergovernmental Panel on Climate Change. [Stocker TF, Qin D, Plattner GK, Tignor M, Allen SK, Boschung J, Nauels A, Xia Y, Bex V, Midgley PM (eds)]. Cambridge University Press, Cambridge, United Kingdom and New York, NY, USA

GCOS (2011) Systematic observation requirements for satellite-based data products for climate (2011 Update)—Supplemental details to the satellite-based component of the "implementation plan for the global observing system for climate in support of the UNFCCC (2010 Update)". GCOS-154 (WMO, December 2011)

Hay CC et al (2015) Probabilistic reanalysis of twentieth-century sea level rise. Nature 517(7535):481

Jevrejeva S, Moore JC, Grinsted A, Matthews P, Spada G (2014) Trends and acceleration in global and regional sea levels since 1807. Glob Planet Change 113:11–22

Milne GA et al (2009) Identifying the causes of sea-level change. Nat Geosci 2:471–478

Nicholls RJ, Cazenave A (2010) Sea level change and the impacts in coastal zones. Science 328:1517–1520. doi:10.1126/science.1185782

Spada G et al (2012) The gravitationally consistent sea level fingerprint of future terrestrial ice loss. Geophys Res Lett 1–5:40

Stammer D, Cazenave A, Ponte RM, Tamisiea ME (2013) Causes for contemporary regional sea level changes. Annu Rev Mar Sci. doi:10.1146/annurev-marine-121211-172406

Tamisiea ME, Mitrovica JX (2011) The moving boundaries of sea level change: understanding the origins of geographic variability. Oceanography 24(2):24–39

Von Schuckmann K, Palmer MD, Trenberth KE, Cazenave A, Chambers D, Champollion N, Hansen J, Josey SA, Loeb N, Mathieu PP, Meyssignac B, Wild M (2016) An imperative to monitor Earth's energy imbalance. Nat Clim Change 6:138–144. doi:10.1038/nclimate2876

Part I
Observations & Contributors to Sea Level

Surv Geophys (2017) 38:7–31
DOI 10.1007/s10712-016-9389-8

Satellite Altimetry-Based Sea Level at Global and Regional Scales

M. Ablain[1] · J. F. Legeais[1] · P. Prandi[1] · M. Marcos[2] · L. Fenoglio-Marc[3,4] · H. B. Dieng[5] · J. Benveniste[6] · A. Cazenave[5,7]

Received: 19 May 2016 / Accepted: 8 October 2016 / Published online: 16 November 2016

Abstract Since the beginning of the 1990s, sea level is routinely measured using high-precision satellite altimetry. Over the past ~25 years, several groups worldwide involved in processing the satellite altimetry data regularly provide updates of sea level time series at global and regional scales. Here we present an ongoing effort supported by the European Space Agency (ESA) Climate Change Initiative Programme for improving the altimetry-based sea level products. Two main objectives characterize this enterprise: (1) to make use of ESA missions (ERS-1 and 2 and Envisat) in addition to the so-called 'reference' missions like TOPEX/Poseidon and the Jason series in the computation of the sea level time series, and (2) to improve all processing steps in order to meet the Global Climate Observing System (GCOS) accuracy requirements defined for a set of 50 Essential Climate Variables, sea level being one of them. We show that improved geophysical corrections, dedicated processing algorithms, reduction of instrumental bias and drifts, and careful linkage between missions led to improved sea level products. Regarding the long-term trend, the new global mean sea level record accuracy now approaches the GCOS requirements (of ~0.3 mm/year). Regional trend uncertainty has been reduced by a factor of ~2, but orbital and wet tropospheric corrections errors still prevent fully reaching the GCOS accuracy requirement.

This paper is an outcome of the workshop on "Integrative Study of Sea Level Budget", International Space Science Institute Workshop, Bern, 2–6 February 2015.

✉ A. Cazenave
anny.cazenave@issibern.ch

1 CLS, Toulouse, France

2 IMEDEA (UIB-CSIC), Esporles, Spain

3 Technical University Darmstadt, Darmstadt, Germany

4 University of Bonn, Bonn, Germany

5 LEGOS, Toulouse, France

6 ESA-ESRIN, Frascati, Italy

7 ISSI, Bern, Switzerland

Similarly at the interannual time scale, the global mean sea level still displays 2–4 mm errors that are not yet fully understood. The recent launch of new altimetry missions (Sentinel-3, Jason-3) and the inclusion of data from currently flying missions (e.g., CryoSat, SARAL/AltiKa) may provide further improvements to this important climate record.

Keywords Satellite altimetry · Sea level · Climate Change Initiative

1 Introduction

Sea level is one of the key indicators of climate change because it integrates changes of several components of the climate system in response to anthropogenic forcing as well as natural forcing factors related to natural sources and internal climate variability. Since the beginning of the twentieth century, the global mean sea level (GMSL) has been rising at a mean rate of 1.7 ± 0.3 mm/year as recorded by in situ tide gauges (e.g., Church et al. 2011, 2013; Woppelmann et al. 2009; Jevrejeva et al. 2008). However, values in the range 1.2 to 1.9 mm/year have also been proposed (Hay et al. 2015; Jevrejeva et al. 2014). Since the early 1990s, sea level variations are routinely measured by high-precision satellite altimetry. In terms of global mean, sea level rise over 1993–2014 amounts to 3.4 ± 0.4 mm/year (e.g., Nerem et al. 2010; Cazenave et al. 2014; Ablain et al. 2015). This value is two times larger than that of the previous decades, suggesting an acceleration of the GMSL rise. Present-day GMSL rise primarily reflects ocean warming (through thermal expansion of sea water) and land ice melting, two processes which result from anthropogenic global warming (Church et al. 2013). The Earth is currently in a state of thermal imbalance because of concentrations of anthropogenic greenhouse gases (GHG) in the atmosphere (Von Schukmann et al. 2016). Most of this heat excess is accumulated in the ocean (93 %); the remaining 7 % being used to warm the atmosphere and continents, and melt sea and land ice. GMSL rise is a direct consequence of these processes. Over the course of its five assessments, the Intergovernmental Panel on Climate Change (IPCC) has reported a significant improvement in our understanding of the sources and impacts of GMSL rise. Over the altimetry era, observed sea level rise and sum of contributions (ocean thermal expansion, land ice melt, land water storage change) concur, allowing the closure of the sea level budget for this period within estimated uncertainties (Church et al. 2013). Confidence in projections of future sea level rise has increased, thanks to improved physical understanding and closer agreement between model hinscasts and observations. However, significant problems still remain. The IPCC 5th assessment report—AR5—(Church et al. 2013) reported a 0.4 mm/year difference between the observed GMSL rate and the sum of contributions over the 1993–2010 time span. Yet uncertainties of components of the sea level budget equation (including sea level) are still large, in the order of 1 mm/year (2-sigma) (Church et al. 2013). The challenge is thus to reduce the components' errors, in order to check the statistical significance of the difference between observed sea level and sum of contributions. The satellite altimetry-based sea level record is affected by errors due to the imperfect altimeter corrections applied to the data (with the orbit solution and the wet tropospheric correction displaying the largest uncertainties), geographical averaging process and imperfect linkage between successive altimetry missions. In terms of long-term (decadal) trends, such factors contribute to the 0.4 mm/year difference quoted above (Ablain et al. 2009, 2015). At the interannual time scale, errors in the GMSL record are also significant and amount to 2–4 mm (Ablain et al. 2009; Dieng et al. 2015a, b). Secondly, as far as the contributions are concerned, current estimates of ice sheet and glacier mass balances also display significant uncertainty (Church et al. 2013; Clark et al. 2015). Another issue concerns the land water contribution due to human activities (e.g., ground water depletion and dam building), its quantification being very difficult due to lack of global data (Church et al. 2013; Dieng et al. 2015c). Finally, although the steric contribution (effects of ocean thermal expansion and salinity) was considerably improved since the advent of the Argo project in the early 2000s. But the contribution of the deep ocean (below 2000 m) remains unknown, and prior to Argo, the steric component is quite uncertain due to the poor and heterogeneous distribution of historical hydrographic observations.

A precise estimation of the influence of these factors is crucial to understand processes at work under current climate change and to validate the climate models used for future projections. Over the last decade, the Global Climate Observing System (GCOS), in support of the United Nations Framework Convention on Climate Change (UNFCCC), has put together a set of requirements for satellite data to meet the needs of the climate change community (see GCOS 2011, for the satellite supplement). These requirements are broken down into key parameters of the Earth system, called Essential Climate Variables (ECVs). The goal is to provide accurate and stable values on the long-term, satellite-based ECV data products for researchers. Among the 50 ECVs identified so far by GCOS, 26 are observable from space. Sea level is one of these.

To respond to this need for climate-quality satellite data, the European Space Agency (ESA) has set up a programme, known as the ESA Climate Change Initiative (CCI). The aim of the programme is to realize the full potential of the long-term global Earth Observation archives from satellites as a significant and timely contribution to the ECV databases required by the UNFCCC. The ECVs are derived from multiple satellite data sets via international collaboration and include specific information on the possible biases and uncertainties of the data set. The CCI provides a unique opportunity to set up dialogue and cooperation between Earth observation and climate research communities.

In this overview article, we focus on the sea level record computed in the context of the ESA CCI project. Contributions and sea level budget issues are discussed in other papers in this Special Issue. Section 2 summarizes the high-precision satellite altimetry missions and their characteristics. Section 3 briefly presents the CCI sea level (SL_cci) project. In Sects. 4 and 5, we discuss how multi-mission altimetry-based sea level products are built and what the current level of uncertainties of the global and regional products are. Validation procedures are discussed in Sect. 6. Section 7 provides a summary of the main accomplishments. Conclusions are presented in Sect. 8.

2 Brief History of Satellite Altimetry Missions

Satellite altimetry has revolutionized the research in ocean dynamics by providing high-precision, high-resolution measurements of the ocean surface topography with global coverage and a revisit time of a few days or weeks. The concept of (nadir) satellite altimetry measurement is rather straightforward. The onboard radar altimeter transmits a short pulse of microwave radiation with known power towards the nadir. Part of the incident radiation reflects back to the altimeter. Measurement of the round-trip travel time provides the height of the satellite above the instantaneous sea surface (called altimeter range R). The quantity of interest in oceanography is the height of the instantaneous sea surface above a fixed reference surface (typically a conventional reference ellipsoid). This quantity (called SSH) is simply the difference between the height H of the satellite above the reference ellipsoid and the altimeter range R: SSH = H-R. H is computed through precise orbit determination, a long-tested approach in space geodesy, which combines accurate modelling of the dynamics of the satellite motion and tracking measurements (Global Positioning System-GPS, Doppler Orbitography and Radiopositioning Integrated by Satellite—DORIS, or Satellite Laser Ranging) between the satellite and observing stations on Earth or on other observing satellites (Rudenko et al. 2012; Couhert et al. 2015). The range from the satellite to the sea, R, must be corrected for various components of the atmospheric refraction as well as for biases between the mean electromagnetic scattering surface and mean sea surface at the air–sea interface in the footprint of the radar. Other corrections due to a number of geophysical effects must also be applied. Chelton et al. (2001) describe the principle of satellite radar altimetry and details of the estimation of the SSH. They also discuss all corrections to be applied to the SSH measurements, including drifts and bias from onboard instruments.

Satellite altimetry was envisaged in the 1960s, was recognized as a high priority measurement at the Williamstown Symposium in 1969 (Kaula 1970), and the first objective was to measure the shape of the Earth. The development was pursued during the 1970s with an experiment onboard Skylab III, which in 1973 produced the first measurements of undulations in the marine geoid. GEOS3 (NASA) was the very first altimetry mission, launched in 1975 and providing data until 1979 (Agreen 1982). It was followed by Seasat (1978; NASA) and Geosat (1985; US Navy). These pioneering missions led to important discoveries (e.g., Lillibridge et al. 2006). They revealed, in particular, that the mean sea surface is not flat but mimics the oceanfloor topography (Fig. 1). In effect, the sea surface consists of two parts: (1) a static (i.e. time invariable) component that coincides with the geoid, an equipotential surface of the Earth's gravity field, and (2) a time-variable component due to ocean dynamics (e.g., ocean tides, currents, waves). At short and medium wavelengths ($\sim$<1000 km), the mean sea surface reflects the topographic features of the ocean floor.

The amplitude of the static component anomalies ranges from a few decimetres to several tens of meters. This explains why the first altimetry missions easily detected these features in spite of their lesser SSH measurement accuracy (uncertainty of several

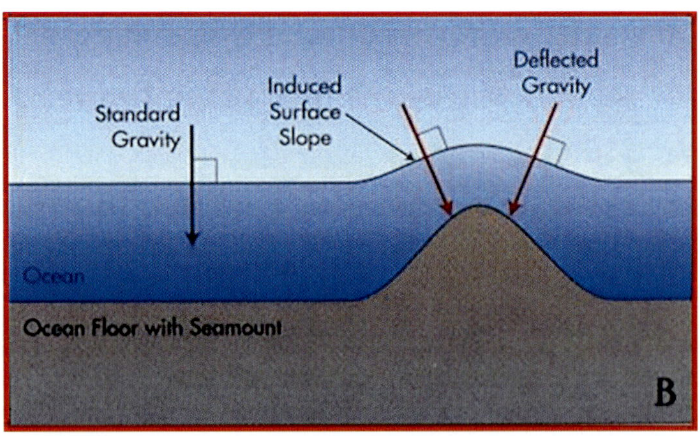

Fig. 1 Sketch of the mean sea surface deformations in response to the gravitational attraction of the sea floor topography

decimetres, mostly due to the orbit error, e.g., Fu and Cazenave 2001) and high instrumental noise.

Nevertheless, these early missions clearly demonstrated the high potential of satellite altimetry to study the dynamics of the world ocean.

The launch of the ERS-1 and TOPEX/Poseidon satellites in 1991 and 1992 opened the era of high-precision altimetry, allowing mapping of the SSH for the first time within a few centimetre accuracy for a single measurement. TOPEX/Poseidon was particularly precise at ocean basin scale and was used as the reference mission while ERS-1 provided the chance to explore the mesoscale variability. They have been followed by several other high-precision altimetry missions with different instrumental characteristics leading to an ever-increasing precision in the SSH measurement. Jason-1, -2, and -3 continued on the tracks of TOPEX/Poseidon, to be succeeded by Sentinel-6/Jason-CS in 2020. Meanwhile ERS-2, Envisat, CryoSat, SARAL/AltiKa and Sentinel-3 supplied and are supplying complementary observations. CryoSat was designed to reveal changing ice fields, but turned out to be an excellent oceanographic mission (Labroue et al. 2012). We now have at our disposal a 25-year-long multi-mission altimetry data set of very high value for studying ocean circulation (because of geostrophy, SSH measurements can be translated in terms of ocean circulation), ocean dynamics and sea level variations.

3 The ESA Climate Change Initiative and the Sea Level ECV

As noted in the introduction, sea level has been identified as a key marine ECV within the CCI programme. Indeed, precise monitoring of changes in the mean level of the oceans is crucial for understanding not just the climate, but also the socio-economic consequences of any rise in sea level.

The sea level project conducted in the CCI programme (SL_cci) gathers a consortium of 15 European partners including experts on altimeter standards as well as a climate research group dedicated to the quality assessment of the products. The first phase of the project (2011–2013) was the opportunity to involve the climate research community and define the user requirements for climate applications. The estimation of the SSH requires not only the knowledge of the altimeter range, but also the instrumental corrections, the satellite orbit and different geophysical corrections to the altimeter range (tides, troposphere and ionosphere corrections, sea state bias, dynamic atmospheric correction; see Table 1) that have to be selected for each altimeter mission. Note that in the following, terms sea level, SSH and SL_cci ECV are used interchangeably.

From the perspective of the production of a sea level ECV, evolutions of these altimeter standards and algorithms were central to the project since they affect the physical content of the SL_cci ECV. The strategy was thus to focus on the improvement of the altimeter corrections which constitute the most important sources of errors at climate scales. Following this strategy, new altimeter algorithms have been developed and tested for all altimeter missions within the phase 1 of the project. Other algorithms from external projects have also been included in the process. A formal validation protocol has been developed (Round Robin approach) for the estimation and the validation of their performances. The evaluation of these standards has been performed, distinguishing different spatial (global and regional) and temporal (long-term, interannual and seasonal) scales. All validation reports are available at www.sea-level-cci.org. A panel of international experts contributed to the selection of the best algorithms for climate

Table 1 Altimeter standards selected for the sea level calculation for the SL_cci products (release 1.1): the choice of these corrections can change with time or from a project to another

Corrections	ERS-1	ERS-2	Envisat	Jason-1	Jason-2	T/P	GFO
Orbit	Reaper combined orbit (Rudenko et al. 2015)		CNES POE-D (Couhert et al. 2015)			GSFC POE (09/2008)	GSFC POE
Sea state bias	BM3 (Gaspar and Ogor 1994)	Nonparametric SSB (Mertz et al. 2005)	Nonparametric SSB V2.1 release	Nonparametric SSB (Tran et al. 2012)	Nonparametric SSB GDR-D release	Nonparametric SSB (Tran et al. 2010)	Nonparametric SSB (Tran et al. 2010)
Ionosphere	NIC09	BENT + GIM	From dual frequency altimeter range measurements (cycles 1–64) and GIM afterwards	From dual frequency altimeter range measurements	From dual frequency altimeter range measurements	From dual frequency altimeter range measurements (TOPEX) and DORIS (Poseidon)	GIM model
Wet troposphere	GPD corrections (Fernandes et al. 2015)						
Dry troposphere	ERA-interim based (Carrere et al. 2016)						
Dynamical atmospheric corrections	ERA-interim based (Carrere et al. 2016)						
Ocean tide	GOT 4.8 (Ray 2013)						
Mean sea surface	DTU 2010 (Andersen 2010)						
Pole tide	(Wahr 1985)						
Solid earth tide	Elastic response to tidal potential (Cartwright and Tayler 1971; Cartwright and Edden 1973)						
Loading tide	GOT4v8 (Ray 2013)						

applications. This has led to a level 2 altimeter database representing more than 50 years of cumulated data from 7 altimeter missions (TOPEX/Poseidon, Jason-1 and 2, ERS-1 and 2, Envisat, and GFO).

The SL_cci ECV consists of monthly maps of sea level anomalies (SLA), and the multi-mission mapping technique used to produce these maps (optimal interpolation) has been optimized for climate scales. The first version of the SL_cci ECV was disseminated in 2012, and the time series has benefited from regular temporal extensions so that the SL_cci v1.1 ECV covers the period 1993–2014 (DOI:10.5270/esa-sea_level_cci-1993_2014-v_1.1-201512). In addition to the monthly SLA maps, the ECV products also include ocean indicators computed over the total period. This includes the temporal evolution of the GMSL and its trend, regional mean sea level trends, and amplitude and phase of the annual signal. The products are available upon request at info-sealevel@esa-sealevel-cci.org, and the Product User Guide can be found on the project website: www.esa-sealevel-cci.org. A full description of the SL_cci v1.1 ECV is provided in Ablain et al. (2015).

The following section provides a detailed discussion of the sea level processing performed in the SL_cci project.

4 The Sea Level Record from High-Precision Satellite Altimetry Missions

This section reviews the altimeter standards and gridding processes used to build the sea level record. One important output of the CCI Programme is the error characteristics of the ECVs, which are described in this section.

4.1 Geophysical Corrections Applied to the SSH Measurements

In this section, we describe the geophysical corrections that are applied to the SSH measurements (hereinafter called 'altimeter standards'). The processing to provide the mean sea level record depends on the altimeter standards selected to derive the sea level from 1-Hz altimeter measurements, and on the gridding process applied to average the along-track measurements and calculate the GMSL time series. Before describing further this processing, it is worth noting that there are some processing differences between the different groups producing GMSL records. The impact of these differences has been described and quantified in several studies (Masters et al. 2012; Henry et al. 2014). The GMSL trend can be modified by few sub-millimetres per year (0.1–0.2 mm/year) due to these differences.

As briefly described in the previous section, corrections need to be applied to the SSH measurement: propagation corrections as the altimeter radar wave is delayed during atmosphere travel (ionospheric correction, wet tropospheric correction, dry tropospheric correction), ocean surface correction for the sea state which directly affects the radar wave (electromagnetic bias), geophysical corrections for the tides (ocean, solid Earth and polar tides as well as loading effects), and atmospheric corrections for the ocean response to atmospheric dynamics (inverse barometer correction for low frequency, atmospheric dynamics correction for high frequency). Furthermore, SSH is calculated for each altimetric measurement considered as valid according to criteria (e.g., threshold, spline, statistics on the ground track) applied either to the main altimetric parameters, the geophysical corrections or the SSH directly. These criteria may vary from one mission to another depending on the altimeter instrumental characteristics. The precise references for the corrections and orbits used when calculating the mean sea level are given in Table 1 for the SL_cci project. Most of these corrections are not contained in the altimeter level-2 products (e.g., TOPEX M-GDR, Jason-1/Jason-2 GDR). They have been calculated and updated in a multi-mission altimetry database in order to calculate homogenous sea level for all altimetry missions.

4.2 Gridding Process

The recommended method by the SL_cci project in order to produce mean sea level grids has been developed for the SSALTO DUACS (Segment Sol Multimission Altimetrie et Orbitographie, Data Unification and Altimeter Combination System) System (Dibarboure et al. 2011). The main advantages of this method are, first of all, use of TOPEX/Poseidon, Jason-1, Jason-2 and Jason-3 as reference missions in order to obtain the most accurate long-term stability (Ablain, et al. 2009) and use of all other complementary missions (ERS-1, ERS-2, Envisat, Geosat Follow-On, CryoSat, SARAL/AltiKa and Sentinel-3A/B) to increase the spatial resolution of mean sea level grids.

The gridding process is composed of the following steps:

1. Calculation of the along-track sea level over all the altimeter period (1993–2014) for all the altimeter missions with homogenized corrections (as listed in Table 1), after removing spurious data (e.g., impacted by rain cells, sea ice).
2. Calculation of the mean sea level biases between the reference missions, both at global and regional scales. The verification phases between two consecutive missions (i.e. satellites on the same ground track apart from each other by few seconds; e.g., TOPEX/Jason-1, Jason-1/Jason-2, Jason-2/Jason-3) allow estimates of global biases with an accuracy close to 0.5–1 mm in terms of mean sea level (Zawadzki and Ablain 2016). It is worth noting that the absolute GMSL bias is arbitrarily set to 0 at 1993.
3. Reduction of the orbit errors between all the missions through a global minimization of the crossover differences observed within the reference mission and between reference and complementary missions (Dibarboure et al. 2011).
4. Computation of SSH grids (with a spatial resolution of 0.25° using a rectangular projection and a temporal resolution of 1 month) combining data from all missions using an objective analysis approach (Ducet et al. 2000; Le Traon et al. 2003).

The GMSL time series (Fig. 2) is easily deduced from the sea level grids by a area weighting averaging (taking into account the box area dependence with latitude) over the oceanic domain observed by the altimetry data (82°S to 82°N).

Other research teams (University of Colorado, AVISO, CSIRO, NOAA, NASA) only use the reference missions (TOPEX/Poseidon, Jason-1 and Jason-2) to provide GMSL time series. Their method is more simple since steps (3) and (4) described above are replaced by a simple averaging on a cycle basis of each mission (e.g., 1° along the latitudinal axis, 3° along the longitudinal axis for AVISO). Then all the time series are linked together during the verification phases (TOPEX/Jason-1 and Jason-1/Jason-2). The main advantage of this approach is the reduction of the computing time (fewer altimetry missions and no multi-mission adjustments). On the other hand, the GMSL is only estimated between 66°S and 66°N, and the regional sea level variations are not as well-represented as in the SL_cci. Furthermore, errors in altimetry measurements, such as long wavelength orbit errors or oceanic tide errors, are not removed and can impact the mean sea level estimate up to 1–2 mm at each cycle. However, the differences between these gridding process approaches, which each have their own limitations, only slightly impact

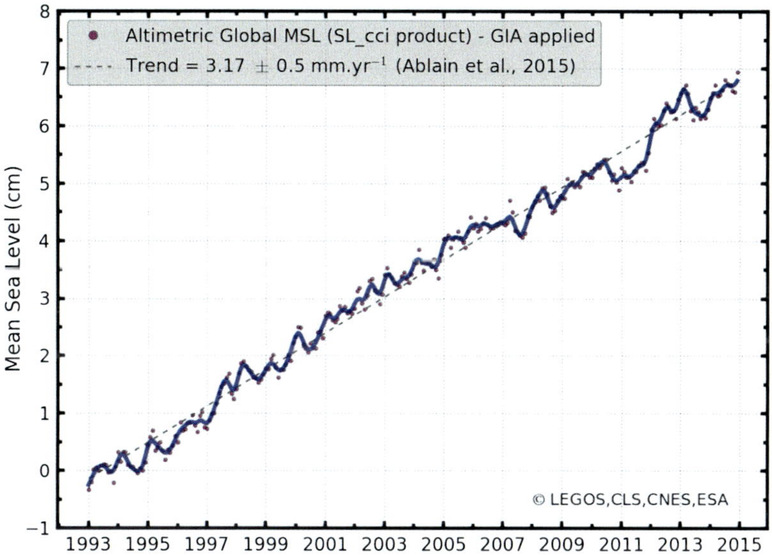

Fig. 2 Global mean sea level evolution over the period 1993–2014 from SL_cci project (DOI:10.5270/esa-sea_level_cci-1993_2014-v_1.1-201512). Annual and semi-annual signals have been removed from the monthly estimates (*red dots*), and a 6-month filter has been applied to produce the *blue curve*

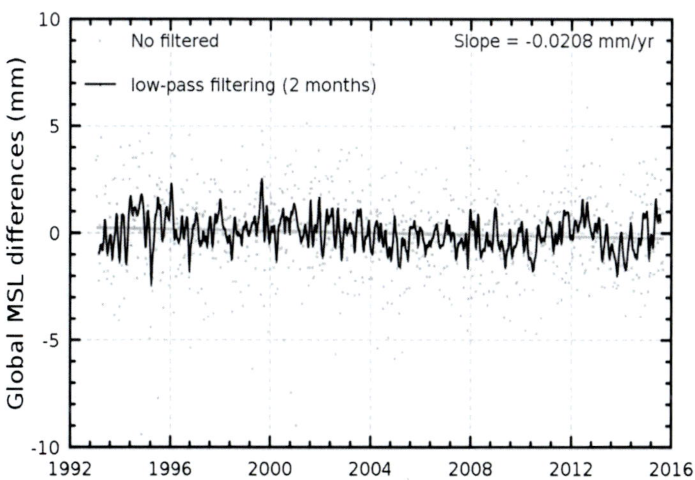

Fig. 3 Global mean sea level differences comparing the method applied to SL_cci products (based on SSALTO/DUACS system) and to AVISO global mean sea level time series. Same altimeter standards are used in both cases

the GMSL trend or the interannual signals (Fig. 3). The differences are lower than 0.05 mm/year over the whole altimetry period for the trend and reach 1–2 mm over shorter periods between 1 and 3 years (Henry et al. 2014).

4.3 Global Mean Sea Level Rise Characteristics

GMSL rises between 3.2 and 3.4 mm/year over the 1993–2014 period, according to the different groups (SL_cci project, AVISO, University of Colorado, NASA, NOAA, CSIRO). Although the global evolution is nearly linear over the period (the linear error adjustment provided by the least squares method is about 0.02 mm/year), interannual variations are also observed. Removing the trend from GMSL time series highlights these variations over a 1-year to 3-year period (Fig. 4). Their magnitudes depend on the period (+3 mm in 1998–1999, -5 mm in 2011–2012, and +10 mm in 2015–2016) and are well-correlated with El Niño Southern Oscillation (ENSO) events. In Fig. 4, the Multivariate ENSO Index (MEI) has been shown to better represent this temporal correlation.

4.4 Global Mean Sea Level Uncertainties

GMSL data contain remaining errors at different time scales. In the SL_cci project, an error budget dedicated to the main temporal scales (i.e. long term—5–10 years or more, inter-annual—<5 years—and seasonal) has been established (see Table 2). Regarding the GMSL trend, an uncertainty of 0.5 mm/year was estimated over the whole altimetry era (1993–2015) within a confidence interval of 95 % (2-sigma). The main source of error is the radiometer wet tropospheric correction with a drift uncertainty in the range of 0.2–0.3 mm/year (Legeais et al. 2014). To a lesser extent, the orbit error (Couhert et al. 2015) and the altimeter parameters (range, sigma-0, significant wave height) instabilities (Ablain et al. 2012) add additional uncertainty, of the order of 0.1 mm/year. It is worth noting that for these two corrections, the uncertainties are higher in the first altimetry decade (1993–2002) when TOPEX/Poseidon, ERS-1 and ERS-2 measurements display larger errors. Furthermore, imperfect links between TOPEX-A and TOPEX-B (February 1999), TOPEX-B and Jason-1 (April 2003), Jason-1 and Jason-2 (October 2008) lead to errors of 2, 1 and 0.5 mm, respectively (Ablain et al. 2009; Zawadzki and Ablain 2016). They cause a GMSL trend uncertainty of about 0.1 mm/year over the 1993–2014 period. It is relevant to

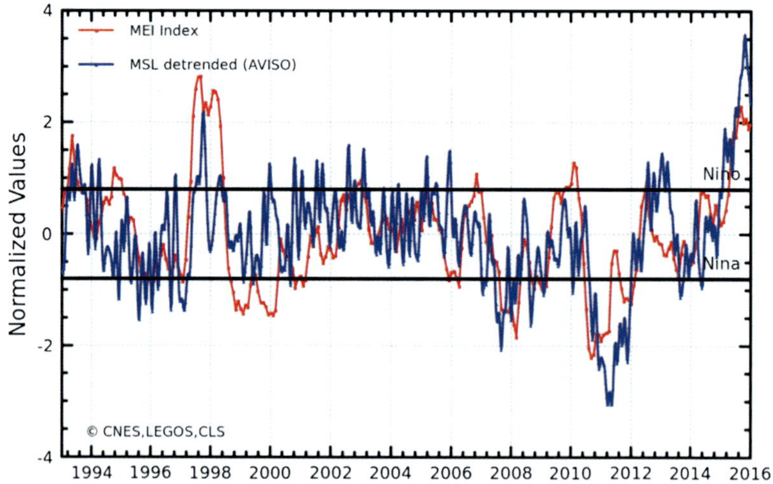

Fig. 4 Comparison of the Multivariate ENSO Index (MEI) and the global mean sea level time series (from AVISO) after removing the global mean trend

Table 2 Mean sea level error budget for the main climate scales (Ablain et al. 2015)

Spatial scales	Temporal scales	Altimetry errors	User requirements
GMSL	Long-term evolution (>10 years)	<0.5 mm/year	0.3 mm/year
	Interannual signals (<5 years)	<2 mm over 1 year	0.5 mm over 1 year
	Annual signals	<1 mm	Not defined
Regional MSL	Long-term evolution (>10 years)	<3 mm/year	1 mm/year
	Annual signals	<1 cm	Not defined

note that the remaining uncertainty of ~0.5 mm/year on the GMSL trend remains 0.2 mm/year higher than the GCOS requirements (of 0.3 mm/year, see GCOS 2011).

All sources of errors described above and the gridding process, already described in Sect. 4.2, also have an impact at interannual time scale (<5 years). The level of error is still 1.5 mm higher than the GCOS requirement of 0.5 mm. This may have consequences on the sea level closure budget studies at interannual time scale. For the annual signal, the amplitude error is estimated lower than 1 mm. Knowing that the annual amplitude of the GMSL is in the order of 9 mm, this error can be considered low.

5 Regional Sea Level

5.1 Spatial Trend Patterns in Sea Level

The regional mean sea level trends (Fig. 5) are directly deduced from the gridded mean sea level time series. As mentioned above, the gridding process applied in the SL_cci project

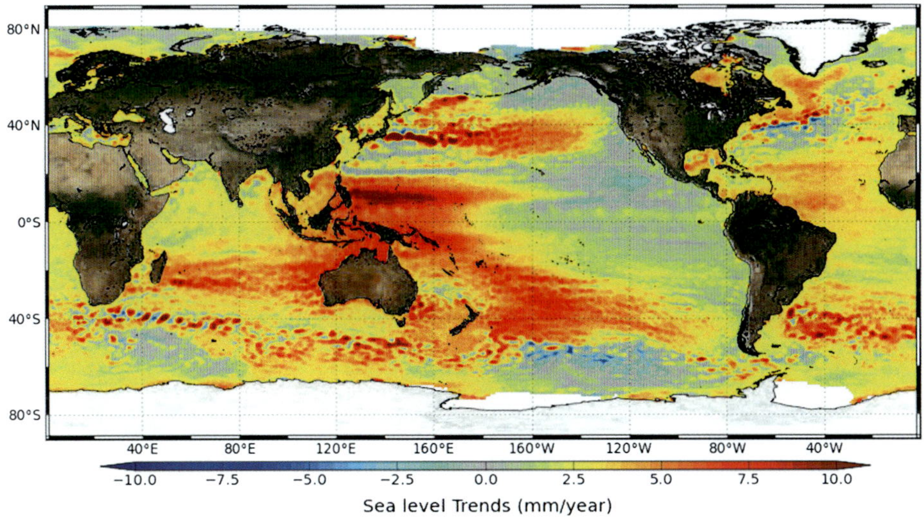

Fig. 5 Regional mean sea level trends over the 1993–2014 period from SL_cci products (release 1.1)

(derived from the SSALTO/DUACS system) provides a spatial resolution of 0.25° between 82°S to 82°N. The results discussed below only apply to the SL_cci products.

5.2 Uncertainties at Regional Scale

At regional scale, trend uncertainty is of the order of 2–3 mm/year (see below). Although the orbit error has been significantly reduced for this spatial scale during the last few years, it remains the main source of uncertainty (in the range of 1–2 mm/year; Couhert et al. 2015) with large spatial patterns at hemispheric scale. The Earth gravity field model errors explain an important part of these uncertainties (Rudenko et al. 2014). Furthermore, errors are higher in the first decade (1993–2002) for which the Earth gravity field models are less accurate due to the unavailability of the Gravity Recovery And Climate Experiment (GRACE) data. Additional errors are still observed, e.g., for the radiometer-based wet tropospheric correction in tropical areas, other atmospheric corrections in high latitudes, and high frequency corrections in coastal areas. The combined errors give rise to an uncertainty of 0.5–1.5 mm/year. Finally, the 2–3 mm/year uncertainty on regional sea level trends remains a significant error compared to the 1 mm/year GCOS requirement, even if this project has led to a 0.5 to 1.5 mm/year error reduction.

In a recent study (Prandi et al., in preparation), uncertainties on sea level trends have been produced. The method to estimate spatial trend uncertainties is based on generalized least-squares (also called inverse method). With this approach, we can separately estimate the errors and the long-term trends, taking into account the natural variability of ocean dynamics (mesoscale circulation, interannual variability). Results (Fig. 6) show that even with no error covariance, trends are not significant in areas of high oceanic variability (Fig. 6, left). When considering measurement errors with 95 % confidence intervals, trend errors generally range from 1 to 3 mm/year (Fig. 6 middle). Adding serial correlation due to natural ocean variability shifts the confidence interval to larger values, from 1 to 4–5 mm/year (Fig. 6, right). In all cases, a large fraction of sea level trends is significant (67 and 52 %, respectively) and cannot be explained by natural variability. It is worth noting that these results rely on numerous assumptions about error covariance shapes and amplitudes.

5.3 New Arctic Products

In this section, a specific focus is performed on the Arctic mean sea level evolution. This is an area of great interest for climate studies with rapid climatic changes, such as the dramatic reduction of sea ice extent. Models also predict that the Arctic Ocean will be

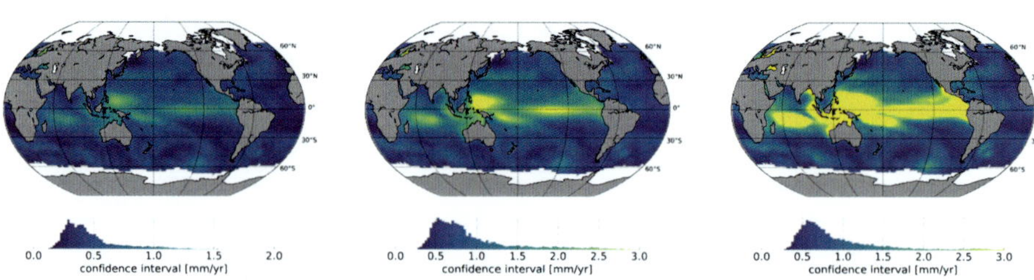

Fig. 6 Uncertainty maps of regional sea level trends. *Left* no error covariance (usual least-squares fit); *middle* measurement error covariance only; *right* measurement error and natural variability error covariance

experiencing large changes in the future (IPCC AR5). However, to date, the Arctic Ocean remains poorly observed by satellite altimetry, mainly due to sea ice cover that prevents accurate sea level measurements.

In recent years, several teams have been working towards a better knowledge of Arctic SSH (e.g., Prandi et al. 2012; Giles et al. 2012; Cheng et al. 2015). Recently, improvements on the processing of altimetry measurements in this area based on a new waveform classification and retracking algorithm (Poisson et al., in preparation) have allowed us to derive improved mean sea level maps with increased data coverage and higher mean sea level accuracy from the ice-covered Arctic (Fig. 7).

6 Validation and Error Assessment of CCI Products at Global and Regional Scales

In situ measurements are used to validate altimeter sea level records. Two types of in situ sensors are generally used: tide gauges and Argo floats. Both provide independent SSH measurements that are very valuable to detect anomalies in the altimeter records.

6.1 Validation with Tide Gauges

Tide gauges are instruments, generally set at the coast, which measure SSH relative to a local datum. There are two methods to compare tide gauges measurements with altimetry data: absolute calibration at dedicated sites, and regional or global comparisons. Absolute calibration requires a carefully monitored tide gauge, along with a precise positioning device (e.g., GPS), placed under or near altimeter ground tracks. There are three such sites in Harvest (Haines et al. 2010), Corsica (Bonnefond et al. 2015) and Bass Strait (Watson et al. 2011), which provide very valuable SSH differences time series from the beginning of the altimeter record. The other approach is to use a much wider network of tide gauges,

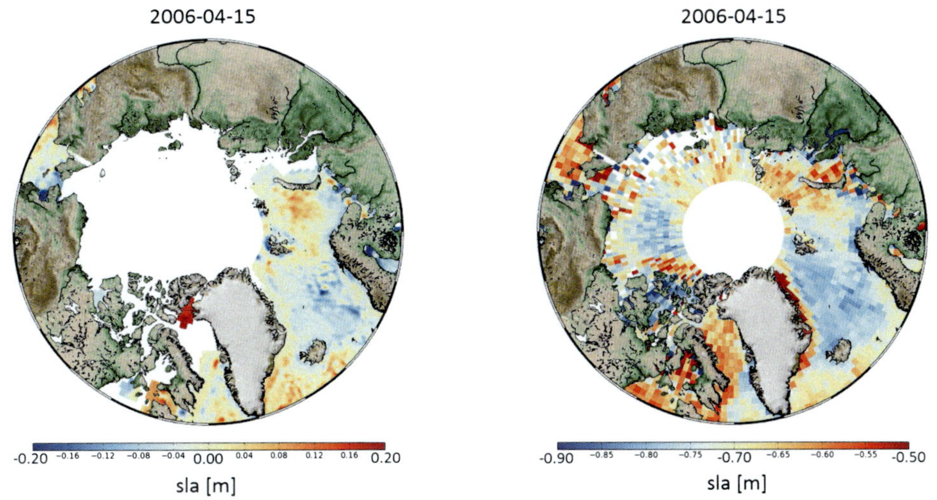

Fig. 7 Maps of sea level anomalies (SLA) in the Arctic Ocean on 15 April 2006. *Left* panel: map derived from global SL_cci products with no specific processing in the Arctic region. *Right* panel: map derived from Envisat data with improved data processing

which are individually less accurate but provide a larger ensemble, to build regional or global biases between an altimetry mission and tide gauges (Nerem et al. 2010; Mitchum et al. 1998, 2010). As the differences between absolute sea level measured by altimetry and relative sea level measured by tide gauges could also partly arise from vertical motion of the land on which the tide gauge is grounded, stations to be used need to be carefully selected and also corrected by vertical land motion if known (Fenoglio-Marc et al. 2004; Santamaría-Gómez et al. 2014).

Tide gauges unevenly sample the global coastlines, and comparisons do not cover the deep open ocean. All comparison methods rely on a similar processing, which is briefly described here. A complete description of the comparison method is available, for example, in Valladeau et al. (2012) and Wöppelmann and Marcos (2016). First, relative SSH measurements from tide gauges are corrected for various effects (tides, atmospheric pressure, vertical land motions) so that the physical content is comparable to absolute SSH measurements from altimetry. Then, altimetry measurements are collocated to tide gauges stations (using different approaches such as bilinear interpolation, area average, etc.), and a time series of altimetry minus tide gauge sea level is extracted at each in situ station. Eventually a global average is estimated from the ensemble of the different times series.

While all groups use similar methods, processing details may vary and result in slightly different estimates of altimeter minus tide gauges biases (Mitchum 1998; Watson et al. 2015). Methods also vary depending on the focus of the comparisons, for example whether it is aimed at obtaining calibrated altimetry records (Watson et al. 2015) or evaluating vertical crustal motions (Wöppelmann and Marcos 2016). But in any case, the advantage of the large number of stations is that a global bias time series can be computed and can then be used to characterize the level of agreement between altimetry and in situ records.

Figure 8 displays two examples of metrics derived from global differences between altimetry and tide gauges. The left panel shows the evolution of global differences between altimetry and tide gauges for the SL_cci and DUACS-DT products (Pujol et al. 2016). In both cases, no significant drift is detected, and differences are generally below 1 cm. The right panel focuses on the residual annual signal observed in the differences. The SL_cci product is found to be in better agreement with tide gauges than the DUACS-DT product regarding the annual signal amplitude.

One objective of such comparisons is to ensure that the altimeter record is not drifting over time. Meanwhile, it is essential to determine the accuracy of tide gauges comparisons. Mitchum (1998) and Watson et al. (2015) claim that the error of the method is about

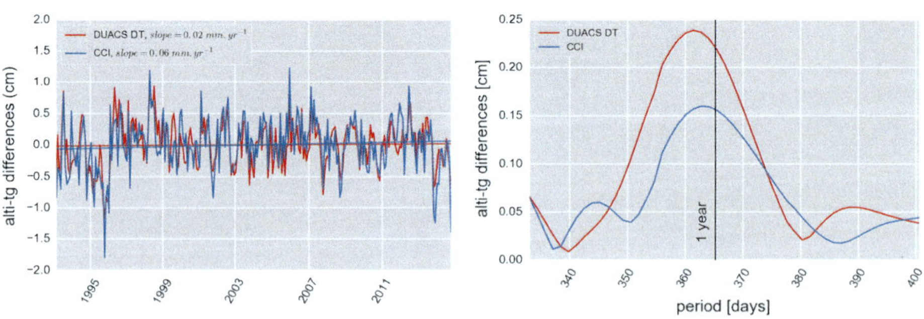

Fig. 8 *Left* time series of global differences between CCI (*red*) or DUACS-DT (*blue*) products and tide gauges. *Right* amplitude of the annual signal in differences between CCI (*red*) or DUACS-DT (*blue*) and tide gauges

0.4 mm/year while Valladeau et al. (2012) provide a 0.7 mm/year estimate and Santamaría-Gómez et al. (2012) found a value of 0.6 mm/year due to vertical land motions. This is actually one important source of error affecting relative SSH measurements by tide gauges. When no precise positioning at the stations exists, these corrections rely on Glacial Isostatic Adjustment (GIA) models (Peltier 2004) that do not account for contemporary vertical land motion sources (present-day ice melt, surface loading, ground water extraction, etc.). Wöppelmann and Marcos (2016) quantified vertical land movements not linked to the GIA process that may reach up to 10 mm/year, although on average they cancel out (0.01 ± 0.27 mm/year) if the number of tide gauge stations used is large enough.

6.2 Validation Using Argo Floats

Data from Argo floats (Roemmich et al. 2009) are another source of in situ information about the state of the ocean. They do not directly measure SSH but vertical profiles of temperature and salinity. These can be converted into density anomalies and integrated over depth to provide dynamic height anomalies (DHA), which can then be compared to altimetry-based SSH data (Valladeau et al. 2012; Legeais et al. 2016). DHA and altimeter SSH do not have consistent physical contents, as DHA are only the steric part of total sea level as measured by altimetry and, unlike tide gauges, cannot be used for absolute calibration of altimeter data but are rather used as a reference to compare two altimeter products or standards. If needed, the mass component can be derived from GRACE measurements. Argo floats are deployed at sea and, since 2005, provide a homogeneous sampling of the upper 2000 m of the global ocean, thus complementing tide gauges stations (Roemmich et al. 2009). Altimeter SSH measurements are interpolated at the time and position of Argo profiles to form an ensemble of SSH minus DHA differences from which various metrics are drawn.

Figure 9 contains a Taylor diagram that compares the CCI and DUACS-DT sea level products to a reference formed by the sum of Argo DHA and GRACE mass component. The diagram visualizes the closeness of altimetry to the reference in terms of correlation and RMS of the differences. Figure 9 shows the total signal separated into different frequencies. The results indicate that at low frequencies the SL_cci product is more consistent with Argo data than DUACS-DT (similar correlation but lower RMS). When all frequencies are considered, differences between the two products are low. A comprehensive review of uncertainty sources for Argo/altimetry comparisons can be found in Legeais et al. (2016).

6.3 Regional Validation

In addition to the global validation described above, a regional validation of the SL_cci products is performed based on the comparison with in situ data for selected regions: North Sea and Mediterranean Sea. These regions have been chosen for the availability of dense and accurate in situ measurements and ocean model data.

Regional closure of the sea level budget was investigated in the Mediterranean Sea. Figure 10 shows the smoothed monthly SL_cci series over January 2003–December 2014 and the sum of the steric and ocean mass components estimated from Argo temperature and salinity data of the EN4 database (Good et al. 2013) and GRACE data (Fenoglio-Marc et al. 2012). The sea level derived from the SL_cci products is in agreement with sea level derived from the sum of steric and mass

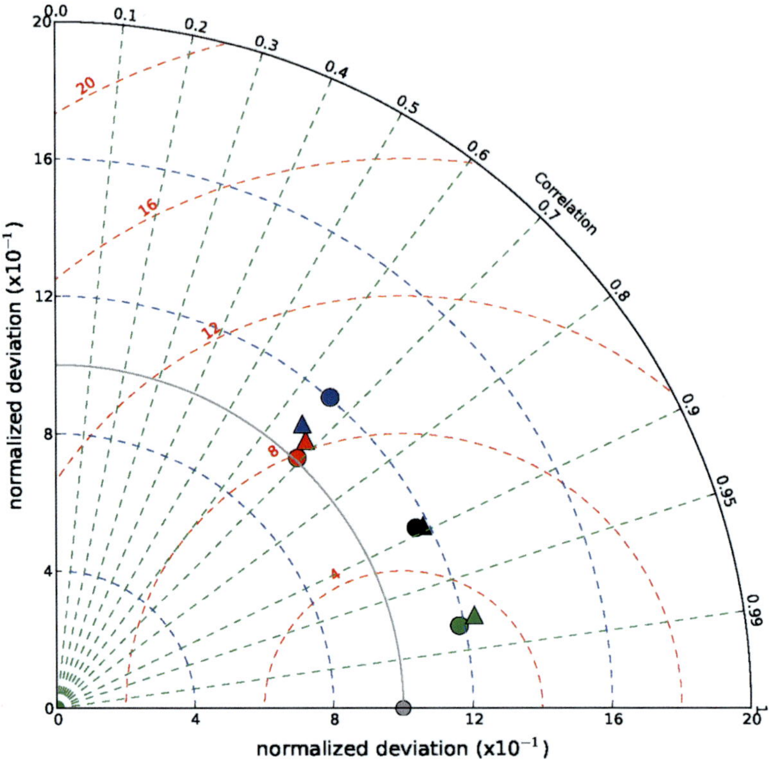

Fig. 9 Taylor diagram of the SL_cci (*triangles*) and SSALTO/DUACS-DT (*circles*) time series compared with the sum of Argo DHA (referenced to 900 dbar) and ocean mass from GRACE GRGS RL03v1 times series (*grey dot*) over 2005–2014. Total time series are in *black* and annual signals in *green*. High (in *red*) and low (in *blue*) frequencies are first adjusted from annual signal and detrended. Taylor diagrams are used to quantify the degree of correspondence between modeled and observed parameters according to 3 variables: correlation coefficient, root-mean-squares error and standard deviation

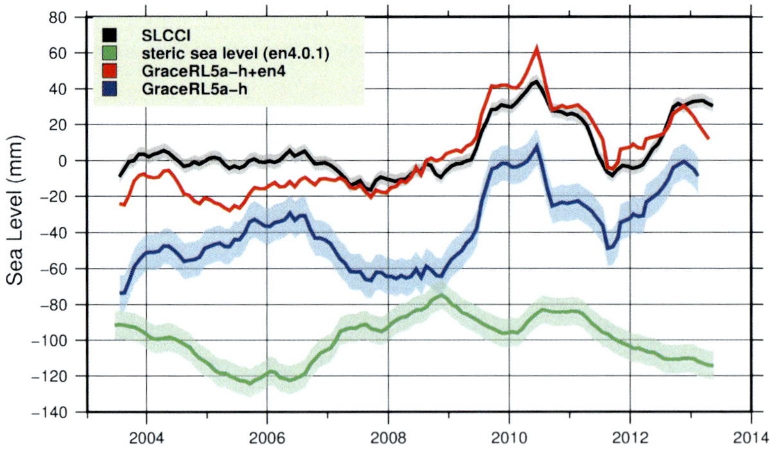

Fig. 10 SL_cci-based regional mean sea level (2004–2014) in the Mediterranean Sea (*black*) and sea level computed as sum of steric and mass components (*red*), with steric (*green*) and mass (*blue*) components

components, with a difference in trend and interannual signals of 0.8 m/year and 10.9 mm, respectively.

In the North Sea, SL_cci data are validated by comparing with tide gauges, quality-controlled with geodetic-referenced data in the German Bight and at a few other stations in the North Sea. In this case, the primary goal is to validate products and estimate errors for the along-track altimetric SSH, to verify their regional mission-long sea level trends and errors, and to compare signals and errors with the gridded sea level solutions. The same analysis has been performed for CryoSat-2 data processed with the SAMOSA model and retracker (Ray et al. 2014) in the ESRIN/GPOD SARvatore service. In the along-track comparison, the uncorrected sea level from tide gauges, expressed in ellipsoidal heights above the reference ellipsoid GRS80, are compared to the SL_cci products corrected as described in Fenoglio-Marc et al. (2015). Figure 11 shows a standard deviation of the differences. It amounts about 7 cm, which reduces to 4 cm when the tidal model TPX08 is used (http://volkov.oce.orst.edu/tides/global.html). The impact of the choice of the improved GNSS Path Delay (GPD+) wet tropospheric correction (Fernandes et al. 2015) in the coastal zone is not significant in this area. We have compared in the same region the monthly time series SL_cci gridded products and tide gauges. They agree well in terms of annual amplitude (differences <1.0 cm) and phase, with statistically significant correlations at all stations. Altimetry and tide gauge sea level trends are not statistically different at any station. The comparison of GPS-derived vertical land motion with the trend of the difference between altimetry and tide gauge shows differences in the order of 1 mm/year, which is within the trend uncertainty (Fig. 12). This uncertainty appears large for an accurate computation of vertical land motion rates from tide gauge and altimetry data. However, we notice a better agreement between altimeter and tide gauge (correlation, standard deviation and difference of trends) when SL_cci data are used, which indicates a higher quality of the SL_cci compared to other altimeter products.

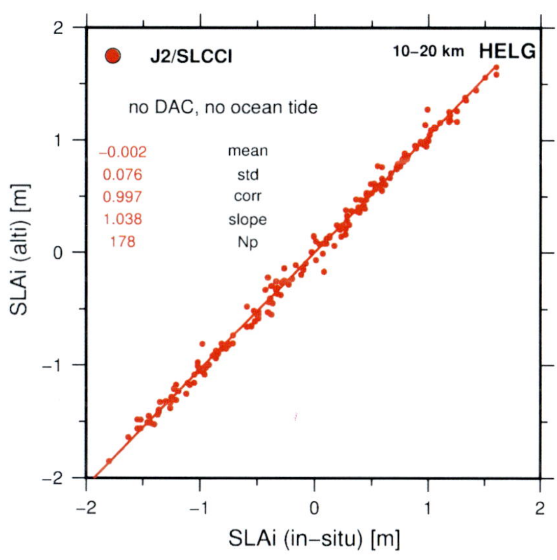

Fig. 11 Scatterplot of instantaneous SLA and statistic of differences for the complete Jason-2 SL_cci along-track data and from in situ data at the Helgoland tide gauge. Data are selected with spatial distance from the station between 10 and 20 km and temporal difference of 30 min. Np is the number of data points

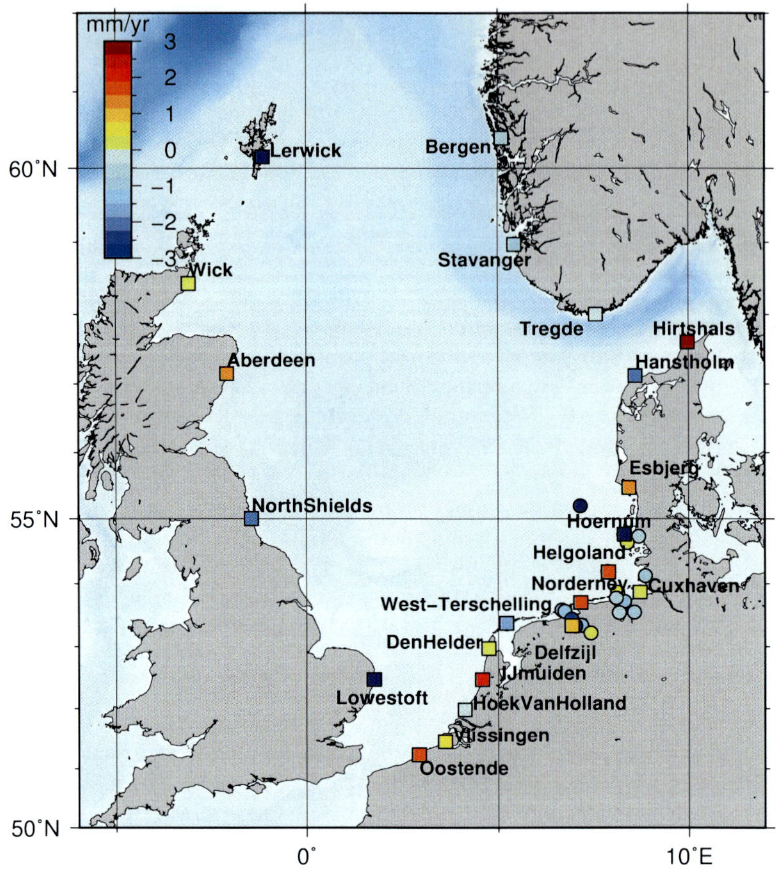

Fig. 12 Vertical land motion from GPS (*circle*) and from SL_cci altimetry minus tide gauges (*square*) in the North Sea, in mm/year

7 The CCI Sea Level Project: A Summary

Compared with previously existing products, the major evolutions of the SL_cci product are related to the following parameters. First, the orbit solutions of the different altimeter missions have been chosen so that the homogeneity of the regional sea level trends has been improved. Secondly, the GPD altimeter wet troposphere correction allows an improved estimation of the wet troposphere path delay in coastal areas. It also improves the sea level estimation in the open ocean, at high latitudes, correcting for invalid observations due to land, ice and rain contamination, and instrument malfunction. This correction exploits the data from various sources, including the Global Navigation Satellite Systems (GNSS). In addition, new dynamic atmospheric corrections computed with the ERA-Interim reanalysis lead to a strong sea level error reduction (Fig. 13) and strong improvement of the regional sea level trends over the early altimetry years.

The most impressive result of the SL_cci project is obtained using a new instrumental correction for the Envisat mission (Garcia and Roca 2010; Thibaut et al. 2012). It is illustrated in Fig. 14 by separating the ERS-1 and 2/Envisat and TOPEX/Jason-1 and 2

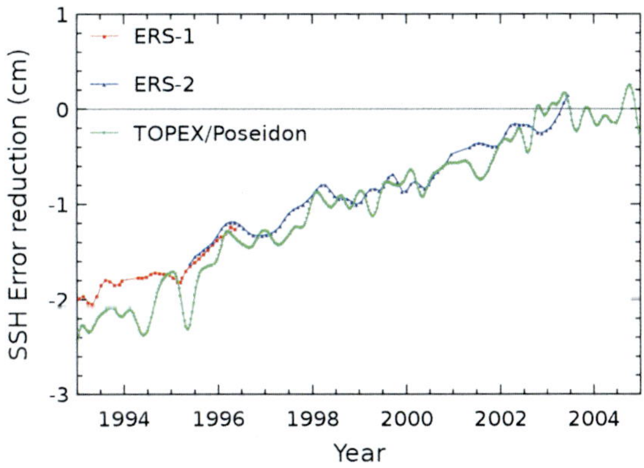

Fig. 13 Sea surface height error reduction for ERS-1 and 2 and TOPEX/Poseidon missions using a dynamic atmospheric correction forced by the ERA-Interim reanalysis compared with the operational ECMWF atmospheric fields

GMSL time series using alternately the old and new altimeter corrections: the trend difference between both time series has been significantly reduced thanks to the new instrumental correction (by 0.9 mm/year). The work performed contributed to better characterize and reduce altimetry errors at climate scales.

New level 2 altimeter algorithms have been developed, focusing on improving the ECV homogeneity and reducing the errors. Compared with the v1.1 SL_cci ECV, the major improvements that can be found in the reprocessed version are associated with the following aspects:

- New GFZ and CNES orbit solutions (Rudenko et al. 2015; Jalabert et al. 2015) have been selected for the SSH calculation of past and present altimeter missions. Compared with the previous POE-D version, the POE-E solution improves the sea level estimation and has a significant impact on the regional sea level trends (Fig. 15, left).
- The FES 2014 ocean tide model (Carrere et al. 2015) is used in the SLA calculation. Compared with other model (GOT 4.8), it leads to a reduced variance of the sea level in many coastal areas and at high latitudes (Fig. 15, right).
- An enhanced radiometer wet troposphere correction, called GPD+ (Fernandes et al. 2015), was selected for the SLA calculation of all altimeter missions. External independent measurements have been used to ensure the stability of this new correction. It significantly impacts the global decadal signals and also the sea level estimation in coastal areas.

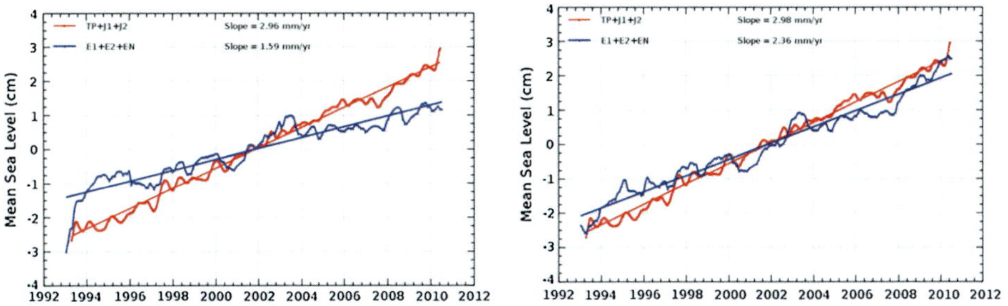

Fig. 14 GMSL evolution and associated trends computed with the TOPEX/Poseidon, Jason 1 and 2 (*red*), and ERS-1, ERS-2 and Envisat (*blue*) altimetry missions

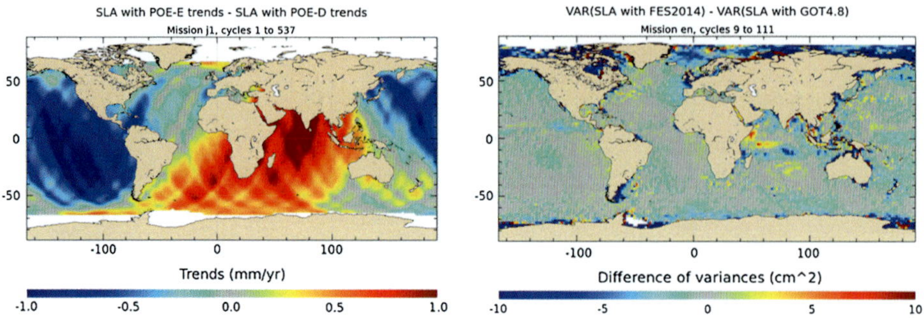

Fig. 15 *Left* map of the difference of Jason-1 (cycles 1–537) mean sea level trends computed successively with POE-D and POE-E orbit solutions. *Right* map of the difference of variance of the SSH computed successively with GOT 4.8 and FES 2014 ocean tide model for the Envisat mission (cycle 9–111)

The SL_cci products benefit from a quality control that includes internal validation, consistency check, and comparison with in situ data. In addition to this validation process, the scientific quality assessment of the ECV is an important ongoing task of the SL_cci project. Two types of assessments are investigated: (1) comparison of ocean model-based sea level with the CCI products and (2) study of the sea level budget. In (1), different methodologies are developed: (a) study of the sensitivity of an ocean reanalysis (the GECCO general ocean circulation model with data assimilation) to the new CCI sea level data via inclusion of these data in the assimilation procedure; (b) comparison with ocean-only simulation at different resolutions and with existing ocean reanalyses, which assimilate subsurface data; (c) assessment of sea level changes at high altitudes and in the Arctic Ocean by comparison of the SL_cci products with simulation runs of the Norwegian Earth System Model (NorESM).

The sea level budget approach consists in computing the sea level components using different observing systems, and comparing their sum to the SL_cci GMSL (Dieng et al. 2015a, b). Figure 16 shows the globally averaged SL_cci time series over January 2003–December 2014 with the sum of the steric and ocean mass components (estimated from Argo temperature and salinity data down to 2000 m depth and GRACE data). Over this time span, there is a very good agreement between the CCI sea level and sum of components, both in terms of trend and interannual variability. Therefore, the SL_cci data lead to quasi-closure of the sea level budget.

Within the second phase of the SL_cci project (2014–2016), updated altimeter standards and corrections are developed in the perspective of a full reprocessing of the sea level ECV (delivered end 2016). By the end of the project, this v2.0 time series will cover the period 1993–2015. Nine altimeter missions will be included, with SARAL/AltiKa and CryoSat-2 missions being new in the dataset.

8 Conclusions

Sea level, a climate variable that integrates changes of several components of the climate system, was identified by GCOS as an ECV and was further selected by ESA to be included in the first phase of the CCI programme. In this paper, we have reported how altimetry-based sea level products from different missions are built, what the current levels of uncertainties of the global and regional products are and how they have been validated.

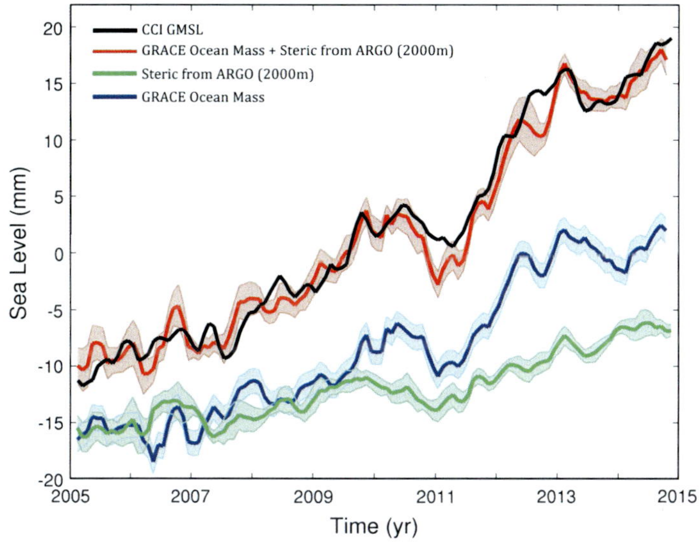

Fig. 16 SL_cci-based GMSL (*black*); Argo-based steric sea level (*green*) GRACE-based ocean mass (in equivalent of sea level, *blue*) over January 2005–December 2014 (update from Dieng et al. 2015b). The *red curve* is the sum of the steric and ocean mass components. An arbitrary vertical offset was applied to the *green* and *blue curves* for clarity

SL_cci products have been significantly improved, revisiting a myriad of instrumental and geophysical corrections. An important outcome of the CCI is that the SL_cci products are now well-characterized in terms of errors. Accounting for the ESA altimetry missions, which have a high-inclination orbit, a specific effort was dedicated to the Arctic region where the sea level evolution was poorly known until recently.

Despite the important effort invested so far, the sea level products provided within the CCI programme still do not fully satisfy the GCOS requirements, in particular at inter-annual time scales. Thus, further improvements of the altimetric standards are needed. The next inclusion of Jason-3 and Sentinel-3A (100 % SAR mode) data to the nearly 25-year-long time series will certainly lead in the near future to more accurate sea level time series, provided that long-term drifts of these new missions are carefully accounted for.

The implementation of the ESA CCI programme has led to the coordination of the Earth Observation and the Climate Research communities. This is a valuable outcome of the programme, and the CCI framework should be sustained in the future, conquering new space-based ECVs, improving existing ECVs, further assimilating ECVs in models and closing imbalances involving climate variables.

Finally, concerning the closure of the sea level budget, efforts are still needed to further improve the accuracy and to characterize the remaining uncertainties of components contributing to sea level, such as glaciers and ice sheet mass balances, ocean heat content and salinity changes, and land water storage changes.

Scientific analysis of the long-term sea level evolution and its societal impacts requires the implementation of an operational and sustainable production of the sea level Climate Data Record (CDR). Regular updates of the time series are also necessary so that the period covered by the dataset is always current. Such challenge has been addressed by the Copernicus Climate Change Service (C3S), which aims at combining observations of the climate system with the latest science to produce a consistent, comprehensive and credible description of the past and present-day climate in Europe and worldwide. It will become a major contribution from the European Union to the WMO Global Framework for Climate Services and its Climate Monitoring Architecture.

The C3S will ensure the production of the sea level CDR. The sea level record is highly dependent on the altimetry data used as input of the production system. First, the maintenance of the historical altimetry databases is required since the reprocessing of the measurements of past missions will lead to an improved quality of the whole CDR. Secondly, the integration of recent (CryoSat-2, Jason-3, Sentinel-3A) and future (Sentinel-3B, Sentinel-6, SWOT) altimetry missions are of crucial importance to guarantee the future of the sea level record.

Acknowledgments We thank G. Woppelmann and an anonymous reviewer for helpful comments on the original manuscript. M. Ablain, J.F.Legeais, P. Prandi, J. Benveniste, L. Fenoglio-Marc, H.B. Dieng and A. Cazenave acknowledge the support by ESA in the frame of the CCI project. The authors also thank all partners of the SL_cci project. M. Marcos acknowledges a 'Ramon y Cajal' contract funded by the Spanish Ministry of Economy. This work was supported by the research project CLIMPACT (CGL2014-54246-C2-1-R) funded by the Spanish Ministry of Economy.

References

Ablain M, Cazenave A, Valladeau G, Guinehut S (2009) A new assessment of the error budget of global mean sea level rate estimated by satellite altimetry over 1993–2008. Ocean Sci 5:193–201

Ablain M, Philipps S, Urvoy M, Tran N, Picot N (2012) Detection of long-term instabilities on altimeter backscattering coefficient thanks to wind speed data comparisons from altimeters and models. Mar Geod 35(S1):42–60. doi:10.1080/01490419.2012.718675

Ablain M, Cazenave A, Larnicol G, Balmaseda M, Cipollini P, Faugère Y, Fernandes MJ, Henry O, Johannessen JA, Knudsen P, Andersen O, Legeais J, Meyssignac B, Picot N, Roca M, Rudenko S, Scharffenberg MG, Stammer D, Timms G, Benveniste J (2015) Improved sea level record over the satellite altimetry era (1993–2010) from the Climate Change Initiative project. Ocean Sci 11:67–82. doi:10.5194/os-11-67-2015

Agreen RW (1982) The 3.5-year GEOS-3 data set. NOAA Technical Memorandum NOS NGS 33, NOAA, Rockville, MD

Andersen OB (2010) The DTU10 Gravity field and mean sea surface (2010) Second international symposium of the gravity field of the earth (IGFS2), Fairbanks, Alaska, 20–22 September 2010. http://www.space.dtu.dk/english/~/media/Institutter/Space/English/scientific_data_and_models/global_marine_gravity_field/dtu10.ashx. Access 20 June 2014

Bonnefond P, Exertier P, Laurain O, Guillot A, Picot N, Cancet M, Lyard F (2015) SARAL/AltiKa absolute calibration from the multi-mission Corsica facilities. Mar Geod 38(S1):171–192

Carrere L, Lyard F, Cancet M, Guillot A, Picot N, Dupuy S (2015) FES 2014: a new global tidal model. Ocean Surface Topography Science Team, Reston, Virginia, USA, October 2015. http://meetings.aviso.altimetry.fr/fileadmin/user_upload/tx_ausyclsseminar/files/OSTST2015/TIDE-01-Carrere.pdf

Carrere L, Faugère Y, Ablain M (2016) Major improvement of altimetry sea level estimations using pressure derived corrections based on ERA-interim atmospheric reanalysis. Ocean Sci Discuss. doi:10.5194/os-2015-112

Cartwright DE, Tayler RJ (1971) New computations of the tide-generating potential. Geophys J Int 23(1):45–73

Cartwright DE, Edden AC (1973) Corrected tables of tidal harmonics. Geophys J Int 33(3):253–264

Cazenave A, Dieng H, Meyssignac B, von Schuckmann K, Decharme B, Berthier E (2014) The rate of sea level rise. Nat Clim Change 4:358–361. doi:10.1038/NCLIMATE2159

Chelton DB, Ries JC, Haines BJ, Fu LL, Callahan PS (2001) Satellite altimetry. In: Fu L-L, Cazenave A (eds) Satellite altimetry and earth sciences, a handbook of techniques and applications. Academic Press, London. Int Geophys Ser 69:1–131

Cheng Y, Andersen O, Knudsen P (2015) An improved 20-year Arctic Ocean altimetric sea level data record. Mar Geod 38(2):146–162

Church JA, White NJ, Konikow LF, Domingues CM, Cogley JG, Rignot E, Gregory JM, van den Broeke MR, Monaghan AJ, Velicogna I (2011) Revisiting the earth's sea-level and energy budgets from 1961 to 2008. Geophys Res Lett. doi:10.1029/2011gl048794

Church JA, Clark PU, Cazenave A, Gregory JM, Jevrejeva S, Levermann A, Merrifield MA, Milne GA, Nerem RS, Nunn PD, Payne AJ, Pfeffer WT, Stammer D, Unnikrishnan AS (2013) Sea level change. In: Stocker TF, Qin D, Plattner G-K, Tignor M, Allen SK, Boschung J, Nauels A, Xia Y, Bex V, Midgley PM (eds) Climate change 2013: the physical science basis. Contribution of working group I to the fifth assessment report of the intergovernmental panel on climate change. Cambridge University Press, Cambridge

Clark PU et al (2015) Recent progress in understanding and projecting regional and global mean sea level. Curr Clim Change. doi:10.1007/s40641-015-0024-4

Couhert A, Luca Cerri L, Legeais JF, Ablain M, Zelensky NP, Haines BJ, Lemoine FG, Bertiger WI, Desai SD, Michiel Otten M (2015) Towards the 1 mm/y stability of the radial orbit error at regional scales. Adv Space Res 55:2–23

Dibarboure G, Pujol M-I, Briol F, Le Traon PY, Larnicol G, Picot N, Mertz F, Ablain M (2011) Jason-2 in DUACS: updated system description, first tandem results and impact on processing and products. Mar Geod 34(3–4):214–241

Dieng H, Palanisamy H, Cazenave A, Meyssignac B, von Schuckmann K (2015a) The sea level budget since 2003: inference on the deep ocean heat content. Surv Geophys 36:1. doi:10.1007/s10712-015-9314-6

Dieng H, Cazenave A, von Shuckmann K, Ablain M, Meyssignac B (2015b) Sea level budget over 2005–2013: missing contributions and data errors. Ocean Sci 11:789–802. doi:10.5194/os-11-789-2015

Dieng H, Champollion N, Cazenave A, Wada Y, Schrama E, Meyssignac B (2015c) Total land water storage change over 2003–2013 estimated from a global mass budget approach. Environ Res Lett 10:124010. doi:10.1088/1748-9326/10/12/124010

Ducet N, Le Traon PY, Reverdin G (2000) Global high resolution mapping of ocean circulation from the combination of TOPEX/POSEIDON and ERS-1/2. J Geophys Res (Oceans) 105(C8):19477–19498

Fenoglio-Marc L, Groten E, Dietz C (2004) Vertical land motion in the Mediterranean Sea from altimetry and tide gauge stations. Mar Geod 27(3–4):683–701

Fenoglio-Marc L, Becker M, Rietbroeck R, Kusche J, Grayek S, Stanev E (2012) Water mass variation in Mediterranean and Black Sea. J Geodyn. doi:10.1016/j.jog.2012.04.001

Fenoglio-Marc L, Dinardo S, Scharroo R, Roland A, Dutour M, Lucas B, Becker M, Benveniste J, Weiss R (2015) The German Bight: a validation of CryoSat-2 altimeter data in SAR mode. Adv Space Res. doi:10.1016/j.asr.2015.02.014

Fernandes MJ, Lázaro C, Ablain M, Pires N (2015) Improved wet path delays for all ESA and reference altimetric missions. Remote Sens Environ 169(2015):50–74. doi:10.1016/j.rse.2015.07.023

Fu L-L, Cazenave A (2001) Satellite altimetry and earth sciences; a handbook of techniques and applications. Academic Press, San Diego. Int Geophys Ser 69

Garcia P, Roca M (2010) ISARD_ESA_L1B_ESL_CCN_PRO_064, issue 1.b, 1 November 2010, "On-board PTR processing analysis: MSL drift differences"

Gaspar P, Ogor F (1994) Estimation and analysis of the sea state bias of the ers-1 altimeter. Rapport technique, Report of task B1-B2 of IFREMER Contract n° 94/2.426016/C. 84

GCOS (2011) Systematic observation requirements for satellite-based data products for climate (2011 update)—supplemental details to the satellite-based component of the "Implementation plan for the global observing system for climate in support of the UNFCCC (2010 update)". GCOS-154 (WMO, December 2011)

Giles KA, Laxon SW, Ridout AL, Wingham DJ, Bacon S (2012) Western Arctic Ocean freshwater storage increased by wind-driven spin-up of the BeaufortÂ Gyre. Nat Geosci 5(3):194–197

Good SA, Martin MJ, Rayner NA (2013) EN4: quality controlled ocean temperature and salinity profiles and monthly objective analyses with uncertainty estimates. J Geophys Res Oceans 118:6704–6716. doi:10.1002/2013JC009067

Haines BJ, Desai SD, Born GH (2010) The harvest experiment: calibration of the climate data record from TOPEX/Poseidon, Jason-1 and the ocean surface topography mission. Mar Geod 33(S1):91–113

Hay CC et al (2015) Probabilistic reanalysis of twentieth-century sea level rise. Nature 517(7535):481

Henry O, Ablain M, Meyssignac B, Cazenave A, Masters D, Nerem S, Garric G (2014) Effect of the processing methodology on satellite altimetry-based global mean sea level rise over the Jason-1 operating period. J Geod 88:351–361. doi:10.1007/s00190-013-0687-3

Jalabert E, Couhert A, Moyard J, Mercier F, Houry S, Rios-Bergantinos S (2015) Jason-2, SARAL and CryoSat-2 status. Ocean surface topography science team meeting, Reston, Virginia, USA, October 2015. http://meetings.aviso.altimetry.fr/fileadmin/user_upload/tx_ausyclsseminar/files/OSTST2015/POD-01-Jalabert.pdf

Jevrejeva S, Moore JC, Grinsted A, Woodworth PL (2008) Recent global sea level acceleration started over 200 years ago? Geophys Res Lett 35:L08715. doi:10.1029/2008GL033611

Jevrejeva S et al (2014) Trends and acceleration in global and regional sea levels since 1807. Glob Planet Change 113:11–22

Kaula W (1970) The terrestrial environment: solid earth and ocean physics. Williamstown report, M.I.T., Cambridge, MA, NASA CR-1579, April 1970. http://ilrs.gsfc.nasa.gov/docs/williamstown_1968.pdf

Labroue S, Boy F, Picot N, Urvoy M, Ablain M (2012) First quality assessment of the CryoSat-2 altimetric system over ocean. Adv Space Res 50(8):1030–1045. doi:10.1016/j.asr.2011.11.018

Le Traon PY, Faugère Y, Hernandez F, Dorandeu J, Mertz F, Ablain M (2003) Can we merge GEOSAT follow-on with TOPEX/Poseidon and ERS-2 for an improved description of the ocean circulation? J Atmos Ocean Technol 20:889–895. doi:10.1175/1520-0426(2003)020<0889:CWMGFW>2.0.CO;2

Legeais J-F, Ablain M, Thao S (2014) Evaluation of wet troposphere path delays from atmospheric reanalyses and radiometers and their impact on the altimeter sea level. Ocean Sci 10:893–905. doi:10.5194/os-10-893-2014

Legeais J-F, Prandi P, Guinehut S (2016) Analyses of altimetry errors using Argo and GRACE data. Ocean Sci 12:647–662. doi:10.5194/os-12-647-2016

Lillibridge J, Smith WHF, David Sandwell D, Scharroo R, Frank G, Lemoine FG, Zelensky NP (2006) 20 years of improvements to GEOSAT altimetry. Symposium: 15 years of progress in radar altimetry, Venice, Italy, March 13–18, 2006. http://earth.esa.int/workshops/venice06/participants/509/paper_509_lillibridge.pdf

Masters D, Nerem RS, Choe C, Leuliette E, Beckley B, White N, Ablain M (2012) Comparison of global mean sea level time series from TOPEX/Poseidon, Jason-1, and Jason-2. Mar Geod 35(1):20–41

Mertz F, Mercier F, Labroue S, Tran N, Dorandeu J (2005) ERS-2 OPR data quality assessment; long-term monitoring—particular investigation. CLS.DOS.NT-06.001. http://www.aviso.altimetry.fr/fileadmin/documents/calval/validation_report/E2/annual_report_e2_2005.pdf. Access 11 May 2016

Mitchum GT (1998) Monitoring the stability of satellite altimeters with tide gauges. J Atmos Ocean Technol 15(3):721–730

Mitchum GT, Nerem RS, Merrifield MA, Gehrels WR (2010) Modern estimates of sea level changes. In: Church J, Woodworth P, Aarup T, Wilson WS (eds) Understanding sea level rise and variability. Wiley-Blackwell, New York, pp 122–142

Nerem RS, Chambers DP, Choe C, Mitchum GT (2010) Estimating mean sea level change from the TOPEX and Jason altimeter missions. Mar Geod 33(1):435–446

Peltier WR (2004) Global glacial isostasy and the surface of the ice-age earth: the ICE-5G (VM2) model and GRACE. Annu Rev Earth Planet Sci 32:111–149

Prandi P, Ablain M, Cazenave A, Picot N (2012) A new estimation of mean sea level in the arctic ocean from satellite altimetry. Mar Geod 35(1):61–81

Pujol M-I, Faugère Y, Taburet G, Dupuy S, Pelloquin C, Ablain M, Picot N (2016) DUACS DT2014: the new multimission altimeter dataset reprocessed over 20 years. Ocean Sci Discuss. doi:10.5194/os-2015-110 **(in review)**

Ray RD (2013) Precise comparisons of bottom-pressure and altimetric ocean tides. J Geophys Res Oceans 118:4570–4584. doi:10.1002/jgrc.20336

Ray C, Martin-Puig C, Clarizia MP, Ruffini G, Dinardo S, Gommenginger C, Benveniste J (2014) SAR altimeter backscattered waveform model. IEEE Trans Geosci Remote Sens 53(2):911–919. doi:10.1109/TGRS.2014.2330423

Roemmich D, Johnson GC, Riser S, Davis R, Gilson J, Owens WB, Garzoli SL, Schmid C, Ignaszewski M (2009) The Argo program: observing the global ocean with profiling floats. Oceanography 22:34–43

Rudenko S, Otten M, Visser P, Scharroo R, Schöne T, Esselborn S (2012) New improved orbit solutions for the ERS-1 and ERS-2 satellites. Adv Space Res 49(8):1229–1244

Rudenko S, Dettmering D, Esselborn S, Schöne T, Förste Ch, Lemoine J-M, Ablain M, Alexandre D, Neumayer K-H (2014) Influence of time variable geopotential models on precise orbits of altimetry satellites, global and regional mean sea level trends. Adv Space Res 54(1):92–118. doi:10.1016/j.asr.2014.03.010

Rudenko R, Neumayer K-H, Dettmering D, Esselborn S, Schöne T (2015) Improvements in precise orbit determination of altimetry satellites. Ocean Surface Topography Science Team meeting, Reston, Virginia, USA, October 2015. http://meetings.aviso.altimetry.fr/fileadmin/user_upload/tx_ausyclsseminar/files/OSTST2015/POD-04-Rudenko_OSTST2015_20151021new.pdf

Santamaría-Gómez A, Gravelle M, Collilieux X, Guichard M, Martín Míguez B, Tiphaneau P, Wöppelmann G (2012) Mitigating the effects of vertical land motion in tide gauge records using a state-of-the-art GPS velocity field. Glob Planet Change 98–99:6–17

Santamaría-Gómez A, Gravelle M, Wöppelmann G (2014) Long-term vertical land motion from double-differenced tide gauge and satellite altimetry data. J Geod 88:207–222. doi:10.1007/s00190-013-0677-5

Thibaut P, Poisson J-C, Roca M, Nilo Garcia P (2012) WP2100 altimeter instrumental processing: RRDP and validation reports, sea level climate change initiative project, phase I. Algorithm selection meeting, Toulouse, 2 May 2012. http://www.esa-sealevel-cci.org/webfm_send/77

Tran N, Labroue S, Philipps S, Bronner E, Picot N (2010) Overview and update of the sea state bias corrections for the Jason-2, Jason-1 and TOPEX missions. Mar Geod 33(S1):348–362. doi:10.1080/01490419.2010.487788

Tran N, Philipps S, Poisson J-C, Urien S, Bronner E, Picot N (2012) Oral: impact of GDR-D standards on SSB corrections. Aviso, OSTST. http://www.aviso.altimetry.fr/fileadmin/documents/OSTST/2012/oral/02_friday_28/01_instr_processing_I/01_IP1_Tran.pdf

Valladeau G, Legeais JF, Ablain M, Guinehut S, Picot N (2012) Comparing altimetry with tide gauges and argo profiling floats for data quality assessment and mean sea level studies. Mar Geod 35(1):42–60

Von Schukmann K, Palmer MD, Trenberth KE, Cazenave A, Chambers D, Champollion N (2016) An imperative to monitor Earth's energy imbalance. Nat Clim Change 6:138–144

Wahr JM (1985) Deformation induced by polar motion. J Geophys Res 90(B11):9363–9368. doi:10.1029/JB090iB11p09363

Watson C, White N, Church J, Burgette R, Tregoning P, Coleman R (2011) Absolute calibration in bass strait, Australia: TOPEX, Jason-1 and OSTM/Jason-2. Mar Geod 34(3–4):242–260

Watson CS, White NJ, Church JA, King MA, Burgette RJ, Legresy B (2015) Unabated global mean sea-level rise over the satellite altimeter era. Nat Clim Change. doi:10.1038/nclimate2635

Wöppelmann G, Marcos M (2016) Vertical land motion as a key to understanding sea level change and variability. Rev Geophys. doi:10.1002/2015RG000502

Woppelmann G, Letetrel C, Santamaria A, Bouin MN, Collilieux X, Altamimi Z, Williams SDP, Miguez BM (2009) Rates of sea-level change over the past century in a geocentric reference frame. Geophys Res Lett. doi:10.1029/2009gl038720

Zawadzki L, Ablain M (2016) Accuracy of the mean sea level continuous record with future altimetric missions: Jason-3 vs. Sentinel-3a. Ocean Sci 12:9–18. doi:10.5194/os-12-9-2016

Surv Geophys (2017) 38:33–57
DOI 10.1007/s10712-016-9392-0

Monitoring Sea Level in the Coastal Zone with Satellite Altimetry and Tide Gauges

Paolo Cipollini[1] · Francisco M. Calafat[2] · Svetlana Jevrejeva[2] · Angelique Melet[3,4] · Pierre Prandi[5]

Received: 19 June 2016 / Accepted: 17 October 2016 / Published online: 18 November 2016

Abstract We examine the issue of sustained measurements of sea level in the coastal zone, first by summarizing the long-term observations from tide gauges, then showing how those are now complemented by improved satellite altimetry products in the coastal ocean. We present some of the progresses in coastal altimetry, both from dedicated reprocessing of the radar waveforms and from the development of improved corrections for the atmospheric effects. This trend towards better altimetric data at the coast comes also from technological innovations such as Ka-band altimetry and SAR altimetry, and we discuss the advantages deriving from the AltiKa Ka-band altimeter and the SIRAL altimeter on CryoSat-2 that can be operated in SAR mode. A case study along the UK coast demonstrates the good agreement between coastal altimetry and tide gauge observations, with root mean square differences as low as 4 cm at many stations, allowing the characterization of the annual cycle of sea level along the UK coasts. Finally, we examine the evolution of the sea level trend from the open to the coastal ocean along the western coast of Africa, comparing standard and coastally improved products. Different products give different sea level trend profiles, so the recommendation is that additional efforts are needed to study sea level trends in the coastal zone from past and present satellite altimeters. Further improvements are expected from more refined processing and screening of data, but in particular from the constant improvements in the geophysical corrections.

Keywords Sea level · Coastal zone · Radar altimetry · Coastal altimetry · Tide gauge

✉ Paolo Cipollini
cipo@noc.ac.uk

1 National Oceanography Centre, European Way, Southampton SO14 3ZH, UK

2 National Oceanography Centre, Brownlow Street, Liverpool L5 3DA, UK

3 LEGOS, Université de Toulouse, CNES, CNRS, IRD, UPS, Toulouse, France

4 Mercator Océan, 8-10 rue Hermès, 31520 Ramonville Saint-Agne, France

5 Space Oceanography Division, CLS, 8-10 rue Hermès, 31520 Ramonville Saint-Agne, France

1 Introduction

In the previous paper in this issue, Ablain et al. explain in detail the importance of sea level rise, a measure of the increase in ocean volume, as a clear indicator of climate change and one of its main effects. Satellite altimeter-derived sea level rise is now well quantified both as a global mean and in its geographical distribution (as, respectively, visible in Figs. 2, 5 of Ablain et al., 2016) thanks to multiple efforts that include the sea level climate change initiative (Ablain et al. 2015). Satellite-based measurements of sea level compare well with the measurements from the global tide gauge network (Fig. 8 of Ablain et al., 2016). The requirement prescribed by the Global Climate Observing System (GCOS) of an accuracy better than 0.3 mm/year in the altimeter-derived rate of global mean sea level rise is still not fully met; however, the improvements seen in the last few years make that target a realistic one. The future looks promising for the precise determination of global and regional sea level rise from the integration of altimetry and tide gauges.

However, there is still an observational gap in our knowledge, and this gap is in the region that represents the main interface between our society and the ocean, i.e. the coastal zone. While tide gauges are usually located at the coast therefore providing coastal sea level measurements (those are relative to a local datum as discussed in Sect. 6 of Ablain et al., 2016, but can be made absolute in the presence of accurate GPS positioning), altimeters have difficulties in the coastal zone. Altimeter data are normally flagged as bad and discarded shorewards of 10–50 km from the coast depending on the particular instrument and the local coastal morphology. As a consequence, all studies combining or contrasting altimeter-derived and tide gauge-derived sea level rise have been essentially comparisons between rates in different locations, and to our knowledge no study has filled this observational gap as yet. Filling the gap, which may at first sound like a purely academic exercise, becomes important when the following two factors are considered:

1. All impacts of sea level rise on society and ecosystems are going to be suffered entirely at the coast. As an example, a recent study (Hauer et al. 2016) found that a rise of only 90 cm by 2100 would put 4.2 million people at risk of inundation in the coastal zone for the continental USA alone. When projected globally and considering highly vulnerable areas such as low-lying island and deltas, the number of people that would be flooded if not relocated quickly rises to reach the order of 100 million or more (Hinkel et al. 2014).
2. Many stretches of the world's coast still do not possess in situ sea level measuring devices, and those stretches include many vulnerable regions in developing countries. Altimetry is at present the only way of obtaining measurements of sea level variations in those regions and can already offer 24 years of observations from the TOPEX/Jason-1/2/3 'reference' series and from the ERS–Envisat–AltiKa series,[1] so it will remain valuable in order to extend the sea level record back in time also when tide gauges will eventually start to be installed in those regions.

It is therefore of great importance to be able to link the satellite altimeter measurements of sea level rise with the tide gauge measurements, by bridging the open-ocean measurements

[1] The TOPEX/Jason-1/2/3 series of satellite altimeters is on a 9.92-day repeat orbit and has continuous measurements since September 1992. The ERS-1/ERS-2/Envisat/AltiKa series is on a 35-day repeat orbit and has measurements since April 1992 with some gaps, the longest of which is between Envisat's change of orbit in October 2010 and the start of the AltiKa data in March 2013. From July 2016, AltiKa has definitely left the 35-day orbit, and it is in a drifting orbit phase with no more orbit keeping manoeuvres, so the time series on that orbit have ended.

with those in close proximity to the coast. This has been one of the motivations that since the early 2000s have led to the development of the field of coastal altimetry (Vignudelli et al. 2011), which aims at recovering more numerous and better measurements of sea surface height in the coastal strip. Processing enhancements and improvements in the corrections is at the basis of a number of improved coastal altimetry datasets now becoming available, than can potentially be used for sea level studies but will need a thorough validation for that purpose.

In this contribution, we first review the status of the measurements of sea level in the coastal zone from tide gauges (Sect. 2), and then in Sect. 3 we examine the same issue from the point of view of coastal altimetry, with an overview of the improvements in the retrieval of sea surface height from altimetry in the coastal zone which have been made possible by algorithmic improvements and development of better corrections and data editing. This section also presents the datasets available for altimetry and coastal altimetry, for the benefit of the potential users of those products. Some particularly promising prospects for the monitoring of sea level in the coastal zone come from the advent of new technologies, whose coastal performance is discussed in Sect. 4: these technologies are Ka-band altimetry from AltiKa, and SAR altimetry from CryoSat-2 (and now Sentinel-3). We then present two specific examples of coastal sea level monitoring: in Sect. 5, a case study showing how local sea level can be monitored with altimetry and tide gauges around the coast of the UK and, in Sect. 6, some results on the variations of sea level rise rate as a function of distance from shore along the West African coast. The concluding section translates the science reviewed and results presented into recommendations for refinements to the processing, which should lead to further progress of the field.

2 Monitoring Sea Level with Tide Gauges

Tide gauges have been used since ancient times to measure sea level changes at the coast. In Amsterdam, an historical record of observations of sea level changes using a pole provides evidence of sea level rise and variability since 1700 (Van Veen 1945). Several tide gauge records from locations in Europe, for instance Liverpool since 1768 (Woodworth 1999) and Stockholm since 1774 (Ekman 1988) greatly contribute to our understanding of sea level changes over the eighteenth century. Since the 1830s, automatic or self-registering tide gauges were developed. The first automatic tide gauges are often credited to those installed in the Thames Estuary, England (Matthäus 1972). The first automatic tide gauge outside Europe was installed in San Francisco, USA in 1851. By the end of the nineteenth century, automatic tide gauges had been installed at many ports (Baltic and Mediterranean Seas, the USA coast). A network global in scope was starting to appear. Sea level rise during the twentieth century is estimated using almost 1000 tide gauge locations (e.g., Douglas 1997; Jevrejeva et al. 2006, 2014; Church and White 2006, 2011; Hay et al. 2015), with techniques accounting for the fact that most of those tide gauges do not cover the entire twentieth century.

Tide gauge data provide valuable information about sea level changes from a few seconds to centuries at the locations where they are installed; however, these observations suffer from several limitations:

- the geographical distribution of tide gauges is naturally confined to the continental margins and some ocean islands, which provides poor sampling of the ocean basins; in addition, most tide gauges are located in the Northern Hemisphere (Europe, Japan and the USA);

- available tide gauge records do not all cover the same time period, and their number decreases rapidly as we go back in time, especially prior to 1960;
- tide gauges are attached to the land, providing measurements relative to the Earth's crust, which could move. Vertical land movement is the one of the main difficulties to interpret tide gauge measurements (Wöppelmann and Marcos 2016; Jevrejeva et al. 2014). Some extreme examples (Fig. 1) of the effect that vertical land motion can have on local sea level are Fort Phrachula, near Bangkok, where relative sea level has been rising at a rate of about 15 mm/year since the 1960s due to subsidence caused by increased groundwater extraction, or Stockholm where it is falling at a rate of 3.8 mm/year due to crustal uplift associated with glacial isostatic adjustment.
- there is no common reference level for the individual tide gauge records, despite some clear recommendations for it (for instance Woodworth et al. 2013), and this creates a problem of stacking records together.

It should be noted that for long-term (climate) applications the temporal variation of the tidal constituents must also be understood and accounted for; these are best studied in detail with a high-frequency (i.e. hourly or more frequent) tidal datasets such as the Global Extreme Sea Level Analysis (GESLA) dataset and its follow-on GESLA-2: this issue is examined in detail in Woodworth (2010).

Since 1933, tide gauge records have been collected and distributed by the Permanent Service for Mean Sea Level (PSMSL). There are records from more than 2000 locations (Fig. 2) available from PSMSL via the webpage www.psmsl.org (Holgate et al. 2012). An international programme, the Global Sea Level Observing System (GLOSS), is established under the Joint Technical Commission for Oceanography and Marine Meteorology

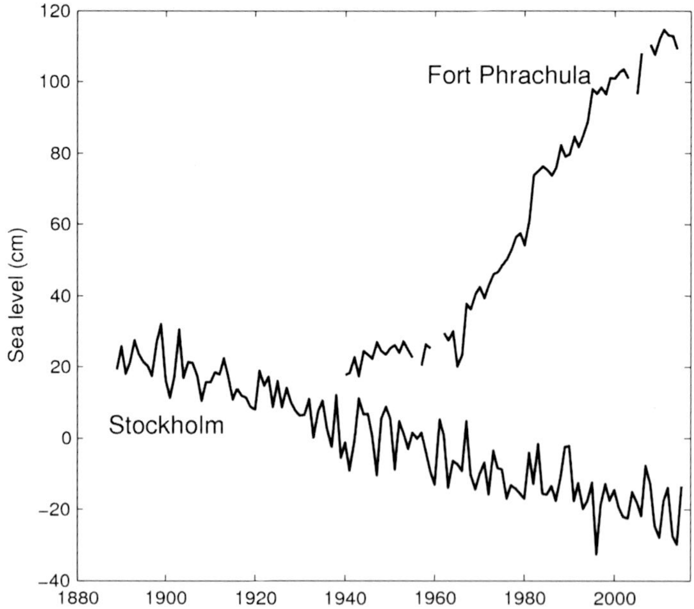

Fig. 1 Annual values of sea level from the tide gauge records at Fort Phrachula (Bangkok, Thailand) and Stockholm (Sweden)

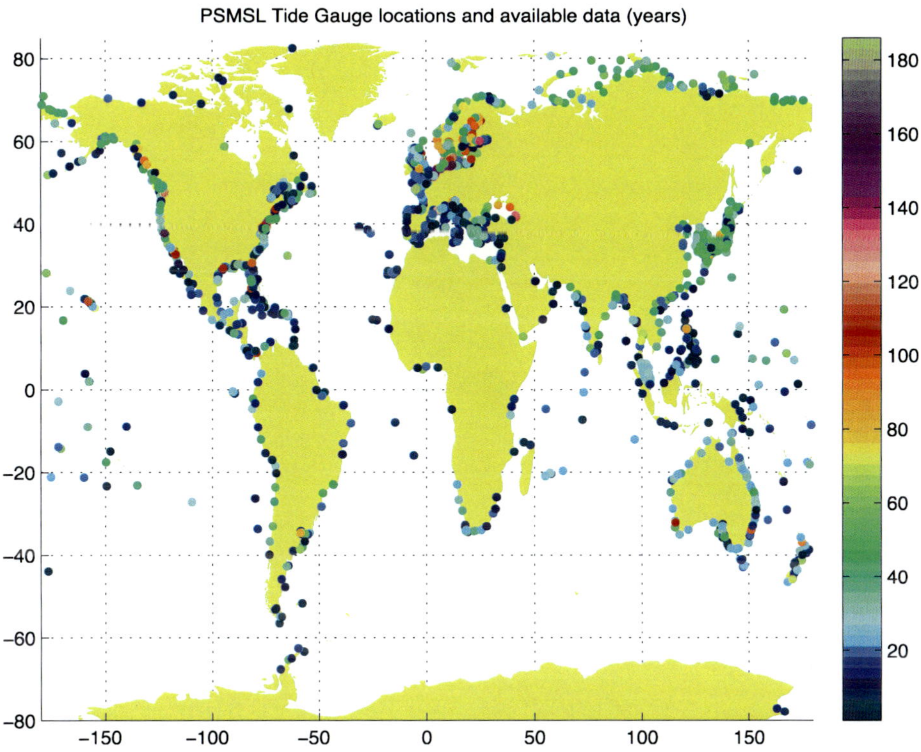

Fig. 2 Locations of tide gauges in the PSMSL database for which annual mean sea level information is available and the number of available annual records in each station

(JCOMM) of the World Meteorological Organization (WMO) and the Intergovernmental Oceanographic Commission (IOC), with the goal of creating high-quality global and regional sea level networks for application to climate, oceanographic and coastal sea level research. There are 289 sea level stations (global core network) around the world designed to provide an approximately evenly distributed sampling of global coastal sea level variations. GLOSS sites, which include Global Positioning System (GPS) receivers to monitor vertical land movements, contribute to long-term climate change studies such as those of the WMO-UNEP Intergovernmental Panel on Climate Change (IPCC).

Tide gauge observations are used to measure and predict tides, quantify the size of tsunamis and storm surges. Tide gauge records are widely used in coastal engineering for design of coastal infrastructure. Sea level datasets from tide gauges are utilized in many scientific disciplines, e.g., geodesy, oceanography, geology, paleo-oceanography studies and climatology. Selected tide gauges, in particular those located on islands, are also used for altimeter calibration. The most familiar application of tide gauge data is global and regional sea level rise and variability, providing information on long-term changes in global sea level during the last two centuries. Individual tide gauge observations (Douglas 1997), global sea level reconstructions using tide gauge data (Gornitz et al. 1982; Jevrejeva et al. 2006, 2008, 2014; Grinsted et al. 2007; Merrifield et al. 2009; Ray and Douglas 2011) and reconstructions that jointly use satellite altimetry and tide gauge records (Church and White 2006, 2011) or apply Bayesian fingerprinting techniques to tide gauge observations (Hay et al. 2015) show the evolution of sea level rise for the past 50–100 years. Three of

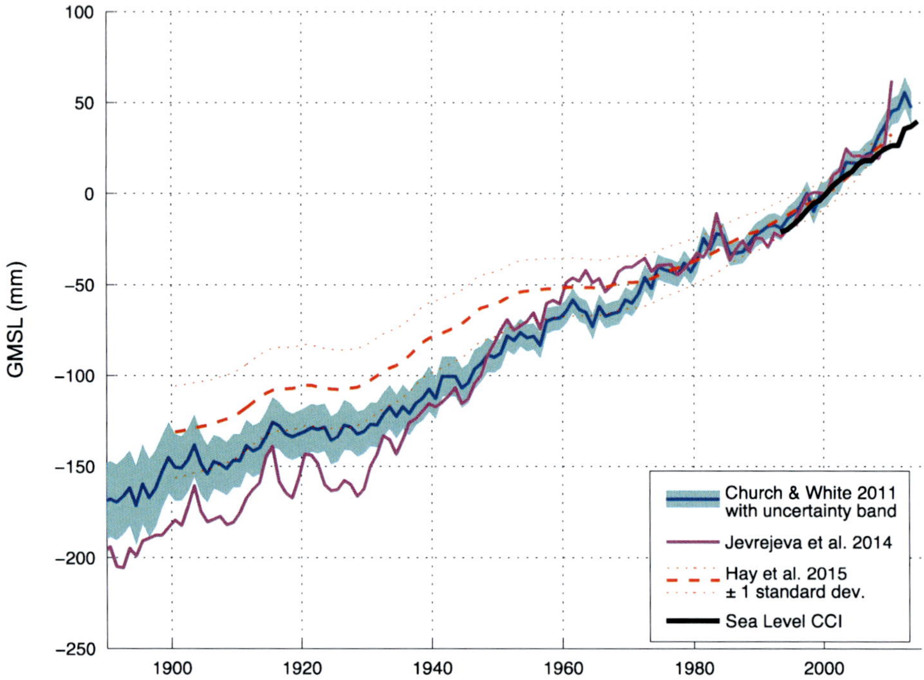

Fig. 3 Global mean sea level (GMSL) from various reconstructions cited in the text and altimetry

these reconstructions, i.e. Church and White (2011), Jevrejeva et al. (2014) and the 'Kalman smoother' reconstruction by Hay et al. (2015) are shown in Fig. 3 alongside the altimeter-derived global mean sea level from the ESA Sea Level Climate Change Initiative (CCI) (Ablain et al. 2015). This illustrates the encouraging agreement between in situ-based and satellite-based quantifications of the global mean sea level.

3 Monitoring Sea Level with Coastal Satellite Altimetry

Satellite altimetry has been one of the workhorses of open-ocean operational oceanography and global sea level monitoring, so efforts are naturally being made to use it also in the coastal zone. The main motivation is its spatial and temporal coverage: altimetry is global in space, covering even the most remote areas of the oceans (and the polar oceans with some satellites), and we have already 24 years of data from missions with accuracies of the order of just a few cm, starting with the ERS-1 launch in 1991.[2] Moreover, in addition to sea level, it provides measurements of significant wave height (SWH) and wind.

A simple illustration of the goal of coastal altimetry is in Fig. 4, showing the profile of sea level anomaly (SLA, i.e. the anomaly of the sea surface height w.r.t. its temporal mean) along Jason-1 pass #003, cycle 130 as it crossed the south-west coast of India on 31

[2] A further extension backwards of the altimetric data record amenable to long-term sea level research could in principle be achieved with ad hoc reprocessing of the GEOSAT mission (1985–1989), and is highly recommended. This will, however, require careful intercalibration with the post-1991 missions, which can be achieved by using carefully selected tide gauges as calibration transfer standards.

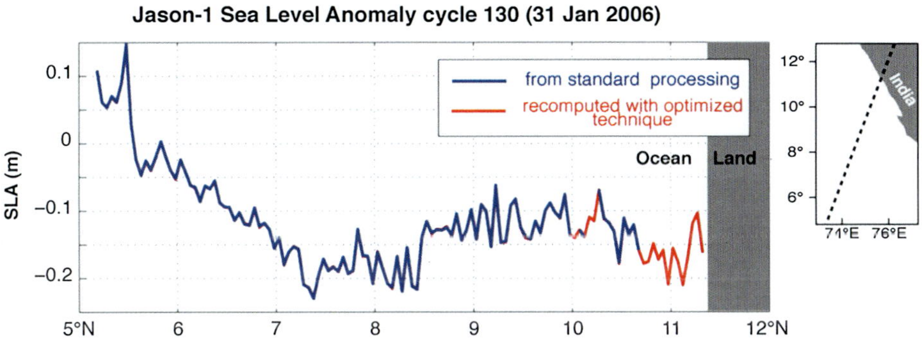

Fig. 4 Example to illustrate the concept of coastal altimetry: profile of sea level anomaly along Jason-1 pass #003, cycle 130 crossing the south-west coast of India. In *blue,* the portion of the profile where all the data flags are set to 'valid' in the conventional data products. In *red,* the portion of the profile that can be recovered with optimized processing as described in the text

January 2006. The portion of the profile where all the data flags are set to 'valid' in the conventional data products is plotted in blue and excludes several tens of km close to the coast, i.e. all the portion north of 10.7°N. Further gaps are seen in the data are between 10° and 10.2°N, possibly due to problems with some of the corrections or to problems with waveform retracking which could be due for instance to the presence of surface slicks: often, coastal altimetry techniques allow the recovery of meaningful measurements also in such problematic open-ocean circumstances. The portion in red is what can be recovered with an improved processing, which in this particular example consisted of de-flagging combined with customized screening of the corrections (i.e. using different validity ranges).

A lively international community of scientists has been involved in research and development of techniques for coastal altimetry in the last 10 years, with substantial support from space agencies such as the European Space Agency (ESA) and the Centre National d'Études Spatiales (CNES) as well as other research institutions, via several projects, amongst which:

- CNES' PISTACH (2007–2011) for Jason-2 and ESA's COASTALT (2008–2012) for Envisat, both developing specialized waveform retracking (see 3.1) and corrections.
- The X-TRACK initiative by the Centre of Topography of the Oceans and the Hydrosphere (CTOH) at LEGOS—Laboratoire d'Etudes en Géophysique et Océanographie Spatiales. X-TRACK products are derived for all precise altimeter missions by means of coastal-oriented screening of altimeter data and corrections (see Birol et al. 2016) and are used for the example on variability of sea level trends in the coastal zone in Sect. 6.
- The PEACHI products for AltiKa (Valladeau et al. 2015) currently produced by CLS (Collecte Localisation Satellites) under CNES funding. An assessment of these products is in Sect. 4.1.

Table 1 lists the available products for open-ocean and coastal and highlights those products that are provided at a higher post rate (20 or 40 Hz, corresponding to along-track distances of approximately 350 and 175 m, respectively) and are therefore more amenable to coastal altimetry applications. The coastal altimetry community (http://www.coastalt.eu/

Table 1 Available products for open-ocean and coastal altimetry as of October 2016

ID	Produced by	Altimeter	Product level	Posting rate	Coverage	Download from	Comments
AVISO	CLS, CNES CNES	e1, tx, e2, en, j1, j2, c2 (LRM/PRLM), sa, h2	L2, L3, L4 also L4	1 Hz	Global + European regions + Arctic + SW Indian	AVISO+	Widely used reference dataset processed with standard techniques. Distribution of global, Mediterranean Sea, Black Sea products is migrating to CMEMS during 2016
CMEMS	CLS CNES	e1, tx, e2, en, j1, j2, c2 (LRM/PRLM), sa (s3a to be added soon)	L3 L3 for assimilation	1 Hz	Global + European regions	marine.copernicus.eu	Marine environment monitoring service of the EC/ESA Copernicus programme, providing products and services for all marine applications
PISTACH	*CLS* *CNES*	*j2*	*L2*	*20 Hz*	*Global*	*AVISO+*	*Experimental Jason-2 products for hydrology and coastal studies with specific processing. Will be discontinued at the end of 2016 in favour of PEACHI*
PEACHI	*CLS* *CNES*	*sa, (j2 to be added soon)*	*L2*	*40 Hz*	*Global*	*AVISO+*	*Experimental SARAL/AltiKa products including dedicated retracking and corrections leading to more accurate products for coastal zones, hydrology and ice. From 2017 expected to generate also j2 products*
X-TRACK	*LEGOS-CTOH*	*tx, j1, j2, gfo, en*	*L2, L3*	*1 Hz* *20 Hz (test)*	*23 regions*	*CTOH* *AVISO+*	*Specific processing using improved data screening and latest corrections available (see Sect. 6)*
RADS	EUMETSAT, NOAA, TUDelft	gs, e1, tx, pn, e2, gfo, j1, n1, j2, c2, sa	L2	1 Hz	Global	TUDelft	Widely used dataset, mirrored by tens of sites worldwide, with continuously updated corrections, but no specific coastal processing
ALES	*NOC*	*j2, n1, (j1, j3 to be added soon)*	*L2*	*20 Hz*	*Global, <50 km from coast*	*PODAAC*	*Experimental products from the ALES processor included in SGDR-type files alongside the standard products and corrections*

Table 1 continued

ID	Produced by	Altimeter	Product level	Posting rate	Coverage	Download from	Comments
SARvatore	*ESA-ESRIN*	*c2 (SAR only)*	*L2*	*20 Hz*	*SAR mode regions*	*ESA GPOD*	*On-demand Processing service for the CryoSat-2 SAR mode data where the user can configure some processing parameters to meet specific requirements (for istance for the coastal zone)*
COP	*ESA*	*c2 (LRM/PLRM)*	*L2*	*20 Hz*	*Global*	*ESA*	*Global products for CryoSat-2 from an ocean processor (output is in PLRM over the SAR mode regions)—but no specific coastal processing*

The italic text highlights those products that are provided at a higher post rate (20 or 40 Hz, corresponding to along-track distances of approximately 350 and 175 m, respectively) and are therefore more amenable to coastal altimetry applications

The abbreviations used for the altimeters are gs, Geosat (1985–1989); e1, ERS-1 (1991–1996); tx, TOPEX (1992–2002); pn, Poseidon (1992–2002), e2, ERS-2 (1995–2011); gfo, Geosat Follow-On-1 (2000–2008); j1, Jason-1 (2002–2013); en, Envisat (2002–2012); j2, Jason-2 (2008-present); c2, CryoSat-2 (2010-present); sa, SARAL/AltiKa (2013-present); h2, HY-2A (2014-present); j3, Jason-3 (2016-present); s3a, Sentinel-3A (2016-present). For CryoSat-2 (c2), a further specification is added when data are only available from the low-resolution mode and pseudo-low-resolution mode (LRM/PLRM) or only from the SAR mode regions. The abbreviations used for product levels are L2, along-track data with corrections; L3, data gridded on regular grids in space and time; L4, products derived from analysis of multiple measurements, such as climatologies community) holds regular workshops where the science and techniques of coastal altimetry are reviewed and the various applications are showcased and discussed.

3.1 Strategies for Improving the Coastal Altimetry Data

Let us first recall the basic equation by which the fundamental measurement taken by a radar altimeter, i.e. a measurement of range from the instrument to the sea surface, is converted into an accurate measurement of the surface height:

$$Surface\_height = orbital\_altitude - (range + corrections) \tag{1}$$

where *orbital_altitude* is the height of the satellite centre of mass with respect to a reference surface (typically a reference ellipsoid) which is normally modelled to an accuracy of 2–3 cm by using a combination of GPS positioning, laser ranging and radio positioning using ground stations. The *Surface_height* so obtained contains the geoid variations as well as the oceanographic signal; subtraction of a mean sea surface[3] removes the time-invariant geoid and the mean of the dynamic topography and yields the SLA (provided also the tidal signal and the atmospheric signals are corrected for).

In the coastal zone, in addition to the refinement of the statistical techniques for screening and filtering of the various data and corrections (such as in X-TRACK), there are two complementary courses of actions for improving the quality of the retrieved data: (1) applying specialized retracking (i.e. improving the estimation of the *range* term in Eq. 1) and (2) applying improved corrections for the atmospheric, surface or geophysical effects (i.e. improving the *corrections* term).

Reflected radar pulses returning to the altimeter receiver are recorded against time in the so-called waveforms. These are sent to the receiving station on the ground and 'retracked', i.e. fitted with a waveform functional form (waveform model) to yield the fundamental measurements of range (from which sea level is measured), SWH and radar backscatter (in turn related to wind). The fitting is usually carried out via least squares or maximum likelihood algorithms. Over the open ocean, waveforms normally conform well to the Brown model (Brown 1977; Hayne 1980). In a band typically extending ~ 10 km from the coastline, a significant portion of the radar waveforms depart from the Brown model (this portion gets larger approaching the coast, as shown in Fig. 3 of Halimi et al. 2013), calling for modified retracking strategies. The factors that impact on the waveforms are not only the presence of land in the altimetric footprints, but also the occurrence of 'bright targets' in the footprint such as patches of very calm water in sheltered areas (Gómez-Enri et al. 2010). A summary of the various strategies proposed in recent years for coastal retracking is in Passaro et al. (2014); these strategies include the use of a modified functional form (as in Halimi et al. 2013), the pre-classification of waveforms and the retracking of 'sub-waveforms', i.e. a portion of the waveform unaffected (or less affected) by coastal artefacts (as in Yang et al. 2012).

A solution recently proposed to improve the retrieval of sea level, SWH and wind in the coastal zone is the Adaptive Leading-Edge Subwaveform (ALES) algorithm (Passaro et al. 2014). ALES is a two-pass retracking algorithm, based on the Brown model, where the second-pass sub-waveform window is selected based on the first-pass estimates of SWH, in a way that

[3] A mean sea surface (MSS) is the level of the sea due to all those contributions that can be assumed constant in time and can be computed as the temporal mean of sufficiently long time series. More accurate MSS models (for instance DTU15) are built from combinations of multiple altimetric and gravimetric missions. The choice of a specific MSS over another is critical in the coastal zone, and there may be biases between datasets if they are referred to different MSS.

minimizes the performance degradation w.r.t. an ideal, uncorrupted full-waveform case. This algorithm has been validated for sea level and SWH and used for different altimeters in number of case studies (Passaro et al. 2014, 2015a, b, 2016; Gómez-Enri et al. 2016). Figure 5 shows an overpass of Jason-2 tangent to the coast of Elba Island in the Mediterranean Sea, where standard retracking algorithms yield unrealistic metre-level variations of the sea level while the adoption of ALES visually improves the height profile. This kind of qualitative assessment of course needs to be followed up by a quantitative validation of the data, whose scope and methods depend to some extent on the intended application, and some validation has been carried out already in the papers cited above, while further comparison between ALES data and tide gauges around the coast of the UK is presented in Sect. 5.

The improvements in retracking have been accompanied by equally important improvements in some of the corrections that need to be applied to altimetry data to account for atmospheric path delays and other geophysical effects. The two major improvements are in the correction of the path delay due to tropospheric water vapour ('wet tropospheric' correction, see Obligis et al. 2011) and in the tide models that are needed for all those applications where the tidal component is not part of the observed signal and need to be removed (Ray et al. 2011). Here we will briefly summarize the main advances in the wet tropospheric correction, while for the improvement in tidal models we refer to the comprehensive review by Stammer et al. (2014).

The wet tropospheric correction is almost proportional to the integrated water vapour content of the atmosphere. Over the open ocean, it is either directly measured by a 2- or 3-channel passive microwave radiometer on board some altimeters, or can be estimated with good accuracy using meteorological models, which however lack spatial structure. On approaching

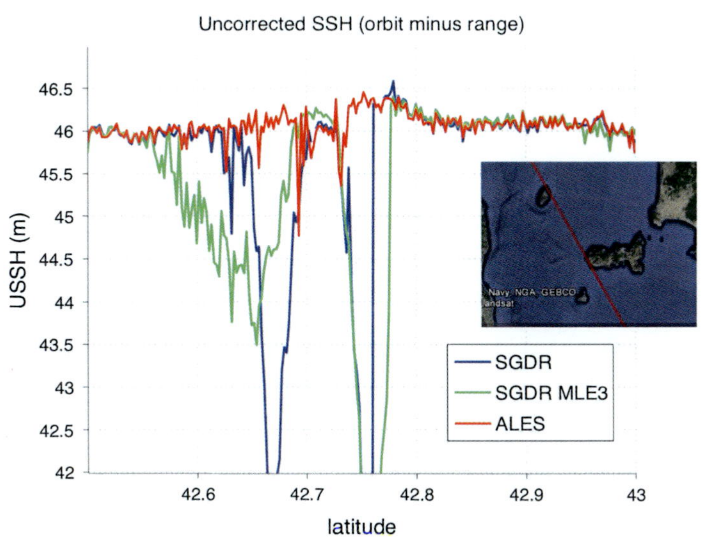

Fig. 5 Example of improved retrieval of sea level (here labelled as sea surface height or SSH) close to the coast via a specialized algorithm. The *inset* shows a map of Jason-2 descending pass 0044 (*red* track) in the vicinity of Elba Island in the Mediterranean Sea. The *line* plots show the 20-Hz uncorrected SSH (i.e. orbital altitude of the satellite minus retracked range) measured during the overpass of that ground track during orbital cycle 252 (at 13:40 on 7 May 2015) using three different retrackers: the standard Brown 4-parameter available in the sensor geophysical data records ('SGDR', in *blue*), the 3-parameter maximum likelihood estimator also in the SGDR ('SGDR MLE3', in *green*), and the ALES retracker (Passaro et al. 2014). The ALES estimates are much less affected by the proximity to the Elba Island coast within 42.65 and 42.8°N

the coastal zone, the radiometer-based measurements degrade rapidly when land enters the radiometer footprint, which has a 10–50 km diameter depending on the particular channel. Models, on the other hand, are still not particularly able to capture the shorter-scale changes in water vapour in the coastal zone and lack accuracy. The need for an improved correction has been apparent since the inception of coastal altimetry and several solutions have been proposed (Obligis et al. 2011). Notable contributions include the improved algorithm proposed by Brown (2010) and applied to the advanced microwave radiometer on the Jason-2 mission, with an estimated error less than 1.2 cm within 5 km from land. Another successful improvement is the GPD correction by Fernandes et al. (2015), built by combining passive microwave measurements from altimetric missions with path delays measured by a network of coastal GNSS stations, and being extended to include measurements from other imaging microwave radiometers. This has been applied globally to 8 missions in the ESA Sea Level CCI project and has yielded a significant impact on regional sea level trends with particular relevance to the coastal and polar regions, due to an efficient correction for land and ice contamination in the radiometer footprint (Fernandes et al. 2015).

The impact of the wet tropospheric correction in the coastal zone is well illustrated by Fig. 6, which shows the mean value of this correction as a function of distance to coast for Jason-2, Envisat and SARAL/AltiKa data. For Jason-2 the wet tropospheric correction has been computed with an improved algorithm, i.e. Brown (2010), showing a much smaller, more realistic variation in the last few kilometres to the coast than the 'standard' corrections derived from the AltiKa and Envisat dual-channel radiometers (red and blue solid curves, respectively), which both start degrading shorewards of 10 km from the coast.

4 The Potential of New Altimetric Technologies in the Coastal Zone

The launch of two satellites—CryoSat-2 and AltiKa with two important technological improvements, i.e. SAR mode altimetry and Ka-band altimetry— has opened new prospects for altimetry. In this section, we describe the potential of these two technologies for

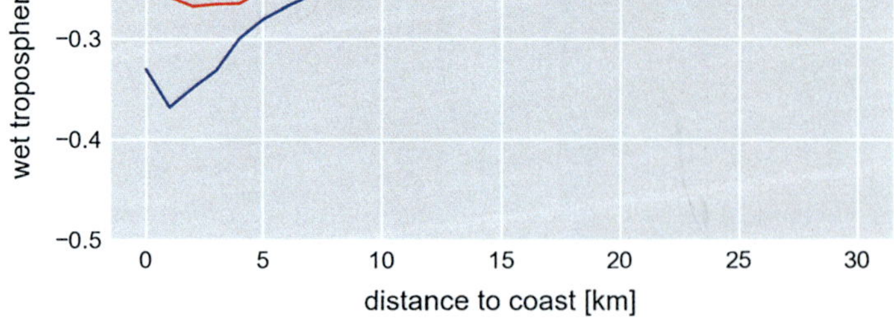

Fig. 6 Mean wet tropospheric correction as a function of distance to coast for SARAL/AltiKa, Jason-2 and Envisat data. A specific coastal algorithm is applied on Jason-2 data, and the dotted red line corresponds to an equivalent coastal processing applied on SARAL/AltiKa data

monitoring sea level in the coastal zone, where those two missions perform particularly well.

4.1 Ka-Band Altimetry: AltiKa

The SARAL/AltiKa French/Indian satellite altimetry mission was launched on 25 February 2013. The platform embarks a DORIS antenna and GPS receivers for precise orbit determination, a dual frequency radiometer for wet tropospheric path delay retrieval, and is the first mission to carry a Ka-band (36.5 GHz) altimeter providing data at the high posting rate of 40 Hz, corresponding to ~180 m along the ground track of the satellite. Compared to previous altimeters that are using the Ku-band at 13.6 GHz (and 20-Hz posting rate), SARAL/AltiKa is expected to provide better vertical resolution of the range thanks to a larger bandwidth, and improved horizontal resolution thanks to a narrower antenna beam (footprint diameter is only 8 km compared to 20 km on Jason-2), at the cost of higher sensitivity to rain events (Vincent et al. 2006; Steunou et al. 2015). The high precision measurements provided by the altimeter are very valuable for the characterization of coastal sea level and dynamics, which is one of the main scientific objectives of the SARAL/AltiKa mission (Verron et al. 2015). The current geophysical data record (GDR) products (GDR-T patch 2 version) are dedicated to open ocean, but the CNES PEACHI prototype (Valladeau et al. 2015) processes high rate data with up-to-date algorithms for different surfaces (coastal, ice, hydrology). These products are available through the ODES portal (http://odes.altimetry.cnes.fr).

Calibration and validation activities demonstrated the excellent performance of SARAL/AltiKa data over the open ocean (Prandi et al. 2015). At the coast, data quality remains very good with little land contamination of the altimeter measurements up to 5 km from the coast. Figure 7 illustrates that by showing the evolution of the standard deviation

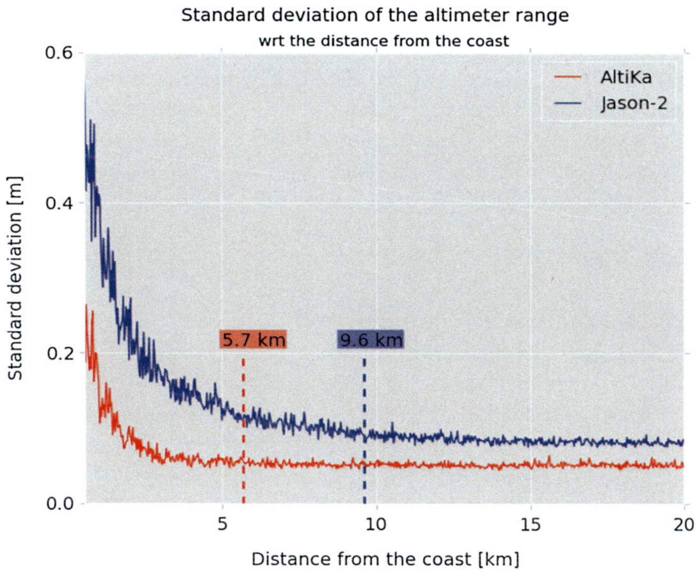

Fig. 7 Standard deviation of the altimeter range as a function of the distance to the coast for Jason-2 and SARAL/AltiKa data

of the altimeter range with respect to distance from the coast, which is directly related to the instrumental noise level. For Jason-2, the noise level rises at about 10 km from the coast. For SARAL/AltiKa, the noise level remains fairly constant, and at the same level as over the open ocean, down to 5 km from the coast or less. Another issue with coastal measurements is the wet tropospheric path delay retrieval from radiometer brightness temperatures, which can be impacted by land contamination as far as 50 km from the coast. Figure 6 displays the wet tropospheric correction derived from the radiometer measurements as a function of distance to coast for different missions. Envisat data show a drop by about 20 cm from 10 km onwards while Jason-2 remains stable thanks to the dedicated processing applied (Brown 2010). On SARAL, with no dedicated processing applied, the drop is reduced with respect to Envisat (about 10 cm, solid red line), while a simple coastal processing (extrapolation of the last uncontaminated brightness temperature) allows the wet tropospheric correction to remain very stable near the coast (dotted red line).

These altimetry-based assessments of instrumental performances can be completed by comparisons to in situ measurements to demonstrate the coastal capabilities of the SARAL/AltiKa mission. Several studies have used SARAL/AltiKa data in coastal zones, and their results tend to confirm what instrumental quality assessment suggests. Troupin et al. (2015) compared ocean currents derived from SARAL/AltiKa altimetry, HF radar and glider measurements and found a good agreement between altimeter and glider currents, as close as 10 km from the coast. Similar results were also found by Pascual et al. (2015), with further improvements foreseen from dedicated near-coast instrumental algorithms and geophysical corrections. Birol and Niño (2015) compared Jason-2 and SARAL/AltiKa data in coastal areas of the north-west Mediterranean Sea. They got a much better sampling from SARAL/AltiKa (more data available), and comparisons with local tide gauges showed a better agreement than with Jason-2, both for correlation (0.7 vs. 0.54) and RMS error (3.3 vs. 4.2 cm). Hareef Baba Shaed et al. (2015) compared significant wave heights (SWH) from SARAL/AltiKa with wave buoy measurements along the coasts of India and found correlations ranging from 0.85 to 0.98 and RMS errors lower than 0.4 m.

All the above examples demonstrate the capabilities of Ka-band altimetry for the monitoring of coastal ocean dynamics. With a complete reprocessing foreseen in 2017, and with its long-term stability validated with comparison with other altimeters and tide gauges as is being done in the Sea Level CCI project, the 3+ years long SARAL/AltiKa record will provide even more valuable data for coastal studies, providing extended observations of sea level over the same 35-day repeat set of orbits sampled by ERS-1, ERS-2 and Envisat between 1992 and 2010.

4.2 SAR Mode Altimetry: CryoSat-2

ESA's CryoSat-2 satellite was launched on 8 April 2010, with the primary mission role of monitoring the cryosphere by measuring variations in ice thickness, but has been proven of exceptional utility also for the monitoring of the oceans (see for instance Dibarboure et al. 2012). A technological innovation of CryoSat-2 Synthetic Aperture Interferometric Radar Altimeter (SIRAL) is the delay-Doppler mode of measurement (Raney 1998) which we will refer to as 'SAR mode' as it involves an unfocused along-track synthetic aperture radar (SAR) processing of the radar echoes. When in SAR mode, SIRAL exploits the Doppler information in the returned pulse to achieve a much finer resolution in the along-track direction (the width of the along-track SAR resolution cell is ~350 m), virtually independent of the sea state. The size of the altimeter footprint in the across-track direction is the same of a conventional altimeter, i.e. 2–20 km depending on the sea state. By averaging independent measurement from adjacent cells, the augmented along-track resolution can be traded off, all or in part, to achieve lower noise on the estimated parameters. Due to power and data downlink constraints, SIRAL on CryoSat-2 can only be operated in SAR mode (and in another experimental mode, SAR interferometry, of primary use over ice surfaces) over a small portion of the Earth's surface, and is instead in conventional low-resolution mode over most of the surface. In practice, this means that SAR mode data are available over a number of 'patches', one of which covers the entire European coastal sector. The importance of CryoSat-2 is magnified by the fact that SIRAL is a precursor of the Synthetic aperture Radar ALtimeter (SRAL) on the Copernicus Sentinel-3 satellites due to provide systematic oceanographic observations for the next 20 years (Donlon et al. 2012). Sentinel-3A was launched on 16 February 2016 and is being operated in SAR mode over the entire global ocean.

The performance of SAR mode altimetry in the coastal zone has been recently assessed using CryoSat-2 data around the coast of the British Isles within the ESA CP4O project (Cotton et al. 2015; Cipollini and Calafat 2016). The noise level in the altimeter measurements has been computed as the absolute value difference between consecutive values (at 20-Hz posting rate) of the total water level envelope or TWLE (i.e. the SLA inclusive of tides and atmospheric forcing; the results would be virtually the same by using SLA), and its variation in the coastal strip has been studied. Figure 8 shows a representative result of this analysis, with a scatterplot of noise and its statistics in 1-km bins of distance from coast. The median of the distribution in particular is a good indicator of the 'typical' level of noise, and it can be seen that this median stays flat at 5 cm or less from the open ocean up to 3 km from the coast and it is still relatively low (6 cm) at 2 km and 9 cm at 1 km.

Results such those presented are extremely encouraging in terms of demonstrating the low noise level in SAR altimetry data due to the excellent performance of the radar, but for

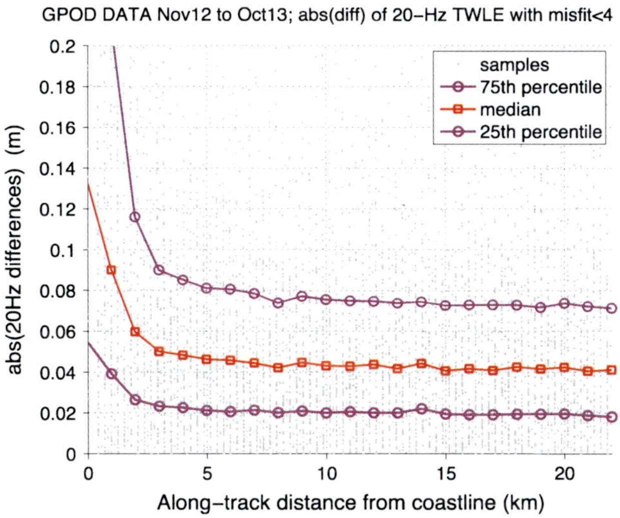

Fig. 8 Example of coastal performance of SAR altimetry. Scatterplot of noise values (estimated as the absolute value difference between consecutive Total Water Level Envelope (TWLE) measurements) against along-track distance from coast, and the statistics of its distribution in 1-km distance bins, for CryoSat data around the coast of the British isles reprocessed with the GPOD SARvatore processor (Dinardo 2014) within the ESA CP4O project (Cotton et al. 2015). The data have been screened based on a threshold on retracking misfit. From Cipollini and Calafat (2016)

the particular application to long-term monitoring of sea level what is paramount is the stability of the whole measurement system, including the corrections. As for AltiKa, this is currently being investigated in the ESA Sea Level CCI project.

5 A Case Study Around the Coast of the UK

As an example of how the local sea level can be monitored with altimetry and tide gauges, which also allows investigating the link between deep-ocean and coastal sea level variability, we present here the results of an analysis of sea level around the UK coastline, which was conducted within the framework of the Sea Level SpaceWatch project. This project was funded by the UK Space Agency within the Space for Smarter Government Programme to design and prototype an operational service delivering systematically updated sea level observations around the UK, from a combination of satellite altimeter observations and tide gauge measurements (Cotton 2016). The focus of the analysis presented here is on the annual cycle of sea level over the period 2002–2015, but first we also present a comparison between the altimetry and tide gauge observations on interannual timescales, as a form of validation.

Here we use along-track altimetry data from Jason-1 and Jason-2 as reprocessed by the coastally adapted ALES retracker (Passaro et al. 2014) described in Sect. 3.1 and covering the period 2002–2015. We use a total of 58 tide gauge records, of which 46 were obtained from the data archives of the British Oceanographic Data Centre (BODC), 11 from the UK Coastal Channel Observatory (CCO), and 1 from the Port of London Authority (PLA). The temporal resolution of the tide gauge data is 15 min for records stored at the BODC and 10 min for those stored at the CCO and PLA. For consistency with the satellite altimetry data, the atmospheric correction was applied to the tide gauge data. In particular, we used the dynamic atmospheric correction (DAC) provided by AVISO (ftp.aviso.altimetry.fr), which consists of the barotropic response of the ocean to wind forcing and atmospheric pressure as estimated by the Mog2D-G model for periods shorter than 20 days and the inverse barometer (IB) approximation for longer periods. The DAC data are provided in the form of 6-h sea level fields on a 1/4° × 1/4° regular grid covering the global oceans. The atmospheric correction at each tide gauge is taken from the nearest DAC grid point to the tide gauge.

The amplitude, A, and phase, ϕ, of the annual cycle are estimated by first fitting the following linear model to the sea level data:

$$y_t = a + bt + c\cos\left(\frac{2\pi}{365.24}t\right) + d\sin\left(\frac{2\pi}{365.24}t\right) + e_t$$

where e_t is the error term. And then, once the regression coefficients have been estimated, by

$$A = \sqrt{c^2 + d^2}$$

$$\phi = \arctan(d/c)$$

As an initial validation test, we compare de-seasoned and de-trended time series of sea level from satellite altimetry and tide gauges at each station. Because altimetry measurements are not, in general, taken at the exact location of a tide gauge, but at some point nearby, nor are they collocated in time with the tide gauge observations, some processing is necessary to obtain consistent altimeter–tide gauge pairs for the comparison. It is clear that altimetry measurements taken too far from a tide gauge station may not be relevant to the sea level measured by the tide gauge and thus a search radius centred on each tide gauge needs to be imposed. The approach taken here is to test for different values of the radius within the range 0–200 km and find that minimizing the root-mean-square difference (RMSD) between the tide gauge and altimetry observations. The advantage of this approach is that the selected radius represents the optimum radius for each tide gauge individually and thus different radiuses are used for different tide gauges. Then for each altimetry pass, an altimetry value is obtained by computing the median (which is more robust against outliers) of all records falling within the selected radius. The corresponding tide gauge matching value is obtained by linearly interpolating the tide gauge observations to the time of the altimetry pass. The resulting time series were then converted to monthly values of sea level by averaging all available values within each month.

The correlation between the altimetry and tide gauge time series is significant at all 58 stations except Portbury, Severn Bridge and Teignmouth (Fig. 9a). Higher correlations are found at stations in the Irish Sea, the North Channel, the Atlantic Ocean and the North Sea, with an average value of about 0.75, whereas lower correlations are found at stations in the English Channel with an average value of about 0.45. Regarding the RMSD (Fig. 9b), there is a clearly defined spatial pattern consistent with that found for the correlation, with the northern stations showing lower RMSD values ($\sim$3.8 cm) and stations in the English channel showing larger differences ($\sim$5.8 cm). Residual tidal errors (after the correction) may play a role in those differences, but the issue warrants further investigation.

There is also a good agreement between satellite altimetry and tide gauge observations in terms of the annual cycle, for both its amplitude (Fig. 10a) and phase (Fig. 10b). The amplitude of the annual cycle ranges from 5 to 9 cm at all station with the exception of the Barmouth tide gauge, which due to its location seems to be affected by river Mawddach and shows a very large annual amplitude of 85 cm. From Fig. 10a, we note that the

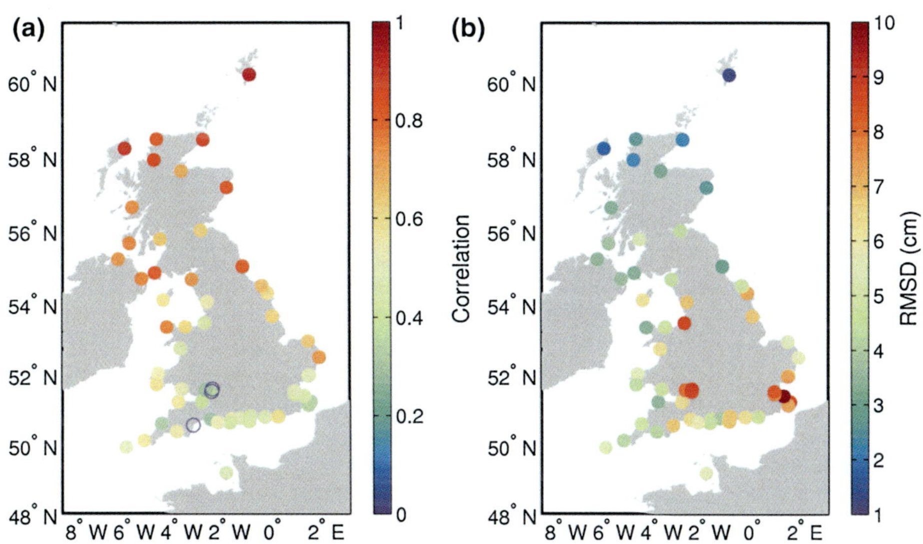

Fig. 9 Correlation (**a**) and root-mean-square difference (RMSD) (**b**) between de-seasoned and de-trended sea level from altimetry and tide gauge observations. *Empty circles* in **a** denote non-significant correlation

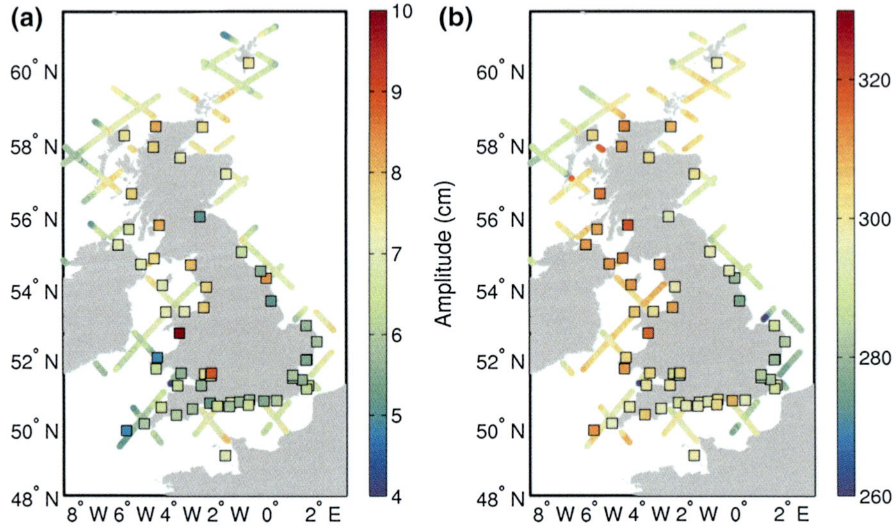

Fig. 10 Annual amplitude (**a**) and phase (**b**) of the mean annual cycle from along-track satellite altimetry measurements and from tide gauge observations (*squares*)

amplitude of the annual cycle tends to be larger in the northern coasts of the Great Britain and smaller in the English Channel. Regarding the phase of the annual cycle (Fig. 10b), we see that the cycle peaks in early October along the south-east coast and over a month later in mid-November along most of the west coast of Great Britain. Such geographical non-uniformity in the phase of the annual cycle is observed both in the tide gauge observations and the altimetry data.

6 Evolution of Sea Level Trend from the Open to the Coastal Ocean

As seen in the previous sections, altimetry already provides unique observational datasets to characterize how sea level variability evolves from the open ocean to the coastal ocean, and new technologies promise further improvements. An obvious scope arises to characterize the statistics of long-term sea level variations as functions of the distance from the coast, which would help to establish a quantitative link between open-ocean and coastal sea level variations at temporal scales relevant to climate. One particularly intriguing aspect is the coastward evolution of sea level trends. No conclusive causes supporting a difference of the coastal trend from the open ocean one have been identified so far when those trends are averaged over the entire globe (see Scharroo et al. 2009). So, the global mean sea level rise at the coast is expected to coincide with the rate from open-ocean altimetry like that from the Sea Level CCI in Fig. 3, but this agreement will have to be confirmed by coastal altimetry. However, differences are certainly possible at a regional level, due to changes in strength and location of coastal currents and changes in the coastal winds. For those reasons, the issue of the variation of sea level with distance from the coast remains worth investigating. To provide some insights into this issue, we present in this section a case study over a region off the western coast of Africa. This region has been chosen both because it is particularly vulnerable to sea level rise (Cazenave and

LeCozannet 2014), and because it has a relatively simple coastline, so it is close to an idealized situation. We use along-track nadir altimetry data from the TOPEX/Poseidon, Jason-1 and Jason-2 satellites, which share the same ground tracks and provide the longest sea level anomaly time series, over the 1993–2012 period. Three datasets are analysed and compared:

- the standard AVISO dataset (version 2014 of delayed-time data,) distributed by the Copernicus Marine Environment Monitoring Service (CMEMS). These are the 'vfec' data, i.e. validated, filtered, sub-sampled and corrected for long wavelength errors; the spatial sub-sampling of data results in a spatial resolution of 14 km.
- two datasets provided by LEGOS/CTOH, in which along-track altimetric data have been reprocessed using the 2011 and 2016 versions of the X-TRACK algorithms, respectively (Birol et al. 2016). These are more adapted to coastal regions and provide data at a spatial resolution of 7 km. The editing and extrapolation of geophysical corrections has been improved in X-TRACK2016 compared to X-TRACK2011, on two aspects: (1) the removal of land contamination in the radiometer-derived wet tropospheric correction is now performed in X-TRACK2016 with an algorithm computing the proportion p of land within the radiometer footprint of each altimeter mission and rejecting the wet tropospheric correction values for which $p > 0$. Possible other outliers away from the coast are discarded by an algorithm that detects large differences between the correction values at two consecutive points along the track for each cycle. The resulting data gaps are either interpolated (away from the coast) or filled with the closest values qualified as valid (in the land/sea transition areas). A technique based on the discrete wavelet transform is then used to compute a cleaned and noise-free wet tropospheric correction; (2) the ionospheric correction filtering has been updated to more efficiently detect outliers: the filter is a median absolute deviation (MAD) threshold in X-TRACK2016, used instead of the 3σ-threshold filter used in X-TRACK 2011 (σ being the standard deviation). The sea-state bias (SSB) correction was smoothed in X-TRACK2011 using Bezier curves. In X-TRACK2016, the SSB correction is filtered in the along-track direction with a Loess low-pass filter, and missing values are replaced by the nearest interpolated data. Regarding the other altimeter corrections, usually derived from models, very few values are discarded by the editing process. We choose to replace flagged corrections by their nearest valid neighbours.

Along-track data from the three datasets were post-processed with various filters to remove the remaining erroneous data, using a strategy similar to that used in Melet et al. (2010). As the coastal shelf is narrow off Western Africa (a few tens of km), we selected the sections of the altimetry tracks that are located less than 200 km off the coast (referred to as coastal sections hereafter) to characterize the evolution of sea level trends from the open to the coastal ocean. For each coastal section, the changes in sea level trend were calculated relative to the 'open ocean trend' for that section (defined as the mean trend for the part of the section located in the open ocean from 160 to 200 km off the coast; this choice was made to exclude the continental shelf).

The number of valid independent measurements for which the sea level trend was computed in the coastal sections is larger in the X-TRACKv2016 than in the X-TRACKv2011, especially in the last few kilometres towards the coast, due to a better editing of the data and to a more refined processing of geophysical corrections applied to altimetric data in X-TRACKv2016 than in X-TRACKv2011 (Birol et al. 2016). The number of measurements is much lower in the AVISO dataset (Fig. 11, right panels, in red)

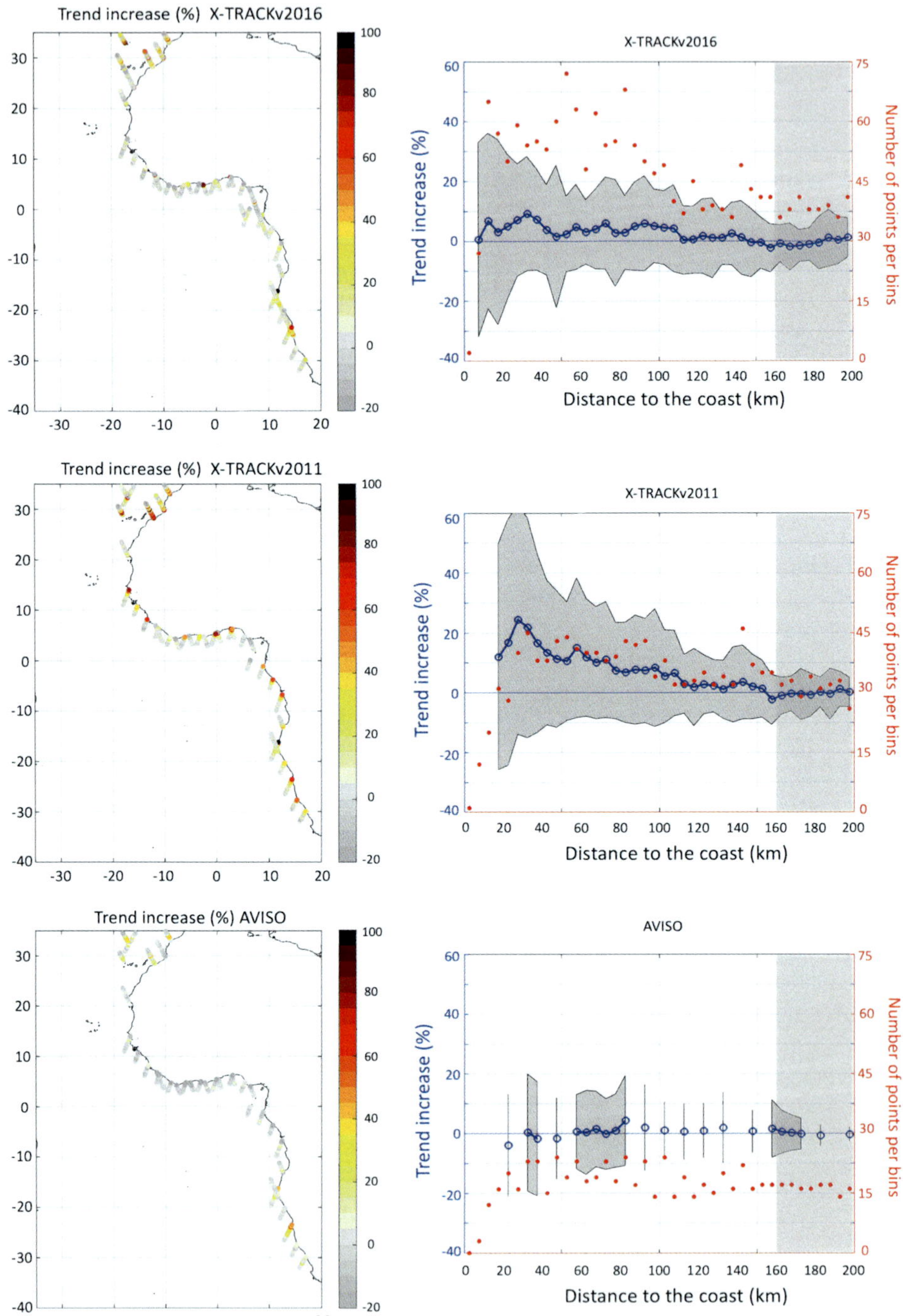

and decreases significantly in the last 30 km off the coast due to both the spatial subsampling of AVISO data, less adapted geophysical corrections in the coastal zone, and a more conservative editing of the data in AVISO than in X-TRACK datasets.

◀ **Fig. 11** (*Left*): Relative changes in SLA trend (mm/year) over 1993–2012 in percentages along coastal sections of altimetry tracks offshore Western Africa for the (*top*) X-TRACKv2016, (*middle*) X-TRACKv2011 and (*bottom*) AVISO along-track datasets. Only the sections of altimetry tracks located less than 200 km off the African coast are studied. Changes are relative to the open-ocean trend defined here as the trend of sea level anomalies averaged over the sections of altimetry tracks located from 160 to 200 km offshore the African coast. (*Right*): Relative changes in SLA trend (in %) were averaged as a function of the distance to the coast, using 5 km wide bins (*blue line, left axis*) for the (*top*) X-TRACKv2016, (*middle*) X-TRACKv2011 and (*bottom*) AVISO along-track datasets. The *grey* envelope shows plus and minus one standard deviation from the average. The number of valid points used to compute the mean and standard deviation of SLA trend changes for each bin is shown in *red* (*right axis*). As in the *left panels*, changes are relative to the trend over the 160–200 km off the coast band (this reference part is shown in *light grey shading*). Results are only shown for bins in which at least 15 coastal sections had valid points for this bin for the calculation of the mean and standard deviation of the trend. Results are based on the TOPEX/Poseidon, Jason-1 and Jason-2 missions over 1993–2012

The changes in sea level trend from the open to the coastal ocean are shown in terms of percentage in Fig. 11 (left panels). In the AVISO dataset, the trend changes from the open to the coastal ocean mostly range from −20 to +20% for the different coastal sections, with a coastward decrease of the sea level trend equatorward of 10°S/10°N, and a coastward increase of the sea level trend poleward of 25°N/25°S. When sea level trend changes are bin-averaged across all the coastal sections as a function of the distance to the coast (Fig. 11, bottom-right panel), no robust evolution of the sea level trend from the open to the coastal ocean can be seen off Western Africa in the AVISO dataset.

In the X-TRACKv2011 dataset, a substantial increase in the sea level trend can be seen for most coastal sections (Fig. 11, middle-left panel). As more valid data are available in the X-TRACKv2011 dataset than in the AVISO dataset, the averaged coastward evolution of sea level trend can be studied closer to the coast (up to 15 km off the coast). On average across the coastal sections, the sea level trend over 1993–2012 steadily increases from the open to the coastal ocean off Western Africa in this dataset, reaching an increase of 25% 25–30 km off the coast (Fig. 11, middle-right panel). The increase weakens in the last few kilometres off the coast and is less robust across the different coastal sections (grey envelope in Fig. 11, middle-right panel).

Results from the X-TRACKv2016 dataset are in between these from the AVISO and X-TRACKv2011 datasets. The sea level trend increases coastward for coastal sections located south of 15°S and north of 25°N (a result qualitatively robust for the 3 datasets), but less so than in the X-TRACKv2011 dataset. Equatorward of 10°N/10°S, no robust coastward evolution of the sea level trend is seen across the different coastal sections. The greater number of valid data in the X-TRACKv2016 dataset allows quantifying the mean coastward evolution of sea level trend up to 5 km of the coast. On average, the sea level trend only slightly increases coastward in the X-TRACKv2016 dataset (by less than 10%), but this is not robust across the coastal sections (Fig. 11, upper-right panel, grey envelope).

These results show that efforts made to improve satellite nadir altimetry products in the coastal ocean allow recovering more data and obtaining more coherent long-term signals in the coastal zone. In particular, the editing and extrapolation of geophysical corrections have been updated and improved in X-TRACKv2016 compared to X-TRACKv2011. Yet, better geophysical corrections themselves are needed to improve the accuracy and reliability of altimetry data in the coastal zone. The analysis performed here highlights that efforts are still needed on the processing of data and on the geophysical corrections applied to satellite data for studying the sea level trend in the coastal zone over the last two decades more robustly.

It should also be noted that the contribution from waves to sea level variability and trend is not considered here as it is removed from satellite altimetry data since the primary focus of satellite altimetry is to study ocean circulation and dynamics. Yet, wave-induced set-up and run-up can contribute to the sea level trend at the coast (Melet et al. 2016).

7 Summary and Conclusions

In this contribution, we have reviewed the status of the measurements of sea level in the coastal zone. First we have summarized the centennial-scale observations that we get from tide gauges, and then we have aimed at showing how improved satellite nadir altimetry products in the coastal ocean are now complementing those observations by allowing meaningful measurements and the retrieval of coherent long-term signals in the coastal zone. The trend towards better altimetric data at the coast comes not only from improved processing and corrections, but also because of technological innovations such as Ka-band altimetry and SAR altimetry, and we have discussed the main advantages deriving from those two innovations that can be now appreciated thanks to the AltiKa altimeter on SARAL and the SIRAL altimeter on CryoSat-2 (in turn a precursor of the Sentinel-3 altimeter, which has recently been launched and is being operated in SAR mode over the entire ocean).

We have then illustrated the use of altimetry for coastal sea level studies with two examples. First, in a case study conducted along the UK coast, we have found a very good agreement between coastal altimetry and tide gauge observations, with RMSDs as low as 4 cm at many stations. This has given us confidence to use the combination of altimetry and tide gauges to characterize the annual cycle of sea level along the UK coasts. We found amplitudes ranging from 5 to 9 cm, with larger amplitudes found in the northern coasts of the Great Britain, and peaks between early October in the south-east coast and mid-November in most of the west coast. Then, we have examined the evolution of sea level trend from the open to the coastal ocean along the western coast of Africa, comparing standard and coastally improved products. We observed that different products give different answers regarding the coastward evolution of the sea level trend, and we cannot yet robustly deduce the quantitative evolution of sea level trend from the open to the coastal ocean.

The clear recommendation stemming from what we have presented is that further efforts are still needed to study sea level trends in the coastal zone from past and present satellite missions. Further improvements are expected from more refined processing and screening of altimetric data, but in particular from the constant improvements in the geophysical corrections applied to them, such as wet tropospheric, tides and dynamical atmospheric corrections, which all become noisier when coming near shore. It is worth noting that such improvements in corrections should enable the full coastal exploitation of the data now flowing in from Ka- and SAR altimetry, and in particular the global SAR altimetry data now coming from Sentinel-3. This growing coastal altimetry field is going to support the monitoring of sea level in the coastal zone as well as other complementary applications such as the study of extreme events (storm surges—see for example Fenoglio-Marc et al. 2015) and the validation of coastal wave models.

Finally, it is important to remark that the advances in coastal altimetry detailed in this paper prepare the modelling community for the flux of higher resolution data—not only those now starting to flow in from Sentinel-3A (and that will be continued by Sentinel-3B/

C/D due for launch over the next 5 years, and then by the two satellites of the Sentinel-6 mission), but also the wide-swath high-resolution observations expected from the surface water and ocean topography (SWOT) mission, due for launch in 2021. The advent of SWOT should hopefully complete the process of 'closing the gap' between altimetric observations and tide gauge observations of sea level and hopefully confirm the full consistency of those two sets of measurements.

Acknowledgements The AVISO altimeter products were produced by SSALTO/DUACS and distributed by AVISO, with support from CNES (http://www.aviso.altimetry.fr/). The X-TRACK altimetry data used in this study were developed, validated and distributed by the CTOH/LEGOS, France. This work has been partially supported by the Natural Environment Research Council (NERC) National Capability funding. Some results were produced in the framework of the ESA Sea Level CCI project (ESRIN Contract No. 4000109872/13/I-NB) and ESA CryoSat Plus for Oceans (CP4O) project (ESRIN Contract No. 4000106169/12/I-NB). Results presented in Sect. 5 were produced in the framework of the Sea Level SpaceWatch project, funded by the UK Space for Smarter Government Programme through a grant from the UK Space Agency. The tide gauge data used in the Sea Level SpaceWatch study were provided by BODC http://bodc.ac.uk, Channel Coastal Observatory http://www.channelcoast.org/ and by the Port of London Authority. The authors want to thank Florence Birol and Benoit Meyssignac for helpful discussions and Angela Hibbert for the processing of the tide gauge data. This paper represents an outcome of the Workshop on 'Integrative study of sea level' at the International Space Science Institute, Bern, Switzerland, 2–6 February 2015.

References

Ablain M, Cazenave A, Larnicol G, Balmaseda M, Cipollini P, Faugère Y, Andersen O (2015) Improved sea level record over the satellite altimetry era (1993–2010) from the Climate Change Initiative project. Ocean Sci 11(1):67–82

Ablain M, Legeais JF, Prandi P, Marcos M, Fenoglio-Marc L, Dieng HB, Benveniste J, Cazenave A (2016) Satellite altimetry-based sea level at global and regional scales. Surv Geophys. doi:10.1007/s10712-016-9389-8

Birol F, Niño F (2015) Ku and Ka-band Altimeter data in the Northwestern Mediterranean Sea: impact on the observation of the coastal ocean variability. Mar Geod 38(S2):313–327

Birol F, Fuller N, Lyard F, Cancet M, Niño F, Delebecque C et al (2016) Coastal applications from nadir altimetry: example of the X-TRACK regional products. Adv Space Res (in press)

Brown GS (1977) The average impulse response of a rough surface and its applications. IEEE Trans Antennas Propag AP-25(1):67–74. doi:10.1109/TAP.1977.1141536

Brown S (2010) A novel near-land radiometer wet path-delay retrieval algorithm: application to the Jason-2/OSTM advanced microwave radiometer. IEEE Trans Geosci Remote Sens 48(4):1986–1992

Cazenave A, LeCozannet G (2014) Sea level rise and its coastal impacts. Earth's Future 2:15–34

Church JA, White NJ (2006) A twentieth century acceleration in global Sea-level rise. Geophys Res Lett 33:L01602. doi:10.1029/2005GL024826

Church JA, White NJ (2011) Sea-level rise from the late ninenteenth to the early twenty-first century. Surv Geophys 32:585–602. doi:10.1007/s10712-011-9119-1

Cipollini P, Calafat FM (2016) Extended evaluation of CryoSat-2 SAR data in the Coastal Zone, CryoSat Plus 4 Oceans—CCN1 Technical Note, vol. 1.2, pp 30. http://www.satoc.eu/projects/CP4O/docs/CP4O_CCN_Coastal_NOC_v1.2.pdf

Cotton D, Andersen O, Boy F, Cancet M, Dinardo S, Gommenginger C, Moreau T (2015) Improved oceanographic measurements from SAR altimetry: results and scientific roadmap from the ESA CryoSat Plus for oceans project. Geophysical Research Abstracts, 17, EGU General Assembly, Vienna, Austria, 12–17 April 2015; EGU2015-4613

Cotton PD, Cipollini P, Calafat FM, Hibbert A, Williams S (2016) Sea level space watch: a service for monitoring UK sea level, wales coastal practitioners conference, Llandudno, April 2016

Dibarboure G, Renaudie C, Pujol MI, Labroue S, Picot N (2012) A demonstration of the potential of Cryosat-2 to contribute to mesoscale observation. Adv Space Res 50(8):1046–1061

Dinardo S (2014) GPOD CryoSat-2 SARvatore Software Prototype User Manual. https://wiki.services.eoportal.org/tiki-index.php?page=GPOD+CryoSat-2+SARvatore+Software+Prototype+User+Manual

Donlon C, Berruti B, Buongiorno A, Ferreira MH, Féménias P, Frerick J, Nieke J (2012) The global monitoring for environment and security (GMES) Sentinel-3 mission. Remote Sens Environ 120:37–57

Douglas BC (1997) Global sea rise: a redetermination. Surv Geophys 18:270–292

Ekman M (1988) The world's longest continuous series of sea level observations. Pure Appl Geophys 127:73–77

Fenoglio-Marc L, Scharroo R, Annunziato A, Mendoza L, Becker M, Lillibridge J (2015) Cyclone Xaver seen by geodetic observations. Geophys Res Lett 42(22):9925–9932

Fernandes MJ, Lázaro C, Ablain M, Pires N (2015) Improved wet path delays for all ESA and reference altimetric missions. Remote Sens Environ 169:50–74

Gómez-Enri J, Vignudelli S, Quartly GD, Gommenginger CP, Cipollini P, Challenor PG et al (2010) Modeling ENVISAT RA-2 waveforms in the coastal zone: case study of calm water contamination. IEEE Geosci Remote Sens Lett 7:474–478

Gómez-Enri J, Cipollini P, Passaro M, Vignudelli S, Tejedor B, Coca J (2016) Coastal Altimetry products in the Strait of Gibraltar. IEEE Trans Geosci Remote Sens. doi:10.1109/TGRS.2016.2565472

Gornitz V, Lebedeff S, Hansen J (1982) Global sea level trend in the past century. Science 215(4540):1611–1614

Grinsted A, Moore JC, Jevrejeva S (2007) Observational evidence for volcanic impact on sea level and the global water cycle. PNAS 104(50):19730–19734

Halimi A, Mailhes C, Tourneret JY, Thibaut P, Boy F (2013) Parameter estimation for peaky altimetric waveforms. IEEE Trans Geosci Remote Sens 51(3):1568–1577

Hauer ME, Evans JM, Mishra DR (2016) Millions projected to be at risk from sea-level rise in the continental United States. Nat Clim Change 6:691–695

Hay CC, Morrow E, Kopp RE, Mitrovica JX (2015) Probabilistic reanalysis of twentieth-century sea-level rise. Nature 517(7535):481–484

Hayne GS (1980) Radar altimeter mean return waveforms from near-normal-incidence ocean surface scattering. IEEE Trans Antennas Propag 28(5):687–692. doi:10.1109/TAP.1980.1142398

Hinkel J, Lincke D, Vafeidis AT, Perrette M, Nicholls RJ, Tol RS, Levermann A (2014) Coastal flood damage and adaptation costs under twenty-first century sea-level rise. Proc Natl Acad Sci 111(9):3292–3297

Holgate SJ, Matthews A, Woodworth PL, Rickards LJ, Tamisiea ME, Bradshaw E, Pugh J (2012) New data systems and products at the permanent service for mean sea level. J Coastal Res 29(3):493–504

Jevrejeva S, Grinsted A, Moore JC, Holgate S (2006) Nonlinear trends and multi-year cycle in sea level records. J Geophys Res 111:2005JC003229

Jevrejeva S, Moore JC, Grinsted A, Woodworth PL (2008) Recent global sea level acceleration started over 200 years ago? Geophys Res Lett. doi:10.1029/2008GL033611

Jevrejeva S, Moore JC, Grinsted A, Matthews AP, Spada G (2014) Trends and acceleration in global and regional sea levels since 1807. Global Planet Change 113C(2014):11–22. doi:10.1016/j.gloplacha.2013.12.004

Matthäus W (1972) On the history of recording tide gauges. Proc R Soc Edinb B 73:26–34. doi:10.1017/S0080455X00002083

Melet A, Gourdeau L, Verron J (2010) Variability in Solomon sea derived from altimeter sea level data. Ocean Dyn 60:883–900

Melet A, Almar R, Meyssignac B (2016) What dominates sea level at the coast: a case study for the Gulf of Guinea. Ocean Dyn 66(5):623–636

Merrifield MA, Merrifield ST, Mitchum GT (2009) An anomalous recent acceleration of global sea level rise. J Clim 22:5772–5781

Obligis E, Desportes C, Eymard L, Fernandes MJ, Lázaro C, Nunes AL (2011) Tropospheric corrections for coastal altimetry. In: Vignudelli S et al (eds) Coastal altimetry. Springer, Berlin, pp 147–176

Pascual A, Lana A, Troupin C, Ruiz S, Faugère Y, Escudier R, Tintoré J (2015) Assessing SARAL/AltiKa data in the coastal zone: comparisons with HF radar observations. Mar Geod 38(S2):260–276

Passaro M, Cipollini P, Vignudelli S, Quartly G, Snaith H (2014) ALES: a multi-mission subwaveform retracker for coastal and open ocean altimetry. Remote Sens Environ 145:173–189

Passaro M, Cipollini P, Benveniste J (2015a) Annual sea level variability of the coastal ocean: the Baltic Sea-North Sea transition zone. J Geophys Res Oceans 120(4):3061–3078

Passaro M, Fenoglio-Marc L, Cipollini P (2015b) Validation of significant wave height from improved satellite altimetry in the German Bight. IEEE Trans Geosci Remote Sens 53(4):2146–2156

Passaro M, Dinardo S, Quartly GD, Snaith HM, Benveniste J, Cipollini P, Lucas B (2016) Cross-calibrating ALES Envisat and CryoSat-2 Delay–Doppler: a coastal altimetry study in the Indonesian Seas. Adv Space Res 58:289–303

Prandi P, Philipps S, Pignot V, Picot N (2015) SARAL/AltiKa global statistical assessment and cross-calibration with Jason-2. Mar Geod 38(S2):297–312

Raney RK (1998) The delay/doppler radar altimeter. IEEE Trans Geosci Remote Sens 36(5):1578–1588

Ray RD, Douglas BC (2011) Experiments in reconstructing twentieth-century sea levels. Prog Oceanogr 91:496–515. doi:10.1016/j.pocean.2011.07.021

Ray RD, Egbert GD, Erofeeva SY (2011) Tide predictions in shelf and coastal waters: status and prospects. In: Vignudelli S et al (eds) Coastal altimetry. Springer, Berlin, pp 191–216

Scharroo R, Leuliette E, Miller L (2009) Is sea level rise larger near the coast? 3rd coastal altimetry workshop, Frascati, Italy, 17–18 September 2009

Shaed KHB, Anand A, Joshi AK, Bhandari SM (2015) Comparison of near coastal significant wave height measurements from SARAL/AltiKa with wave rider buoys in the Indian Ocean region. Mar Geodesy 38(S2):422–436

Stammer D, Ray RD, Andersen OB, Arbic BK, Bosch W, Carrère L, Erofeeva SY (2014) Accuracy assessment of global barotropic ocean tide models. Rev Geophys 52(3):243–282. doi:10.1002/2014RG000450

Steunou N, Desjonquères J-D, Picot N, Sengenes P, Noubel J, Poisson J-C (2015) AltiKa altimeter: instrument description and in flight performance. Mar Geod 38(S1):22–42

Troupin C, Pascual A, Valladeau G, Pujol I, Lana A, Heslop E, Ruiz S, Torner M, Picot N, Tintoré J (2015) Illustration of the emerging capabilities of SARAL/AltiKa in the coastal zone using a multi-platform approach. Adv Space Res 55(1):51–59. doi:10.1016/j.asr.2014.09.011. **ISSN 0273-1177**

Valladeau G, Thibaut P, Picard B, Poisson J-C, Tran N, Picot N, Guillot A (2015) Using SARAL/AltiKa to improve Ka-band altimeter measurements for coastal zones, hydrology and ice : the PEACHI prototype. Mar Geod 38(S2):124–142

Van Veen J (1945) Bestaat er een geologische bodemdaling te Amsterdam sedert 1700? Tijdschr. K. Ned. Aardrijksk. Genoot LXII:2–36

Verron JP, Sengenes J, Lambin J, Noubel N, Stenou A, Guillot N, Picot S, Coutin-Faye R, Sharma RM, Gairola DVA, Raghava Murty JG, Richman D, Griffin A, Pascual F Rémy, Gupta PK (2015) The SARAL/AltiKa altimetry satellite mission. Mar Geod 38(S1):2–21

Vignudelli S, Kostianoy A, Cipollini P, Benveniste J (eds) (2011) Coastal altimetry. Springer, Berlin, p 578. doi:10.1007/978-3-642-12796-0

Vincent P, Steunou N, Caubet E, Phalippou L, Rey L, Thouvenot E, Verron J (2006) AltiKa: a Ka-band altimetry payload and system for operational altimetry during the GMES period. Sensors 6:208–234

Woodworth PL (1999) High waters at liverpool since 1768: the UK's longest sea level record. Geophys Res Lett 26(11):1589–1592

Woodworth PL (2010) A survey of recent changes in the main components of the ocean tide. Cont Shelf Res 30(15):1680–1691

Woodworth PL, Hughes CW, Bingham RJ, Gruber T (2013) Towards worldwide height system unification using ocean information. J Geod Sci 2(4):302–318. doi:10.2478/v10156-012-0004-8

Wöppelmann G, Marcos M (2016) Vertical land motion as a key to understanding sea level change and variability. Rev Geophys 54:64–92. doi:10.1002/2015RG000502

Yang L, Lin M, Liu Q, Pan D (2012) A coastal altimetry retracking strategy based on waveform classification and subwaveform extraction. Int J Remote Sens 33:7806–7819

Surv Geophys (2017) 38:59–87
DOI 10.1007/s10712-016-9387-x

Uncertainties in Steric Sea Level Change Estimation During the Satellite Altimeter Era: Concepts and Practices

C. R. MacIntosh[1,2] · C. J. Merchant[1,2] · K. von Schuckmann[3]

Received: 19 May 2016 / Accepted: 23 September 2016 / Published online: 17 October 2016

Abstract This article presents a review of current practice in estimating steric sea level change, focussed on the treatment of uncertainty. Steric sea level change is the contribution to the change in sea level arising from the dependence of density on temperature and salinity. It is a significant component of sea level rise and a reflection of changing ocean heat content. However, tracking these steric changes still remains a significant challenge for the scientific community. We review the importance of understanding the uncertainty in estimates of steric sea level change. Relevant concepts of uncertainty are discussed and illustrated with the example of observational uncertainty propagation from a single profile of temperature and salinity measurements to steric height. We summarise and discuss the recent literature on methodologies and techniques used to estimate steric sea level in the context of the treatment of uncertainty. Our conclusions are that progress in quantifying steric sea level uncertainty will benefit from: greater clarity and transparency in published discussions of uncertainty, including exploitation of international standards for quantifying and expressing uncertainty in measurement; and the development of community "recipes" for quantifying the error covariances in observations and from sparse sampling and for estimating and propagating uncertainty across spatio-temporal scales.

Keywords In situ observations · Satellite altimetry · Uncertainty · Steric sea level · Global sea level

✉ C. R. MacIntosh
c.r.macintosh@reading.ac.uk

[1] Department of Meteorology, University of Reading, Earley Gate, PO Box 243, Reading RG6 6BB, UK

[2] National Centre for Earth Observation, University of Reading, Reading RG6 6BB, UK

[3] Mercator Ocean, 8-10 rue Hermès, 31520 Ramonville-Saint-Agne, France

1 Introduction

Global mean sea level (GMSL) change integrates all the volume changes of the world ocean (Church et al. 2013). Thermal expansion of sea water is a major driver of change and is highly correlated with global scale ocean heat content (OHC) (Domingues et al. 2008). Over the last 50 years, it is estimated that about 90 % of human-induced heat accumulation in the Earth's climate system (e.g., Trenberth et al. 2014; von Schuckmann et al. 2016) has penetrated into the ocean through subduction and mixing processes, leading to an observed increase in upper OHC (Abraham et al. 2013; Church et al. 2013). The corresponding observed increase in specific volume is steric change (e.g., Levitus et al. 2005). The remaining excess heat from planetary warming goes into melting of both terrestrial and sea ice, and warming the atmosphere and land surface (Trenberth 2009; Hansen et al. 2011; Church et al. 2011; 2013). Thus, quantifying the effect of the sea water density changes on sea level variability is of crucial importance for climate change studies, as the cumulative sea level rise can be regarded as an important climate change indicator, as well as being of direct societal importance.

The dominant component of steric sea level change is temperature-related, i.e. the thermosteric component. Salinity variations associated with freshwater tendencies at the sea surface and redistributed in the ocean's interior have a negligible effect on sea water density and thus on sea level changes on the global scale (e.g., Lowe and Gregory 2006). On regional to basin scales, the role of halosteric effects through the addition and subtraction of freshwater or mixing processes can be large and should not be neglected in sea level studies (e.g., Durack et al. 2014; Gille 2004). Regional freshwater changes are found to have an important imprint on global mean sea level (Boening et al. 2012), but their relation to global halosteric sea level changes has not yet been quantified.

Three approaches to evaluate steric sea level from observations are available. The first is direct estimation from the global ocean in situ observing system, from data available back to the 1950s (e.g., Levitus et al. 2005, 2012). Observations have been mostly limited to the upper ocean (700 m) before the year 2005 due to data sampling issues (Abraham et al. 2013). From 2005 onwards, data sampling has strongly increased (Roemmich et al., 2009), and improved global scale estimates of steric sea level down to 2000 m are now possible (e.g., von Schuckmann et al. 2009). Hydrographic observations from sparse and irregular in situ sampling of the deep ocean exist and show that deep ocean layers (>2000 m) contribute around 0.1 mm per year to global steric sea level change (Purkey and Johnson 2010). Also analyses based on the Coupled Model Intercomparison Project Phase 5 (CMIP5) model simulations highlight the fundamental role of deeper ocean layer temperature changes (below 2000 m depth) to global ocean warming (Palmer and McNeall 2014; Cheng et al. 2016) and thermosteric sea level increase (Lorbacher et al. 2014).

A second method to obtain steric sea level estimates is to use results from ocean reanalyses, which are the combination of ocean models, atmospheric forcing fluxes and ocean observations via data assimilation methods. Ocean reanalyses can, in principle, provide more accurate information than observation-only- or model-only-based ocean estimations (Trenberth et al. 2014). However, methodological uncertainty, deficiencies in the observing system and model biases are major obstacles for the reliable reconstruction of the past ocean climate (Balmaseda et al. 2015; von Schuckmann et al. 2016). Intercomparisons of ocean reanalyses deliver insights into the performance of data assimilation systems, the underlying physical models and adequacy of the ocean observing system, and results suggest that upper layer (<700 m) global thermosteric sea level from ocean reanalyses is comparatively well constrained by observations, in contrast to large uncertainty in the deep ocean (>700 m) and halosteric contributions (Storto et al. 2015). These results emphasise the need to better observe the deep ocean, both for providing observational constraints for future ocean state estimation efforts and also to develop improved models and data assimilation methods (Palmer et al. 2015).

The third method is based on global sea level budget studies by using remote sensing data (e.g., von Schuckmann et al. 2014). Full depth global steric sea level change can be derived from the difference of total sea level change from satellite altimetry (Cazenave and Llovel 2010) and the change from ocean mass change measured using gravimetry (Chambers et al. 2010). Results based on this method underpin the significant role of deep ocean warming (Rietbroek et al. 2016), but uncertainties in the different observing systems are too large to quantify the contribution below 2000 m depth (von Schuckmann et al. 2014). Indirect steric estimates through the sea level budget are still restricted to the period from the year 2002 onwards where satellite gravimetry data are available. However, this indirect method is ideal to monitor the quality of global observing systems in the context of physical budget constraints.

This paper is a review of direct estimation of the steric component of global sea level (i.e. of the first of the three methods described) paying particular attention to questions of uncertainty. It is written in the context of a Special Issue of the journal addressing related topics about sea level budget. First, we introduce some basic concepts of steric sea level in Sect. 2. The main focus of this review is how such differences can be accounted for in terms of the inherent uncertainties present in constructing datasets representing steric changes in global sea level. Section 3 presents the theoretical principles applicable to constructing within a dataset rigorous estimates of those uncertainties, highlighting the practical challenges. In Sect. 4, practices of dataset construction and uncertainty estimation that have been reported in the literature are reviewed, illustrating the variety of approaches. In the light of these practices, some conclusions are drawn in the final section.

2 Basic Concepts

2.1 Calculating Steric Sea level

The integration of the hydrostatic balance equation gives the vertical depth ("thickness") of water between two pressure levels, p_1 and p_2:

$$Z = -\int_{p_1}^{p_2} \frac{1}{g\rho} dp \tag{1}$$

Density [ρ] is a function of temperature T, salinity S and pressure p. To find a form for calculating steric effects on the thickness, Z, between these pressures, a common approach is to recast this expression for a profile (T, S) using a first-order expansion of the specific volume, θ (inverse of density) around a reference profile (T_0, S_0), which can be either a standard sea water reference or, commonly, a climatological background field.

$$Z = -\frac{1}{g}\int_{p_1}^{p_2} \left(\theta_0(T_0, S_0, p) + \frac{\partial \theta}{\partial T}(T - T_0) + \frac{\partial \theta}{\partial S}(S - S_0) \right) dp \tag{2}$$

Defining the "steric thickness", h, as the difference from the thickness of the reference profile (i.e. $h = Z - Z_0$) and transforming the vertical integration into depth coordinates z, the steric thickness can be expressed in terms of the thermal expansion coefficient ($\alpha = \frac{1}{\theta}\frac{\partial\theta}{\partial T}$) and haline expansion coefficient ($\beta = \frac{1}{\theta}\frac{\partial\theta}{\partial S}$). The total steric effect separates cleanly into a thermosteric component, h_T, and a halosteric component, h_S:

$$h = h_T + h_S = \int_{z1}^{z2} \alpha(T - T_0)dz + \int_{z1}^{z2} \beta(S - S_0)dz \tag{3}$$

(adapted from Antonov et al. 2002). The α and β coefficients quantify the fractional change in density (or, equivalently, in specific volume) per unit increase in temperature and salinity, respectively. A temperature increase causes an increase in specific volume (expansion). A salinity increase decreases specific volume; hence, the positive values shown in Fig. 1 are coefficients of contraction (negative expansion).

2.2 Equation of State

Steric sea level change is not fully correlated with OHC change in that the former reflects the ability of water to expand and contract, rather than its heat capacity; a given heat uptake can produce different steric height changes depending on the initial conditions.

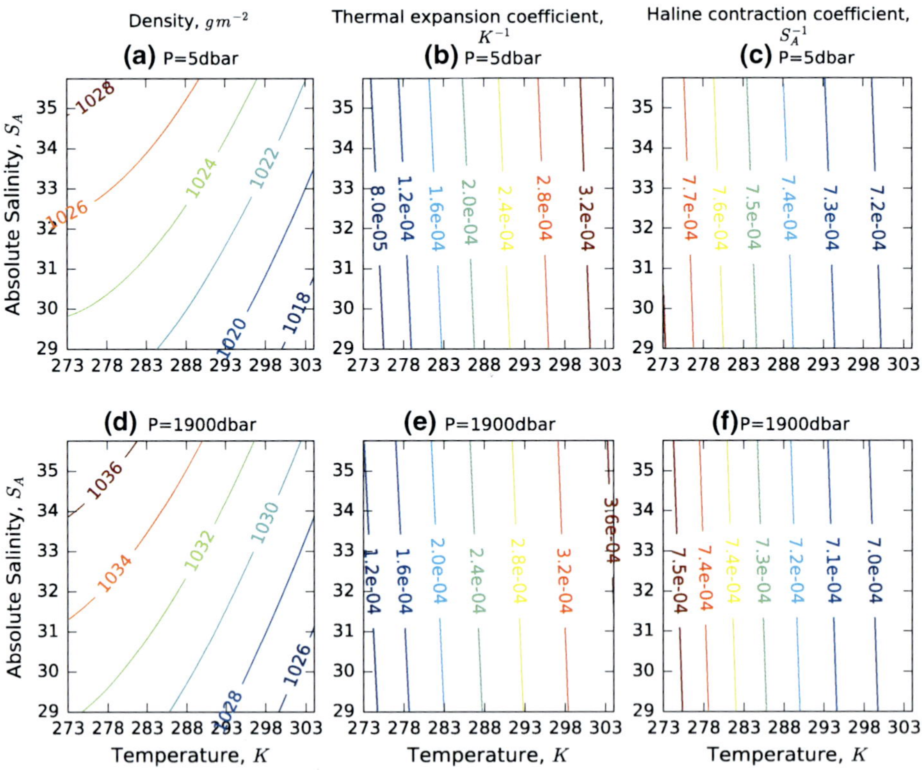

Fig. 1 Dependence of density (**a**, **d**) on temperature and salinity at two pressures (*upper*, 5 db; *lower*, 1900 db). *Centre*: thermal expansion (**b**, **e**, α, K^{-1}) and right: haline contraction (**c**, **f**, $-\beta$, g^{-1} kg) coefficients, which derive from the partial derivatives of the density surface

Density of sea water is a function of temperature and salinity at any given pressure and is described through the Thermodynamic Equation of Seawater, referred to as "TEOS-10". (Pawlowicz et al. 2012, IOC, SCOR and IAPSO 2010). TEOS-10 is the SI-traceable standard for relating temperature, salinity and density relationships in sea water in the limit to the thermodynamic properties of other substances (especially ice and humid air) and replaces the previous standard, EOS-80 (UNESCO 1981a, b).

TEOS-10 uses absolute salinity, S_A (units: g kg^{-1}), which is the mass fraction of dissolved salts in a parcel of water. This replaces the previous standard practical salinity unit, psu, which is simply a measure of the conductivity of the water. Thus, absolute salinity cannot be directly measured and is related to the conductivity (which is measured) via a reference standard composition in the TEOS-10 equations. Absolute salinity is defined as $S_R = 35.16504$ g kg^{-1} (from North Atlantic surface waters) for the standard reference composition sea water (Millero et al. 2008). Natural sea water contains many other substances in somewhat variable proportions, e.g., dissolved silicates. Where there is known to be systematic deviation from the standard relationship between conductivity and S_A, empirical corrections to salinity are used (particularly in the Pacific). These correction factors are outlined in McDougall et al., (2012). As a single measurement of salinity must always be a simplification of the true composition of sea water, the measurement, representation and even the definition of salinity continue to evolve (e.g., Wright et al. 2011).

The dependence of density on temperature and salinity shows that the thermal expansion coefficient α (units: K^{-1}) depends on both salinity and pressure, with effects of up to $\sim$3 % and $\sim$30 % across the particular range (Fig. 1a, d). However, the most influential factor on α is temperature itself, which causes a variation of a factor of 3 (Fig. 1b). Likewise, the haline contraction coefficient $-\beta$ (units of g^{-1} kg) is dependent on salinity itself and pressure, but it most strongly dependent on temperature (Fig. 1c, f). This means, for example, that for the most accurate estimation of thermosteric effects, knowledge of the salinity profile as well as the temperature profile is required.

Given these dependencies, the climatological variation of thermal expansion and haline contraction coefficients can be examined, as in Fig. 2. An increase in ocean temperature in the tropical ocean at depths down to $\sim$500 m is seen to be much more effective in causing sea level rise than the same increase in the colder waters poleward of 60° of latitude or at greater depths. This is mainly determined by the temperature-dependence of α, and the variation of β has a similar spatial shape because its variation is also temperature-determined; however, the dependence has the opposite sign, with least sensitivity of specific volume to salinity in the upper tropical ocean, as expected from Fig. 1c.

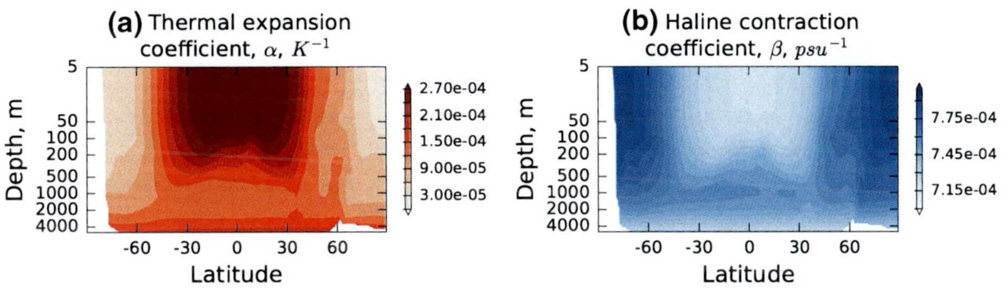

Fig. 2 Zonal average of α (units: K^{-1}) and $-\beta$ (units of g^{-1} kg) as a function of depth, from EN4 (Good et al. 2013)

3 Uncertainty in Steric Sea Level Change

One can readily find different usages for the terms "error" and "uncertainty" in the scientific literature, but these terms have unambiguous, internationally agreed definitions (JCGM 2008), adherence to which brings conceptual precision. "Error" is the difference between the true value and the measured value of a quantity; since the true value is unknown and unknowable, errors in measured values must also be unknowable. "Uncertainty" is the degree of doubt about a measured value: given the result of a measurement or calculation, it quantifies within what dispersion of values around that result is it reasonable to assume that the (unknown) true value lies (JCGM 2008). Uncertainty is usually quantified as the standard deviation of the estimated probability distribution of error in the measured value.

3.1 Importance of Uncertainty

Estimates from multiple groups of global mean steric sea level may show substantial spread in terms of the observed trend and its uncertainty (e.g., Table 2). However, the discrepancies can give an indication of the level of uncertainty. The magnitude of disagreement, and of uncertainty, increases with increasing spatio-temporal resolution, in general. It is important also to attempt to estimate uncertainty within an analysis ("internal estimates") and not only to look at the disagreement between different results ("external analysis"). If internal and external estimates of uncertainty are inconsistent, it is a sign that the origins of errors in the analyses are not adequately understood. For example, agreement between groups of estimates can be unrealistically good (external uncertainty estimates are biased small), if all or many groups adopt approaches that lead to a degree of commonality of errors across analyses. This can be revealed if rigorous internal uncertainty estimates suggest that uncertainty is much greater than disagreement across the group.

This section of the paper discusses principles of "internally" estimating uncertainty in steric sea level variations. The aim is to outline what is involved in developing a comprehensive estimate. This is intended as background information for any readers who may be only partly aware of methods of uncertainty estimation, to inform the discussion of practices found in the literature in Sect. 4.

3.2 Key Uncertainty Concepts

Fully to understand what we can and cannot infer about GMSL rise, calculations of the steric contribution need to be associated with estimates of their uncertainty.

For complex datasets, uncertainty estimation is generally challenging since it requires significant effort to gather fundamental uncertainty information and significant computation to correctly propagate and combine uncertainties to give a final estimate. Faced with this challenge, the temptation is to make simplifying assumptions (such as assuming uncorrelated errors) and/or address only those sources of error that are reasonably well understood. As a result, uncertainty is more likely to be underestimated than overestimated.

The overall uncertainty in a climate dataset can be decomposed into structural uncertainty and value uncertainty (Thorne et al. 2005). In constructing a climate dataset, many choices have to be made. In the case of an analysis of steric sea level change, choices include: which sources of *T–S* profile data to include; what quality control is applied to profiles; what bias corrections to apply to profiles; which auxiliary data to use, such as a reference climatology; what method of data gridding and/or interpolation to apply; what method of vertical integration to use; and what choice of density state function to assume? The dispersion of outcomes that could arise for all well-justified choices reflects the structural uncertainty. Thorne et al. (2005) noted that dataset producers traditionally do not estimate structural uncertainty at all.

Value uncertainty is the total remaining uncertainty given the structural choice made for constructing the dataset. It includes the uncertainty due to sampling, i.e. the facts that the available measurements are finite and not ideally (or even adequately) distributed. Uncertainty in parameters used in calculations (e.g., α) contributes to value uncertainty. Most obviously, the uncertainty in individual measured values of temperature, salinity and pressure propagates to uncertainty in the steric sea level and their estimate is challenging in practice.

Decisions regarding the construction of a dataset may influence both structural and value uncertainty. For example, the uncertainty associated with forming a gridded product from limited sampling as individual profiles has a structural component, which depends on the choice of method (kriging vs. simple binning, and the method used to fill gaps), and a value component, which could in principle be formally propagated (for example, uncertainty associated with random and systematic measurement errors in the profile, and associated with the estimated unsampled variability).

Any calculation of steric SL involves the combination of many measured values, typically profiles of T, S and p. A key question is therefore how uncertainty in these variables propagates to give the steric SL uncertainty. It is not only the magnitude of uncertainty in each measured value that affects the combined uncertainty: any correlation between errors in measured values greatly affects the combined uncertainty, as will be discussed in the next section.

3.3 Uncertainty of Steric Sea Level Estimate from a Single Profile

Using the formulation of steric SL of Eq. (3) as a starting point, we consider what would be involved in a detailed estimation of the uncertainty in h_T, which is the thermosteric component, from a single profile of T against depth. The same principles would apply to the halosteric component and the S profile.

First, note that in formulating Eq. (3), we have already made choices that could be made differently, and which contribute to structural uncertainty. These include: the means of evaluating the expansion coefficient at each level; and the numerical integration scheme used, which makes implicit assumptions about the variation of the integrand between the levels at which measured values are available.

Second, we consider estimating the value uncertainty. In general, the numerical implementation of the integral can be written (exactly or approximately) as a linear combination of measured values at N measurement levels:

$$f = \sum_{n=1}^{N} w_n \alpha_n (T_n - T_{0n}) + 0 \tag{4}$$

where w_n expresses the weight of the nth set of measurements in the profile gets in the integrated result, which depends on the separation between levels and the nature of the numerical scheme used. The relevance of the "+ 0" term will be made clear below. Now, let all the parameter and measured values contributing to f be collected in a column vector $\mathbf{x}$, for example:

$$\mathbf{x}^{\mathrm{T}} = [\, T_1 \quad \cdots \quad T_N \quad S_1 \quad \cdots \quad S_N \quad z_1 \quad \cdots \quad z_N \quad T_{01} \quad \cdots \quad T_{0N} \quad S_{01} \quad \cdots \quad S_{0N} \,] \tag{5}$$

The ordering within $\mathbf{x}$ is arbitrary. It contains all measured values and parameters—including the reference profiles (which may affect α_n) and the depth estimates (which affect w_n).

A full evaluation of the uncertainty, under a first-order approximation, is given by

$$u(h_T) = \sqrt{\mathbf{c}^{\mathrm{T}}\mathbf{U}\mathbf{c}} + u(0) \tag{6}$$

where

$$\mathbf{c} = \frac{\partial f}{\partial \mathbf{x}} \tag{7}$$

which is a vector of sensitivity coefficients, corresponding term-by-term to the contents of $\mathbf{x}$. This vector can be evaluated straightforwardly by differentiation. So, for example,

$$c_1 = \frac{\partial f}{\partial T_1} = w_1 (T_1 - T_{01}) \frac{\partial \alpha_1}{\partial T_1} + w_1 \alpha_1 \cong w_1 \alpha_1 \tag{8}$$

assuming the thermal expansion coefficient is evaluated using only the reference profile and is insensitive, therefore, to T_1.

The matrix $\mathbf{U}$ is the error covariance matrix for the elements in $\mathbf{x}$. This has to be developed from an understanding of the measurements and parameters. The element of U for row i and column j is

$$U_{ij} = u(x_i) u(x_j) r(x_i, x_j) \tag{9}$$

where $u(x_i)$ is the magnitude of uncertainty in value x_i and $r(x_i, x_j)$ is the correlation coefficient between errors in x_i and x_j. Note that it is errors (not uncertainties) that can be correlated. This happens, for example, when a common effect (source of error) contributes to the total errors in both x_i and x_j.

Estimating $\mathbf{U}$ therefore involves developing an understanding of the magnitude of the uncertainty in every measured value and parameter, and having a model for the degree of correlation between errors in different elements of $\mathbf{x}$.

The 0 terms in the definitions of f and the $u(0)$ terms remind us that the total uncertainty, $u(h_T)$, is more than the propagation of the value uncertainties. Discrete data have been vertically integrated using a numerical scheme, which itself is a source of numerical uncertainty. A well-chosen integration scheme will be unbiased, hence the "+ 0" formulation, and will provide an estimate of the numerical uncertainty, $u(0)$.

The above principles are very general and well established. To make them concrete, consider a simplified case of a profile of temperature measurements obtained by a particular sensor. This temperature sensor records digitised output, which effectively acts as a source of noise in individual measurements. The sensor is calibrated to a stated accuracy, and the calibration error is independent of temperature. Let us assume that these are the only error effects. (This scenario is just for illustration: a real sensor's error structure would likely be rather more complex.)

If the temperature data are quantised in bins of width t, the standard uncertainty that this introduces is

$$u_{\text{digit}}(T_i) = \frac{t}{2\sqrt{3}} \tag{10}$$

from considering the standard deviation of a top-hat distribution of full width t. Where vertical gradients in temperature are very gradual relative to t and the vertical sampling interval, there could be some correlation in the error in measured values, but in general for this sensor we assume that this is not the case and treat the digitisation error as random and independent between measured temperatures. So,

$$r_{\text{digit}}(T_i, T_j) = \delta_{ij} \tag{11}$$

where $\delta_{ij} = 0$ when $i \neq j$ and is 1 when $i = j$.

In contrast, the calibration error here is a constant bias. This is an example of a systematic effect, which means an effect that could be corrected for in principle if better information were available (a broad definition that includes the case of constant bias). Here, if calibration were performed again more precisely before deployment, a correction could be estimated. But all corrections are imperfect, and a smaller systematic error would remain, after this correction, associated with a smaller residual uncertainty.

Estimating the magnitude of uncertainty for an error effect can be done variously by statistical means (e.g., repeated laboratory evaluations), by simulation of the measurement process, by sourcing relevant information from published literature, by physical reasoning, etc. In this case, the uncertainty from calibration might be estimated from the manufacturer's stated goal for the calibration accuracy, corresponding to an uncertainty, $u_{calib}(T_i) = u_t$. Given the assumed nature of the error (constant bias), $r_{\text{digit}}(T_i, T_j) = 1$ and so

$$U_{ij} = \frac{t^2}{12}\delta_{ij} + u_t^2 \tag{12}$$

In a more realistic case, there may be several error-causing effects to combine with more complex correlation structures. Modelling the uncertainty contribution of all significant effects to **U** is a significant effort, but a full error covariance model would need to be developed only once for each source of profile data. While **U** is profile-specific (for example, the depths of measurement levels are different for each profile, and therefore, vertical correlation coefficients may differ), developing a community "recipe" for the error covariance matrix is feasible. Several studies address measurement error covariances (e.g., Kaplan et al. 2000; Levitus et al. 2012, hereafter L12, Gaillard et al. 2016), although in practice diagonal error matrices are usually implemented to reduce the substantial computational cost. Neglecting the correlation terms (off-diagonal terms in the matrix) results in an underestimation of steric thickness uncertainty that is potentially large.

Since the error in measured values is a priori independent of errors in parameters, the same recipe for the observation error covariance matrix would apply to a variety of structural/methodological choices, since the full covariance matrix can be constructed of a "data" block and a "parameter" block:

$$\mathbf{U} = \begin{bmatrix} \mathbf{U_{data}} & 0 \\ 0 & \mathbf{U_{param}} \end{bmatrix} \tag{13}$$

This subsection has presented the general equations for uncertainty estimation in a calculated quantity given the input measured and parameter values. Their application has been illustrated by a simplified case of calculating the uncertainty in the thermosteric influence on the thickness of a layer of sea water arising from the T profile uncertainties. Similar steps would address the uncertainty from the S and z values, as well as from the assumed parameters. The concepts behind the uncertainty analysis are well established, but systematic estimation of the necessary parameters, describing the dispersion and correlation of errors from the major effects, requires a significant level of effort.

3.4 Propagating Uncertainty Through Gridding, Interpolation and Integration

To create a full in situ-based estimate of the steric contribution to GMSL from a dataset of profiles involves gridding and/or interpolation, and spatio-temporal integration. One approach is to aggregate the steric thickness estimates from profiles onto a spatio-temporal grid, propagating the uncertainty results for individual thickness estimates to the gridded product. Alternatively, T and S observations may be gridded/interpolated, and steric sea level found by integrating Eq. 3 for those fields, the most commonly used method in the literature (Sect. 4). Gridded datasets aggregate available estimates for selected layers within the ocean on a spatio-temporal grid, typically 1 deg in latitude/longitude and monthly, or coarser. In many cells, there will be no or few data, the proportion being greater the less coarse the spatio-temporal grid. Nonetheless, aggregation of more than one set of observations will be necessary in some cells. The aggregation of two or more observations should account for their relative uncertainty, for example, by weighting more uncertainty data less heavily, and this is straightforward if uncertainty estimates have been associated with each observation. In addition, the limited sampling within the grid implies that, even for cells where data are available, there is a statistical uncertainty from having subsampled the natural variability over the bounds of the cell.

In principle, a full uncertainty model for gridded, interpolated or integrated SL estimates can be built using the principles and equations presented for a single profile in the previous subsection.

The same principles and equations for uncertainty propagation discussed in the previous section also apply to these transformations. Each of the transformations used in averaging to a grid, interpolating to give a complete field and integrating over ocean volumes can be viewed as scaling and reweighting the influence of individual observations in the final result(s). Each transformation can be recast in a form similar to Eq. (4), and Eqs. (6) and (7) can be used to propagate uncertainty to the final result, at least in principle. For gridded/interpolated/integrated products, the number of individual variables rapidly becomes large. The difficulties of propagating observation uncertainty are therefore to associate uncertainty and error correlation information with each observation and deal with the practicalities of organising the computation efficiently.

Two additional considerations arise: the modelling of uncertainty from sparse sampling and the use of "background fields" in interpolation.

The sampling uncertainty, when gridding observations, is an example where "u(0)" can be very significant. The number of vertical profiles at a grid location is usually small, and a given profile gives only a snapshot of the subgrid variability within that column. Thus, sampling uncertainty depends on the magnitude and nature of variability within the space–time box defined by the grid and the time–space pattern of sampling in the observational data. The observations themselves may have limited ability to provide an estimate of the variability, and additional information from ocean models and reanalyses may be useful in constructing a model for variability, such as magnitude of variations in T and S, the correlation length scales of variability in time and space and the degree of covariability between T and S, e.g., Cheng and Zhu (2016). The uncertainty from different patterns of observational sampling can then be estimated via simulation (Monte Carlo) methods.

Background fields may be used in interpolation. Interpolation procedures for sparse observations rely on the geophysically justified assumption that variability is correlated in space and time, so that the observations are informative about variability beyond the time and location of observation. Where no observations are sufficiently close to be informative, interpolation methods generally rely on a background field, such as climatology, to provide the most highly weighted estimate. In such areas, the interpolation uncertainty is larger, tending towards the uncertainty of the background field—e.g., the estimate of climatological variability. In terms of uncertainty estimation, the background field may be treated the same way as actual observations, albeit that the uncertainty will be relatively large. Note that use of a static climatology as a background field can cause bias (underestimation of change) in the face of a real geophysical trend, as has been discussed in reference to GMSL and OHC (e.g., Lyman and Johnson 2008; Boyer et al. 2016).

The weight of observations in determining the estimate for a particular analysis cell can be output from the interpolation method. Where the weight tends to zero, the interpolated value reflects only the background estimate. The fundamental limitation of data in estimating GMSL from in situ data alone is illustrated in Fig. 3. This figure relates to the EN4 dataset (Good et al. 2013) and shows the weight of profile observations for the surface layer of the interpolated analysis, which is an interpolation at 1 × 1 degree resolution in latitude and longitude and is monthly in time. The observation weight has been averaged globally for two depth ranges (see legend), and thus, 1 minus this weight indicates the overall dependence on the background field. Since the ocean volume gaining observational weight from the presence of one or more profiles in the grid box depends on vertical, horizontal and temporal correlation length scale parameters, it is principally the relative change in observation weight over time and variation in space that is instructive here. (With different estimates of the length scale parameters, the observation weights would have a different value, but would show similar spatial and temporal changes.) The steep rise in global mean observation weight over time between the year 2001 and 2005 principally shows how effective the programme of deployment of Argo profiling floats has been at improving the sampling of ocean profiles.

Several studies have discussed the impact of sparse sampling on trends (e.g., Cheng et al. 2015; Abraham et al. 2013). On average, the choice of the mapping method for irregularly sampled in situ measurements is the largest source of uncertainty (Boyer et al. 2016). Despite the tremendous technical developments of the ocean in situ observing system through the implementation of the Argo array, coverage is not yet truly global. The deep ocean below 2000 m (nearly half the volume) has very few measurements. The few that are available are from sparse, but very precise, hydrographic sections from research vessels (L12; Desbruyères et al. 2014). There are also gaps in the geographic coverage, with almost no floats in marginal seas (such as the Indonesian Sea; von Schuckmann et al. 2014), under sea ice or polewards of 60° (von Schuckmann et al. 2016). As a consequence, steric sea level estimates still differ at subseasonal to interannual timescale (Trenberth et al. 2016; Dieng et al. 2015; Abraham et al. 2013; VS11) and even show significant disagreement at decadal scale (von Schuckmann et al. 2016).

Another recent analysis (Good 2016) has analysed quantitatively the impact that sparse sampling can have on the reconstructed temperature trends over the global ocean. The study subsamples a spatially complete model ocean to construct a pseudo-observational dataset and then compares global mean temperature estimates from this pseudo-observational set with the model "truth". The study concludes that there is substantial scope for

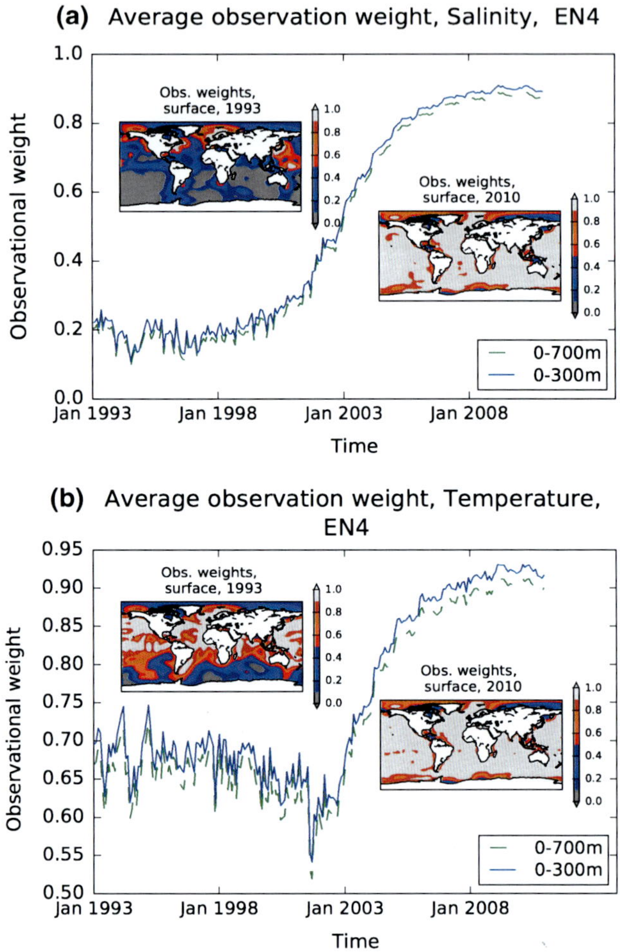

Fig. 3 Global mean observational weights from EN4, for the 0–300 m (*blue*) and 0–700 m (*green, dashed*) layers, salinity (**a**) and temperature (**b**). *Inset panels* illustrate the spatial variability of observational weights at the surface at the beginning (*top left*) and end (*lower right*) of the time period

background climatology to poorly represent the true temperature fields (see also Lyman and Johnson 2014) and for this influence of this climatology to cause systematic errors in reconstructed time series. The study also found that poor spatial coverage in the pre-Argo era also had the potential to introduce spurious variability comparable in magnitude to the true variability. The modelling of sampling uncertainty is thus seen to be critical to fairly estimating the uncertainty in steric sea level (see also Boyer et al. 2016).

A fundamental challenge is the sparseness of observations below 2 km, throughout the period to the present. Uncertainty about deep ocean (>2 km depth) heat content change and associated steric effects is profound. For example, Purkey and Johnson (2010) provide only basin-scale estimates of SL change below 2000 m, as sampling density is insufficient to resolve trends at higher spatial resolution. Their estimate of the uncertainty on the global mean trend is of similar order to the trend itself (0.113 ± 0.100 mm year^{-1}) where the uncertainty estimate is a measure of the 2σ variability in the temperature trend in each region, area weighted, converted to SL coordinates and combined in quadrature. Thus, although plausibly a modest component of steric sea level rise (seemingly between 0 and 20 % of the total), the uncertainty in this deep ocean change is a non-negligible component of the overall trend uncertainty.

4 Data and Uncertainty Methods in the Literature

4.1 Global Mean Estimates

While the focus of this review is the estimate of steric SL change from in situ measurements, here we discuss briefly other methods of estimation. Estimates of SL change that are truly independent can confirm or refute each other and also be used as diagnostic tools to highlight areas where our understanding may be limited.

Statistically robust upper ocean warming directly related to steric rise was found in both current in situ and indirect estimates (Fig. 4), but both of these approaches nonetheless face important challenges or limitations. The indirect estimate is bounded by data system availability, starting in the year 2002 (beginning of GRACE time series), and associated uncertainties are still too large for the extraction of warming trends given by Purkey and Johnson (2010) of about 0.1 ± 0.1 mm year^{-1} below 2000 m depth (von Schuckmann et al. 2014, Fig. 4). For the direct approach, uncertainties in OHC estimates arise from calculating global fields from temporally and spatially irregular data (mapping method), instrument bias corrections and the definitions of a baseline climatology from which anomalies are calculated (Boyer et al. 2016; Lyman and Johnson 2014; Abraham et al. 2013).

Fundamental advances in observing systems (e.g., Argo, gravimetry), in reanalysis systems (Balmaseda et al. 2015), in in situ data bias correction methods (Boyer et al. 2016) and in estimates of deep ocean contributions from in situ measurements (e.g., Purkey and Johnson 2010) have led to significant improvements in global steric sea level estimates over time. The warming of the upper 700 m during 1970–2014 (1993–2014) caused an estimated mean thermosteric rate of rise of 0.8 ± 0.2 mm year^{-1} (0.9 ± 0.2 mm year^{-1})

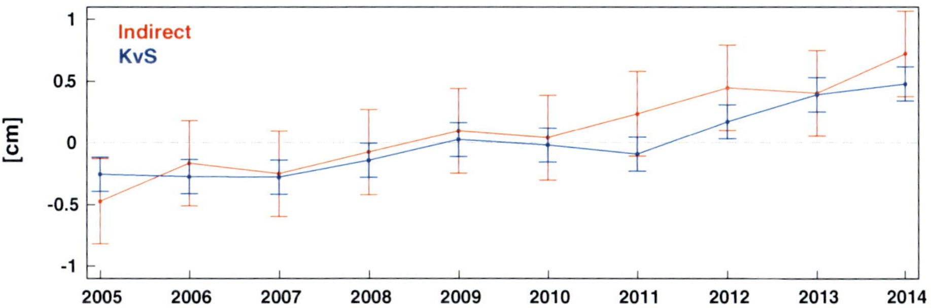

Fig. 4 Annual mean estimates of global thermosteric sea level during 2005–2014 from the surface down to 2000 m depth based on Argo measurements (updated after von Schuckmann and Le Traon 2011 (KvS, *blue*), and the indirect estimate through the sea level budget (*red*). Method, data use and uncertainty estimates for the KvS time series are described in von Schuckmann and Le Traon (2011). For the indirect approach, the evaluation method and the use of GRACE data are described in von Schuckmann et al. 2014; for estimates of total sea level, the gridded product from ESA CCI is used (product version V1.1_20151113, see http://www.esa-sealevel-cci.org/products and Ablain et al. 2015 for more details)

(90 % confidence), which is 30–40 % of the observed rate of GMSL rise (Cheng et al., 2015, Chambers et al. 2016). Steric contributions in the 700–2000 m depth layer account for 0.2 ± 0.1 mm $year^{-1}$ over the altimeter era and about 0.1 ± 0.1 mm $year^{-1}$ below 2000 m depth (Purkey and Johnson 2010, Chambers et al. 2016). Decadal steric trends for the Argo era (2005–2014) are of similar size and amount to 0.8 ± 0.2 mm $year^{-1}$ for the upper 2000 m depth (Table 1; Fig. 3) and 0.9 ± 0.2 mm $year^{-1}$ for the entire ocean depth when taking into account the Purkey and Johnson (2010) estimate. In summary, the most recent steric trend estimates based on the direct and indirect approach are in considerable agreement with associated error bars (Fig. 4).

The indirect estimate as derived from the sea level budget is an updated version of the von Schuckmann et al. (2014) methodology, except that the ESA CCI dataset is used (see figure caption of Fig. 3 for more information on the data used), taking into account error bars of both satellite altimetry (Ablain et al. 2015) and ocean mass from GRACE (see von Schuckmann et al. 2014 for more details on error estimates). A review on data processing and uncertainty methods in the literature is further discussed in Sect. 4.2 and 4.3.

4.2 Temperature and Salinity In Situ Measurements

In constructing a SL analysis and uncertainty budget, investigators will generally undertake quality control and correct for known problems with the data. Biases in the temperature measurement systems have been extensively studied and can be both widespread and systematic in nature (e.g., Abraham et al. 2013, Cheng et al. 2015). As discussed in Sect. 3, there remains post-correction uncertainty that ideally should be estimated. More generally, the selection, quality control and bias correction of input data all contribute to differences SL analyses, reflecting structural uncertainty (e.g., Cheng et al., 2015). A basic choice is whether to build the analysis from raw profile data, or a dataset that has already undergone some processing to remove unreliable profiles and/or correct for biases. Table 3 summarises the input data, additional data sources and interpolation method for several studies.

The in situ observing system is a complex and constantly evolving network of instruments. Substantive discussion of the system requires careful and complex analysis and is outside the scope of this study; however, here we highlight two issues that are areas of substantial ongoing research. The first is the sparse and constantly evolving nature of the in situ observing system. This presents substantial challenges to any potential analyst. Spatially and temporally sparse sampling is not spatially uniform, and temporal coverage (together with quality of coverage) can vary substantially by region. The construction of a spatially and temporally uniform representation of the Earth's oceans from such input data is addressed in many studies (e.g., Good 2016; Ishii and Kimoto 2009; von Schuckmann and Le Traon (2011) and can be the major source of structural uncertainty in estimates of thermosteric SL change (Boyer et al. 2016).

Table 1 Most recent estimates from the GRACE/Argo "golden" era starting in 2005 (note that Argo programme has started in the year 2000, but has reached near global coverage from the year 2005 onwards, e.g., von Schuckmann et al., 2009). Trends are quoted as the mean plus or minus the reported uncertainty based on the spread of results for different estimates of a given method

Method	Steric sea level trend (mm $year^{-1}$, 2005–2014)
Direct estimate	0.8 ± 0.2 (0–2000 m) 0.9 ± 0.2 (full depth, this study (Fig. 3) and Purkey and Johnson (2010))
Indirect estimate	1.2 ± 0.4

Second, the evolving nature of the observing system presents its own challenges. Each observing solution has its own strengths, limitations and biases, which often evolve in time, or with instrument design changes. Particularly in the pre-Argo era, systematic removal of fall speed biases has been a major focus of quality control (QC) efforts. The biases may depend on more than one factor, for example, water temperature variations (Abraham et al. 2016) or XBT (eXpendable Bathythermograph) or MBT (Mechanical Bathythermograph) model (Abraham et al. 2013). Several current correction schemes exist and are in widespread use, for example Ishii and Kimoto (2009), Levitus et al. (2012), Wijffels et al. (2008) and Gouretski and Reseghetti (2010), and the selection of a given method contributes to differences between products as structural uncertainty (Boyer et al. 2016). Abraham et al. (2013) present a substantial review of bias correction methods and results for non-Argo type in situ observing instruments, and a comprehensive review of the current state of knowledge of these systems.

A detailed analysis of the in situ observing system, its strengths and limitation and its evolution over time is outside of the scope of this work; however, a recent major review article by Abraham et al. (2013) addresses all of these issues comprehensively.

Any search of the literature will reveal a large number of estimates of steric SL change from in situ data. Table 2 summarises recent estimates of SL change from analysis of temperature and salinity in situ measurements, from Domingues et al. 2008 (the basis for the estimates used in IPCC AR5,[1] Church et al. 2013; note the "short names" in the table which will be used for brevity hereafter), to present day. It does not include estimates from reanalyses, which have a more complex relationship between input data and SL estimate, or indirect estimates from satellite data, which act as independent data that should ideally be explained in terms of the in situ estimates (Table 1).

Most of the estimates are not directly comparable as they cover different time periods and depths. This illustrates a persistent problem in reviewing the literature and assessing the consensus on sea level trends. This highlights the need for more systematic, coordinated efforts. Currently, it is difficult and time consuming to compare methodological differences or improvements as time and depth considerations can strongly affect trends over these short (on ocean circulation timescales) time periods. Nonetheless, on comparing estimates with similar or identical analysis domains, it does appear that methodological differences result in substantially different trends.

The estimates in Table 2 are not independent of each other in terms of source data: common or overlapping source datasets are used, inevitably. Most of the estimates that include the pre-Argo period use their contemporary version of the World Ocean Database (IQuOD) also addresses (Boyer et al. 2013), sometimes supplemented by other data sources. WOD aims to provide the most comprehensive available temperature and salinity profile data and undergoes a continuous programme of update and review. Those estimates which cover only the Argo period, and use information only from Argo floats, may get their information direct from the Argo teams (e.g., VS14). However, data from the Argo floats are also included in the WOD data store; therefore, none of the estimates is wholly independent of others in terms of source data (Table 3).

Two global data centres exist, i.e. NODC (https://www.nodc.noaa.gov) and Coriolis (http://www.coriolis.eu.org). However, each study may also make use of some auxiliary datasets and conduct additional QC on their input database. IK09 use XBT observations from the Global Temperature and Salinity Profile Programme (GTSPP), Japanese Maritime

[1] The Fifth Assessment Report (AR5) of the United Nations Intergovernmental Panel on Climate Change (IPCC).

Table 2 Summary of recent literature estimates of global mean sea level trend

Study	Short name	Trend, mm year^{-1}	Uncertainty estimate mm year^{-1}	Depth(m)	Time	Spatial resolution (degrees)	Temporal resolution
Ishii and Kimoto 2009	IK09	0.294	±0.057	0–700	1951–2005	1 × 1	Monthly
Levitus et al. 2012	L12	0.41	–	0–700	1955–2010	1 × 1	Annual
Levitus et al. 2012	L12	0.54	–	0–2000	1955–2010	1 × 1	Annual
Llovel et al. 2013 (L12)	LL09	0.42	±0.12	0–700	1960–2010	1 × 1	Quarterly
Llovel et al. 2013 (IK09)	IK09	0.39	±0.12	0–700	1960–2010	1 × 1	Monthly
IPCC, Church et al. 2013	AR5	0.6	±0.3	0–700	1961–2003	1 × 1	Monthly
Domingues et al. 2008	D08	0.52	±0.08	0–700	1961–2003	1 × 1	Monthly
Storto et al. 2015 Objective analyses	StOA	1.11	±0.08	Full depth	1993–2010	1 × 1	Monthly
Ishii and Kimoto 2009	IK09	1.23	±0.295	0–700	1993–2005	1 × 1	Monthly
Cabanes et al. 2013	CORA	0.64	±0.12	10–1500	2005–2010	5° Lat, 10° Lon	Monthly
Cabanes et al. 2013	CORA, Argo only	0.58	±0.1	10–1500	2005–2010	5° Lat, 10° Lon	Monthly
Von Schuckmann and Le Traon 2011	VS14	0.5	±0.1	10–1500	2005–2012	5° Lat, 10° Lon	Quarterly

Self-Defence Force (JMSDF) and sea surface temperature data (Ishii et al. 2006). L12 use data to extend the World Ocean Atlas (WOA) from 2009 to the end of 2010. EN3, which forms the input to two of the studies (D08 and Storto et al. 2015), comprises WOD05 plus data from GTSPP, Argo and the Arctic Synoptic Basin Wide Oceanography (ASBO) project (Ingleby and Huddleston 2007), CORA comprises data from the Coriolis data centre, comprising European ship observations, XBT and other profiling systems, and Argo data. The data are global from 1990 onwards. ARMOR use all T and S profiles from the EN3 dataset, with the exception of those labelled as Argo. Argo data are from the Coriolis data centre up to 2009 and combine these data with SST and satellite altimeter data.

Additional quality assurance (QA) procedures may be used to retain only the "best" data, or, as in IK09 and EN3, to thin the data where many profiles exist in a single location. Thinning the data substantially reduces the cost of optimal interpolation (OI) in the Ishii et al. (2006) data set, allowing several estimates of trends using different QA procedures to be evaluated.

These five distinct starting datasets are used to construct seven gridded products. (D08, L12, IK09, VS11, EN3, CORA and ARMOR) and, from these, twelve estimates of steric

Table 3 Construction of SL trend estimates from temperature and salinity in situ profiles—summary of methods from the literature

	Domingues et al. 2008	Levitus et al. 2012	Ishii and Kimoto 2009	Von Schuckmann and Le Traon 2011	EN3 (Ingleby and Huddleston 2007)	CORA (Cabanes et al. 2013)	ARMOR (Guinehut et al. 2012)
Short name	D08	L12	IK09	VS14	EN3	CORA	ARMOR
Input data	EN3	WOD 2009	WOD 2005	Argo only	WOD 2005	Coriolis	SST, altimeter, EN3, Coriolis
Additional data	No	Yes, see text	Yes, see text	No	Yes, see text	No	No
XBT bias correction	Wijffels et al. 2008	L12	IK09	N/A	Wijffels et al. 2008	Hamon et al. 2012	N/A
MBT	No	Yes	Yes	N/A	Yes	No	N/A
Argo bias correction	Yes their own QC	Yes Argo teams' corrections	Yes Ignore biased data	Yes Argo teams' corrections	No	Yes Argo teams' corrections	Yes Argo teams' corrections
Background climatology	Alory et al. 2007	Locarnini et al. 2010 (WOA)	Locarnini et al. 2010 (WOA)	Von Schuckmann et al. 2009	WOA 98	Not stated	N/A
OI/binning method	Reduced space optimal interpolation	Objective mapping	Objective mapping	Binning, box averaging	Bell et al. 2000	VS11	N/A
Sparse data treatment	Kaplan et al. 2000	Relax to climatology	Relax to climatology	Gap filled from climatology	Relax to climatology	VS11	N/A
Time-varying salinity	No	Yes	No	Yes	Yes	Yes	Yes

SL trends (Table 2). Of the studies reviewed here, IK09 and D08 do not include salinity and therefore strictly provide an estimate of thermosteric SL rise. The other estimates do include time-varying salinity.

4.3 Published Sea Level Trends

During the construction of their estimates of SL change, researchers must make decisions at each step regarding data and methods. Many decisions reflect reasoned judgements, where alternative conclusions could also be defended. This is precisely the value of multiple groups addressing themselves to such work: the different approaches adopted ensure there is some exploration of structural uncertainty across different attempts. The degree to which the structural uncertainty is explored is difficult to assess where no systematic intercomparison of methods has been attempted. There are also procedures that may suppress some structural uncertainty: results of new methods are inevitably compared during development with published estimates, which may lead to reduced diversity of outcome. Nonetheless, a useful step in understanding the degree to which structural uncertainty is explored is reviewing the range of data and method choices present in the current literature. This section is a contribution to such a review and summarises the nature of several key choices about data and method. These are presented against the background of the uncertainty concepts introduced in Sect. 3; we discuss, to the degree possible from a literature review, how each operation might contribute to a formal uncertainty budget and how uncertainty is handled at each stage of the process.

There is substantial community effort to address these apparent discrepancies, by systematically comparing different methods for construction (e.g., Llovel et al. 2013), although these often focus on ocean heat content (e.g., Lyman and Johnson 2014; Cheng et al. 2016; Boyer et al. 2016). A particular community effort had been developed to quantify uncertainties in OHC estimates which arise from calculating global fields from temporally and spatially irregular data (mapping method), instrument bias corrections and the definitions of a baseline climatology from which anomalies are calculated (Boyer et al. 2016). In this context, the International Quality controlled Ocean Database also addresses this issue through a systematic quality analysis of the historical record aiming to define internationally agreed standards and guidance.

4.4 Spatial Structures of Steric Sea Level Change

The global mean estimates of steric SL change are associated with substantial regional variability. This is important when discussing interpolation and data manipulation as sparse sampling in some locations may have a disproportionately large effect, for example, if the water is warm or salinity effects are important. These regional effects are illustrated using data from the EN4 analysis (Good et al. 2013).

Strong warming trends can be observed at northern mid-latitudes to a depth of several hundred metres (Fig. 5) and are known to be located in the North Atlantic (Rhein et al. 2013). These are accompanied at very high latitudes by freshening trends (Fig. 5b), although salinity sampling in this region is sparse, in particular before the Argo era, so the relative strength of this feature may be poorly constrained (Good et al. 2013).

The monthly, gridded data are converted to contributions to sea level (SL) change relative to the time mean via the expansion and contraction coefficients (Eq. (3), Fig. 2). Both α and β are, primarily, functions of temperature (rather than salinity) given pressure (Fig. 1), so trends in temperature and salinity have the most effect in the tropics and at

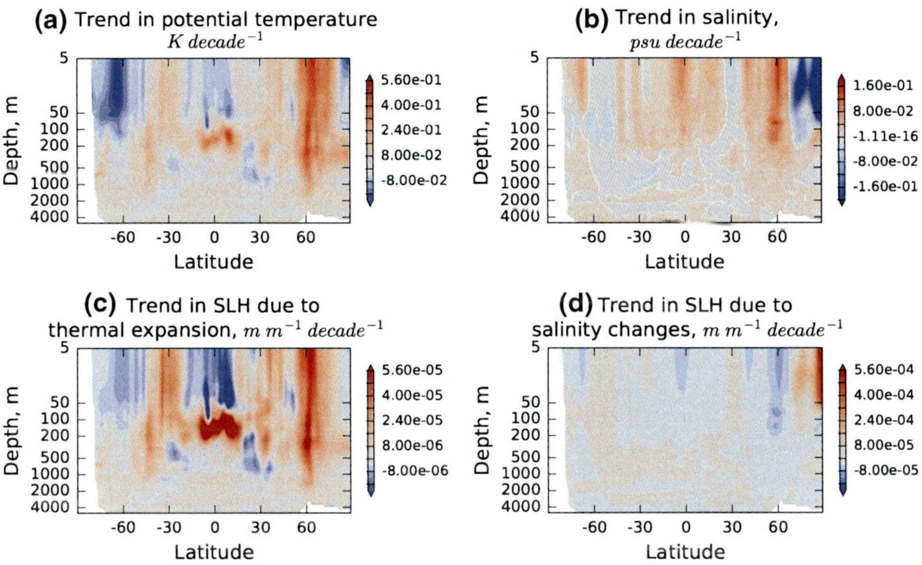

Fig. 5 Temperature and salinity trends for 1993 to 2010 (**a**, **b**) and the resulting trend in thermosteric and halosteric sea level change (**c**, **d**), expressed in metres of sea level rise per metre depth of ocean, per decade

depths of the order of a few hundred metres or less. The larger relative impact of the subsurface tropical warming on SL change is very clear in Fig. 5c. Note that the halosteric trends (Fig. 5d) for this period have a smaller effect on regional SL than do the observed temperature trends; however, there are some areas where they are not negligible, for example around 60 °N. The total trend in SL is calculated by integrating the gridded product vertically (Fig. 6a–c). For comparison, the total observed SL trend, from ESA's Climate Change Initiative (CCI) Sea Level project (Ablain et al. 2015), was shown (Fig. 6d). The regional impact of salinity in the North Atlantic reduces the total steric trend substantially relative to the thermosteric trend alone, in common with previous studies (e.g., Durack et al. 2014)

Overall, mean estimates of global mean steric sea level from different methods have reasonable consistency, although there is considerable spread in individual estimates. Such spread reflects uncertainties in data and methods that are not fully understood and quantified. The global mean reflects the net effect of trends that have considerable spatio-temporal structure and, thus, good understanding of uncertainty in key areas such as the North Atlantic and Pacific oceans is required.

4.5 Interpolation and Regridding/Rebinning

Interpolation and gridding sparse and generally inhomogeneous data are a source of structural and sampling uncertainty, additional to the uncertainty from propagation of value uncertainties. This section discusses briefly the methods used to produce gridded products.

Interpolation methods can generally estimate the statistical uncertainty (which may be referred to as "error" estimate) in interpolated values, based on some underlying model of spatio-temporal error covariance in the input data. Value uncertainty estimates are often used in interpolation to influence the relative weights of observations and the background field, for example. This doesn't in itself ensure full propagation of value uncertainty

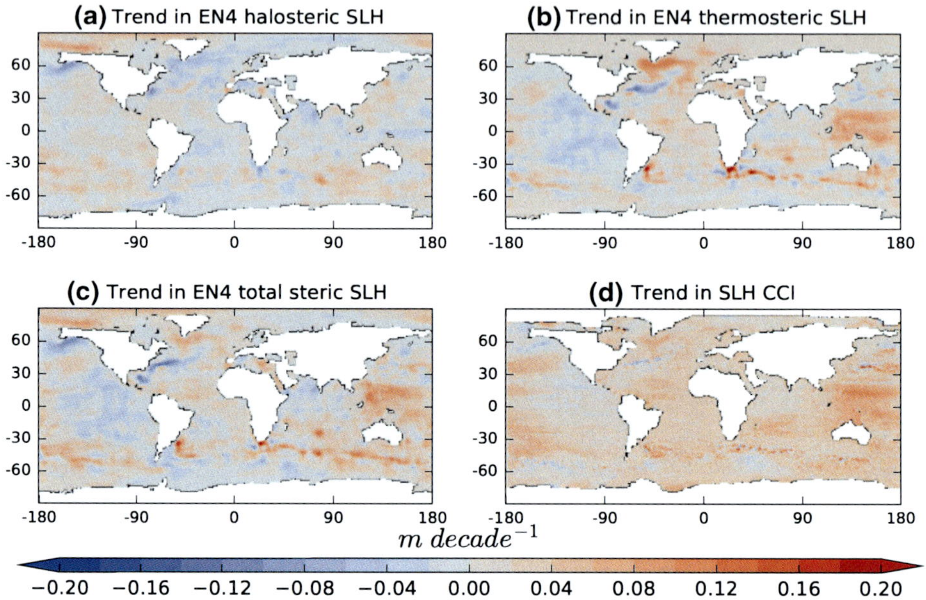

Fig. 6 Spatial distribution of **a** halosteric trend, **b** thermosteric trend, **c** halo- and thermosteric (total steric) trend and **d** total sea level trend (SLH) from satellite altimetry from the ESA CCI project. All trends are evaluated over the period 1993–2010

through the interpolation procedure, if value error correlation structures are simplified or neglected. Uncertainty from limited sampling may be accounted for in the interpolation process using statistics of variability that are estimated to account for unsampled variability on scales of more than one grid cell; sampling uncertainty associated with the relationship between individual profile measurements and grid cell means (so called "intra-box" effects, see Kaplan et al. 2000) should be explicitly recognised (e.g., as in Gaillard et al. 2016).

L12 conduct an objective analysis as described in Locarnini et al. (2010) to produce a gridded dataset. IK09 also use an objective analysis technique, from Derber and Rosati (1989), as do EN3 (described in Bell et al. 2000).

These techniques must all use some estimate of time and length scales across which profiles may be correlated, in order to weight the observations. EN3 use a simple assumption that length scales are invariant across the globe and use values of 300 km latitude and 400 km longitude. The method used in the L12 study comprises several iterations of their objective analysis method using different length scales of influence, as a method of capturing different scales of variability, from 888 to 444 km (Locarnini et al. 2010). IK09 use spatially varying decorrelation scales of 300 km in the horizontal and 10 m in the vertical at the sea surface, and which increase linearly by 30 km/100 m and 6 m/100 m away from the surface.

Neither L12 nor EN3 use an explicit temporal correlation window, although the EN3 for any given month are constructed using the previous month's estimate as input, so some correlation is implicitly included. The decorrelation timescale in Ishii et al. (2006) is 15 days at the sea surface and increases linearly as 2 days/100 m.

D08 use an optimal interpolation technique described in Kaplan et al. (2000), which explicitly calculates a space covariance matrix over the analysis time period. This method includes instrumental error in the analysis and also attempts to account for intra-box sampling uncertainty by calculating the box variance, under the assumption that the error [sic] in the box value is related to high-frequency variability within the box.

ARMOR also use an optimal interpolation method, but merge in situ data with satellite products of SST and altimetry. This study optimally interpolates temperature and salinity fields which have been synthesised from satellite data, with the T and S in situ measurements. The synthetic fields are generated using statistical relationships between surface and subsurface measurements, compiled using in situ data from the well-sampled Argo period.

CORA and VS14 use the method of von Schuckmann and Le Traon (2011, hereafter VS11). Both studies use a box averaging method rather than interpolation. Within each 5 degree latitude × 10 degree longitude × 3 month box, profiles are averaged using a weighted average. The weights are represented by a covariance matrix between each pair of observations, where the values are calculated using correlation scales between pairs of observations of 15 days and 150 km. This addresses the variable sampling density across the dataset by ensuring that more comprehensively sampled regions are not over-represented. For reported uncertainties, VS11 use the method of Bretherton (1976) to weight the grid box variance by the inverse of the covariance matrix. The resulting error is, if the correlation terms are non zero, larger than an estimate in which each observation is assumed to be independent.

4.6 Considerations Regarding Background Climatology

The majority of the studies construct a background climatology to be used in dataset construction and also commonly for gap-filling sparse data. The construction and choice of the ocean climatology used for interpolation requires care (Gaillard et al. 2016). For example, the rapidly increasing density of observations over the past two decades means that a simple average over the available observations would be biased, and the extent of the bias would be spatially varying, as some areas of the globe have been proportionately better sampled for longer. Various methods are used to construct climatologies to avoid this and similar pitfalls.

L12 and IK09 use annual climatologies from the WOA (Locarnini et al. 2010). The climatology is the average of five 10-year climatologies, which gives a reweighting that approximately accounts for the much greater volume of data in later years. In addition, a constraint of vertical stability has been imposed on the temperature/salinity climatologies—i.e. potential density of one level is nowhere lower than the density of the next shallowest level. Unstable climatological profiles can arise because there are many more temperature than salinity measurements and because of errors in measured values. The climatologies are often used to fill data gaps in regions where observations are sparse. L12 and previous estimates by Levitus et al. deal with sparse data by constructing running 5-year estimates of sea level at 1 × 1 degree horizontal resolution. Temperature anomalies are constructed for each 1 degree mean temperature value by subtracting the WOD climatology (Locarnini et al. 2010). Despite these anomalies being composited over a running 5-year period, some gaps in the record still remain. This method uses the WOD background climatology described above to fill gaps and assumes zero anomaly where there is no direct information.

EN3 combine the WOD 98 climatology with their objective analysis from the previous month. The resulting analysis therefore contains some signal persistence through the record. The EN3 background field also relaxes to climatology in the absence of data profiles, reducing the weight of the observations relative to the climatology gradually over 300–400 km. This background field is then used as input to the objective analysis for the following month, effectively widening the temporal window over which observations can have an influence.

Some studies construct their own climatology from the input data. D08 use an optimal interpolation technique developed by Ridgway et al. (2002), which accounts for variable sampling densities at different locations via a weighting function, which also takes into account local bathymetry and land barriers.

VS11 use reference climatology from a previous study (von Schuckmann et al. 2009) to fill gaps in individual Argo profiles. As this study deals only with the Argo era, changes in sampling density need not be a consideration. However, the effect of the choice of the background climatology has been taken into account in the error bar of the global VS11 estimate.

Both IK09 and VS11 make some investigation of the effect of the choice of background climatology. VS11 use a second climatology and calculate the standard deviation of the difference between the two resulting global mean time series. The final estimate of uncertainty is calculated as the quadratic sum of the global mean, area-weighted uncertainty and the uncertainty associated with the climatological time series.

IK09 compare the use of the WOA climatology for interpolation with using their own climatology, as the long-term time mean from the XBTs after they have applied their correction is quite different (0.236 ± 0.066 and 0.294 ± 0.077 mm year^{-1}—see their Table 5 for the full comparison of their estimates exposing structural uncertainty).

4.7 Uncertainty in Trends and Global Means

Uncertainty in a geophysical trend comprises two elements: the statistical uncertainty in fitting a trend line to a finite length of time series in the face of geophysical variability and errors in measured points, plus the uncertainty in the stability of the observing system. Stability is the constancy of the systematic error between the measured values and the truth. The latter component is often neglected, not so much because it is known to be negligible, but because stability is difficult to assess. In the case of steric sea level, the changes in the observing system described earlier do raise the likelihood that systematic effects in the observing system have changed over time. The most obvious instability is the switch from dominance of XBT to Argo profiles, given that different corrections to which these different sources are subject. Moreover, global sampling density had significantly decreased during this transition period of changing observing systems (e.g., Cabanes et al. 2013). To estimate the observational stability, the temporal correlations of errors across the time series could be quantified, although in practice this is complex. This is an area where further research will be beneficial.

It is often unclear exactly what, in terms of uncertainty, is accounted for in quoted uncertainty bounds. Estimates of uncertainty in the published literature may not be directly comparable as they are not representing the same uncertainties. This can lead to apparent contradictions within or between studies, unless the precise nature of the bounds is explicitly made clear.

In some cases, quoted uncertainties in trends are based upon the statistical uncertainty. Here, the question arises as to whether auto-correlation along the time series is properly accounted for: not doing so will tend to underestimate trend uncertainty.

This highlights the need to enhance the discussions of uncertainties for global estimates and to find a common way to directly compare the different approaches and their related scientific interpretations.

In this review, L12, IK09, VS11, LL12, D08 and CORA propagate in some way the uncertainties of the gridded product to the final, global mean estimate. Each study propagates uncertainty into its final estimate differently, in addition to the range of methods used to incorporate uncertainty information discussed above and the types of uncertainty that are accounted for in each estimate.

For example, L12, using the method described in Antonov et al. (2002), combine standard errors from the gridded product as a weighted sum (where the weights are the partial derivatives with respect to the dimension being combined, or as standard RMS if the errors are considered independent) to give global mean error estimates. However, the uncertainties quoted for the trends reflect the statistical uncertainty in the straight line fit and do not exploit any of the previously calculated uncertainty information explicitly. The evolution of the observing system over time raises the likelihood of observational instability (temporal covariance of errors) in steric sea level time series, as well as decreasing uncertainty for the recent decade. It is relatively common for uncertainty estimates in the global mean to be used to provide weights when fitting a trend line. VS11 and CORA use a weighted least squares fit for their trend uncertainty estimation, in which the weights are the propagated global mean uncertainty weights as discussed in Sect. 5. LL09 use the Levitus et al. (2009) gridded temperature uncertainties and weight them according to vertical and horizontal correlation length scales. As in VS11, these weights are used to calculate weighted least squares fit for the final quoted trend. D08 also propagate gridded errors into the global mean using their chosen interpolation method (Kaplan et al. 2000), although it is unclear whether these estimates are propagated into the fitted trends in the final estimate. Generalised least squares (GLS) methods (Aitken 1934) yield trend estimates that account for estimated data uncertainty and temporal error covariance. Although more complex to implement, application of GLS may reveal trend uncertainty to be greater than previously found using (weighted) ordinary least squares.

Iterative approaches to trend uncertainty also are used. Storto et al. (2015) do not provide information on grid box level or time series uncertainty. Rather, they provide as uncertainty estimates the 95 % confidence limits on their trend using a bootstrap (i.e. subsampling of variability) method. Intercomparison of a range of trend estimation methods would increase understanding of their applicability to the case of steric sea level change.

5 Discussion

The present study reviews estimates of steric SL with a particular focus on uncertainty estimates. Resolving the problem of representing global mean steric SL change using point measurements, which provide a subsample of the full ocean state, requires considerable effort and scientific rigour. Many scientifically sound methodological choices are possible. Within the current literature, the diversity of approach is substantial (e.g., compare VS11, L12, D08) and provides a useful method of exposing structural uncertainty. Diversity of approach is strength. Nonetheless, in any area of science, structural uncertainty can be underestimated. Thus, there is a need for realistic "internal" assessment of uncertainty, as well as by looking at the diversity of results.

Sampling uncertainty is a major contributor to total uncertainty and surely is dominant for certain epochs and ocean domains. Nonetheless, value uncertainty should be assessed via rigorous uncertainty propagation and accounting for the correlated nature of errors, not least as a contribution to constraining the steric change in parts of the ocean that are inadequately observed. Community efforts are underway with the Global Ocean Data Assimilation Experiment (GODAE) Ocean View (https://www.godae.org/OSSE-OSE-home.html) and the European initiative AtlantOS (https://www.atlantos-h2020.eu/) through observing system evaluations and observing system simulation experiments (e.g., Halliwell et al. 2014). However, these efforts should, in future, be supported by a community effort to quantify error covariance estimates for classes of profile observation, in addition to the extensive existing body of work on observational bias (e.g., Abraham et al. 2013)

Historically, there has been enormous community effort to maintain, improve and understand the in situ record (e.g., Wijffels et al. 2008; Hamon et al. 2012; Abraham et al. 2013; Boyer et al. 2016 and references therein). Large-scale corrections to the in situ record can have a substantial impact on the evolution of estimated SL change (e.g., D08, IK09). As the true state of the ocean can never be known, it is important that not only is the impact of the corrections assessed (IK09), but that inevitable post-correction uncertainties are also acknowledged and, where possible, estimated.

Recently, as recommended under the CLIVAR[2] research focus CONCEPT-HEAT (http://www.clivar.org/research-foci/heat-budget), a systematic comparison of different methods on common data is encouraged. For the pre-Argo era, such an activity has started under the project IQuOD (www.iquod.org), and a community paper is under way (Boyer et al. 2016). Still, efforts are needed for Argo era data—this, and the underlying international collaboration are of particular importance to support observing system's development into the future.

The development of the Argo observing network has undoubtedly been the single largest modifier to our understanding of the global oceans in recent years. There is of course the need of continued international maintenance, and, if possible extend (e.g., to the deep ocean), the in situ observing system; but there is also scope to improve our exploitation of existing data; through the systematic comparisons outlined above, through formal methodological assessment and through comparison with independent data such as that from the ESA CCI projects (Ablain et al. 2015; Merchant et al. 2015)

Other systematic approaches can include the use of modelled "test" data, to compare methods and aspects of methods by testing their ability to reproduce "known" model results under realistic sampling (Good 2016). Such approaches should be expanded into systematic benchmarking studies (Chandler et al. 2012) to build maximum trust in steric sea level products.

In the absence of such systematic benchmarking comparisons, it can be difficult to assess whether observed differences in SL estimates and their associated uncertainty estimates are consistent. While some individual studies do discuss some aspects of structural uncertainty (e.g., VS11, Lyman and Johnson 2014, IK09), systematic comparison across studies is not generally facilitated and this may mask important differences that arise from changes in the state of the global oceans. The time period under consideration can, for example, have a substantial effect not on the trend estimate, but also on the impact of methodological choices and bias corrections. In the case of XBT bias corrections, the

[2] CLImate and ocean: VARiability, predictability and change. This is one of the four core projects of the World Climate Research Programme (WCRP).

effect in D08 and IK09 is, on the face of it, very different (0.262 ± 0.063 to 0.236 ± 0.066 mm year^{-1} in IK09, an increase of "about 50 %" on previous literature estimates in D08). Closer examination of the time periods under discussion, however, reveals that similar features in the global mean time series are resolved in each case when the bias correction is applied, and that the extra 10 years of ocean data in IK09 alter the slope of the trend, but not the shape of the features in the common time period. A true assessment of this and of any similar case is limited; however, as the D08 study does not explicitly compare the same method, merely referring to previous estimates in the literature, and methodological differences cannot be ruled out.

These approaches also require the deployment of clear (unambiguous) and concise language for the discussion of error and uncertainty. In the absence of such a language, even when studies are comparing like time periods, it may not be obvious that different studies mean very different things by their quoted uncertainty estimates. Commonly, but not exclusively, the quoted uncertainty in a trend is provided using a standard least squares fit. However, the fit may (or may not) be weighted by uncertainties propagated from an optimal interpolation method (VS11), may be estimated by a bootstrap subsampling method (Storto et al. 2015) or may simply be a measure of the deviation of the fit from a straight line (L12). Wherever a quoted uncertainty estimate in a trend is based only on the fitting uncertainty in the face of geophysical variability, important aspects of the true trend uncertainty, such as the degree of instability of the observing system, are omitted. Thus, even if the meaning of each quoted uncertainty is clear, different meanings of quoted uncertainties make interpretation and comparisons between studies difficult (e.g., Table 1). It cannot necessarily be assumed that the result with the smallest quoted uncertainty is in reality the most certain trend estimate.

Developing a well-characterised uncertainty estimate is intrinsically beneficial. Clear discussion of uncertainty sources and formal propagation of uncertainties allow better understanding of differences between datasets, improving our understanding of the underlying physical quantities. Systematic assessment of whether differences in estimates are consistent with their quoted uncertainties is a useful diagnostic of the degree of real understanding of the phenomenon. Results that agree too well (within quoted internal uncertainties) can indicate some lack of independence ("herding" effects) in the development of the datasets, raising concern about the true uncertainty of a consensus picture. In a field in which most studies share at least some input data, clear discussion of uncertainty must be necessary to tease out such effects and their sources. Where results diverge more than expected given their quoted uncertainties, this suggests there is significant influence on the result of factors whose implications are not yet well understood (at the level of the community working in the area).

Historically, ocean scientists have pioneered international approaches to global science. Large-scale data collection and data quality initiatives are often internationally coordinated. Significant progress may accrue from coordination for the development of improved measurement uncertainty information, innovative, rigorous systems of uncertainty propagation and systematic approaches to intercomparison, such as benchmarking. Efforts to establish error covariance estimates and uncertainty estimates for both the current observing network, and, where possible the historic record, are already underway (e.g., Guinehut et al. 2012, CLIVAR, IQuOD). Likewise, progress can be made via further coordinated development and comparison of methods of estimating "intra-box" sampling uncertainty (e.g., Kaplan et al. 2000; Ingleby and Huddleston 2007; Good et al. 2013) and large-scale sampling uncertainty, via coordinated experiments in the reconstruction of known model fields (e.g., Good 2016).

There is therefore great scope to develop community discussions around formalising approaches to uncertainty. Progress in quantifying steric sea level uncertainty will benefit from greater clarity and transparency in discussions of uncertainty in the scientific literature. Rigorous, shared understanding across the community of how uncertainty in measurements should be expressed will enable researchers to better understand and build on the results of others and improve comparability of uncertainty estimates. Progress may be accelerated by developing and sharing rigorous community "recipes" for common problems relating to correct quantification of uncertainty. We identify in this review the potential value of such recipes for quantifying the error covariances in observations and from sparse sampling and for estimating and propagating uncertainty across spatio-temporal scales. International standards for the estimation, propagation and expression of measurement uncertainty exist which are applicable to the problem of steric sea level change. The overall conclusion of this review is to emphasise the importance of progress in quantifying and expressing uncertainty with greater rigour.

Acknowledgments The authors acknowledge the International Space Science Institute for supporting their attendance at a workshop hosted in February 2015. Claire MacIntosh and Chris Merchant are supported by the National Centre for Earth Observation, and the UK's Natural Environmental Research Council (NERC).

References

Ablain M, Cazenave A, Larnicol G, Balmaseda M, Cipollini, P, Faugère Y, Fernandes MJ, Henry O, Johannessen J A, Knudsen P, Andersen O, Legeais J, Meyssignac B, Picot N, Roca M, Rudenko S, Scharffenberg M G, Stammer D, Timms G, Benveniste J (2015) Improved sea level record over the satellite altimetry era (1993–2010) from the Climate Change Initiative project, Ocean Sci 11:67–82, doi:10.5194/os-11-67-2015, dataset doi:10.5270/esa-sea_level_cci-MSLA-1993_2014-v_1.1-201512

Abraham JP, Baringer M, Bindoff NL, Boyer T, Cheng LJ, Church JA, Conroy JL, Domingues CM, Fasullo JT, Gilson J, Goni G, Good SA, Gorman JM, Gouretski V, Ishii M, Johnson GC, Kizu S, Lyman JM, Macdonald AM, Minkowycz WJ, Moffitt SE, Palmer MD, Piola AR, Reseghetti F, Schuckmann K, Trenberth KE, Velicogna I, Willis JK (2013) A review of global ocean temperature observations: implications for ocean heat content estimates and climate change. Rev Geophysics 51:450–483. doi:10.1002/rog.20022

Abraham JP, Cowley R, Cheng L (2016) Quantification of the effect of water temperature on the fall rate of eXpendable bathythermographs. J Atmos Ocean Technol 6:1271–1284

Aitken AC (1934) On least squares and linear combinations of observations. Proc R Soc Edinburgh Sec A 55:42–47

Antonov JI, Levitus S, Boyer TP (2002) Steric sea level variations during 1957–1994: importance of salinity. J Geophys Res 107(C12):8013. doi:10.1029/2001JC000964

Balmaseda MA, Hernandez F, Storto A, Palmer MD, Alves O, Shi L, Smith GC, Toyoda T, Valdivieso M, Barnier B, Behringer D, Boyer T, Chang Y-S, Chepurin GA, Ferry N, Forget G, Fujii Y, Good S, Guinehut S, Haines K, Ishikawa Y, Keeley S, Köhl A, Lee T, Martin MJ, Masina S, Masuda S, Meyssignac B, Mogensen K, Parent L, Peterson KA, Tang YM, Yin Y, Vernieres G, Wang X, Waters J, Wedd R, Wang O, Xue Y, Chevallier M, Lemieux J-F, Dupont F, Kuragano T, Kamachi M, Awaji T, Caltabiano A, Wilmer-Becker K, Gaillard F (2015) The ocean reanalyses intercomparison project (ORA-IP). J Oper Oceanogr 8(sup1):s80–s97. doi:10.1080/1755876X.2015.1022329

Bell MJ, Forbes RM, Hines A (2000) Assessment of the FOAM global data assimilation system for real-time operational ocean forecasting. J Mar Syst 25:1–22

Boening C, Willis JK, Landerer FW, Nerem RS, Fasullo J (2012) The 2011 La Niña: so strong, the oceans fell. Geophys Res Lett 39:L19602. doi:10.1029/2012GL053055

Boyer T, et al (2016) Sensitivity of global ocean heat content estimates to mapping methods, XBT bias corrections, and baseline climatology. JClim (submitted)

Boyer TP, Antonov JI, Baranova OK, Coleman C, Garcia HE, Grodsky A, Johnson DR, Locarnini RA, Mishonov AV, O'Brien TD, Paver CR, Reagan JR, Seidov D, Smolyar IV, Zweng MM (2013) World Ocean Database 2013, NOAA Atlas NESDIS 72, S. Levitus, Ed., A. Mishonov, Technical Ed.; Silver Spring, MD, p 209. doi:10.7289/V5NZ85MT

Bretherton FP, Davis RE, Fandry CB (1976) A technique for objective analysis and design of oceanographic experiments applied to MODE-73. Deep Sea Res 23:559–582

Cabanes C, Grouazel A, von Schuckmann K, Hamon M, Turpin V, Coatanoan C, Paris F, Guinehut S, Boone C, Ferry N, de Boyer Montégut C, Carval T, Reverdin G, Pouliquen S, Le Traon P-Y (2013) The CORA dataset: validation and diagnostics of in situ ocean temperature and salinity measurements. Ocean Sci 9:1–18

Cazenave A, Llovel W (2010) Contemporary sea level rise. Ann Rev Mar Sci 2:145–173. doi:10.1146/annurev-marine-120308-081105

Chambers DP, Wahr J, Tamisiea ME, Nerem RS (2010) Ocean mass from GRACE and glacial isostatic adjustment. J Geophys Res 115:B11415. doi:10.1029/2010JB007530

Chambers DP, Cazenave A, Champollion N, Dieng H, Llovel W, Forsberg R, von Schuckmann K, Wada Y (2016) Evaluation of the global mean sea level budget between 1993 and 2014. Surv Geophys (accepted, this issue)

Chandler RE, Thorne P, Lawrimore J, Willett K (2012) Building trust in climate science: data products for the 21st century. Environmetrics 23:373–381. doi:10.1002/env.2141

Cheng L, Zhu J (2016) Benefits of CMIP5 multimodel ensemble in reconstructing historical ocean subsurface temperature variations. J Clim. doi: 10.1175/JCLI-D-15-0730.1 (in press)

Cheng L, Zhu J, Abraham J (2015b) Global upper ocean heat content estimation: recent progress and the remaining challenges. Atmos Ocean Sci Lett 8(6):333–338. doi:10.3878/AOSL20150031

Cheng L, Trenberth KE, Palmer MD, Zhu J, Abraham JP (2016) Observed and simulated full-depth ocean heat-content changes for 1970–2005. Ocean Sci 12:925–935. doi:10.5194/os-12-925-2016

Church JA, White NJ, Konikow LF, Domingues CM, Graham Cogley JG, Rignot E, Gregory JM, van den Broeke MR, Monaghan AJ, Velicogna I (2011) Revisiting the earth's sea-level and energy budgets from 1961 to 2008. Geophys Res Lett 38:L18601. doi:10.1029/2011GL048794

Church JA, Clark PU, Cazenave A, Gregory JM, Jevrejeva S, Levermann A, Merrifield MA, Milne GA, Nerem RS, Nunn PD, Payne AJ, Pfeffer WT, Stammer D, Unnikrishnan AS (2013) Sea level change. In: Stocker TF, Qin D, Plattner G-K, Tignor M, Allen SK, Boschung J, Nauels A, Xia Y, Bex V, Midgley PM (eds) Climate change 2013: The physical science basis. Contribution of Working Group I to the fifth assessment report of the intergovernmental panel on climate change. Cambridge University Press, Cambridge, United Kingdom and New York, NY, USA

Derber JC, Rosati A (1989) A global oceanic data assimilation technique. J Phys Oceanogr 19:1333–1347

Desbruyères DG, McDonagh EL, King BA, Garry FK, Blaker AT, Moat BI, Mercier H (2014) Full-depth temperature trends in the Northeastern Atlantic through the early 21st century. Geophys Res Lett 41:7971–7979. doi:10.1002/2014GL061844

Dieng HB, Palanisamy H, Cazenave A, Meyssignac B, von Schuckmann K (2015) The sea level budget since 2003: Inference on the deep ocean heat content. Surv Geophys 36:209–229. doi:10.1007/s10712-015-9314-6

Domingues C, Church J, White N, Gleckler P, Wijffels S, Barker P, Dunn J (2008) Improved estimates of upper-ocean warming and multi-decadal sea-level rise. Nature 453:1090–1093

Durack PJ, Wijffels SE, Gleckler PJ (2014) Long-term sea level change revisited: the role of salinity. Environ Res Lett 9:114017. doi:10.1088/1748-9326/9/11/114017

Gaillard F, Reynaud T, Thierry V, Kolodziejczyk N, von Schuckmann K (2016) In situ-based reanalysis of the global ocean temperature and salinity with ISAS: variability of the heat content and steric height. J Clim 29:1305–1323. doi:10.1175/JCLI-D-15-0028.1

Gille ST (2004) How nonlinearities in the equation of state of seawater can confound estimates of steric sea level change. J Geophys Res 109:C03005. doi:10.1029/2003JC002012

Good SA (2016) The impact of observational sampling on time series of global 0–700 m ocean average temperature: a case study. Int J Climatol. doi:10.1002/joc.4654

Good SA, Martin MJ, Rayner NA (2013) EN4: quality controlled ocean temperature and salinity profiles and monthly objective analyses with uncertainty estimates. J Geophys Res Oceans 118:6704–6716. doi:10.1002/2013JC009067

Gouretski V, Reseghetti F (2010) On depth and temperature biases in bathythermograph data: development of a new correction scheme based on analysis of a global ocean database. Deep Sea Res I 57:812–833. doi:10.1016/j.dsr.2010.03.011

Guinehut S, Dhomps A, Larnicol G, Le Traon P-Y (2012) High resolution 3D temperature and salinity fields derived from in situ and satellite observations. Ocean Sci 8:845–857

Halliwell GR, Srinivasan A, Kourafalou V, Yang H, Willey D, Le Hénaff M, Atlas R (2014) Rigorous evaluation of a fraternal twin ocean OSSE system for the open gulf of Mexico. J Atmos Ocean Technol 31:105–130. doi:10.1175/JTECH-D-13-00011.1

Hamon M, Reverdin G, Le Traon P-Y (2012) Empirical correction of XBT data. J Atmos Ocean Tech 29:960–973. doi:10.1175/JTECH-D-11-00129.1

Hansen J, Sato M, Kharecha P, von Schuckmann K (2011) Earth's energy imbalance and implications. Atmos Chem Phys 11:13421–13449. doi:10.5194/acp-11-13421-2011

Ingleby B, Huddleston M (2007) Quality control of ocean temperature and salinity profiles—historical and real-time data. J Mar Syst 65:158–175

IOC, SCOR, IAPSO (2010) The international thermodynamic equation of seawater—2010: calculation and use of thermodynamic properties, Manual and Guides No. 56, Intergovernmental Oceanographic Commission, UNESCO (English). http://www.TEOS-10.org

Ishii M, Kimoto M (2009) Reevaluation of historical ocean heat con-tent variations with time-varying XBT and MBT depth bias corrections. J Oceanogr 65:287–299

Ishii M, Kimoto M, Sakamoto K, Iwasaki SI (2006) Steric sea level changes estimated from historical ocean subsurface temperature and salinity analyses. J Oceanogr 62:155–170

Joint Committee for Guides in Metrology (2008) Evaluation of measurement data—guide to the expression of uncertainty in measurement. Bureau International des Poids et Mesures. p 130. www.bipm.org/en/publications/guides/

Kaplan A, Kushnir Y, Cane MA (2000) Reduced space optimal interpolation of historical marine sea level pressure. J Clim 13:2987–3002

Levitus S, Antonov J, Boyer T (2005) Warming of the world ocean, 1955–2003. Geophys Res Lett 32:L02604. doi:10.1029/2004GL021592

Levitus S, Antonov JI, Boyer TP, Locarnini RA, Garcia HE, Mishonov AV (2009) Global ocean heat content 1955–2008 in light of recently revealed instrumentation problems. Geophys Res Lett 36:L07608. doi:10.1029/2008GL037155

Levitus S, Antonov JI, Boyer TP, Baranova OK, Garcia HE, Locarnini RA, Mishonov AV, Reagan JR, Seidov D, Yarosh ES, Zweng MM et al (2012) World ocean heat content and thermosteric sealevel change (0–2000 m), 1955–2010. Geophys Res Lett 39:L10603. doi:10.1029/2012GL051106

Llovel W, Fukumori I, Meyssignac B (2013) Depth-dependent temperature change contributions to global mean thermosteric sea level rise from 1960 to 2010. Glob Planet Chang 101:113–118. doi:10.1016/j.gloplacha.2012.12.011

Locarnini RA, Mishonov AV, Antonov JI, Boyer TP, Garcia HE, Baranova OK, Zweng MM, Johnson DR (2010) World Ocean Atlas 2009. Volume 1: Temperature. In: Levitus S (ed) NOAA Atlas NESDIS 68. U.S, Government Printing Office, Washington, DC, p 184

Lorbacher K, Nauels A, Meinshausen M (2014) Complementing thermosteric sea level rise estimates. Geosci Model Dev 8:2723–2734. doi:10.5194/gmd-8-2723-2015

Lowe JA, Gregory JM (2006) Understanding projections of sea level rise in a Hadley Centre coupled climate model. J Geophys Res 111:C11014. doi:10.1029/2005JC003421

Lyman JM, Johnson GC (2008) Estimating annual global upper ocean heat content anomalies despite irregular in situ ocean sampling. J Clim 21:5629–5641

Lyman JM, Johnson GC (2014) Estimating global ocean heat content changes in the upper 1800 m since 1950 and the influence of climatology choice. J Clim 27:1945–1957. doi:10.1175/JCLI-D-12-00752.1

McDougall TJ, Jackett DR, Millero FJ, Pawlowicz R, Barker PM (2012) A global algorithm for estimating Absolute Salinity. Ocean Sci 8:1123–1134. doi:10.5194/os-8-1123-2012

Merchant CJ, Embury O, Roberts-Jones J, Fiedler EK, Bulgin CE, Corlett GK, Good S, McLaren A, Rayner, NA, Donlon C (2015) ESA sea surface temperature climate change initiative (ESA SST CCI): analysis long term product version 1.1. Centre for Environmental Data Analysis, 2016. http://catalogue.ceda.ac.uk/uuid/c65ce27928f34ebd92224c451c2a8bed

Millero FJ, Rainer Feistel R, Wright DG, McDougall TJ (2008) The composition of Standard Seawater and the definition of the Reference-Composition Salinity Scale. Deep Sea Res Part I 55:50–72. doi:10.5670/oceanog.2010.21

Palmer MD, McNeall DJ (2014) Internal variability of Earth's energy budget simulated by CMIP5 climate models. Environ Res Lett. doi:10.1088/1748-9326/9/3/034016

Palmer MD et al (2015) Ocean heat content variability and change in an ensemble of ocean reanalyses. Clim Dyn. doi:10.1007/s00382-015-2801-0
Pawlowicz R, McDougall T, Feistel R, Tailleux R (2012) An historical perspective on the development of the Thermodynamic Equation of Seawater—2010. Ocean Sci 8:161–174. doi:10.5194/os-8-161-2012
Purkey SG, Johnson GC (2010) Warming of global abyssal and deep Southern Ocean waters between the 1990s and 2000s: contributions to global heat and sea level rise budgets. J Clim 23:6336–6351. doi:10.1175/2010JCLI3682.1
Rhein M, Rintoul SR, Aoki S, Campos E, Chambers D, Feely RA, Gulev S, Johnson GC, Josey SA, Kostianoy A, Mauritzen C, Roemmich D, Talley LD, Wang F (2013) Observations: ocean. In Climate change 2013: the physical science basis. Contribution of working group I to the fifth assessment report of the intergovernmental panel on climate change, Cambridge University Press, Cambridge.
Ridgway KR, Dunn JR, Wilkin JL (2002) Ocean interpolation by four-dimensional weighted least squares—application to the waters around Australasia. J Atmos OceanTechnol 19(9):1357–1375
Rietbroek R, Brunnabenda S-E, Kuschea Ju, Schröter J, Dahle C (2016) Revisiting the contemporary sea-level budget on global and regional scales. PNAS. doi:10.1073/pnas.1519132113
Roemmich D, A.S. Team (2009) Argo: the challenge of continuing 10 years of progress. Oceanography Soc 22(3):46–55
Storto A et al (2015) Steric sea level variability (1993–2010) in an ensemble of ocean reanalyses and objective analyses. Clim Dyn. doi:10.1007/s00382-015-2554-9
Thorne PW, Parker DE, Christy JR, Mears CA (2005) Uncertainties in climate trends: lessons from upper-air temperature records. BAMS 86:1437–1442. doi:10.1175/BAMS-86-10-1437
Trenberth KE (2009) An imperative for adapting to climate change: tracking Earth's global energy. Curr Opin Environ Sustain 1:19–27. doi:10.1016/j.cosust.2009.06.001
Trenberth KE, Fasullo JT, Balmaseda MA (2014) Earth's energy imbalance. J Clim 27:3129–3144. doi:10.1175/JCLI-D-13-00294
Trenberth KE, Fasullo JT, von Schuckmann K, Cheng L (2016) Insights into Earth's energy imbalance from multiple sources. J Clim. doi:10.1175/JCLI-D-16-0339
UNESCO (1981a) Background papers and supporting data on the International Equation of State of Seawater 1980, Unesco Technical Papers in Marine Science 38, Paris
UNESCO (1981b) Background papers and supporting data on the Practi- cal Salinity Scale 1978, Unesco Technical Papers in Marine Science 37, Paris
von Schuckmann K, Le Traon P-Y (2011) How well can we derive Global Ocean Indicators from Argo data? Ocean Sci 7:783–791. doi:10.5194/os-7-783-2011
von Schuckmann K, Gaillard F, Le Traon P-Y (2009) Global hydrographic variability patterns during 2003–2008. J Geophys Res 114:1–17. doi:10.1029/2008JC005237
von Schuckmann K, Sallée J-B, Chambers D, Le Traon P-Y, Cabanes C, Gaillard F, Speich S, Hamon M (2014) Monitoring ocean heat content from the current generation of global ocean observing systems. Ocean Sci 10:547–557. doi:10.5194/os-10-547-2012
von Schuckmann K, Palmer MD, Trenberth KE, Cazenave A, Chambers D, Champollion N, Hansen J, Josey SA, Loeb N, Mathieu P-P, Meyssignac B, Wild M (2016) An imperative to monitor Earth's energy imbalance. Nat Clim Change 6:138–144. doi:10.1038/nclimate2876
Wijffels SE, Willis J, Domingues CM, Barker P, White NJ, Gronell A, Ridgwayand K, Church JA, (2008) Changing expendable bathythermograph fall rates and their impact on estimates of thermosteric sea level rise. J Clim 21(21):5657–5672. doi:10.1175/2008JCLI2290.1
Wright DG, Pawlowicz R, McDougall TJ, Feistel R, Marion GM (2011) Absolute Salinity, "Density Salinity" and the Reference-Composition Salinity Scale: present and future use in the seawater standard TEOS-10. Ocean Sci 7:1–26. doi:10.5194/os-7-1-2011

Surv Geophys (2017) 38:89–104
DOI 10.1007/s10712-016-9398-7

Greenland and Antarctica Ice Sheet Mass Changes and Effects on Global Sea Level

Rene Forsberg[1] · Louise Sørensen[1] · Sebastian Simonsen[1]

Received: 29 August 2016 / Accepted: 31 October 2016 / Published online: 12 January 2017

Abstract Thirteen years of GRACE data provide an excellent picture of the current mass changes of Greenland and Antarctica, with mass loss in the GRACE period 2002–2015 amounting to 265 ± 25 GT/year for Greenland (including peripheral ice caps), and 95 ± 50 GT/year for Antarctica, corresponding to 0.72 and 0.26 mm/year average global sea level change. A significant acceleration in mass loss rate is found, especially for Antarctica, while Greenland mass loss, after a corresponding acceleration period, and a record mass loss in the summer of 2012, has seen a slight decrease in short-term mass loss trend. The yearly mass balance estimates, based on point mass inversion methods, have relatively large errors, both due to uncertainties in the glacial isostatic adjustment processes, especially for Antarctica, leakage from unmodelled ocean mass changes, and (for Greenland) difficulties in separating mass signals from the Greenland ice sheet and the adjacent Canadian ice caps. The limited resolution of GRACE affects the uncertainty of total mass loss to a smaller degree; we illustrate the "real" sources of mass changes by including satellite altimetry elevation change results in a joint inversion with GRACE, showing that mass change occurs primarily associated with major outlet glaciers, as well as a narrow coastal band. For Antarctica, the primary changes are associated with the major outlet glaciers in West Antarctica (Pine Island and Thwaites Glacier systems), as well as on the Antarctic Peninsula, where major glacier accelerations have been observed after the 2002 collapse of the Larsen B Ice Shelf.

Keywords Greenland ice sheet · Antarctica mass loss · GRACE · Envisat · CryoSat

✉ Rene Forsberg
rf@space.dtu.dk

[1] National Space Institute, Technical University of Denmark (DTU Space), Elektrovej 328, 2800 Lyngby, Denmark

1 Introduction

The melting of the polar ice sheets is a major global concern, especially due to the direct effects on global sea level rise. Although estimation of ice sheet mass balance has been a main goal of glaciological research for decades, reliable observations of current mass changes have only been available since the advent of space observations. Three types of Earth observation methods are in use for this purpose: *satellite altimetry*, where direct measurements of height changes by laser (ICESat) or radar (ERS-1/-2, Envisat, CryoSat-2) altimetry, in combination with climatological/glaciological models for firn (snow) density and compaction, are used to estimate mass loss; *input–output methods*, where measurements of ice flow velocities from synthetic aperture radar data (ERS, Envisat, RADARSAT) over outlet glaciers are combined with glacier thickness data, and models for accumulation and surface mass balance in the interior are used to give a net mass balance; and *gravity field change* missions (GRACE), where the mass changes are measured directly.

Numerous scientific papers on ice sheet change estimation have been published in the recent decade, and a review and intercomparison of all methods were done in the ESA/NASA supported IMBIE study (Ice Sheet Mass Balance Intercomparison Experiment 2011–12), resulting in a landmark paper with nearly 50 co-authors (Shepherd et al. 2012). We therefore refer readers to this paper, and extensive references therein, for background on the ice sheet mass loss estimation methods. The main IMBIE conclusions were that the three space-based methods give consistent results, when properly applied for common periods and that combinations of all methods yield more reliable estimates, with overall mass change estimates closely mirroring GRACE-only estimations.

All the above space-based estimation schemes are affected by various types of errors. Conventional pulse-limited radar satellite altimetry does not cover the most sloping and rugged parts of the ice sheets, which are the most rapidly changing parts; furthermore, radar altimetry is strongly affected by radar penetration into snow and melt events forming ice lenses in the snowpack, especially in Greenland (Nilsson et al. 2015). The input–output method is limited by lack of information on outlet glacier ice thickness, especially in Antarctica, as well as uncertainty in interior surface mass balance models derived from regional climate models. GRACE accuracy is limited by signal leakage from adjacent ice caps, land hydrology, unmodelled ocean mass changes, and—especially for Antarctica—large uncertainty in glacial isostatic adjustment (GIA) models (Wahr et al. 1998).

The mass loss of the ice sheets is typically expressed in units of GT/year, with 100 GT of mass change corresponding to an average global sea level rise of 0.272 mm/year. Examples of recent estimates of mass loss and corresponding sea level rise are outlined in Table 1, mainly extracted from a compilation of Dieng et al. (2015). A large variability in results is found, mainly due to use of linear trends over different time intervals; the use of such trends is not very suitable for longer time series, where decadal changes in climate and ocean temperatures (a major cause of outlet glacier melt, both in Greenland and Antarctica; Holland et al. 2012) would be expected to change ice sheet mass loss correspondingly. It should be noted that the mass loss of the ice sheets does not result in a uniform sea level rise; due to changes in gravitation, land uplift, and earth rheology, Greenland melt will mainly affect tropical and southern latitudes, while Antarctica melt will mainly affect the northern hemisphere, as illustrated in Fig. 1.

Table 1 Some selected mass loss estimates of the ice sheets by different methods

Area	Period	Method	Mass loss		Error estimate		References
			GT/year	mm/year	GT/year	mm/year	
Greenland	2000–2011	Combination	217	0.59	36	0.1	Shepherd et al. (2012)
	2005–2006	Input–output	210	0.57	40	0.11	Rignot and Kanagaratnam (2006)
	2003–2007	Altimetry	176	0.48	4	0.01	Zwally et al. (2011)
	2003–2008	Altimetry	239	0.65	29	0.08	Sørensen et al. (2010)
	2003–2008	Alt. + GRACE	180	0.5	29	0.08	Ewert et al. (2012)
	2003–2009	Combination	265	0.72	58	0.16	Sasgen et al. (2012)
	2003–2010	GRACE	230	0.63	12	0.03	Luthcke et al. (2013)
	2003–2012	GRACE	265	0.72	40	0.11	Velicogna and Wahr (2013)
	2003–2012	GRACE	235	0.64	25	0.07	Groh et al. (2014)
Antarctica	2000–2011	Combination	88	0.24	44	0.12	Shepherd et al. (2012)
	2003–2009	Altimetry	63	0.17	44	0.12	Helm et al. (2014)
	2010–2013	Altimetry	159	0.43	48	0.13	McMillan et al. (2014)
	2003–2011	GRACE	85	0.23	36	0.1	Barletta et al. (2013)
	2003–2010	GRACE	80	0.21	26	0.07	Luthcke et al. (2013)
	2003–2012	GRACE	118	0.32	66	0.18	Velicogna and Wahr (2013)
	2003–2012	GRACE	110	0.3	29	0.08	Groh et al. (2014)

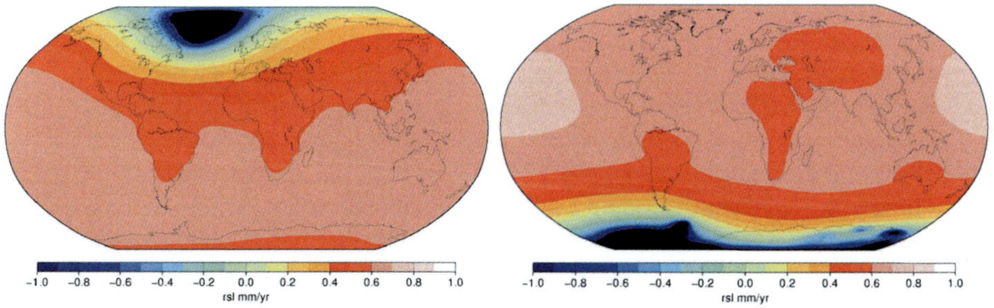

Fig. 1 Relative sea level rise due to Greenland melt (*left*) and Antarctica melt (*right*) for IceSat period 2003–2008, unit mm/year. Figures courtesy of V. Barletta (DTU Space) produced as part of EU Ice2Sea project

2 GRACE Measurements of Ice Sheet Changes

For the investigations of this paper, we will focus on the GRACE mission mass loss estimates, reprocessing GRACE data up to early 2016 based on new, improved GRACE data, with the primary goal to expand the sea level rise curves of the IMBIE study to 2016.

The successful GRACE gravity field mission (Tapley et al. 2004) has since late 2002 provided measurements of the temporal variations of the gravitational field of the earth in the form of monthly expansions of the gravitational field in spherical harmonics (Bettadpur 2003). These data have provided a unique way to monitor the changing ice sheets of the earth, and their link with climate change. GRACE measurements have clearly documented the accelerating mass loss of the Greenland ice sheet, with early results of analysis (e.g., Velicogna and Wahr 2006; Horwath and Dietrich 2006; Chen et al. 2006; Luthcke et al. 2006; Forsberg and Reeh 2007) giving quite variable results in the 150–250 GT/year range, depending on the analysed period (for examples see Table 1).

Recent comprehensive multiple-method "reconciled" estimates (Shepherd et al. 2012; Sasgen et al. 2012) give results in a more narrow band around −225 GT/year for Greenland, and −85 GT/year for Antarctica, for the period 2003–2011. Depending on the time periods analysed, the mass loss trend estimates tend to increase with time, as the melt of both Greenland and especially Antarctica shows accelerating signals.

Differences in the published GRACE mass loss estimates are mainly due to method differences in how to convert the monthly gravity field solutions (expressed as spherical harmonic expansions, also termed "Level-2" GRACE data) into surface mass changes. The limited spatial resolution of GRACE (around 300–350 km), differences in processing methods adopted by the different Level-2 data centres, differences in data corrections, and leakage of GRACE signals between different mass bodies, are major sources of differences; in particular, critical is the glacial isostatic adjustment (GIA) correction. The nature of most of these error sources is explained in details in the fundamental pre-launch GRACE methodology paper (Wahr et al. 1998). The same paper also outlined spherical harmonic filtering and conversion methods for converting GRACE spherical harmonic data into mass change estimates.

The point mass inversion methods used in this paper are fundamentally different from the spherical harmonic conversion methods of the above pre-launch paper. In the point mass (or mascon) methods, a direct inversion of the measured signals at orbit altitude is converted into associated mass distributions at the earth's surface. This can either be done by direct global analysis of "raw" GRACE satellite to satellite ranging data (termed "Level-1" data), solving for global mascons (e.g., Luthcke et al. 2006, 2013), or by solving for regional distributions of mascons, e.g., representing ice covered areas, as done by, e.g., Forsberg and Reeh (2007), Baur and Sneeuw (2011) and Barletta et al. (2013). In the regional point mass estimation methods, a priori knowledge of the source region of the mass changes can be taken into account in a simple intuitive manner, but with the risk of increasing "leakage" from unmodelled land and ocean mass changes close to the ice sheet. The IMBIE project has, however, confirmed that estimates by either methods in Greenland and Antarctica provide similar results, when applied to the same periods and using the same auxiliary geophysical and environmental corrections.

We present in the sequel a reanalysis of a new 13.5-year GRACE data set from an improved unconstrained Kalman filter processing scheme at ITSG/TU Graz (Mayer-Gürr et al. 2014; Klinger et al. 2016), giving a clear picture of the year-to-year mass change signals, the regions of greater mass loss, and providing an extension of the IMBIE (Shepherd et al. 2012) estimates of sea level rise due to ice sheet melt. We additionally supplement the GRACE analysis for Greenland and Antarctica, with mass change results from satellite altimetry (Envisat and CryoSat), using a joint GRACE/altimetry inversion method, to pinpoint with greater resolution where current mass changes are taking place. The joint inversion method improves the limited resolution of GRACE. This improved resolution is relatively more important for Greenland rather than for Antarctica, where the role of resolution is less dominant due to the larger area, and where leakage from adjacent ice caps is not an issue.

3 Basics of Analysis of GRACE Data

All GRACE Level-2 data are provided as monthly spherical harmonic geopotential expansions of form

$$V(r,\phi,\lambda) = \frac{GM}{R}\sum_{n=2}^{N}\left(\frac{R}{r}\right)^{n} \times \sum_{m=0}^{n}(C_{nm}\cos m\lambda + S_{nm}\sin m\lambda)\bar{P}_{nm}(\sin\phi) \tag{1}$$

where V is the geopotential, G is the gravitational constant, M is Earth's mass, R is the Earth radius, and the fully normalized spherical harmonic coefficients C_{nm} and S_{nm}, provided by a processing centre, such as CSR (Center for Space Research, University of Texas) or GFZ (Geoforschungs Center, Potsdam), the primary GRACE mission data providers.

We have in our computations, however, used new reprocessed Level-2 data provided by Institute for Theoretical and Satellite Geodesy, TU Graz (ITSG), to harmonic degree and order 90 (Klinger et al. 2016). The ITSG GRACE data appear to be improved relative to corresponding CSR and GFZ data and have recently been adopted for use in the ESA Climate Change Initiative projects for both the Greenland and Antarctica ice sheets, see http://www.esa-icesheets-cci.org/.

The ITSG spherical harmonic coefficients have been supplemented with C_{20}-terms from satellite laser ranging derived from the International Laser Ranging Service data (available at https://podaac.jpl.nasa.gov/), and corrections for geocentre mass variations (C_{10}, C_{11}, S_{11}) by the method of Swenson et al. (2008), as provided at http://grace.jpl.nasa.gov/. The spherical harmonic coefficients are further modified for the elastic response of the Earth's crust to a mass load change, which gives an indirect effect on the gravity response. We take this into account in the mass inversion method by modifying the GRACE coefficients for elastic loading by

$$C^{*}_{nm} = \frac{1}{1+k_n}C_{nm} \quad S^{*}_{nm} = \frac{1}{1+k_n}S_{nm} \tag{2}$$

where k_n are the elastic Love numbers, using Preliminary Reference Earth Model (PREM) numbers as listed in Wahr et al (1998).

The GRACE spherical harmonic data, as corrected above, should then in principle represent the land ice, land hydrology, GIA mass changes, and other geodynamic/earthquake effects, as atmosphere and ocean mass changes are modelled and subtracted from the original Level-2 coefficient data. In the case of Greenland and Antarctica, the role of land hydrology is small to negligible, especially on temporal trends, and land hydrology effects from the ice-free part of Greenland and Antarctica have thus been ignored in this study.

An important further correction is the GIA effect. Many different past deglaciation history and Earth models have been used to infer the GIA effects in Greenland and Antarctica. We use here two "standard" models, namely the ICE-5G model for Greenland (Peltier 2004) and the W12 model for Antarctica (Whitehouse et al. 2012), as also used in the IMBIE study. The GIA models are also given as spherical harmonic models. We will not discuss further the possible errors in these models, but only note that the relatively large error quoted in the IMBIE estimates (±25 GT for Greenland and ±50 GT for Antarctica) is predominantly due to the GIA effects.

The results for the observed GRACE gravity trends at satellite altitude for Greenland and Antarctica over the period April 2002 to January 2016 (with a total of 148 available monthly epochs) are shown in Fig. 2. Because of the monthly "stacking" of data to estimate a trend, the use of "destriping" filters (Kusche et al. 2009), commonly in use for attenuating the dominant north–south trending error patterns in GRACE data (due to the near-polar orbit), seems not to be needed. No further filtering has thus been applied to the gravity trend data. Figure 2 shows major signals associated with the margins and major outlet glaciers in Greenland, and for Antarctica the overall dominating signal over the Amundsen Sea Embayment glacier systems in West Antarctica.

4 Generalized Inverse Point Mass Inversion for Greenland and Antarctica Changes

From the fitted GRACE trends, cf. Fig. 2, a trend of mass change at the Earth's surface can be obtained by a linear, albeit ill-posed inversion procedure. The formula for the response function for this inversion can be found in Heiskanen and Moritz (1967). For a point mass j at the surface, the gravity attraction at orbit height location i is of form

$$\delta g_i = Gm_j \frac{R^2 r - R^3 \cos\psi}{\left[r^2 + R^2 - 2Rr\cos\psi\right]^{3/2}} \tag{3}$$

where ψ is the spherical distance, R is the Earth's radius, and $r \sim R + 480$ km is the GRACE orbit altitude (the GRACE orbit has decayed slightly over the years, so r is not constant).

In the inversion method, the observed gravity δg_i values (relative to a mean value over the GRACE period) over the region at GRACE orbit altitude are combined in the observation vector y

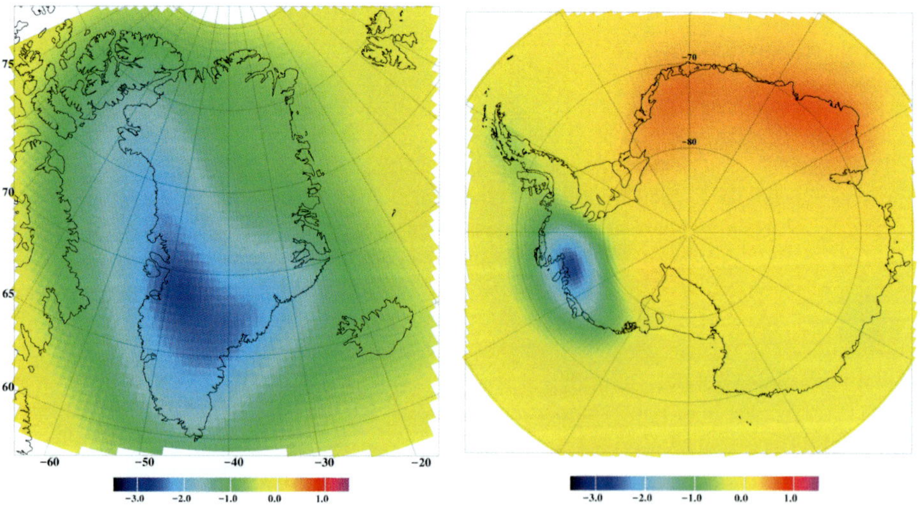

Fig. 2 GRACE gravity change signals 2002–2016 over Greenland (*left*) and Antarctica (*right*), at orbit altitude. GRACE ITSG-2016 solution to degree $n = 90$, corrected for elastic response. Unit μGal/year

$$\underline{y} = \{dg_i\}, \quad i = 1, \ldots, n \tag{4}$$

and modelled by a dense set of point masses m_j in a solution vector x

$$\underline{x} = \{m_j\}, \quad j = 1, \ldots, m \tag{5}$$

located at the surface of the ice covered region to be studied. The key element of the method is that point masses are *only* located where the ice sheet changes are assumed to take place (defined by an "indicator" grid, cf. Fig. 3); ocean and ice-free land gravity field changes are assumed to be removed by GIA, ocean, and hydrological corrections prior to applying the inversion.

Using an ice-only set of point mass (mascon) elements has both advantages and drawbacks. An advantage is that a priori knowledge of the ice sheet location is applied, but a drawback is that any change signals from nearby non-glacial sources (e.g., land hydrology or ocean model errors) are "leaking" into the ice mass change, likely to a larger degree than in the spherical harmonics approach. To avoid such leakage, neighbouring ice caps in Greenland (especially the Canadian ice caps) need to be modelled simultaneously, otherwise the Canadian ice cap changes would "leak" into Greenland and give too high mass loss estimates there.

The elements of the basic point mass equation (3) form a response matrix A, where the linear observation equations

$$\underline{y} = A\underline{x} \tag{6}$$

may be solved by Tychonoff generalized inverse by

$$\underline{x} = [A^{\mathrm{T}}A + \lambda I]^{-1}A^{\mathrm{T}}\underline{y} \tag{7}$$

Here I is the unit matrix, and λ is a regularization factor, needed to obtain a non-singular inversion problem. The λ-factor determines the necessary trade-off between model smoothness and residuals; the total mass change of the Greenland ice sheet will only to a

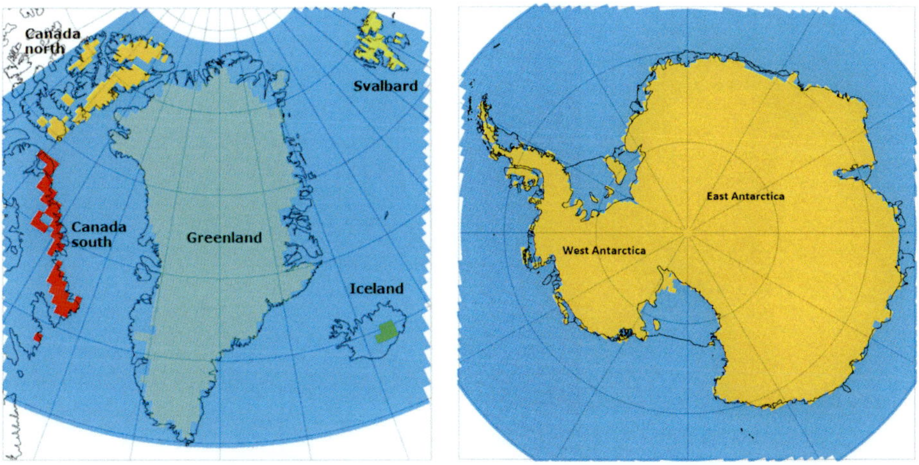

Fig. 3 Indicator grids for the Greenland ice sheet and adjacent ice caps, and for the grounded ice regions of Antarctica. The mascons are distributed at approx. 50 km resolution. For Greenland, only the Ellesmere and Devon Island ("Canada north") are solved for to avoid excess leakage

small degree be affected by the choice of λ; the areal shape of the modelled mass change is, however, stronger affected by this choice. The selection of the optimal λ-parameter has been discussed at length in Baur and Sneeuw (2011) and will in the end be up to a subjective trade-off.

Linear equations like (7) are readily and quickly solved by the Cholesky method for positive definite linear equation systems. The equations may either be solved for gravity *trends*, or—on a month by month basis—by gravity *residuals* relative to the mean of the period. The latter will give a time series of the mass balance. Because of the geometry of the input (a regular grid, covering the region of interest) and the output (a fixed set of space domain masses, covering the ice sheets), the normal equation matrix of (7) will be constant, except for the "right hand side" ($A^T\underline{y}$) observations. Processing many epochs is therefore extremely fast, as the Cholesky factorization needs to be done only once.

Figure 4 shows the results of the selected point mass solution by the inversion method, using a relatively weak regularization (small λ-factor), and a 200–250 km cut-away zone for GRACE observations beyond the ice sheets (to limit leakage from oceans and land areas); the resolution of the mass cells is around 50 km (0.5° × 1° for Greenland and 0.5° × 2° for Antarctica). It is seen that the mass balance, expressed in mm water-equivalent change, is negative in a quite narrow band along the ice margins in Greenland, in good accordance with in situ and airborne observations (Krabill et al. 2000; Sørensen et al. 2010; Helm et al. 2014) and that the Antarctica changes are dominated by the West Antarctica sector of the Amundsen Sea, with the major changes in the region of the Pine Island/Thwaites Glacier systems.

Figure 5 shows the time series for the Greenland and Antarctic ice sheets, with no time-domain low-pass filtering done on the monthly estimates. Included in the Greenland plot are also results from the northern Canadian ice caps (Ellesmere and Devon Islands), estimated jointly with Greenland. It should be noted, though, that the estimates from the northern Canadian ice caps are further affected by leakage error from the southern Canadian ice caps and glaciers of Baffin and Bylot Island, so the quoted estimates for

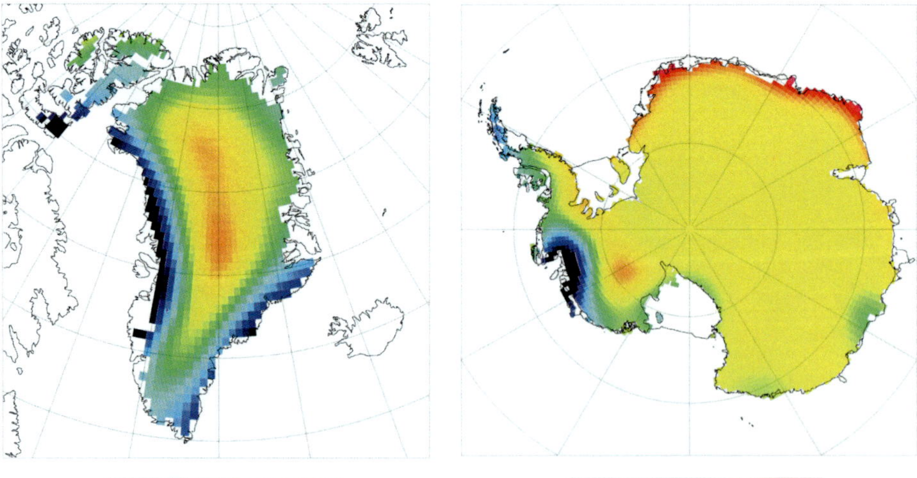

Fig. 4 Mass change trends 2002–2016 for Greenland and Antarctica. Unit: mm water-equivalent/year. Note difference in *colour scale* (and size of the ice sheets, Antarctica map scale only half of Greenland)

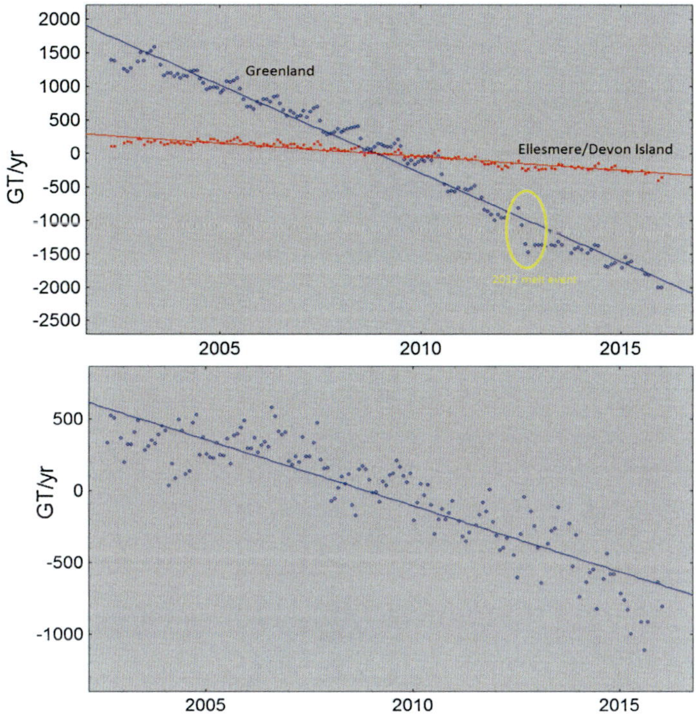

Fig. 5 Unfiltered GRACE time series of mass balance for Greenland and Ellesmere/Devon Island (*upper*), and Antarctica (*lower*), with monthly solution values (*dots*) and linear trend fit. The *Yellow ellipse* shows the 2012 record melt event in Greenland

Ellesmere and Devon Islands are probably too large (and not really a topic for this paper). The estimated mass trends and associated global sea level rise are additionally listed in Table 2.

It should be noted that the Greenland ice loss estimates include minor outlying ice caps and glaciers; the mass loss of the outlying ice bodies has been estimated at −28 ± 11 GT/year from ICESat altimetry in the period 2003–2008 (Bolsch et al. 2013); this estimate

Table 2 Mass change and global sea level rise for Greenland and Antarctica from GRACE

Time period	April 2002–2015 (13.7 years)		2006–2015 (10 years)		2011–2015 (5 years)	
	GT/year	mm/year	GT/year	mm/year	GT/year	mm/year
Greenland, including outlying ice caps (±25 GT/year)	−264	−0.72	−295	−0.80	−265	−0.72
Ellesmere and Devon Island[b] (±20 GT/year)	−41	−0.11	−48	−0.13	−45	−0.12
Antarctica[a] (±50 GT/year)	−92	−0.25	−120	−0.33	−147	−0.40

[a] Two anomalous first epochs (April–May 2012) in GRACE time series deleted

[b] Estimates include leakage from Baffin Island

might be slightly overestimated, as it includes a major part of the ice sheet classified as "outlying" (the central East Greenland nunatak zone); the estimate will most likely be an underestimate for more recent periods, where many outlying ice caps and outlet glaciers have experienced rapid melt.

Figure 5 shows that the mass loss of both Greenland and the northern Canadian ice caps are accelerating, with summer 2012 seeing an exceptionally large melt event in Greenland, and the Canadian ice caps experiencing rapid acceleration since 2008. But it is also seen that taking a trend over only the last 5 years gives an apparent slowdown of the Greenland ice sheet melt, following the anomalous large melt event of 2012 (Nghiem et al. 2012). In July 2012, the entire Greenland ice sheet for the first time in recent times experienced melt and rain even at the highest elevations, due to unusually warm meteorological conditions; using a 5-year trend interval does not make too much sense because of the highly unusual events of recent years, where the 2012 summer was followed by an unusually cold 2013 summer.

For Antarctica, a major apparent acceleration is seen continent-wide, especially due to the West Antarctica ice streams, with major snow accumulation effects in East Antarctica offsetting to some degree the acceleration in West Antarctica, as studied in more detail in Shepherd et al. (2012) and Lenaerts et al. (2013).

5 Satellite Altimetry Used to Outline Detailed Melt Regions in Greenland and Antarctica

In this section, we use an extension of the inversion method to include satellite altimetry in the mass loss determination. In Greenland, the relatively smaller size of the region (compared to Antarctica) makes the lack of GRACE resolution issue more serious, in terms of pinpointing the true regions of mass loss. Identifying *where* the mass loss is coming from is useful to understand changes; to use satellite altimetry for a stand-alone estimate of mass changes, i.e. to validate GRACE results or vice versa, is not the aim of this section, but rather the goal of IMBIE (Shepherd et al. 2012), and the upcoming IMBIE-2 project. We will therefore here solely focus on the altimetry height changes used as a tool to enhance GRACE resolution and—for Greenland—as an effective tool to control leakage error, e.g., relative to the Canadian ice caps.

For the Greenland elevation changes, we use data from the ESA Envisat radar altimetry mission for the period 2002–2010 (www.esa.int/envisat) and CryoSat-2 data for the period 2010–2015 (www.esa.int/cryosat). The Envisat elevation changes are estimated by a repeat-track algorithm (Sørensen et al. 2015), while CryoSat-2 data (SARIn and LRM modes) are retracked by a novel threshold retracker, with the central ice sheet LRM mode data further relocated by an updated Greenland Digital Elevation Model (DEM) (Nilsson et al. 2016). CryoSat elevation change estimates are computed by a "binning and stacking" method (Forsberg et al. 2013), where monthly elevation residuals relative to an initial Greenland DEM are analysed for bias and trend in 5 km resolution grid cells after gross-outlier rejection. The biases of these residuals are subsequently used to update the used DEM, and the binning and stacking process repeated for the final d*h*/d*t* results. Both the Envisat and CryoSat changes are spatially filtered at resolutions around 15–20 km for the final results.

For Antarctica, a similar CryoSat binning and stacking has been performed for a 5-year period July 2010–June 2016, starting from the BEDMAP-2 DEM (Fretwell et al. 2012), using the newest Baseline C ESA Level 2 data (OCOG retracker), with DEM-relocation corrections for LRM mode data. The Antarctica elevation changes from the altimetry are shown in Fig. 6, along with the corresponding GRACE change data.

The inversion method outlined in Sect. 3 may also include height change estimates from altimetry. In the joint inversion method, an additional set of observation equations for $\underline{x}$ may be formed simply for the trends over the designated period by

$$dg_i/\mathrm{d}t = \rho_{\mathrm{model}(i)}\mathrm{d}h_i/\mathrm{d}t + f_i \tag{8}$$

where f_i is a firn (upper snow layer) compaction correction, and ρ_{model} is a modelled surface density (for more details see Sørensen et al. 2010, or Simonsen et al. 2013). The surface density would typically be ice density of 0.92 g/cm^3 in the margin-near ablation zone and have lower values of 0.3–0.4 g/cm^3 for uncompacted snow at the higher elevations (Sørensen et al. 2010).

The firn model for Greenland is derived from the regional climate model HIRHAM, run by the Danish Meteorological Institute (R. Mottram, pers.comm.; Simonsen et al. 2013). An example of the firn compaction correction (for the Envisat period) is shown in Fig. 7 (left), along with the results of the "optimal" combined altimetry/GRACE combination. The GRACE/altimetry inversion in Greenland is performed using the CryoSat-2- or Envisat-derived mass change estimates at a 25-km resolution UTM (Universal Transverse Mercator) grid, covering the entire Greenland ice sheet and peripheral ice caps and glaciers (the UTM projection is superior to the frequently used Polar Stereographic projection in Greenland, giving fewer observation cells and smaller scale distortion corrections). The degrees of freedom in the inversion process include the λ-factor (7) as well as the a priori standard deviations of both GRACE and altimetry data, with a "best" weighting scheme reproducing the overall GRACE mass loss estimate, while keeping the detailed spatial resolution of the altimetry.

For Antarctica, the inversion method has been applied with ice density 0.92 g/cm^3 only; then, the inversion method will effectively estimate on overall mass product consistent with GRACE, within the *shape* parameter of CryoSat. A derived a posteriori regional scale

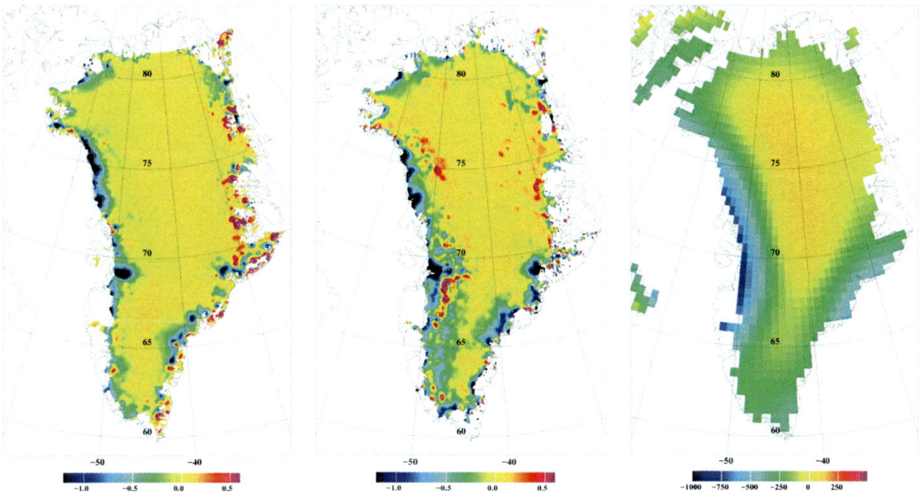

Fig. 6 Elevation changes of the Greenland ice sheet from Envisat 2002–2010 (*left*), CryoSat 2010–2015 (*centre),* and GRACE 2010–2015 (*right*). Units: m/year for height changes, and mm/year water equivalent for GRACE

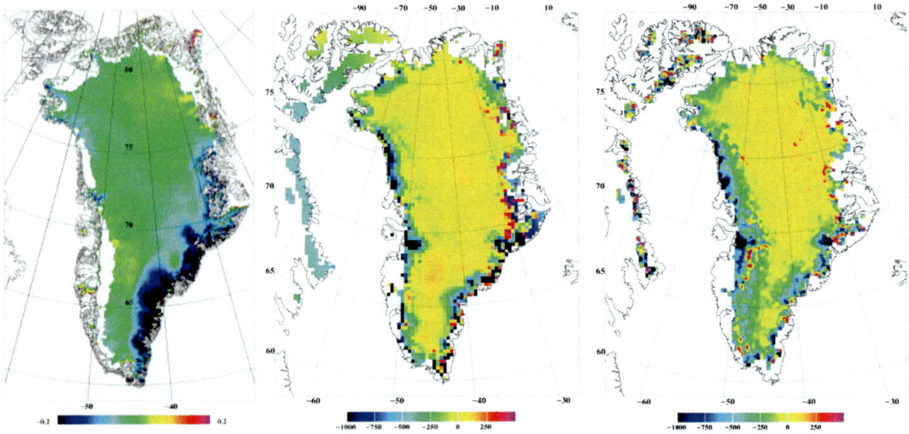

Fig. 7 *Left*: average Greenland firn compactions corrections from HIRHAM (unit: m/year). *Centre/right*: mass solutions by constrained GRACE inversion with Envisat (*centre*) and CryoSat (*right*, including Canadian ice caps), unit mm water equivalent/year

correction to d*h*/d*t* would thus represent a kind of proxy for a composite of density variations and firn compaction parameters. The Antarctica inversion was done on a 40-km resolution polar stereographic grid. The results are shown in Fig. 8, and it is seen that due to larger size of the Antarctica continent relative to Greenland, the difference between the "pure" and "combined" GRACE solutions is relatively smaller.

The overall mass change of the joint estimation methods for both Greenland and Antarctica is within a few GT/year of the overall GRACE-only mass change estimate. The small variation is due to the strong adjustment constraint from GRACE (the firn model corrections to the Greenland altimetry are by themselves around 40 GT/year). The strong

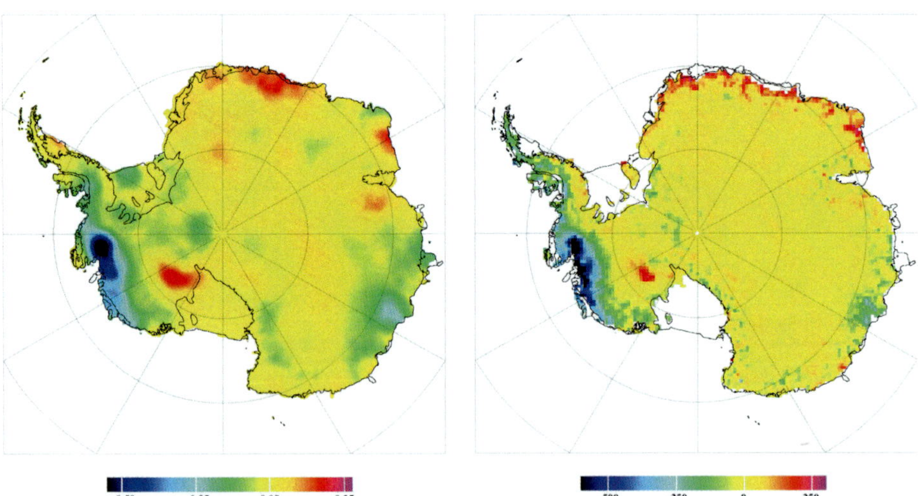

Fig. 8 Antarctica elevation changes from CryoSat 2010–2015 (*left*, unit: m/year), and jointly estimated GRACE/CryoSat mass changes (*right*, unit mm water equivalent/year). CryoSat data smoothed to 0.3° resolution

GRACE constraint also overcomes the lack of data in altimetry solutions over the most sloping parts of the ice sheets.

6 Discussion and Conclusions

In this paper, we have outlined current estimates of total and regional mass loss for the Greenland and Antarctica ice sheets, based on GRACE data alone, and detailed combined GRACE/altimetry mass change estimates. Although specific methods and data sources have been used, overall mass change results are representative for many similar investigations and thus confirm the values of the associated global sea level rise presented here.

The mass changes in Greenland are clearly seen to be associated with relatively narrow marginal ice zones, especially in West and South-East Greenland, and major outlet glaciers such as Jakobshavn and Helheim glaciers, most clearly seen in the joint GRACE/altimetry inversion results. The overall mass loss of Greenland is accelerating, but the last 6 years (2010–2015) have seen large year-to-year interannual variations, with 2010 and especially 2012 record melt years. The period 2010–15 has an apparent decreasing trend relative to 2005–2010, which might not be significant due to the short period, the large melt events, and the lack of some key monthly GRACE data (the GRACE satellite pair is now operating way beyond its original design lifetime).

The large Greenland melt event in 2012 has made the use of radar altimetry for height and mass change estimation more complicated, with inherent changes in melt-generated ice layers inside the upper layers of the firn, leading to "noisy" CryoSat elevation change data, apparent in Fig. 6 (Nilsson et al. 2015; Forsberg et al. 2013).

For Antarctica, the GRACE time series shows a clear acceleration between 2002 and present, dominated by the West Antarctica outlet glaciers of the Amundsen Embayment (Thwaites and Pine Island glacier systems), but also mass loss in the Antarctic Peninsula and in smaller outlet glacier regions of coastal East Antarctica near 120°E (Totten Glacier region). The larger size of the continent gives a larger degree of similarity of the jointly estimated altimetry/GRACE mass changes, relative to the GRACE-alone estimates. A major apparent mass increase in Queen Maud Land, East Antarctica, are most likely expressions of the major snowfall events in the region after 2009–2011, which has been estimated to give a mass gain on the order of 350 GT in East Antarctica (Boening et al. 2012; Lenaerts et al. 2013).

The overall GRACE-derived sea level change from the ice sheets is plotted alongside the results of the IMBIE 2012 reconciled estimates of Greenland and Antarctica mass changes in Fig. 9. The IMBIE results are from combinations of altimetry, GRACE and input–output SAR interferometry methods, and therefore do not agree exactly with the GRACE-only solutions of this paper. Furthermore, the reprocessed GRACE results 2002–2016 have improved significantly since the IMBIE study. The accumulated global sea level rise during the IMBIE period seems to agree reasonably well for Greenland, albeit with a slightly lower trend in Antarctica (which might be due to the difference in GIA models used). Beyond the IMBIE period, Fig. 9 shows the extension of the global sea level curve, giving a present (2016) accumulated sea level rise of 12 mm for Greenland and 5 mm for Antarctica since the IMBIE starting time of 1992, continuing currently at rates around -0.8 mm/year for Greenland and -0.4 mm/year for Antarctica. These values are expected to be updated and verified in the near term, as a new, broader international NASA/ESA intercomparison effort is reporting its results (IMBIE-2).

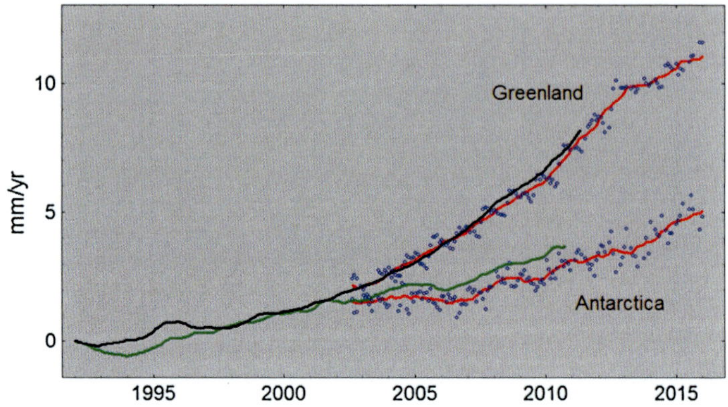

Fig. 9 Accumulated global sea level rise from the ITSG GRACE 2002–2016 monthly solutions (*blue*), with yearly GRACE moving average filtered values (*red*), plotted together with IMBIE 1992–2011 Greenland (*black*) and Antarctica (*green*) reconciled mass change estimates (Shepherd et al. 2012)

Acknowledgements This paper is based on a presentation at the Workshop of the International Space Science Institute (ISSI), Bern, Switzerland, February 2015. The sea level "fingerprinting" plots in Fig. 1 was provided by V. Barletta. Johan Nilsson, former PhD student at DTU Space, now at NASA-JPL, provided the retracked CryoSat-2 data for Greenland. Torsten Mayr-Gürr, TU Graz, provided early access to the new ITSG-processed GRACE L2 data. Comments by Anny Cazenave and Nicolas Champollion improved the original manuscript.

References

Barletta VR, Sørensen LS, Forsberg R (2013) Variability of mass changes at basin scale for Greenland and Antarctica. Cryosphere 6:3397–3446. doi:10.5194/tcd-6-3397-2012

Baur O, Sneeuw N (2011) Assessing Greenland ice mass loss by means of point-mass modeling: a viable methodology. J Geodesy 85:607–615. doi:10.1007/s00190-011-0463-1

Bettadpur S (2003) GRACE level-2 gravity field product user handbook. CSR Publ. GR-03-01, University of Texas, Austin. http://podaac.jpl.nasa.gov/grace

Boening C, Lebsock M, Landerer F, Stephens G (2012) Snowfall-driven mass change on the East Antarctic ice sheet. Cryosphere. doi:10.1029/2012GL053316

Bolsch T, Sørensen LS, Simonsen SB, Mölg N, MacGuth H, Rastner P, Paul F (2013) Mass loss of Greenland's glaciers and ice caps 2003-8 revealed from ICESat laser altimetry data. Geophys Res Lett 40(5):875–881. doi:10.1002/grl.50270

Chen JL, Wilson CR, Tapley BD (2006) Satellite gravity measurements confirm accelerated melting of greenland ice sheet. Science 313:1958–1960. doi:10.1126/science.1129007

Dieng HB, Champollion N, Wada Y, Schrama E, Meyssignac B (2015) Total land water storage change over 2003–13 estimated from a global mass budget approach. Environ Res Lett 10:124010. doi:10.1088/1748-9326/10/12/124010

Ewert H, Groh A, Dietrich R (2012) Volume and mass changes of the Greenland ice sheet inferred from ICESat and GRACE. J Geodyn 59–60:111–123

Forsberg, R and Reeh N (2007) Mass change of the Greenland ice sheet from Grace. In: Proceedings of the 1st international symposium of the IGFS, Harita Dergisi, Ankara, vol 18, pp 454–458

Forsberg R, Sørensen L, Levinsen J, Nilsson J (2013) Mass loss of Greenland from GRACE, IceSat and CryoSat. In: Proceedings of the CryoSat workshop, Dresden, ESA Special Publication 717 paper S6-4

Fretwell P, Pritchard HD, Vaughan DG, Bamber JL, Barrand NE, Bell R, Bianchi C, Bingham RG, Blankenship DD, Casassa G, Catania G, Callens D, Conway H, Cook AJ, Corr H, Damaske D, Damm V, Ferraccioli F, Forsberg R, Fujita S, Gogineni P, Griggs JA, Hindmarsh R, Holmlund P, Holt J, Jacobel RW, Jenkins A, Jokat W, Jordan T, King EC, Kohler J, Krabill W, Riger-Kusk M, Langley K, Leitchenkov G, Leuschen C, Luyendyk B, Matsuoka K, Nogi Y, Nost O, Popov S, Rignot E, Rippin D, Riviera A, Roberts J, Ross N, Siegert M, Smith A, Steinhage D, Studinger M, Sun B, Tinto B, Welch

B, Young D, Xiangbin C, Zirizzotti A (2012) Bedmap2: improved ice bed, surface and thickness datasets for Antarctica. Cryosphere 6:4305–4361. doi:10.5194/tcd-6-4305-2012

Groh A, Ewert H, Fritsche M, Rülke A, Rosenau R, Scheinert M, Dietrich R (2014) Assessing the current evolution of the Greenland ice sheet by means of satellite and ground-based observations. Surv Geophys 35:1459–1480

Heiskanen W, Moritz H (1967) Physical geodesy. Wheeler, San Franciso

Helm V, Humbert A, Miller H (2014) Elevation and elevation change of Greenland and Antarctica derived from CryoSat-2. Cryosphere 8:1539–1559

Holland DM, Thomas RH, de Young B, Ribergaard MH, Lyberth B (2012) Acceleration of Jakobshavn Isbrae triggered by warm subsurface ocean waters. Nat Geosci. doi:10.1038/ngeo316

Horwath M, Dietrich R (2006) Errors of regional mass variations inferred from GRACE monthly solutions. Geophys Res Lett 33:L07502. doi:10.1029/2005GL025550

Klinger B, Mayer-Gürr T, Behzadpour S, Ellmer M, Kvas A and Zehentner N (2016) The new ITSG-Grace2016 release, EGU General Assembly 2016, Vienna, Austria, doi: 10.13140/RG.2.1.1856.7280

Krabill W, Abdalati W, Frederick E, Manizade S, Martin C, Sonntag J, Swift R, Thomas R, Wright W, Yungel J (2000) Greenland ice sheet: high-elevation balance and peripheral thinning. Science 289:428–430

Kusche J, Schmidt R, Petrovic S, Rietbroek R (2009) Decorrelated GRACE Time-variable gravity solutions by GFZ, and their validation using a hydrological model. J Geodesy 83:903–913. doi:10.1007/s00190-009-0308-3

Lenaerts JTM, Van Meijgaard E, Van den Broeke MR, Ligtenberg SRM, Horwath M, Isaksson E (2013) Recent snowfall anomalies in Dronning Maud Land, East Antarctica, in a historical and future climate perspective. Geophys Res Lett. doi:10.1002/grl.50559

Luthcke SB, Zwally HJ, Abdalati W, Rowlands DD, Ray DD, Nerem RS, Lemoine FG, McCarthy JJ, Chin DS (2006) Recent Greenland ice mass loss by drainage basin from satellite gravity observations. Science 314:1286–1289. doi:10.1126/science.1130776

Luthcke SB, Sabaka TJ, Loomis BD, Arendt AA, Mccarthy JJ, Camp J (2013) Antarctica, Greenland and Gulf of Alaska land-ice evolution from an iterated GRACE global mascon solution. J Glaciol. doi:10.3189/2013JoG12J147

Mayer-Gürr T, Zehentner N, Klinger B, Kvas A (2014) ITSG-Grace2014: a new GRACE gravity field release computed in Graz. In: GRACE Science Team Meeting (GSTM), Potsdam

McMillan M, Shepherd A, Sundal A, Briggs K, Muir A, Ridout A, Hogg A, Wingham D (2014) Increased ice losses from Antarctica detected by CryoSat-2. Geophys Res Lett 41(11):899–3905. doi:10.1002/2014GL060111

Nghiem SV, Hall DK, Mote TL, Tedesco M, Albert MR, Keegan K, Shuman CA, DiGirolamo NE, Neumann G (2012) The extreme melt across the Greenland ice sheet in 2012. Geophys Res Lett. doi:10.1029/2012GL053611

Nilsson J, Vallelonga PT, Simonsen SB, Sørensen LS, Forsberg R, Dahl-Jensen D, Hirabayashi M, Goto-Azuma K, Hvidberg CS, Kjær HA, Satow K (2015) Greenland 2012 melt event effects on CryoSat-2 radar altimetry. Geophys Res Lett 42:3919–3926. doi:10.1002/2015GL063296

Nilsson J, Gardner A, Sørensen LS, Forsberg R (2016) Improved retrieval of land ice topography from CryoSat-2 data and its impact for volume-change estimation of the Greenland ice sheet. Cryosphere 10:2953–2969. doi:10.5194/tc-10-2953-2016

Peltier WR (2004) Global glacial isostasy and the surface of the ice-age earth: the ice-5G (VM2) model and grace. Annu Rev Earth Planet Sci 32:111

Rignot E, Kanagaratnam P (2006) Changes in the velocity structure of the Greenland ice sheet. Science 311:986–990

Sasgen I, van den Broeke M, Bamber JL, Rignot E, Sørensen LS, Wouters B, Martinec Z, Velicogna I, Simonsen SB (2012) Timing and origin of recent regional ice-mass loss in Greenland Earth Planet. Sci Lett 333–334:293–303

Shepherd A, Ivins E, Geruo A et al (2012) A reconciled estimate of ice sheet mass balance. Science 338(6111):1183–1189. doi:10.1126/science.1228102

Simonsen SB, Stenseng L, Adalgeirsdottir G, Fausto R, Hvidberg CS, Lucas-Picher P (2013) Assessing a multilayered dynamic firn-compaction model for Greenland with ASIRAS radar measurements. J Glaciol 59(215):545–558

Sørensen LS, Simonsen SB, Nielsen K, Lucas-Picher P, Spada G, Adalgeirsdottir G, Forsberg R, Hvidberg C (2010) Mass balance of the Greenland ice sheet—a study of ICESat data, surface density and firn compaction modelling. Cryosphere 5:173–186. doi:10.5194/tcd-4-2103-2010

Sørensen LS, Simonsen SB, Meister R, Forsberg R, Levinsen J, Flament T (2015) Envisat-derived elevation changes of the Greenland ice sheet, and a comparison with ICESat results in the accumulation area. Remote Sens Environ 160:56–62. doi:10.1016/j.rse.2014.12.022

Swenson SC, Chambers DP, Wahr J (2008) Estimating geocenter variations from a combination of GRACE and ocean model output. J Geophys Res Solid Earth 113(B8), Article B08410. doi:10.1029/2007JB005338

Tapley BD, Bettadpur S, Watkins M, Reigber C (2004) The gravity recovery and climate experiment: mission overview and early results. Geophys Res Lett 31:L09607. doi:10.1029/2004GL019920

Velicogna I, Wahr J (2006) Acceleration of Greenland ice mass loss in spring 2004. Nature 443:329–331. doi:10.1038/nature05168

Velicogna I, Wahr J (2013) Time-variable gravity observations of ice sheet mass balance: precision and limitations of the GRACE satellite data. Geophys Res Lett 40:3055–3063

Wahr J, Molenaar M, Bryan F (1998) Time variability of the earths gravity field: hydrological and oceanographic effects and their possible detection by GRACE. J Geophys Res 103:30205–30229. doi:10.1029/98JB02844

Whitehouse PL, Bentley MJ, Le Brocq AM (2012) A deglacial model for Antarctica: geological constraints and glaciological modelling as a basis for a new model of Antarctic glacial isostatic adjustment. Quat Sci Rev 32:1

Zwally HJ et al (2011) Greenland ice sheet mass balance: distribution of increased mass loss with climate warming; 2003–2007 versus 1992–2002. J Glaciol 57:88–102

Surv Geophys (2017) 38:105–130
DOI 10.1007/s10712-016-9394-y

Observation-Based Estimates of Global Glacier Mass Change and Its Contribution to Sea-Level Change

B. Marzeion[1] · N. Champollion[2] · W. Haeberli[3] · K. Langley[4] · P. Leclercq[5] · F. Paul[3]

Received: 2 May 2016 / Accepted: 21 October 2016 / Published online: 11 November 2016

Abstract Glaciers have strongly contributed to sea-level rise during the past century and will continue to be an important part of the sea-level budget during the twenty-first century. Here, we review the progress in estimating global glacier mass change from in situ measurements of mass and length changes, remote sensing methods, and mass balance modeling driven by climate observations. For the period before the onset of satellite observations, different strategies to overcome the uncertainty associated with monitoring only a small sample of the world's glaciers have been developed. These methods now yield estimates generally reconcilable with each other within their respective uncertainty margins. Whereas this is also the case for the recent decades, the greatly increased number of estimates obtained from remote sensing reveals that gravimetry-based methods typically arrive at lower mass loss estimates than the other methods. We suggest that strategies for better interconnecting the different methods are needed to ensure progress and to increase the temporal and spatial detail of reliable glacier mass change estimates.

Keywords Glacier mass change · In situ and satellite observations · Glacier modeling · Mean sea level

✉ B. Marzeion
ben.marzeion@uni-bremen.de

1 Institute of Geography, University of Bremen, Postfach 330 440, 28334 Bremen, Germany

2 International Space Science Institute, Hallerstrasse 6, 3012 Bern, Switzerland

3 Department of Geography, University of Zurich, Winterthurerstrasse 190, 8057 Zurich, Switzerland

4 Asiaq Greenland Survey, Qatserisut 8, 3900 Nuuk, Greenland

5 Department of Geosciences, University of Oslo, P.O. Box 1047, Blindern, 0316 Oslo, Norway

1 Introduction

Glaciers are distinctive features of many high-altitude and high-latitude landscapes around the world. Their geometric response to changes in atmospheric conditions is slow enough to filter out high-frequency weather and climate variability, but fast enough to provide humans with visible impressions of systematic changes of the environment, without the need for technical or statistical tools. Because of these properties, glaciers have become one of the key indicators of climate change (e.g., WGMS 2008; Marzeion et al. 2014a). Perhaps more importantly, glaciers are closely linked to the Earth system not only by being shaped by atmospheric conditions and their topographic setting, but by changing the seasonality of water runoff in many large river systems (e.g., Immerzeel et al. 2010; Kaser et al. 2010; Huss 2011), by being central to many geomorphologic processes (e.g., Egholm et al. 2009; Korup et al. 2010; Thomson et al. 2010; Koppes et al. 2015; Haeberli et al. 2016) and by affecting sea level through changes of the terrestrially stored water mass (see e.g., Radić and Hock 2010; Huss and Farinotti 2012; Grinsted 2013 for current estimates of mass stored in glaciers). Of interest here are changes to the mass of water stored in glaciers on the global scale, thus affecting the global mean sea level, over a time scale of several years (i.e., excluding seasonal mass changes, Jansson et al. 2003).[1] Methods to derive glacier mass changes from in situ and remote sensing observations are presented and compared in order to obtain a globally coherent picture from spatially incomplete and temporarily inconsistent datasets.

The importance of glacier mass change for sea-level change is a function of the time scales and spatial scales considered: on decadal and shorter time scales, and on small spatial scales, sea-level variability is dominated by ocean dynamics and wind stress (e.g., Meyssignac and Cazenave 2012; Richter et al. 2012). On longer time scales and small spatial scales, glaciers may contribute to distinct patterns in relative sea-level change through mass redistribution and isostatic effects (e.g., Tamisiea et al. 2003; Larsen et al. 2005; Melini et al. 2015). Finally, long-term, systematic changes to glacier mass will impact their relative importance for the global sea level. That is, in globally cool periods with large glaciers, small temperature changes can lead to relatively large sea-level change from glaciers, while in warm periods with small glaciers, even large temperature change has a limited impact on glacier-related sea-level change (Marzeion et al. 2014b). Because of this nonlinearity, glaciers will likely play a lesser role for sea-level change in the twenty-first century than they did within the twentieth century, but still be a main contributor along with ice sheet mass loss and thermal expansion of sea water (Meier et al. 2007; Church et al. 2013; Allison et al. 2014).

In order to be able to assess our understanding of sea-level change, knowledge of glacier mass change is therefore critical. In recognition of this relevance, glaciers were identified as an Essential Climate Variable (ECV) in the terrestrial component of the Global Climate Observing System (GCOS) in support of the United Nations Framework Convention on Climate Change (UNFCCC). Observations of glaciers are organized within the Global Terrestrial Network for Glaciers (GTN-G), with the corresponding work being coordinated between the World Glacier Monitoring Service (WGMS; mainly responsible for in situ observations), the Global Land Ice Measurement from Space initiative (GLIMS, satellite observations, Raup et al. 2007) and the National Snow and Ice Data Center at Boulder/Colorado (NSIDC, data management and photo collection).

[1] Note that glacier mass loss does not equal ocean mass gain, as not all melt water may end up in the ocean (cf. Sect. 5).

While the observational network is rapidly growing, data on glaciers are characterized by a severe sub-sampling problem on the global scale: whereas there are individual time series of direct observations of glacier mass change reaching back to the ninetieth century (see Sect. 2.1), they necessarily represent a very small sample of the roughly 200,000 glaciers worldwide (Pfeffer et al. 2014). Glacier length change observations (presented in Sect. 2.2) are often available for even longer time periods and valuable for contextualizing the shorter, direct observations of mass change, but the link between glacier length and mass changes is complicated through ice dynamics and is strongly non-linear, and thus needs careful calibration. Remote sensing techniques of glacier area and volume change (see Sects. 3.1, 3.2) are comprehensive and spatially detailed, but available on the global scale for only relatively short time. Moreover, conclusions on associated ice mass change are not simple to draw, as both glacier volume and area change first need to be related to ice mass (by estimating ice density or ice thickness, respectively) or have to be derived by other means (e.g., Glasser et al. 2011). Direct remote sensing of glacier mass change (see Sect. 3.3) is a relatively recent and promising observation method. As of now, time series are short and the spatial resolution is coarse (a few hundred km), complicating the separation of the glacier mass change signal from other mass changes. Finally, it is possible to use observations of the state of the atmosphere in order to force a glacier model to obtain estimates of glacier mass change (see Sect. 4). Modeling of glaciers allows for comprehensive and long-term assessments. However, the quality of the model results critically depends on model reliability and the quality of atmospheric observations, which generally speaking are of questionable quality in glacierized regions because of their remoteness and terrain complexity. In the discussion (Sect. 5), we critically review the efforts to synthesize these different approaches and indicate recent achievements and the most pressing gaps.

2 Long-Term Glacier Monitoring

2.1 Glacier Mass Change

Data on glacier mass balance and length fluctuations document changes in time, often at high temporal resolution (annually), while glacier inventories, repeated at time intervals of a few decades, enable an assessment of how representative this information is in space. Glacier mass balance is commonly considered to be an undelayed response to atmospheric forcing, whereas glacier length change represents a filtered, enhanced and cumulative but also delayed response to a longer-term forcing, taking place at the temporal scale of climate change (i.e., decades to millennia). This delayed response of geometric adjustment to changes in climatic conditions and glacier mass balance ("response time"; decades for steep/mid-size glaciers to centuries for the flat/largest glaciers, see, e.g., Jóhannesson et al. 1989; Haeberli and Hoelzle 1995; Bahr et al. 1998; Harrison et al. 2001) induces a feedback on glacier mass balance (via mass balance-altitude effects). For the current measurement network, emphasis is put on calibrated long-term mass balances of entire glaciers investigated with direct field measurements (stakes and snow pits), enabling a detailed process understanding (accumulation-ablation, comparison with meteorological information) at high resolution in time (seasons, years) and with repeated precision mapping to determine volume/mass changes for the entire glacier.

The study by Zemp et al. (2015) provides the most recent summary of in situ observations collected by the World Glacier Monitoring Service (WGMS). They conclude that rates of early twenty-first century mass loss are historically unprecedented for the worldwide sample of observed glaciers and mainly driven by ablation processes. For the sample of long and continuous mass balance time series, rates of mass loss increased from about 0.2 m water equivalent (w.e.) per year in the decade 1980–1990 to about 0.4 m w.e. per year in the decade 1990–2000 to about 0.8 m w.e. per year in the first decade of the twenty-first century. Besides this strong evidence for accelerated climate change, the worldwide uniformity of the signal is striking: with very few exceptions (e.g., Karakoram, Pamir, see Cogley 2009; Kääb et al. 2012; Gardner et al. 2013), glaciers are shrinking rapidly and at a global scale. It is this clarity which makes glaciers "unique demonstration objects of climate change" (WGMS 2008) and key indicators in global climate system monitoring. Locally and regionally, differential change patterns take place: in case of continued atmospheric warming, glaciers with short response times can adjust quickly and remain relatively close to equilibrium conditions, while glaciers with long response times increasingly depart from equilibrium conditions (Haeberli and Hoelzle 1995). As a result, glaciers with long response times are now far too large as compared to equilibrium with twenty-first century climatic conditions and show massive down-wasting in many regions of the world (Larsen et al. 2007; Paul and Haeberli 2008; Gardelle et al. 2013). Detailed repeat inventories in well-documented regions indeed show a tendency for larger glaciers to have higher mass losses than smaller glaciers, but statistics relating rates of mass loss to geometric and climatic parameters generally exhibit large scatter and complex spatial patterns (Fischer et al. 2015).

The question of representative mass balance time series can be treated on the basis of these results, where one has to distinguish between two types of representativeness: concerning representation of the climate signal at high time resolution, well-calibrated mass balances from relatively small and comparatively steep glaciers with short response times are most significant. However, sea level is predominantly influenced by the largest and comparatively flat glaciers on Earth with long response times (Meier et al. 2007), thus requiring a different set of representative glaciers. Results from one set cannot be directly translated to the other (e.g., Le Bris and Paul 2015), because the strong or even extreme dis-equilibrium of large glaciers exerts a strong feedback on rates of mass loss via the mass balance-altitude feedback (Raymond et al. 2005). For total glacier mass change, Cogley (2009 and updates thereof) developed a scheme for weighting the spatially inhomogeneous observations and combining geodetic and direct mass balance measurements.

The marked dis-equilibrium of (especially large/flat) glaciers means that there is a strong commitment for future mass loss (Mernild et al. 2013; Marzeion et al. 2014b). In the European Alps, for instance, about half of the still existing glacier area will have to disappear in order for glaciers to adjust to climatic conditions during the first decade of the twenty-first century (Carturan et al. 2013a). Also in other similar cases, i.e. in many mid-latitude mountain ranges, the option of "saving the glaciers" hardly exists anymore because most of the ice will have disappeared within the coming decades already, before measures of global climate policy could show significant effects. Mernild et al. (2013) estimate the committed, but not yet realized, mass loss of glaciers globally at 163 ± 69 mm sea-level equivalent (SLE) based on the climate of first decade of the twenty-first century, Marzeion et al. (2014b) estimate this number at 66 ± 2 mm SLE, but for the climate of the earlier (and substantially cooler) period of 1961 to 1990.

As a consequence of ongoing rapid glacier vanishing, many of the mass balance glaciers in the international glacier monitoring network are likely to disintegrate or completely disappear within the next few decades (e.g., Carturan et al. 2013b). New mass balance programs must therefore be established on still larger/higher/thicker glaciers to save continuity. The homogenization of long-term series can then be carried out on the basis of repeat glacier inventories and inter-comparison of past melt rates in order to determine the effect of the changed set of observed glaciers. The same technique can also be used to assess quantitative information on mass changes through time of large glacier samples and entire mountain regions (Paul and Haeberli 2008). New technologies like numerical modeling of detailed glacier-bed topographies (e.g., Clarke et al. 2013; Huss and Farinotti 2012; Linsbauer et al. 2012) that can be used to bridge spatio-temporal gaps in the observational network and low-cost/high-precision measurements with LIDAR (Bhardwaj et al. 2016a) or drones (Bhardwaj et al. 2016b) are now refining observational techniques as well as the frequency and quality of obtained results.

2.2 Glacier Length Change

As direct observations of glacier volume and glacier mass change did not have global coverage prior to before the late twentieth century, other observations of glacier changes are needed to determine the glacier contribution to sea-level rise prior to the second half of the twentieth century. Glacier length fluctuations seem to be the most suited type of measurements to get a representative picture of long-term global glacier changes, as direct measurements of length changes (also called front variations) have a long history. When worldwide glacier monitoring began in 1894 with the founding of the International Glacier Commission, collection and publication of frontal fluctuations was the first focus (Zemp et al. 2014). In addition, frontal positions can be reconstructed from geomorphologic landforms (e.g., moraines), biological evidence such as overridden trees and the lifetime of lichens (e.g., Rabatel et al. 2005; Bushueva and Solomina 2012) and historical evidence such as paintings, photos, and early maps (e.g., Zumbühl 1980; Nussbaumer and Zumbühl 2012). Combining all this information has resulted in a comprehensive overview of global glacier fluctuations, also from periods before monitoring programs started.

In all glacierized regions on Earth measurements of glacier length fluctuations are available through in situ measurements, often in connection with long-term monitoring programs or through remote sensing. The first direct measurements originate from the nineteenth century, and the number of measurements strongly increased in the beginning of the twentieth century. Since 2011, WGMS has also collected reconstructed frontal variations (Zemp et al. 2011), so that at present the database contains approx. 40,000 observations of about 2000 glaciers and covers the period from the sixteenth century until the present.

A qualitative analysis of the data included in the WGMS database shows a dominant retreat of glaciers over the last 100–150 years with some intermittent advances (Zemp et al. 2015). The advances are small (a few 100 m) compared to the overall retreat (up to a few km) and asynchronous in the different glacier regions—except for the 1970s, when about a third of the glaciers included in the dataset at that time advanced.

Leclercq et al. (2014) present a worldwide dataset of long-term glacier length changes for 471 glaciers that includes a large compilation of reconstructed historical changes from a large variety of sources which, at the time, were not included in the WGMS (after submission of the data to the WGMS, they are also included in the WGMS dataset). Leclercq et al. (2014) include only long-term, cumulative glacier length records that start prior to 1950 and cover four decades at least. Like Zemp et al. (2015), Leclercq et al. (2014) found a general trend of glacier retreat since the middle of the nineteenth century, despite some intermittent periods of advance for several glaciers. Based on the global dataset of 471 glaciers, the mean twentieth century glacier length reduction is 1.5 km. There is a large variation in the retreat of individual glaciers ranging from 30 m to 23 km over the twentieth century, and a difference in regional averages of length change of the glaciers in the dataset. For example, the mean retreat of Alaskan glaciers in the dataset is 5.18 ± 0.33 km, while the average twentieth century retreat in Scandinavia is 1.11 ± 0.02 km, and in High Mountain Asia is 1.02 ± 0.05 km. The global average rate of retreat in the first half of the twentieth century, 12.5 m year^{-1} in the period 1921–1960, is larger than the retreat in the second half of the twentieth century (7.4 m year^{-1} in the period 1961–2000), which may reflect the shortening of glaciers themselves, as length change can be expected to scale with glacier length (Jóhannesson et al. 1989). On average, calving glaciers retreated much more than land-terminating glaciers, with a twentieth century average retreat of 4.9 and 1.1 km, respectively. Nevertheless, the average relative retreat of calving glaciers is almost identical to the global mean relative retreat of land-terminating glaciers, again confirming the relation between glacier length and length change.

The global average of relative glacier length change is remarkably consistent despite the large increase in the number of glacier length records and the corresponding improvement of the global coverage of the glacierized regions. Of the 197 glaciers included in Oerlemans et al. (2007), more than half are located in central Europe and Scandinavia and only 12 in the Arctic. The representativeness of the world's glaciers has substantially improved in later versions of the dataset by addition of, among others, more than 100 records in the Arctic and around 30 in both the Southern Andes and Central Asia. Despite the large differences in the characteristics of the three datasets, the global average relative glacier length change is fairly similar (Fig. 1), indicating a worldwide coherent glacier change.

Like glacier volume, glacier terminus fluctuations lag and filter variations in the climatic forcing due to the response time of the dynamical glacier system. The length

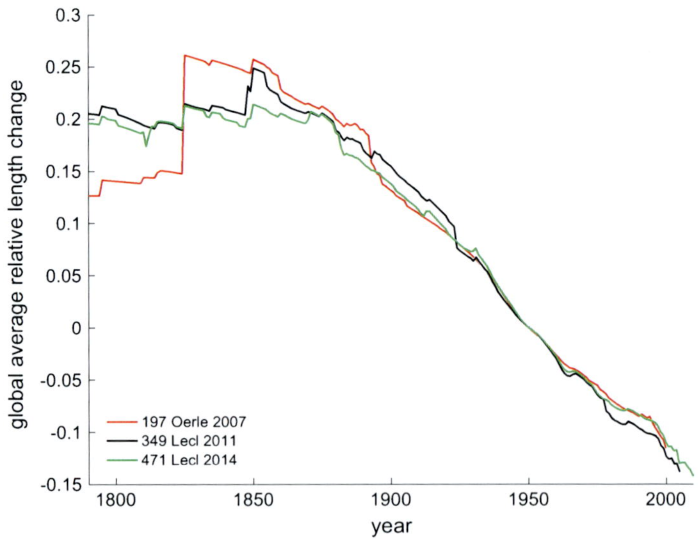

Fig. 1 Global average relative length change (i.e., length change normalized to the glacier length in 1950), based on the glaciers included in three versions of the dataset of worldwide glacier length fluctuations as used in Oerlemans et al. (2007, 197 records, *red*), Leclercq et al. (2011, 349 records, *black*) and Leclercq et al. (2014, 471 records, *green*)

response time is generally a bit larger than the volume response time (Oerlemans 2001), so that the observed length changes cannot directly be translated to volume changes.

Lüthi et al. (2010) use a dynamic equivalent simple model to describe glacier evolution forced by varying equilibrium line altitudes. After calibration of the model parameters, based on the match of modeled and observed length changes, they are able to reproduce time series of observed volume changes for 13 glaciers in Switzerland. In order to produce absolute volumes, the model results have to be scaled with the observed volume change over one time interval for each of the glaciers. However, if applied to a small sample of glaciers, these absolute volume estimates may be relatively uncertain. Oerlemans et al. (2007) proposed a different approach to derive volume change from length observations on a global scale. They combined observed cumulative length changes of 197 glaciers to reconstruct a global length change signal. They used the relative length change, rather than the absolute cumulative length change, in their measure for global length change, as the relative length change gives a globally very coherent signal (see Fig. 1). This global relative length change was scaled to global relative volume change, and the relative volume change is translated into global glacier mass change by calibration against the global glacier mass loss over the period 1961–2003, based on a compilation of glacier mass balance data (Dyurgerov and Meier 2005). Oerlemans et al. (2007) found a glacier contribution to sea-level change of 55 ± 10 mm SLE for the period 1850–2000 (0.37 mm SLE year^{-1}), and 45 ± 70 mm SLE for 1900–2000 (0.45 mm SLE year^{-1}).

Leclercq et al. (2011) estimated a significantly higher glacier contribution to sea-level rise. Their estimate is based on the same method as used by Oerlemans et al. (2007), but they used extended datasets of global glacier length changes and mass balance observations that provided better global coverage. The mass balance dataset also included available geodetic mass balances (Cogley 2009), and the number of length change records is increased from 197 to 349. The estimated glacier contribution to sea-level change resulted in 91 ± 10 mm SLE for the period 1850–2005 (0.59 mm SLE year^{-1}), and 67 ± 16 mm SLE for 1900–2005 (0.64 mm SLE year^{-1}). A more recent update, based on even more geodetic and glaciological mass balance observations and glacier length records, results in an even higher estimate of glacier contribution to sea-level rise of 80 ± 21 mm SLE for 1900–2005 (0.76 mm SLE year^{-1}), excluding the Antarctic peripheral glaciers (Marzeion et al. 2015). Uncertainty analysis shows that the results are most sensitive to the calibration of the scaled global length signal to the global mass balance observations compilation (Leclercq et al. 2011). The large increase in reconstructed sea-level contribution of glaciers as presented in Oerlemans et al. (2007), Leclercq et al. (2011) and Marzeion et al. (2015) can be ascribed mainly to the increase in the glacier mass loss in the consecutive versions of the global mass balance observations (Dyurgerov and Meier 2005; Cogley 2009 and updates thereof, WGMS 2015).

The development of automated flow-line calculations based on glacier inventory data and digital elevation models (e.g., Kienholz et al. 2014; Machguth and Huss 2014) now enables large numbers of length change data to be collected all over the world, based on satellite information. However, the interpretation of such data is becoming increasingly more difficult as more and more glaciers are turning from a dynamically active retreat to down-wasting, and even collapse, with ill-defined glacier margins. In this regard, glaciers with dynamic instabilities and surge-type glaciers with excessive length changes (km scale) over short periods of time (years) need to be excluded to the greatest extent possible using available databases (Sevestre and Benn 2015, see also Roe 2011).

3 Satellite Remote Sensing of Glaciers

The use of space-borne observations for global glacier monitoring has increased dramatically over the past two decades. By all means, satellites provide information that is complementary to field observations, in particular when it comes to complete spatial coverage and remote regions. Moreover, they also provide observations for time periods that are not covered by field observations (Barandun et al. 2015) or products that are difficult to obtain in the field or from airborne remote sensing. For the time period they cover, their particular strength is in global scale observations with consistent and reproducible methods. Key products derived from satellite data are glacier outlines and inventories (in combination with a digital elevation model, DEM, e.g., Andreassen et al. 2012), consistent DEMs of the surface topography with global coverage (e.g., the SRTM DEM or the ASTER GDEM), elevation changes over entire glaciers from differencing DEMs from two epochs or at points from repeat altimetry (e.g., Nuth et al. 2010), surface flow velocities for determination of mass fluxes (e.g., Melkonian et al. 2013, 2016), glacier mass changes from space-borne gravimetry observations (using the GRACE satellites, e.g., Jacob et al. 2012) and glacier facies mapping (ice, firn, snow) that is used as a proxy for mass balance (e.g., Rabatel et al. 2008) and an important input dataset for hydrologic models or for calibration and/or validation of distributed mass balance models (e.g., Immerzeel et al. 2009; Paul et al. 2009).

Apart from elevation changes and flow velocities that are already determined from datasets acquired at two different points in time, area and length changes can also be derived from multi-temporal datasets. Whereas change rates are of interest for a comparison across regions, it can be instructive to look at the developments of trends through time, i.e. acceleration of rates. This has mostly been done with time series of area changes (e.g., Paul et al. 2004b; Klein and Kincaid 2006; Narama et al. 2010; Tennant et al. 2012; Falaschi et al. 2013) but also with changes in velocity (Heid and Kääb 2012; Quincey et al. 2015) or elevation (Holzer et al. 2015). A trend in any of these parameters is most useful for the determination of climate change impacts (e.g., an increase of mass loss or area shrinkage rates through time). However, all of these datasets have to be determined in regard to the area covered by glaciers. Unfortunately, generating precise glacier outlines still requires large amounts of manual work by trained experts, e.g., to delineate their debris-covered parts (see 3.1.2). In consequence, glacier outlines used for calculations are often temporally misaligned with other datasets. A baseline dataset with complete global coverage has only recently been established in the form of the Randolph Glacier Inventory (RGI, Pfeffer et al. 2014),

Since it became available, the RGI was widely applied by the glaciological and hydrological community for a range of applications, including an estimate of the global glacier mass budget as derived from a combination of different methods (Gardner et al. 2013). A key asset of the inventory is the possibility to upscale information that is spatially incomplete (e.g., determined only at points) or heterogeneous (e.g., data voids in DEMs), using relations that can be derived from the individual measurements (e.g., the elevation dependence of mass loss or size class specific area change rates). With careful handling of the different time periods covered by each dataset, it is possible to derive a global picture of glacier mass changes based on field data of selected glaciers (providing the longest time series), geodetic changes from DEM differencing (providing decadal changes for entire mountain ranges), repeat altimetry (point information at sub-annual timescales) and direct mass changes from spaceborne gravimetry (for large regions that are homogenously glacierized). Table 1 gives an overview of some common methods used to derive elevation, volume or mass changes for glaciers. Airborne datasets are not considered in this table, but provide an important means to cover (temporal) data gaps (e.g., Operation IceBridge, Studinger et al. 2010) or calibration data for mass balance time series using the geodetic method and satellite-derived products (e.g., photogrammetric DEMs, Huss et al. 2015; Andreassen et al. 2016).

Owing to the complementary nature of satellite-derived observations, accuracy assessment and validation by field measurements is difficult. It is generally performed by comparing datasets across sources (field, air, space). For example, DEM quality is determined (and adjusted) using laser altimetry data from ICESat, cumulative mass balances measured in the field are calibrated with geodetic balances derived from DEM differencing, snow lines derived from satellite data are validated by field observations, and mass changes from spaceborne gravimetry are spatially constrained using glacier extents. Hence, all source data and methods have to play multiple roles and are equally valuable. Below, we provide a short overview on how glacier outlines, elevation/volume and mass changes are derived from satellite sensors and present the latest results from global scale assessments.

3.1 Glacier Mapping and Area Changes

3.1.1 The Importance of Precise Glacier Extents for the Determination of Glacier Mass Changes

As mentioned above, precise glacier extents are key to determining glacier mass changes and thus their contribution to sea-level change. Apart from technical and methodological challenges related to the mapping itself, a key problem of a precise delineation lies in the very nature of glaciers as climatic indicators: their rapid geometric changes in response to climate change and/or dynamic instabilities. This inevitably results in a temporal mismatch of glacier outlines and the datasets that use them to spatially constrain the analysis. Henceforth, all calculations using glacier outlines from a different point in time have an inherent uncertainty that can be smaller (e.g., in regions where glaciers show limited changes over decades) or larger (e.g., for rapidly changing calving or surging glaciers). In Fig. 2, an example is shown with some rapidly shrinking valley glaciers in northern Patagonia that require a very good temporal match of the outlines and the DEM to precisely derive volume loss.

For several applications, it is sufficient to just know where glaciers are located, i.e. a simple (binary) yes/no mask of glacier coverage. Such a glacier mask can be derived rather quickly from satellite imagery (for clean ice) as it is not required to separate glacier entities with drainage divides derived from a watershed analysis of a DEM. Glacier-specific calculations can only be performed when the glacier mask is divided into entities by drainage divides. In a first step, a single glacier inventory is already highly useful as it allows selecting glaciers by size, using their mean elevation as a proxy for the balanced-budget equilibrium line altitude ELA_0 (Braithwaite and Raper 2010) and henceforth precipitation amounts (e.g., Sakai et al. 2015), or calculating their ice-thickness distribution (Huss and Farinotti 2012; Linsbauer et al. 2012; Frey et al. 2014) or length (Kienholz et al. 2014; Machguth and Huss 2014). When glacier-specific outlines are combined with elevation changes derived from DEM differencing and an appropriate assumption for density (e.g., Huss 2013), it is possible to not only determine the sea level equivalent contribution of an entire region (e.g., Rignot et al. 2003; Schiefer et al. 2007; Berthier et al. 2010), but also

Table 1 Methods that can be used to determine elevation, volume and mass changes of glaciers from satellite data

Survey	Sensor	Change measured	Coverage	Time period	Temporal resolution	Issues	Accuracy
Field data	Stakes and pits, GPR	Elevation, mass	Selected glaciers globally	Since 1947	Annual	Point data, small sample, extrapolation	0.2 m
Repeat altimetry	ICESat, Cryosat2	Elevation	Globally but only along stripes	Since 2003 Since 2012	Sub-annual	Point data, extrapolation	0.5 m
DEM differencing	SRTM, GDEM SPOT/national High-resolution	Volume	Globally, w/o poles Mountain regions Selected glaciers	Around 2000 After 2000, 1960s After 2010	Decadal	Data voids, artifacts	1–2 m Better 1 m Better 0.5 m
Gravimetry	GRACE	Mass	Regional means and globally	Since 2002	Monthly	Coarse resolution	Variable

Field measurements added as reference

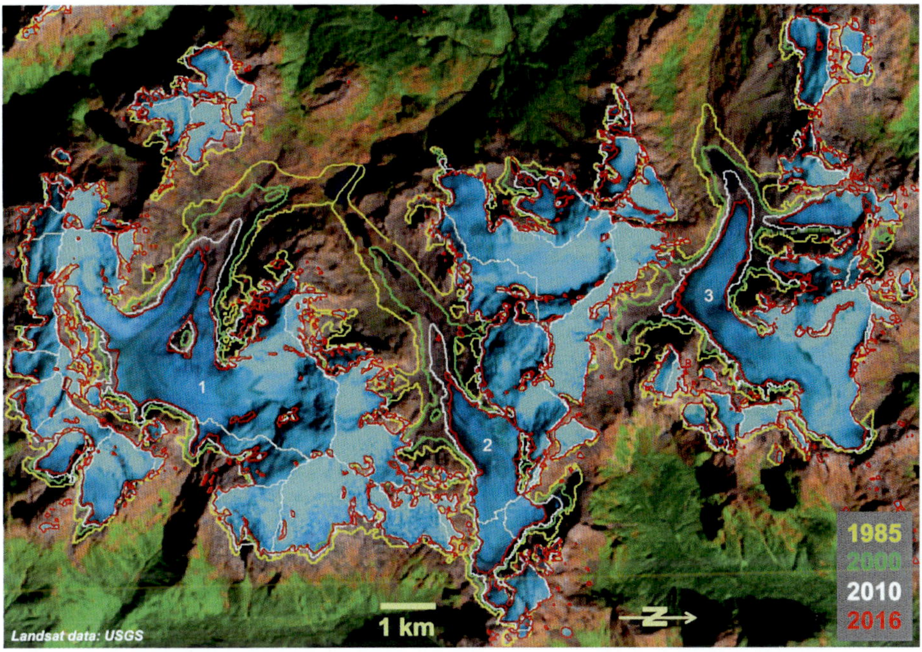

Fig. 2 Dramatic glacier shrinkage in northern Patagonia around the volcano Monte Inexplorado from 1985 to 2016. When glaciers retreat kilometers per decade, outlines need to be obtained for the time of DEM acquisition. Good temporal match of datasets (i.e., outlines and DEMs) is essential for correct change deduction. The three largest glaciers flowing to the west have lost 1/3 of their area from 1985 to 2010

the mean changes for each individual glacier. This so-called geodetic balance can be compared to cumulative field data and is used for the calibration of the in situ measurements. It further provides additional insights to the representativeness of the glaciers selected for field measurements for the entire region (Paul and Haeberli 2008; Le Bris and Paul 2015). On a larger scale, such a separation of contiguous ice masses by hydrologic drainage divides is also required to separate the glaciers and ice caps surrounding the Greenland Ice Sheet from the ice sheet and thus determine their mass change separately (Rastner et al. 2012).

When glacier outlines are available from at least two points in time, all kinds of associated changes can be derived. Whereas area changes themselves might only be indicative of mass changes, there are several further parameters that can be used to estimate corresponding mass changes. One is the change in minimum elevation that is—in a first-order approximation—two times higher than the change in mean elevation and thus the balanced-budget ELA (Haeberli and Hoelzle 1995). Combined with a mass balance gradient and the area-elevation distribution (hypsometry) of a glacier, the resulting mass changes can be determined, albeit only for a decadal time scale where measured changes are significant.

Finally, estimates of the mass balance may be obtained by determining the accumulation area ratio through remote sensing (Rabatel et al. 2008; Barandun et al. 2015). However, because of the required manual delineation of the snow line, this is not done frequently.

3.1.2 *Derivation of Glacier Outlines and Trends from Satellite Data*

Automated glacier mapping with optical satellite images is straightforward when cloud-free scenes from the end of the ablation period or dry season (with minimal snow cover) are available. Snow and ice have a very low reflectance in the shortwave infrared (SWIR), whereas reflectance is much higher in the visible and near-infrared (VNIR). Hence, dividing reflectances in a VNIR band by those in a SWIR band gives high values over glaciers and low ones over all other terrain where lower reflectance values are divided by higher ones when using a red/SWIR band ratio. With a simple threshold this ratio image can be segmented into a binary glacier map and transformed into glacier outlines using raster-vector conversion. When abundant small snow patches are present, a 3-by-3 (or 5-by-5) median filter applied to the binary glacier map is useful for eliminating them and reducing noise (Paul et al. 2002; Rastner et al. 2012). The manually selected threshold value is most sensitive in regions of shadow and should be optimized for this region in a way that workload for corrections is minimized. Manual editing is required for debris-covered glacier parts as these have the same spectral reflectance as the surrounding terrain and can thus not be mapped based on spectral reflectance alone (Paul et al. 2004a, b). Apart from debris cover, clouds and seasonal snow are the main bottleneck for a fast and frequent production of glacier outlines (Paul et al. 2011). All three obstacles will likely improve when Sentinel 2A and 2B provide 10 m resolution images at least every 5 days (more often toward higher latitudes). Whereas the high repetition rate will help in acquiring cloud-free images from the real end of the ablation period, the high spatial resolution will help in identifying snow patches and delineating glaciers with debris more precisely (Fig. 3).

Thanks to the opening of the Landsat archive by USGS (Wulder et al. 2012), numerous studies have mapped glacier extents over large regions with the method described above (Bolch et al. 2010; Rastner et al. 2012; Frey et al. 2012; Guo et al. 2015) and contributed to the global completeness of the RGI (Pfeffer et al. 2014). The body of literature on regional glacier area changes revealed a globally consistent picture of shrinking glaciers over the past 50 years with rates mostly between 0.05 and 1% area loss per year (Vaughan et al. 2013). In some regions (Alps, Low Latitudes), they are even higher (up to 2%) and the majority of investigated regions showed a recent increase of area loss rates.

3.2 Elevation and Volume Changes

Surface elevation changes can be used together with the glacier outline to determine volume changes for a given time period. To convert the measured volume changes to changes in mass, a density must be assigned. It is important to note that surface elevation changes and thereby volume changes do not necessarily mean a loss or gain in mass since they do not consider firn compaction or internal refreezing. Thus, by themselves, these parameters cannot be directly transferred to a sea-level budget.

There are three main methods used for determining surface elevations and thereby surface elevation changes of glaciers from satellite measurements. These are: repeat altimetry (Csathó et al. 2014), DEM differencing (Wang and Kääb 2015), and a combination of these two (Nuth et al. 2010; Helm et al. 2014). As shown in Table 1, each of the methods has advantages and disadvantages with regards to temporal and spatial coverage. The combination of altimetry and DEMs allows for the advantages of both methods to be exploited (e.g., Kääb et al. 2012). Combining multiple data sources where available, be it

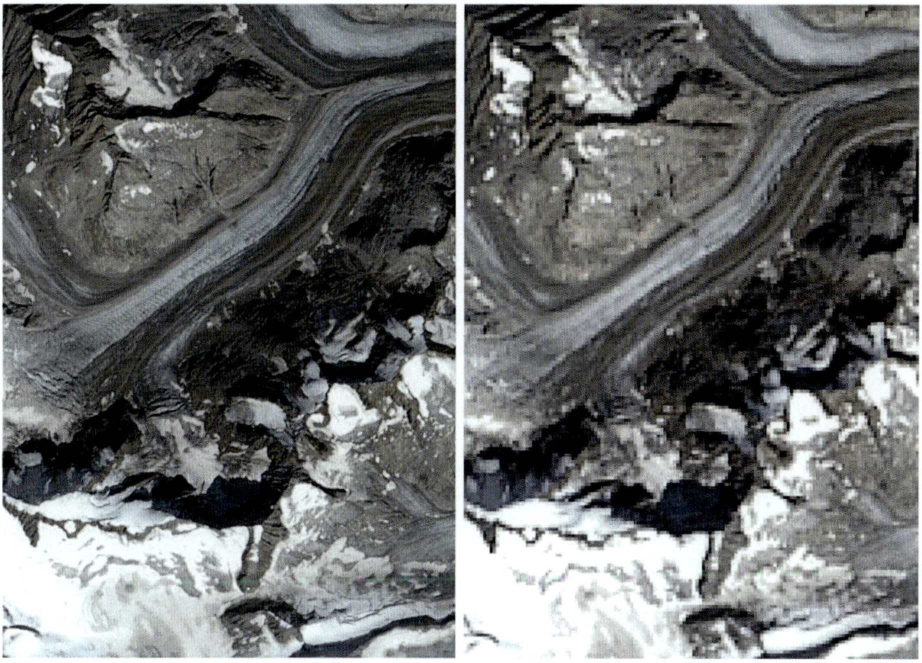

Fig. 3 Comparison of a Sentinel 2 image (*left*) with 10 m resolution (acquired on 29 August 2015) to a Landsat 8 image (*right*) with 30 m resolution (31 August 2015) for the southern part of Unteraarglacier in the Swiss Alps. With Sentinel 2 fine spatial details of glacier flow become visible (ogives) and debris-covered regions can be identified and delineated more precisely. Sentinel data: Copernicus 2015, Landsat data: USGS

satellite, airborne or ground-based, gives the best possible assessment of change (Schenk and Csathó 2012; Csathó et al. 2014).

Both laser and radar have been used in past and current altimetry missions. Laser has a shorter wavelength and thus has the advantage of having higher resolution, but clouds reflect energy at this frequency so that the surface below is not measured when clouds are present. Radar is not affected by clouds, so the surface can always be measured. However, at the typical frequencies used, C-band to Ku-band, the energy penetrates into snow and ice, meaning that the distance measured may correspond to a region below the true surface. The amount of penetration and subsequent contribution of volume scattering to the signal depends both on the electromagnetic frequency of the signal and on the characteristics of the target material, in particular on liquid water content. It is greatest over the dry snow zones of the ice sheets, but most variable and therefore most difficult to correct for over glaciers and ice caps (e.g., Rignot et al. 2001; Müller et al. 2011; Gray et al. 2015).

ICESat was the first laser altimetry mission, operational from 2003 to 2009. Currently, an ICESat2 mission (Abdalati et al. 2010) is planned for launch in late 2017. There are currently 6 radar altimetry satellites in operation (http://www.altimetry.info/missions/). Of these, Cryosat-2 was the first to be dedicated to measuring changes in the cryosphere and is novel in that it has both SAR and Interferometric capabilities to improve both along and across track resolution. Altika is novel in that it operates at Ka-band frequency, presenting less of a surface penetration problem. To derive elevation changes, the commonly used cross-over method and the repeat-track method are suitable for both laser and radar missions (e.g., Moholdt et al. 2010).

Comprehensive coverage of individual glaciers is difficult with the relatively sparse coverage of altimetry missions such ICESat, although with swath processing of Cryosat-2 data this is no longer an issue (Gray et al. 2013). However, using a DEM for the surface elevation at one point in time and an altimetry dataset for the elevation at a second point in time allows all altimetry data points to be used as opposed to just those with repeat or cross-tracks. Use of a radar-based DEM, such as the widely used SRTM DEM, requires consideration of the penetration of the radar signal and the impacts this may have on the deduced surface elevation changes (Kääb et al. 2012).

Prior to differencing two DEMs, it is critical that they are co-registered to ensure the elevation change is calculated for corresponding ground points. Despite good co-registration, biases may remain due to acquisition strategy (Berthier et al. 2007) or DEM creation (Nuth and Kääb 2011), which will propagate to the final difference DEM if they cannot be corrected for. The final difference DEM will only be as good as the poorest input DEM. The most obvious manifestation of this is that no-data pixels, or voids, in one of the DEMs will produce voids in the resultant elevation change grid. Additional errors include noise and biases and can be evaluated by making statistical analyses of the surrounding stable terrain (e.g., Nuth and Kääb 2011).

The time stamps on the DEMs being differenced is a key part of any interpretation of the derived elevation change. Depending on the time difference, the calculated elevation changes can reveal short-term dynamic behavior, seasonal changes, annual or multi-decadal changes. Since the whole glacier system can be assessed, for land-terminating glaciers, the dynamic component of ice submergence and emergence cancels out. Thus, the DEM differencing method provides a possibility to assess climatic-induced elevation changes where sufficiently long time series are available. A time period of at least five years is recommended in order to avoid seasonal and small scale fluctuations, although the ideal period will also depend on the accuracy of the DEMs and the magnitude of the change being measured.

The recent sub-meter resolution optical satellites such as Quickbird, WorldView and Pléiades provide high-resolution spaceborne DEMs that push the spatial and temporal limits of what can be achieved with spaceborne DEM differencing (Kronenberg et al. 2016; Melkonian et al. 2016; Berthier et al. 2014).

3.3 Glacier Mass Change from Spaceborne Gravity Observations

Glacier mass changes can be estimated at the global scale since the launch of the Gravity Recovery And Climate Experiment (GRACE) twin satellites in summer 2002. This mission is devoted to the observation of relative Earth gravity changes. On a monthly timescale, changes in gravity observed from space are mainly caused by changes in the mass of surface water (Wahr and Molenaar 1998). Due to the mission design and satellite orbit (polar orbit at an altitude of 500 km), Earth gravity changes are delivered as monthly spherical harmonic coefficients at the Earth surface with a spatial resolution around 300 km by the three data centers responsible for the GRACE mission (Geoforschungszentrum Postdam, GFZ; Center for Space Research at University of Texas, CSR; Jet Propulsion Laboratory, JPL; Wahr and Molenaar 1998; Jacob et al. 2012; Sakumura et al. 2014). The most recent gravity field from GRACE is the Level 2 Release 05 (RL05), which includes de-aliasing processing. Sakumura et al. (2014) have shown that using the arithmetic mean of all gravity field solutions is the most effective in reducing the noise.

All the methods to estimate glacier mass changes from spaceborne gravimetry observations convert the spherical harmonic coefficients (Stokes coefficients) into surface water thickness equivalent and finally into mass changes using the formulation presented in Wahr and Molenaar (1998). The main challenge remains to attribute the observed mass changes to glaciers and other mass change sources. Two recent methods exist to do this: the first is the mass concentration blocks approach (Jacob et al. 2012; Schrama et al. 2014). This attempts to determine the mass change of glacierized regions by fitting mass values (i.e., a set of Stokes coefficients) obtained from the gravity field measured by GRACE into small regions (called mascons) where the mass is assumed to be uniformly distributed. This approach can be applied to Level 2 GRACE data or also to raw Level 1 measurements. The second method is the forward modeling approach (Chen et al. 2013; Yi et al. 2015) that is based on an iterative and global forward modeling of the Stokes coefficients, in order to separate the terrestrial from the ocean signal. Known locations of terrestrial mass sources can be used to constrain mass changes. In this approach, the total mass on the Earth surface is conserved.

Note that usually during the attribution of mass changes to regions with glaciers and ice caps, the Stokes coefficients of degree 2 and order 0, as well as the degree 1 are, respectively, replaced with those from satellite laser ranging and those estimated based on Swenson et al. (2008). In addition, a correction for glacial isostatic adjustment (GIA) and a Gaussian filter (usually 300 km large or more) is applied.

Despite recent method improvements, uncertainties are still important in glacier mass changes estimated from GRACE measurements, especially due to the heterogeneity and the small size of glaciers compared to the spaceborne gravimetry spatial resolution, which is determined by satellite altitude and distance between the two twin satellites (Wahr and Molenaar 1998). For example, there is significant leakage of the glacier mass signal into the oceans. Excluding coastal regions, or using the forward modeling approach is useful to solve this issue. However, these solutions are not completely satisfying because they are based on questionable assumptions, such as a uniform distribution of mass change in the ocean. Another example of uncertainty is the difficulty to distinguish between mass change originating from land water storage changes and glacier mass changes. Using known spatial patterns (glacier or river basins, mascons, etc.) or hydrological modeling can be a basis for removing the land water contribution from the signal. However, the spatial separation is not clear everywhere, especially in regions where the land water storage signal or annual cycle is large (e.g., Himalaya regions, Andes region) and where hydrological models show large discrepancies. Finally, the necessity to remove the GIA signal from GRACE-based mass changes causes uncertainty (Schrama 2016). The GIA signal does not include more recent effects, e.g., the rebound from the Little Ice Age, which is essential in some glacierized areas like Patagonia or Alaska (Larsen et al. 2005; Jacob et al. 2012).

The most recent global estimates of glacier mass changes include Jacob et al. (2012) with 0.41 ± 0.08 mm SLE year^{-1} (2003–2010 time period) and Schrama et al. (2014) with 0.45 ± 0.03 mm SLE year^{-1} (2003–2013) using the mascon-based approach, Chen et al. (2013) with 0.54 ± 0.10 mm SLE year^{-1} (2005–2011) and Yi et al. (2015) with 0.58 ± 0.04 mm SLE year^{-1} (2005–2014) using the forward modeling-based method, and the four following studies which combined GRACE-based estimates with other datasets (e.g., altimetry, literature assessment, sea-level budget approach): Gardner et al. (2013) with 0.60 ± 0.08 mm SLE year^{-1} (2003–2009), Dieng et al. (2015) with 0.58 ± 0.1 mm

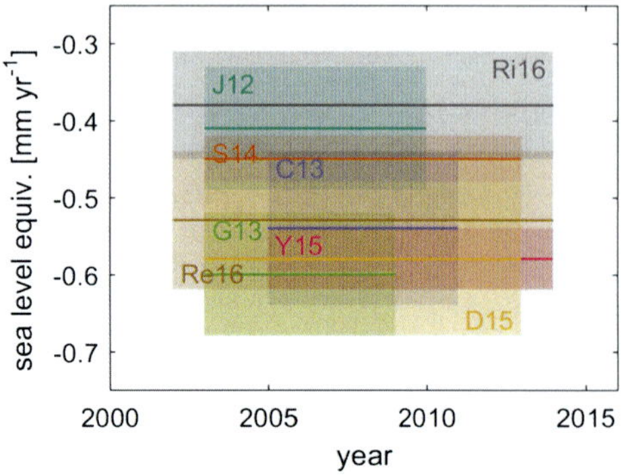

Fig. 4 Glacier mass change estimates based (partly) on GRACE data. Boxes indicate period covered and upper and lower confidence level of estimate. J12 is Jacob et al. (2012), G13 is Gardner et al. (2013), C13 is Chen et al. (2013), S14 is Schrama et al. (2014), D15 is Dieng et al. (2015), Y15 is Yi et al. (2015), Re16 is Reager et al. (2016) and Ri16 is Rietbroek et al. (2016)

SLE year^{-1} (2003–2013), Reager et al. (2016), an update of Gardner et al. (2013), with 0.53 ± 0.09 mm SLE year^{-1} (2002–2014) and Rietbroek et al. (2016) with 0.38 ± 0.07 mm SLE year^{-1} (2002–2014). See Fig. 4 for a comparison of these estimates. Because of the high temporal variability (see Fig. 5) and its influence on the estimated trends (Yi et al. 2015), all these results are difficult to compare with one another. However, most values are consistent and show an average rate of glacier mass losses of 0.51 ± 0.07 mm SLE year^{-1} for the last decade (2002/2005–2013/2015), which represents 184 Gt of fresh water added into the ocean every year. GRACE-based glacier mass losses are smaller than those derived from in situ measurements or modeling estimates. Note that peripheral glaciers in Antarctica and Greenland (excluded from the studies above) were estimated to additionally have contributed 0.02 ± 0.03 mm SLE year^{-1} and 0.1 ± 0.02 mm SLE year^{-1}, respectively, during 2003–2009 (Gardner et al. 2013).

4 Modeling Glaciers Based on Climate Observations

While comprehensive monitoring of glaciers on the global scale only became available with modern remote sensing methods, the state of the global atmosphere has been observed by a dense (at least compared to glacier observations) network of weather stations for many more decades. Based on objective analysis and homogenization of individual records, Harris et al. (2014) present a monthly, globally complete gridded dataset of temperature and precipitation starting with the beginning of the twentieth century. Datasets like this (similarly, global reanalysis data) allow another observation-based estimate of past glacier change: instead of relying on observations of glaciers, observations of the state of the atmosphere may be used as boundary conditions for glaciological modeling. However, most published global glacier models rely on this reconstructive mode of operation only for purposes of model validation and calibration, neglecting any past glacier geometry change (e.g., Radić et al. 2014; Huss and Hock 2015).

However, on time scales of decades and more, glacier geometry change presents several important feedbacks to glacier mass balance (e.g., Paul 2010). On the one hand, the stabilizing, negative feedback of glacier advance and thus progression to a warmer environment during periods of mass gain, and the retreat to higher, cooler altitudes during periods of mass loss, are important. On the other hand, particularly the largest glaciers with flat and thick tongues often only have weakly inclined or even retrograde beds. This leads to down-wasting rather than dynamically active retreat to higher elevations and may

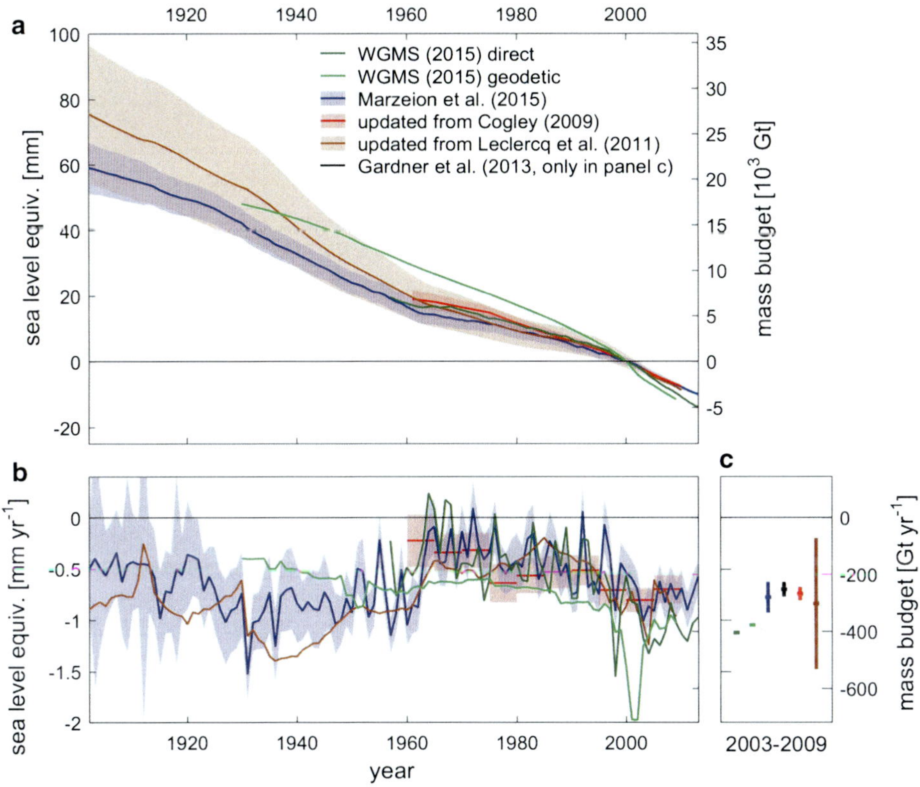

Fig. 5 Globally integrated glacier mass change: **a** accumulated in time, relative to the year 2000; **b** mass change rates; **c** mean mass change rates during 2003 to 2009. *Shading* and *error bars* indicate the 90% confidence interval. Note that all values exclude glaciers in the Antarctic periphery. For the time series labeled WGMS, the global mean of direct and geodetic specific mass balance measurements from Zemp et al. (2015, data from WGMS 2015) were multiplied with the global glacier area of Pfeffer et al. (2014). Figure extended from Marzeion et al. (2015)

prevent a recovery of the glacier tongue once the surface lowering has started. Because of the delayed response of glaciers to climate forcing, these feedbacks are not simple to capture in glacier models in reconstructive mode: to estimate the past glacier geometry, the past mass balance has to be known, which itself is a function of the past geometry. Marzeion et al. (2012, updated in Marzeion et al. 2015), developed an iterative procedure to identify the state of a glacier in the year of model initialization (e.g., 1901) that will result in the observed state of the glacier in the year of observation (e.g., 2005) after a forward model run, using atmospheric conditions from Harris et al. (2014) as boundary condition. They then used the RGI to reconstruct past changes of each of the world's glaciers, using direct glacier observations for model calibration and cross-validation.

While this approach benefits from the relative abundance of atmospheric observations, it suffers from the additional uncertainty introduced by the glacier model. Moreover, the validation with direct glacier observations is of limited value for uncertainty estimates in the earlier period of the reconstruction, as most direct glacier observations were taken at times and in places where the observations of the state of the atmosphere can be assumed to be of above-average quality and quantity compared to other remote and mountainous regions.

5 Synthesis and Discussion

Direct and geodetic measurements of glacier mass change, calibrated reconstructions of mass change based on glacier length change records and modeled reconstructions forced by climate observations are all available for (at least) several decades. Over the past years, progress in data availability and methodological advances has led to greater consistency in the results of the respective methods (Marzeion et al. 2015). In Fig. 5, we compare the most recent results for the twentieth century. In Table 2, we additionally list estimates based on remote sensing for a more recent period.

Over multi-decadal time periods, the greatest deviation is between the estimates based on the arithmetic means of direct and geodetic observations (WGMS 2015) on the one hand, and the estimates based on glacier length change observations (Leclercq et al. 2011), glacier modeling (Marzeion et al. 2015) and based on spatially weighted observations (Cogley 2009) on the other hand. Since the data used by Cogley (2009) and WGMS (2015) are to a large degree identical, this discrepancy can probably be explained by non-representative sampling of the glacier observations. Specifically, glacier observations appear to be made on glaciers that tend to have more negative mass balances than the global, or even regional, mean (Gardner et al. 2013, also found indications of this).

During the more recent years when estimates from remote sensing are available, it becomes apparent that they consistently indicate weaker glacier mass losses, even when taking into account the different regions considered (i.e., results from GRACE generally exclude glaciers in the Greenland and Antarctic peripheries). While the error margins allow a reconciliation of most of these estimates, it needs to be better understood where the systematic differences originate.

It is important to consider that not all meltwater from shrinking or vanishing glaciers directly contributes to sea-level rise (Haeberli and Linsbauer 2013; Loriaux and Casassa 2013). Some glacier parts are below sea level and, because of the ice/water density

Table 2 Comparison of mass change estimates during common periods, in mm SLE year^{-1} and the 90% confidence interval, where given in the source

Source	2003–2009	1961–2010	1901–2010
WGMS (2015) direct	−1.12	−0.57	–
WGMS (2015) geodetic	−1.05	−0.85	–
Marzeion et al. (2015)	−0.78 ± 0.15	−0.49 ± 0.05	−0.62 ± 0.05
Updated from Cogley (2009)	−0.75 ± 0.07	−0.54 ± 0.05	–
Updated from Leclercq et al. (2011)	−0.84 ± 0.64	−0.58 ± 0.15	−0.78 ± 0.19
Gardner et al. (2013)	−0.70 ± 0.07	–	–
Average of GRACE-based studies, see Sect. 3.3 for sources	−0.61 ± 0.07[a]	–	–

[a] Averaged over different time periods (2002/2005–2013/2015) and adding the estimate for Greenland peripheral glaciers from Gardner et al. (2013)

difference, even cause a slight lowering of the sea level when replaced by water. Meltwater from glaciers on land may be held back in lakes, which form in over-deepened parts of glacier beds when becoming exposed. On its way to the ocean, glacier meltwater on land can be lost by natural processes or be used through human activity (Bury et al. 2013; Carey et al. 2014a, b). Some rivers fed by glacier meltwater drain endorheically and do not reach the ocean at all (Kääb et al. 2015; Neckel et al. 2014). This effect may in fact be a partial explanation for the difference between GRACE-based versus other estimates: GRACE integrates large regions and thus includes in its glacier mass loss estimates the increase of water stored in lakes or in the ground. Coastal effects such as shoreline migration, changes in ocean area and isostatic adjustments to land and ocean surfaces must also be considered (Huss and Hock 2015).

Scientific research into the questions mentioned above is at its very beginning and relates to complex systems. In recent years, the development of flux/stress/slope-related approaches to calculate detailed ice thicknesses and glacier-bed topographies (Haeberli 2016) enabled coherent and well-constrained calculations of glacier volumes and first estimates of ice below sea level and glacier overdeepenings. Based on detailed worldwide ice-thickness modeling, Huss and Farinotti (2012) provided a value of 0.43 ± 0.06 m sea-level equivalent for the total ice volume. Haeberli and Linsbauer (2013) used information from modeled glacier-bed topographies for still existing mountain glaciers (Linsbauer et al. 2012 for the Swiss Alps; cf. also Linsbauer et al. 2015 for the Himalaya–Karakoram region) to infer that ice below sea level and in overdeepenings together accounts for a few (probably 1–6) centimeters SLE, with millimeters rather than centimeters in overdeepenings and centimeters rather than millimeters below sea level. This is confirmed by Huss and Hock (2015) who re-calculated the total glacier volume to be 37.4 cm SLE and that 11–14% of the ice volume to be lost in the twenty-first century is already below sea level (glacier-bed overdeepenings at higher elevations not included). Not even order-of-magnitude assessments are available for the amount of meltwater being diverted over land. The evolution in time is likely different for the various components of glacier volume. The amount of ice below sea level is likely to decrease in the twenty-first century due to the retreat of glaciers out of the ocean (Haeberli and Linsbauer 2013; Huss and Hock 2015). However, the amount of meltwater stored in overdeepenings may increase over time even into the twenty-second century.

There is another systematic error likely leading to an underestimation of past glacier mass loss, which is based on the temporal limitation of the glacier inventories used for upscaling measurements or initializing models. These inventories do not necessarily contain information on glaciers that have already disappeared but produced meltwater during their disappearance. As these are necessarily small glaciers, their contribution to the total mass loss is likely also small. But whereas estimates exist on the number and size of small glaciers missing in state-of-the-art inventories (Pfeffer et al. 2014 estimate an upper bound of the missing glaciers at 1.1–1.4% of global glacier area), we are not aware of any estimate of their potential past mass loss. However, Bahr and Radić (2012) find that their contribution to the total glacier mass is probably non-negligible, and it is particularly for these small glaciers, that the option of "saving the glaciers" hardly exists anymore (see Sect. 2.1).

In conclusion, an integrative observational strategy combining in situ measurements, satellite observations and numerical modeling is desirable to provide a comprehensive view of past and ongoing glacier change. To bridge the gap between detailed local investigations and global coverage, a tiered strategy has already been developed (Haeberli et al. 2000; WGMS 2015), including:

- process understanding and model development/calibration/assimilation
    - extensive measurements of energy/mass balance, flow, etc.
- regional indicators
    - mass change (index stakes + photogrammetry, LIDAR)
- regional representativeness
    - cumulative length change, DEM differencing
- global coverage
    - inventories (remote sensing/geoinformatics)

To this end, field measurements of mass balance will remain essential in the future to (a) separate surface effects from effects of glacier flow (submerging/emerging flow, etc.) and, hence, (b) understanding of, and numerical model development for, mass and energy balance processes. The full implementation of the above strategy will likely result in more reliable, spatially and temporally well-resolved data on glacier mass change. Ideally, it could lead to similar agreement between the different methods on regional scales as can now be found on the global scale, which would be an important step to better understanding glacier mass changes at a regional scale.

Acknowledgements Open access funding provided by Austrian Science Fund (FWF). This paper is a result of the ISSI Workshop on Integrative Study of Sea Level, held in Bern, Switzerland 2–6 February 2015. P.W. Leclercq acknowledges funding by the European Research Council under the European Union's Seventh Framework Programme (FP/2007–2013)/ERC Grant Agreement No. 320816. F.P. acknowledges funding by the ESA project Glaciers_cci (4000109873/14/I-NB). B.M. acknowledges funding by the Austrian Science Fund (FWF): P25362-N26. We thank the two reviewers for their constructive comments and suggestions that greatly helped to improve the manuscript.

References

Abdalati W, Zwally HJ, Bindschadler R, Csatho B, Farrell SL, Fricker HA, Harding D, Kwok R, Lefsky M, Markus T, Marshak A, Neumann T, Palm S, Schutz B, Smith B, Spinhirne J, Webb C (2010) The ICESat-2 laser altimetry mission. Proc IEEE 98(5):735–751

Allison I, Colgan W, King M, Paul F (2014) Ice sheets, glaciers and sea level rise. In: Haeberli W, Whiteman C (eds) Snow and ice-related hazards, risks and disasters. Elsevier, Amserdam, pp 713–747

Andreassen LM, Winsvold SH, Paul F, Hausberg JE (2012) Inventory of Norwegian Glaciers, Norwegian Water Resources and Energy Directorate report no. 38-2012

Andreassen LM, Elvehøy H, Kjøllmoen B, Engeset RV (2016) Reanalysis of long-term series of glaciological and geodetic mass balance for 10 Norwegian glaciers. Cryosphere 10:535–552

Bahr DB, Radić V (2012) Significant contribution to total mass from very small glaciers. Cryosphere 6:763–770

Bahr DB, Pfeffer WT, Sassolas C, Meier MF (1998) Response time of glaciers as a function of size and mass balance: 1. Theory. J Geophys Res Solid Earth 103(B5):9777–9782

Barandun M, Huss M, Sold L, Farinotti D, Azisov E, Salzmann N, Usubaliev R, Merkushkin A, Hoelzle M (2015) Re-analysis of seasonal mass balance at Abramov glacier 1968–2014. J Glaciol 61(230):1103–1117

Berthier E, Arnaud Y, Kumar R, Ahmad S, Wagnon P, Chevallier P (2007) Remote sensing estimates of glacier mass balances in the Himachal Pradesh (Western Himalaya, India). Remote Sens Environ 108:327–338

Berthier E, Schiefer E, Clarke GK, Menounos B, Rémy F (2010) Contribution of Alaskan glaciers to sea-level rise derived from satellite imagery. Nat Geosci 3(2):92–95

Berthier E, Vincent C, Magnússon E, Gunnlaugsson ÁÞ, Pitte P, Le Meur E, Masiokas M, Ruiz Pálsson F, Belart JMC, Wagnon P (2014) Glacier topography and elevation changes derived from Pléiades sub-meter stereo images. Cryosphere. doi:10.5194/tc-8-2275-2014

Bhardwaj A, Sam L, Bhardwaj A, Martín-Torres FJ (2016a) LiDAR remote sensing of the cryosphere: present applications and future prospects. Remote Sens Environ 177:125–143

Bhardwaj A, Sam L, Martín-Torres FJ, Kumar R (2016b) UAVs as remote sensing platform in glaciology: present applications and future prospects. Remote Sens Environ 175:196–204

Bolch T, Menounos B, Wheate R (2010) Landsat-based glacier inventory of western Canada, 1985–2005. Remote Sens Environ 114(1):127–137

Braithwaite RJ, Raper SCB (2010) Estimating equilibrium-line altitude (ELA) from glacier inventory data. Ann Glaciol 50(53):127–132

Bury J, Mark BG, Carey M, Young KR, McKenzie JM, Baraer M, French A, Polk MH (2013) New geographies of water and climate change in Peru: coupled natural and social transformations in the Santa River watershed. Ann Assoc Am Geogr 103(2):363–374. doi:10.1080/00045608.2013.754665

Bushueva IS, Solomina ON (2012) Fluctuations of Kashkatash glacier over last 400 years using cartographic, dendrochronologic and lichonometric data. Ice Snow 2:121–130

Carey M, Baraer M, Mark BG, French A, Bury J, Young KR, McKenzie JM (2014a) Toward hydro-social modeling: merging human variables and the social sciences with climate-glacier runoff models (Santa River, Peru). J Hydrol 518(Part A):60–70. doi:10.1016/j.jhydrol.2013.11.006

Carey M, McDowell G, Huggel C, Jackson J, Portocarrero C, Reynolds JM, Vicuña L (2014b) Integrated approaches to adaptation and disaster risk reduction in dynamic Socio-cryospheric systems. In: Haeberli W, Whiteman C (eds) Snow and ice-related hazards, risks, and disasters. Elsevier, Amsterdam, pp 219–261

Carturan L, Filippi R, Seppi R, Gabrielli P, Notarnicola C, Bertoldi L, Paul F, Rastner P, Cazorzi F, Dinale R, Dalla Fontana G (2013a) Area and volume loss of the glaciers in the Ortles-Cevedale group (Eastern Italian Alps): controls and imbalance of the remaining glaciers. Cryosphere 7(5):1339–1359

Carturan L, Baroni C, Becker M, Bellin A, Cainelli O, Carton A, Casarotto C, Dalla Fontana G, Godio A, Martinelli T, Salvatore MC (2013b) Decay of a long-term monitored glacier: careser glacier (Ortles-Cevedale, European Alps). Cryosphere 7:1819–1838

Chen JL, Wilson CR, Tapley BD (2013) Contribution of ice sheet and mountain glacier melt to sea level rise. Nat Geosci 6:549–552. doi:10.1038/NGEO1829

Church JA, Clark PU, Cazenave A, Gregory JM, Jevrejeva S, Levermann A, Merrifield MA, Milne GA, Nerem RS, Nunn PD, Payne AJ, Pfeffer WT, Stammer D, Unnikrishnan AS (2013) Sea level change. In: Stocker TF, Qin D, Plattner G-K, Tignor M, Allen SK, Boschung J, Nauels A, Xia Y, Bex V, Midgley PM (eds) Climate change 2013: the physical science basis. Contribution of working group I to the fifth assessment report of the intergovernmental panel on climate change. Cambridge University Press, Cambridge, UK and New York, NY, USA

Clarke GK, Anslow F, Jarosch A, Radic V, Menounos B, Bolch T, Berthier E (2013) Ice volume and subglacial topography for western Canadian glaciers from mass balance fields, thinning rates, and a bed stress model. J Clim 26(12):4282–4303

Cogley JG (2009) Geodetic and direct mass-balance measurements: comparison and joint analysis. Ann Glaciol 50(50):96–100

Csathó BM, Schenk AF, van der Veen CJ, Babonis G, Duncan K, Rezvanbehbahani S, van den Broeke MR, Simonsene SB, Nagarajan S, van Angelen JH (2014) Laser altimetry reveals complex pattern of Greenland Ice Sheet dynamics. Proc Natl Acad Sci 111(52):18478–18483

Dieng HN, Champollion N, Cazenave A, Wada Y, Schrama E, Meyssignac B (2015) Total land water storage change over 2003-2013 estimated from a global mass budget approach. Environ Res Lett 10(12):124010. doi:10.1088/1748-9326/10/12/124010

Dyurgerov MB, Meier MF (2005) Glaciers and the changing earth system: a 2004 Snapshot. Occasional paper 58, INSTAAR, University of Colorado

Egholm DL, Nielsen SB, Pedersen VK, Lesemann JE (2009) Glacial effects limiting mountain height. Nature 460(7257):884–887

Falaschi D, Bravo C, Masiokas M, Villalba R, Rivera A (2013) First glacier inventory and recent changes in glacier area in the Monte San Lorenzo Region (47°S), Southern Patagonian Andes, South America. Arct Antarct Alp Res 45:19–28

Fischer M, Huss M, Hoelzle M (2015) Surface elevation and mass changes of all Swiss glaciers 1980–2010. Cryosphere 9:525–540. doi:10.5194/tc-9-525-2015

Frey H, Paul F, Strozzi T (2012) Compilation of a glacier inventory for the western Himalayas from satellite data: methods, challenges and results. Remote Sens Environ 124:832–843

Frey H, Machguth H, Huss M, Huggel C, Bajracharya S, Bolch T, Kulkarni A, Linsbauer A, Salzmann N, Stoffel M (2014) Estimating the volume of glaciers in the Himalaya–Karakoram region using different methods. Cryosphere 8:2313–2333

Gardelle J, Berthier E, Arnaud Y, Kääb A (2013) Region-wide glacier mass balances over the Pamir–Karakoram–Himalaya during 1999–2011. Cryosphere 7:1263–1286. doi:10.5194/tc-7-1263-2013

Gardner AS, Moholdt G, Cogley JG, Wouters B, Arendt AA, Wahr J, Berthier E, Hock R, Pfeffer WT, Kaser G, Ligtenberg SRM, Bolch T, Sharp MJ, Hagen JO, van den Broeke MR, Ligtenberg SR (2013) A reconciled estimate of glacier contributions to sea level rise: 2003 to 2009. Science 340(6134):852–857

Glasser NF, Harrison S, Jansson KN, Anderson K, Cowley A (2011) Global sea-level contribution from the Patagonian Icefields since the Little Ice Age maximum. Nat Geosci 4(5):303–307

Gray L, Burgess D, Copland L, Cullen R, Galin N, Hawley R, Helm V (2013) Interferometric swath processing of Cryosat data for glacial ice topography. Cryosphere 7:1857–1867. doi:10.5194/tc-7-1857-2013

Gray L, Burgess D, Copland L, Demuth MN, Dunse T, Langley K, Schuler TV (2015) CryoSat-2 delivers monthly and inter-annual surface elevation change for Arctic ice caps. Cryosphere 9:1895–1913. doi:10.5194/tc-9-1895-2015

Grinsted A (2013) An estimate of global glacier volume. Cryosphere 7(1):141–151

Guo W, Liu S, Xu J, Wu L, Shangguan D, Yao X, Wei J, Bao W, Yu P, Liu Q, Jiang Z (2015) The second Chinese glacier inventory: data, methods and results. J Glaciol 61(226):357–372

Haeberli W (2016) Brief communication: on area- and slope-related thickness estimates and volume calculations for unmeasured glaciers. Cryosphere Discuss. doi:10.5194/tc-2015-222

Haeberli W, Hoelzle M (1995) Application of inventory data for estimating characteristics of and regional climate-change effects on mountain glaciers: a pilot study with the European Alps. Ann Glaciol 21:206–212

Haeberli W, Linsbauer A (2013) Brief communication: global glacier volumes and sea level—small but systematic effects of ice below the surface of the ocean and of new local lakes on land. Cryosphere 7:817–821. doi:10.5194/tc-7-817-2013

Haeberli W, Barry R, Cihlar J (2000) Glacier monitoring within the global climate observing system. Ann Glaciol 31:241–246

Haeberli W, Schaub Y, Huggel C (2016) Increasing risks related to landslides from degrading permafrost into new lakes in de-glaciating mountain ranges. Geomorphology. doi:10.1016/j.geomorph.2016.02.009

Harris IPDJ, Jones PD, Osborn TJ, Lister DH (2014) Updated high-resolution grids of monthly climatic observations—the CRU TS3.10 dataset. Int J Climatol 34(3):623–642

Harrison WD, Elsberg DH, Echelmeyer KA, Krimmel RM (2001) On the characterization of glacier response by a single time-scale. J Glaciol 47(159):659–664

Heid T, Kääb A (2012) Repeat optical satellite images reveal widespread and long term decrease in land-terminating glacier speeds. Cryosphere 6:467–478

Helm V, Humbert A, Miller H (2014) Elevation and elevation change of Greenland and Antarctica derived from CryoSat-2. Cryosphere 8(4):1539–1559

Holzer N, Vijay S, Yao T, Xu B, Buchroithner M, Bolch T (2015) Four decades of glacier variations at Muztagh Ata (eastern Pamir): a multi-sensor study including Hexagon KH-9 and Pléiades data. Cryosphere 9:2071–2088

Huss M (2011) Present and future contribution of glacier storage change to runoff from macroscale drainage basins in Europe. Water Resour Res 47(7):W07511

Huss M (2013) Density assumptions for converting geodetic glacier volume change to mass change. Cryosphere 7:877–887

Huss M, Farinotti D (2012) Distributed ice thickness and volume of all glaciers around the globe. J Geophys Res Earth Surf. doi:10.1029/2012JF002523

Huss M, Hock R (2015) A new model for global glacier change and sea-level rise. Front Earth Sci. doi:10.3389/feart.2015.00054

Huss M, Dhulst L, Bauder A (2015) New long-term mass-balance series for the Swiss Alps. J Glaciol 61(227):551–562

Immerzeel WW, Droogers P, de Jong SM, Bierkens MFP (2009) Large-scale monitoring of snow cover and runoff simulation in Himalayan river basins using remote sensing. Remote Sens Environ 113(1):40–49

Immerzeel WW, van Beek LP, Bierkens MF (2010) Climate change will affect the Asian water towers. Science 328(5984):1382–1385
Jacob T, Wahr J, Pfeffer WT, Swenson S (2012) Recent contributions of glaciers and ice caps to sea level rise. Nature 482(7386):514–518
Jansson P, Hock R, Schneider T (2003) The concept of glacier storage: a review. J Hydrol 282(1):116–129
Jóhannesson T, Raymond C, Waddington ED (1989) Time-scale for adjustment of glaciers to changes in mass balance. J Glaciol 35(121):355–369
Kääb A, Berthier E, Nuth C, Gardelle J, Arnaud Y (2012) Contrasting patterns of early 21st century glacier mass change in the Himalaya. Nature 488:495–498. doi:10.1038/nature11324
Kääb A, Treichler D, Nuth C, Berthier E (2015) Brief communication: contending estimates of 2003–2008 glacier mass balance over the Pamir–Karakoram–Himalaya. Cryosphere 9:557–564. doi:10.5194/tc-9-557-2015
Kaser G, Großhauser M, Marzeion B (2010) Contribution potential of glaciers to water availability in different climate regimes. Proc Natl Acad Sci 107(47):20223–20227
Kienholz C, Rich JL, Arendt AA, Hock R (2014) A new method for deriving glacier centerlines applied to glaciers in Alaska and northwest Canada. Cryosphere 8:503–519
Klein AG, Kincaid JL (2006) Retreat of glaciers on Puncak Jaya, Irian Jaya, determined from 2000 and 2002 IKONOS satellite images. J Glaciol 52(176):65–79
Koppes M, Hallet B, Rignot E, Mouginot J, Wellner JS, Boldt K (2015) Observed latitudinal variations in erosion as a function of glacier dynamics. Nature 526(7571):100–103
Korup O, Montgomery DR, Hewitt K (2010) Glacier and landslide feedbacks to topographic relief in the Himalayan syntaxes. Proc Natl Acad Sci 107(12):5317–5322
Kronenberg M, Barandun M, Hoelzle M, Huss M, Farinotti D, Azisov E, Usubaliev R, Gafurov A, Petrakov D, Kääb A (2016) Mass-balance reconstruction for Glacier No. 354, Tien Shan, from 2003 to 2014. Ann Glaciol 57(71):92–102
Larsen CF, Motyka RJ, Freymueller JT, Echelmeyer KA, Ivins ER (2005) Rapid viscoelastic uplift in southeast Alaska caused by post-Little Ice Age glacial retreat. Earth Planet Sci Lett 237(3):548–560
Larsen CF, Motyka RJ, Arendt AA, Echelmeyer KA, Geissler PE (2007) Glacier changes in southeast Alaska and northwest British Columbia and contribution to sea level rise. J Geophys Rese Earth Surf. doi:10.1029/2006JF000586
Le Bris R, Paul F (2015) Glacier-specific elevation changes in western Alaska. Ann Glaciol 56(70):184–192
Leclercq PW, Oerlemans J, Cogley JG (2011) Estimating the glacier contribution to sea-level rise for the period 1800–2005. Surv Geophys 32:519–535
Leclercq PW, Oerlemans J, Basagic HJ, Bushueva I, Cook AJ, Le Bris R (2014) A data set of worldwide glacier length fluctuations. Cryosphere 8:659–672. doi:10.5194/tc-8-659-2014
Linsbauer A, Paul F, Haeberli W (2012) Modeling glacier thickness distribution and bed topography over entire mountain ranges with GlabTop: application of a fast and robust approach. J Geophys Res Earth Surf. doi:10.1029/2011JF002313
Linsbauer A, Frey H, Haeberli W, Machguth H, Azam MF, Allen S (2015) Modelling glacier-bed overdeepenings and possible future lakes for the glaciers in the Himalaya–Karakoram region. Ann Glaciol. doi:10.3189/2016AoG71A627
Loriaux T, Casassa G (2013) Evolution of glacial lakes from the Northern Patagonia Icefield and terrestrial water storage in a sea-level rise context. Global Planet Change 102:33–40. doi:10.1016/j.gloplacha.2012.12.012
Lüthi MP, Bauder A, Funk M (2010) Volume change reconstruction of Swiss glaciers from length change data. J Geophys Res Earth Surf. doi:10.1029/2010JF001695
Machguth H, Huss M (2014) The length of the world's glaciers—a new approach for the global calculation of center lines. Cryosphere 8:1741–1755
Marzeion B, Jarosch AH, Hofer M (2012) Past and future sea-level change from the surface mass balance of glaciers. Cryosphere 6:1295–1322. doi:10.5194/tc-6-1295-2012
Marzeion B, Cogley JG, Richter K, Parkes D (2014a) Attribution of global glacier mass loss to anthropogenic and natural causes. Science 345(6199):919–921
Marzeion B, Jarosch AH, Gregory JM (2014b) Feedbacks and mechanisms affecting the global sensitivity of glaciers to climate change. Cryosphere 8(1):59–71
Marzeion B, Leclercq PW, Cogley JG, Jarosch AH (2015) Brief communication: global reconstructions of glacier mass change during the 20th century are consistent. Cryosphere 9:2399–2404. doi:10.5194/tc-9-2399-2015
Meier MF, Dyurgerov MB, Rick UK, O'Neel S, Pfeffer WT, Anderson RS, Anderson SP, Glazovsky AF (2007) Glaciers dominate eustatic sea-level rise in the 21st century. Science 317(5841):1064–1067

Melini D, Gegout P, King M, Marzeion B, Spada G (2015) On the rebound: modeling Earth's ever-changing shape. EOS 96:14–17. doi:10.1029/2015EO033387

Melkonian AK, Willis MJ, Pritchard ME, Rivera A, Bown F, Bernstein SA (2013) Satellite-derived volume loss rates and glacier speeds for the Cordillera Darwin Icefield, Chile. Cryosphere 7(3):823–839

Melkonian AK, Willis MJ, Pritchard ME, Stewart AJ (2016) Recent changes in glacier velocities and thinning at Novaya Zemlya. Remote Sens Environ 174:244–257

Mernild SH, Lipscomb WH, Bahr DB, Radić V, Zemp M (2013) Global glacier changes: a revised assessment of committed mass losses and sampling uncertainties. Cryosphere 7(5):1565–1577. doi:10.5194/tc-7-1565-2013

Meyssignac B, Cazenave A (2012) Sea level: a review of present-day and recent-past changes and variability. J Geodyn 58:96–109

Moholdt G, Nuth C, Hagen JO, Kohler J (2010) Recent elevation changes of Svalbard glaciers derived from ICESat laser altimetry. Remote Sens Environ 114(11):2756–2767

Müller K, Hamran SE, Sinisalo A, Hagen JO (2011) Phase center of L-band radar in polar snow and ice. IEEE Trans Geosci Remote Sens 49(11):4572–4579

Narama C, Kääb A, Duishonakunov M, Abdrakhmatov K (2010) Spatial variability of recent glacier area changes in the Tien Shan Mountains, Central Asia, using Corona (similar to 1970), Landsat (similar to 2000), and ALOS (similar to 2007) satellite data. Glob Planet Change 71:42–54

Neckel N, Kropáček J, Bolch T, Hochschild V (2014) Glacier mass changes on the Tibetan Plateau 2003–2009 derived from ICESat laser altimetry measurements. Environ Res Lett. doi:10.1088/1748-9326/9/1/014009

Nussbaumer SU, Zumbühl HJ (2012) The Little Ice Age history of the Glacier des Bossons (Mont Blanc area, France): a new high-resolution glacier length curve based on historical documents. Clim Change 111(2):301–334

Nuth C, Kääb A (2011) Co-registration and bias corrections of satellite elevation data sets for quantifying glacier thickness change. Cryosphere 5:271–290. doi:10.5194/tc-5-271-2011

Nuth C, Moholdt G, Kohler J, Hagen JO, Kääb A (2010) Svalbard glacier elevation changes and contribution to sea level rise. J Geophys Res 115:F01008. doi:10.1029/2008JF001223

Oerlemans J (2001) Glaciers and climate change. A.A. Balkema, Rotterdam

Oerlemans J, Dyurgerov M, van de Wal RSW (2007) Reconstructing the glacier contribution to sea-level rise back to 1850. Cryosphere 1:59–65. doi:10.5194/tc-1-59-2007

Paul F (2010) The influence of changes in glacier extent and surface elevation on modeled mass balance. Cryosphere 4:569–581

Paul F, Haeberli W (2008) Spatial variability of glacier elevation changes in the Swiss Alps obtained from two digital elevation models. Geophys Res Lett. doi:10.1029/2008/GL034718

Paul F, Kääb A, Maisch M, Kellenberger TW, Haeberli W (2002) The new remote-sensing-derived Swiss glacier inventory: I.Methods. Ann Glaciol 34:355–361

Paul F, Huggel C, Kääb A (2004a) Combining satellite multispectral image data and a digital elevation model for mapping of debris-covered glaciers. Remote Sens Environ 89(4):510–518

Paul F, Kääb A, Maisch M, Kellenberger TW, Haeberli W (2004b) Rapid disintegration of Alpine glaciers observed with satellite data. Geophys Res Lett 31:L21402

Paul F, Escher-Vetter H, Machguth H (2009) Comparison of mass balances for Vernagtferner obtained from direct measurements and distributed modeling. Ann Glaciol 50:169–177

Paul F, Andreassen LM, Winsvold SH (2011) A new glacier inventory for the Jostedalsbreen region, Norway, from Landsat TM scenes of 2006 and changes since 1966. Ann Glaciol 52(59):153–162

Pfeffer WT, Arendt AA, Bliss A, Bolch T, Cogley JG, Gardner AS, Hagen J-O, Hock R, Kaser G, Kienholz C, Miles ES, Moholdt G, Mölg N, Paul F, Radić V, Rastner P, Raup BH, Rich J, Sharp MJ (2014) The Randolph Glacier Inventory: a globally complete inventory of glaciers. J Glaciol 60(221):537–552

Quincey DJ, Glasser NF, Cook SJ, Luckman A (2015) Heterogeneity in Karakoram glacier surges. J Geophys Res Earth Surf 120:1288–1300. doi:10.1002/2015JF003515

Rabatel A, Jomelli V, Naveau P, Francou B, Grancher D (2005) Dating of Little Ice Age glacier fluctuations in the tropical Andes: Charquini glaciers, Bolivia, 16°S. CR Geosci 337:1311–1322

Rabatel A, Dedieu JP, Thibert E, Letréguilly A, Vincent C (2008) 25 years (1981–2005) of equilibrium-line altitude and mass-balance reconstruction on Glacier Blanc, French Alps, using remote-sensing methods and meteorological data. J Glaciol 54(185):307–314

Radić V, Hock R (2010) Regional and global volumes of glaciers derived from statistical upscaling of glacier inventory data. J Geophys Res Earth Surf 115:F01010

Radić V, Bliss A, Beedlow AC, Hock R, Miles E, Cogley JG (2014) Regional and global projections of twenty-first century glacier mass changes in response to climate scenarios from global climate models. Clim Dyn 42:37–58

Rastner P, Bolch T, Mölg N, Machguth H, Le Bris R, Paul F (2012) The first complete inventory of the local glaciers and ice caps on Greenland. Cryosphere 6:1483–1495

Raup BH, Kääb A, Kargel JS, Bishop MP, Hamilton G, Lee E, Paul F, Rau F, Soltesz D, Khalsa SJS, Beedle M, Helm C (2007) Remote sensing and GIS technology in the Global Land Ice Measurements from Space (GLIMS) Project. Comput Geosci 33:104–125

Raymond C, Neumann TA, Rignot E, Echelmeyer K, Rivera A, Casassa G (2005) Retreat of Glaciar Tyndall, Patagonia, over the last half-century. J Glaciol 51(173):239–247

Reager JT, Gardner AS, Famiglietti JS, Wiese DN, Eicker A, Lo MH (2016) A decade of sea level rise slowed by climate-driven hydrology. Science 351(6274):699–703. doi:10.1126/science.aad8386

Richter K., Nilsen JE, Drange H (2012) Contributions to sea level variability along the Norwegian coast for 1960–2010. Journal of Geophysical Research: Oceans 117(C5)

Rietbroek R, Brunnabend SE, Kusche J, Schröter J, Dahle C (2016) Revisiting the contemporary sea-level budget on global and regional scales. Proc Natl Acad Sci 113(6):1504–1509. doi:10.1073/pnas.1519132113

Rignot E, Echelmeyer K, Krabill W (2001) Penetration depth of interferometric synthetic-aperture radar signals in snow and ice. Geophys Res Lett 28(18):3501–3504

Rignot E, Rivera A, Casassa G (2003) Contribution of the Patagonia Icefields of South America to sea level rise. Science 302(5644):434–437

Roe GH (2011) What do glaciers tell us about climate variability and climate change? J Glaciol 57:567–578

Sakai A, Nuimura T, Fujita K, Takenaka S, Nagai H, Lamsal D (2015) Climate regime of Asian glaciers revealed by GAMDAM glacier inventory. Cryosphere 9(3):865–880

Sakumura C, Bettadpur S, Bruinsma S (2014) Ensemble prediction and intercomparison analysis of GRACE time-variable gravity field models. Geophys Res Lett 41:1389–1397. doi:10.1002/2013GL058632

Schenk T, Csathó B (2012) A new methodology for detecting ice sheet surface elevation changes from laser altimetry data. IEEE Trans Geosci Remote Sens 50(9):3302–3316

Schiefer E, Menounos B, Wheate R (2007) Recent volume loss of British Columbian glaciers, Canada. Geophys Res Lett 34(16):L16503

Schrama EJO (2016) Glacial isostatic adjustment and the observed sea-level rise—an overview. Surv Geophys **(submitted)**

Schrama EJO, Wouters B, Rietbroek R (2014) A mascon approach to assess ice sheet and glacier mass balances and their uncertainties from GRACE data. J Geophys Res Solid Earth 119:6048–6066. doi:10.1002/2013JB010923

Sevestre H, Benn DI (2015) Climatic and geometric controls on the global distribution of surge-type glaciers: implications for a unifying model of surging. J Glaciol 61(228):646–662

Studinger M, Koenig L, Martin S, Sonntag J (2010) Operation icebridge: using instrumented aircraft to bridge the observational gap between Icesat and Icesat-2. In: IEEE international on geoscience and remote sensing symposium (IGARSS), pp 1918–1919

Swenson S, Chambers D, Wahr J (2008) Estimating geocenter variations from a combination of GRACE and ocean model output. J Geophys Res 113:B08410. doi:10.1029/2007JB005338

Tamisiea ME, Mitrovica JX, Davis JL (2003) A method for detecting rapid mass flux of small glaciers using local sea level variations. Earth Planet Sci Lett 213(3):477–485

Tennant C, Menounos B, Wheate R, Clague JJ (2012) Area change of glaciers in the Canadian Rocky Mountains, 1919 to 2006. Cryosphere 6:1541–1552

Thomson SN, Brandon MT, Tomkin JH, Reiners PW, Vásquez C, Wilson NJ (2010) Glaciation as a destructive and constructive control on mountain building. Nature 467(7313):313–317

Vaughan DG, Comiso JC, Allison I, Carrasco J, Kaser G, Kwok R, Mote P, Murray T, Paul F, Ren J, Rignot E, Solomina O, Steffen K, Zhang T (2013) Observations: Cryosphere. In: Stocker TF, Qin D, Plattner GK, Tignor M, Allen SK, Boschung J, Nauels A, Xia Y, Bex V, Midgley PM (eds) Climate change 2013: the physical science basis. Contribution of working group I to the fifth assessment report of the intergovernmental panel on climate change. Cambridge University Press, Cambridge, United Kingdom and New York, NY, USA, pp 317–382

Wahr J, Molenaar M (1998) Time variability of the Earth's gravity field: hydrological and oceanic effects and their possible detection using GRACE. J Geophys Res 103(B12):30205–30229. doi:10.1029/98JB02844

Wang D, Kääb A (2015) Modeling glacier elevation change from DEM time series. Remote Sens 7(8):10117–10142

WGMS (2008) Global glacier changes: facts and figures. In: Zemp M, Roer I, Kääb A, Hoelzle M, Paul F, Haeberli W (eds) UNEP, World Glacier Monitoring Service. University of Zurich, Switzerland

WGMS (2015) Global glacier change bulletin no. 1 (2012–2013). In: Zemp M, Gärtner-Roer I, Nussbaumer SU, Hüsler F, Machguth H, Mölg N, Paul F, Hoelzle, M (eds) ICSU(WDS)/IUGG(IACS)/UNEP/

UNESCO/WMO, World Glacier Monitoring Service, Zurich, Switzerland, 230 pp. doi: 10.5904/wgms-fog-2015-11

Wulder MA, Masek JG, Cohen WB, Loveland TR, Woodcock CE (2012) Opening the archive: how free data has enabled the science and monitoring promise of Landsat. Remote Sens Environ 122:2–10

Yi S, Sun W, Heki K, Qian A (2015) An increase in the rate of global mean sea level since 2010. Geophys Res Lett 42:3998–4006. doi:10.1002/2015GL063902

Zemp M, Zumbühl HJ, Nussbaumer SU, Masiokas MH, Espizua LE, Pitte P (2011) Extending glacier monitoring into the Little Ice Age and beyond. PAGES News 19(2):67–69

Zemp M, Amstrong R, Gärnter-Roer I, Haeberli W, Hoelzle M, Kääb A, Kargel JS, Khalsa SJS, Leonard GJ, Paul F, Raup BH (2014) Introduction: global glacier monitoring—a long-term task integrating in situ observations and remote sensing. In: Kargel JS, Leonard GJ, Bishop MP, Kääb A, Raup BH (eds) Global land ice measurements from space. Springer, Berlin

Zemp M, Frey H, Gärtner-Roer I, Nussbaumer SU, Hoelzle M, Paul F, Haeberli W, Denzinger F, Ahlstrøm AP, Anderson B, Bajracharya S, Baroni C, Braun LN, Cáceres BE, Casassa G, Cobos G, Dávila LR, Delgado Granados H, Demuth MN, Espizua L, Fischer A, Fujita K, Gadek B, Ghazanfar A, Hagen JO, Holmlund P, Karimi N, Li Z, Pelto M, Pitte P, Popovnin VV, Portocarrero CA, Prinz R, Sangewar CV, Severskiy I, Sigurđsson O, Soruco A, Usubaliev R, Vincent C (2015) Historically unprecedented global glacier decline in the early 21st century. J Glaciol 61(228):745–762. doi:10.3189/2015JoG15J017

Zumbühl HJ (1980) Die Schwankungen der Grindelwaldgletscher in den historischen Bild-und Schriftquellen des 12. bis 19. Jahrhunderts: ein Beitrag zur Gletschergeschichte und Erforschung des Alpenraumes. Birkhäuser, Basel

Surv Geophys (2017) 38:131–152
DOI 10.1007/s10712-016-9399-6

Recent Changes in Land Water Storage and its Contribution to Sea Level Variations

Yoshihide Wada[1,2,3,4] · **John T. Reager**[5] · **Benjamin F. Chao**[6] · **Jida Wang**[7] · **Min-Hui Lo**[8] · **Chunqiao Song**[9] · **Yuwen Li**[6] · **Alex S. Gardner**[5]

Received: 21 June 2016 / Accepted: 1 November 2016 / Published online: 15 November 2016

Abstract Sea level rise is generally attributed to increased ocean heat content and increased rates glacier and ice melt. However, human transformations of Earth's surface have impacted water exchange between land, atmosphere, and ocean, ultimately affecting global sea level variations. Impoundment of water in reservoirs and artificial lakes has reduced the outflow of water to the sea, while river runoff has increased due to groundwater mining, wetland and endorheic lake storage losses, and deforestation. In addition, climate-driven changes in land water stores can have a large impact on global sea level variations over decadal timescales. Here, we review each component of negative and positive land water contribution separately in order to highlight and understand recent changes in land water contribution to sea level variations.

Electronic supplementary material The online version of this article (doi:10.1007/s10712-016-9399-6) contains supplementary material, which is available to authorized users.

✉ Yoshihide Wada
y.wada@uu.nl; yoshihide.wada@nasa.gov

1 NASA Goddard Institute for Space Studies, 2880 Broadway, New York, NY 10025, USA

2 Center for Climate Systems Research, Columbia University, 2880 Broadway, New York, NY 10025, USA

3 Department of Physical Geography, Faculty of Geosciences, Utrecht University, Heidelberglaan 2, 3584 CS Utrecht, The Netherlands

4 International Institute for Applied Systems Analysis, A-2361 Laxenburg, Austria

5 NASA Jet Propulsion Laboratory, California Institute of Technology, Pasadena, CA 91109, USA

6 Institute of Earth Sciences, Academia Sinica, Taipei 11529, Taiwan

7 Department of Geography, Kansas State University, 118 Seaton Hall, Manhattan, KS 66506, USA

8 Department of Atmospheric Sciences, National Taiwan University, Taipei 10673, Taiwan

9 Department of Geography, University of California Los Angeles, 1255 Bunche Hall, Los Angeles, CA 90095, USA

Keywords Land water storage · Sea level rise (SLR) · Groundwater depletion (GWD) · Reservoir impoundment · Climate variability

1 Introduction

Sea level rise (SLR) over the past century is generally attributed to increased ocean heat content (thermal expansion, e.g., Abraham et al. 2013) and increased rates of melt and solid ice discharge (calving) from glaciers and ice sheets (e.g., Gardner et al. 2013; Shepherd et al. 2012). Large-scale anthropogenic and natural changes in land water storage, defined as snow, surface water, soil moisture, and groundwater storage, excluding glaciers, have also contribute to observed rates of SLR on annual to centennial timescales (Milly et al. 2003; Syed et al. 2010; Reager et al. 2016).

Human transformations of Earth's surface have impacted continental patterns of river flow and water exchange between land, atmosphere, and ocean, ultimately affecting global sea level variations. Massive impoundment of water in reservoirs and artificial lakes has reduced the outflow of water to the sea, while river runoff has increased due to excessive groundwater mining, wetland and endorheic lake storage losses, and deforestation (Chao et al. 2008; Wada et al. 2012a, b; Church et al. 2013). In the IPCC Fifth Assessment report (IPCC AR5), anthropogenic changes in terrestrial water storage (primarily filling of reservoirs and groundwater mining) were included in the sea level change budget but natural fluctuations were excluded due to poor knowledge of their change and the expectation that such changes would be small on decadal timescale. Recent work by Reager et al. (2016) showed that climate-driven changes in water stores (e.g., soil moisture and groundwater) can have a large impact on global sea level variations over decadal timescales.

Impoundment of fresh water behind reservoirs constructed in the 1950s through the 1980s resulted in increased storage of water on land resulting in a lowering of sea level. Over the past few decades, the rate of impoundment has been surpassed by increased human mining of groundwater reserves leading to a net increase in sea levels (Wada 2016). Better understanding of the temporal evolution of the contributing processes leading to change in total land water storage is critical to closing decadal sea level budgets and to understanding changes in rates of sea level change.

Around 8000 km^3 of water is presently sequestered behind large reservoirs. More than 90% of this total reservoir capacity was created after the 1950s resulting in an cumulative decrease in the global sea level change of 30 mm (Chao et al. 2008). Other processes such as groundwater mining (or pumping of groundwater at rates exceeding the natural recharge rate), wetland and endorheic lake storage losses, and deforestation have led to an increase in the rates of land water contribution to the oceans. Water stored in wetlands and endorheic lakes (e.g., the Aral Sea and the Lake Urmia) has been heavily used for agricultural production leading to a net transfer of water from land to the ocean resulting in SLR. Deforestation reduces the infiltration capacity of the soil due to compaction by heavy logging, farm machinery, overgrazing and trampling by cattle and increases soil erosion. In newly cleared areas, runoff usually increases, especially in the rainy season, leading to greater chances of flooding. The combined effect of these processes has reduced soil moisture and groundwater reserves, increasing the rate of sea level change by 0.3–0.5 mm $year^{-1}$ during a recent few decades (Church et al. 2013; Gregory et al. 2013; Wada et al. 2016).

Knowledge of how anthropogenic changes in land water storage affect future sea levels is poorly constrained. A simple extrapolation of recent trends is likely unrealistic for several reasons: (1) A major dam-building boom is underway across the developing world. An additional few hundred km^3 water could therefore be sequestered on land, amounting to between 2 and 4% of present reservoir storage. This rate of dam building is likely to slow down in coming decades as suitable sites become scarcer and public opposition to environmental impacts of dams grows. (2) Deforestation may also decrease as forests are cleared from all but inaccessible sites and as stricter environmental regulations aim to preserve remaining forests. (3) Groundwater mining has been increasing due to excessive pumping in many irrigated regions (e.g., India, Pakistan, China, and Iran), and it is unclear whether such rates can be sustained into the future. All these unknowns lead to large uncertainties in projections of anthropogenic changes in land water storage.

Here we review each of the contributing process to changes in land water storage to provide understanding of recent trends in associated land water contribution to sea level change.

2 Mechanisms of Land Water Contribution and Other Sea Level Components

Construction of reservoirs and resulting impoundment for power generation and water resource management results in increased land water storage through reservoir filling and raising of surrounding groundwater tables. A great quantity of water, which has been impounded behind dams (artificial reservoirs), would otherwise reside in the oceans rather than on land. Thus, each reservoir can be thought of as a "one-time" deduction of SLR. Impounded water is ultimately sourced from the ocean, resulting in a lowering of mean sea level (Chao et al. 2008). Groundwater pumping for agriculture (mostly irrigation) and other uses (industrial and municipal) contributes to SLR through the persistent removal of groundwater at rates exceeding natural recharge rates. The mined water finds its way to the ocean as increased surface runoff and/or evapotranspiration that later falls as precipitation over the ocean (Wada et al. 2016). Human water use is also a major driver for decreasing wetland and endorheic lake storage across the globe. In addition, forest clearing releases water stored in both biotic tissues and soil, which leads to positive SLR.

Anthropogenic influences, such as water use and land use change on the Earth's land water storage, are clearly seen in the decreasing volume of the Aral sea (Pala 2006, 2011), decreased flows of the Colorado and Yellow Rivers (Gleick 2003), dwindling groundwater resources over intense irrigated regions such as the Ogallala aquifer (Scanlon et al. 2012a, b), the California's Central Valley (Famiglietti et al. 2011), the North China Plain (Cao et al. 2013), northwest India and northeast Pakistan (Rodell et al. 2009; Tiwari et al. 2009), and the Tigris-Euphrates (Voss et al. 2013). The degree of aquifer depletion is reported at an alarming rate over the Indus, Saudi Arabia, Iran, northeastern China, the southwestern and Central USA, and northern Mexico (Rodell et al. 2009; Tiwari et al. 2009; Famiglietti et al. 2011; Konikow 2011; Döll et al. 2012, 2014; Scanlon et al. 2012a, b; Cao et al. 2013; Voss et al. 2013; Famiglietti 2014).

In addition to anthropogenic influences on land water storage, seasonal to decadal changes in climate can also modify the apparent SLR rate. Changes in the patterns of seasonal precipitation over the continents (e.g., the migration of the intertropical convergence zone during rainy seasons) and winter snow accumulation at high latitudes (e.g., Siberia and North America) can act to modulate global mean sea level and the annual amplitude of global mean sea level (Wouters et al. 2011). Multi-year variability in the distribution of water between the land and oceans due to El Niño–Southern Oscillation

(ENSO) effects on weather and precipitation can result in droughts and/or raised groundwater tables over basin to continental scales that modify rates of ocean mass change (Boening et al. 2012; Syed et al. 2010; Reager et al. 2016).

Many complex processes contribute to generate net changes in global and regional sea levels, and the relative importance of each contributing term depends on the time period being analyzed. Many of the contributing processes are not stationary and can undergo substantial changes in magnitude or even direction at annual to decadal timescales.

3 Groundwater Contribution

The rate of groundwater depletion (GWD) and its contribution to SLR has been subject to much debate (Gregory et al. 2013; Taylor et al. 2013). In the IPCC AR4 (Solomon et al. 2007), the contribution of non-frozen terrestrial waters including GWD to sea level variation is not considered due to its perceived uncertainty (Wada 2016).

GRACE (Gravity Recovery and Climate Experiment) satellite observations (Tapley et al. 2004; Chen et al. 2016) open a path to monitor total water storage changes including groundwater in data scarce regions (hereafter referred to as *"satellite-based method"*) (Strassberg et al. 2007; Jacob et al. 2012; Shamsudduha et al. 2012; Voss et al. 2013). Rodell et al. (2009) and Tiwari et al. (2009) reported substantial GWD over Northwest India (17.7 ± 4.5 km^3 year^{-1}; 0.05 ± 0.013 mm year^{-1}) and adjacent regions (54 ± 9 km^3 year^{-1}; 0.15 ± 0.025 mm year^{-1}) (see also Table 1). Coarse spatial resolution and noise contamination inherent in GRACE data hinder global application of estimating GWD (*satellite-based method*). Van Dijk et al. (2014) used a data assimilation framework to integrate water balance estimates derived from GRACE satellite observations and several hydrological models, which improved the estimate of global GWD derived from a hydrological model from 168 to 92 km^3 year^{-1} (averaged over the 2003–2012 period).

Earlier estimates of GWD contribution to SLR ranges from 0.075 to 0.30 mm year^{-1} (Sahagian et al. 1994; Gornitz 1995, 2001; Foster and Loucks 2006; see also Table 1). These studies evaluate direct groundwater storage changes but only cover a limited number of regions of the world. Using a global-scale hydrological model (hereafter referred to as *"flux-based method"*), Wada et al. (2010) estimated the current rate of global GWD to be 283 ($\pm$40) km^3 year^{-1} (0.8 ± 0.1 mm year^{-1}), responsible for 25 ($\pm$3) % of recently observed SLR, which is increased substantially from 126 ($\pm$32) km^3 year^{-1} (0.35 ± 0.09 mm year^{-1}) in 1960. The *flux-based method* provides an upper bound of GWD as it does not account for increased capture due to decreased groundwater discharge and enhanced recharge from surface waters (Bredehoeft 2002). A subsequent study of Wada et al. (2012b) applied a correction factor to remediate the overestimation and to constrain the original GWD estimate using regionally reported numbers and estimated that the average global GWD rate amounts to 163 ($\pm$28) km^3 year^{-1} during 1990–2000, equivalent to a SLR of 0.46 ($\pm$0.08) mm year^{-1} Wada et al. (2012b) estimated that the contribution of GWD to global sea level increased from 0.035 ($\pm$0.009) to 0.57 ($\pm$0.09) mm year^{-1} during the 20th century and projected that this would increase to 0.82 ($\pm$0.13) mm year^{-1} by 2050. Döll et al. (2014) used hydrological modeling, well observations, and GRACE satellite gravity anomalies to estimate a 2000–2009 global GWD of 113 km^3 year^{-1} (0.314 mm year^{-1}) (see Table 1). Pokhrel et al. (2015) used an integrated hydrological model (*flux-based method*), which explicitly simulates groundwater dynamics and pumping within a global land surface model, to estimate a global GWD of 330 km^3 year^{-1} (0.92 mm year^{-1}) for year 2000. The later study overestimated GWD by calculating the difference of water demand and water

Table 1 Global and regional estimates of groundwater depletion (GWD) contribution to sea level rise (SLR)

		GWD (km^3 $year^{-1}$)	SLR (mm $year^{-1}$)	Year	Notes	Sources
Global	Sahagian et al. (1994)	86.7	0.24	Contemporary	Limited regions (e.g., USA, India, China)	Literature Country statistics
	Postel (1999)	200	0.56	Contemporary	Global	Literature Country statistics
	Gornitz (2001)	36–108	0.1–0.3	Contemporary	Limited regions (e.g., USA, India, China)	Literature Country statistics
	Foster and Loucks (2006)	26.8	0.075	Contemporary	Limited regions (e.g., Middle East, Northern Africa)	Literature Country statistics
	Wada et al. (2010)	126 (±32) 283 (±40)	0.35 (±0.09) 0.79 (±0.11)	1960 2000	Depletion equals abstraction in excess of recharge	IGRAC-GGIS PCR-GLOBWB (0.5°)
	Konikow (2011)	145 (±39)	0.4 (±0.11)	2000–2008	Extrapolated for other than USA, north India, North China Plain, Saudi Arabia, Nubian and Sahara	In situ groundwater level measurements, GRACE satellite observation, calibrated groundwater model, extrapolation (15.4 %; depletion to abstraction ratio of USA)
	Wada et al. (2012)	163 (±28) 204 (±30)	0.45 (±0.07) 0.57 (±0.08)	1990–2000 2000	Corrected against reported regional depletion estimates	IGRAC-GGIS PCR-GLOBWB (0.5°)
	Pokhrel et al. (2012)	455 (±42)	1.27 (±0.12)	2000	Depletion equals water demand in excess of water availability	MATSIRO (1.0°)
	Döll et al. (2014)	113	0.31	2000–2009	Application of deficit irrigation	WaterGAP (0.5°), In situ groundwater level measurements, GRACE satellite observation
	Van Dijk et al. (2014)	92	0.26	2003–2012	Without data assimilation, original depletion equals 168 km^3 $year^{-1}$	Data assimilation with GRACE satellite observation
	Famiglietti (2014)	77.4	0.22	2003–2013	Time periods vary among studies considered Limited regions	Various studies using GRACE-derived total terrestrial water storage changes
	Pokhrel et al. (2015)	330	0.92	2000	570 km^3 $year^{-1}$ for the global groundwater abstraction for 2000	MATSIRO (1.0°)
	Wada et al. (2016)	7.2 (±1.4) 97 (±14)	0.02 (±0.004) 0.27 (±0.04)	1900 2000	Coupled atmosphere-land–ocean model simulation	NCAR CESM-CAM4-CLM4 (1.0°)

Table 1 continued

		GWD (km^3 $year^{-1}$)	SLR (mm $year^{-1}$)	Year	Notes	Sources
Regional Northwest Sahara	Richey et al. (2015)	2.7	0.008	2003–2012	Algeria, Libya, Tunisia	GRACE-derived total terrestrial water storage changes
Middle East and North Africa (MENA)	Foster and Loucks (2006)	26.8	0.0075	Contemporary		Literature Country statistics
	Voss et al. (2013)	13.0 (±1.6)	0.036 (±0.005)	2003–2009	Cumulative 91.3 (± 10.9) km^3 for 2003–2009	GRACE-derived total terrestrial water storage changes
Arabian	Richey et al. (2015)	15.5	0.04	2003–2013	Iraq, Jordan, Oman, Qatar, Saudi Arabia, UAE, Yemen	GRACE-derived total terrestrial water storage changes
Guarani	Richey et al. (2015)	1.0	0.003	2003–2013	Argentina, Brazil, Paraguay, Uruguay	GRACE-derived total terrestrial water storage changes
North China Plain (NCP)	Cao et al. (2013)	4.0 2.5 4.0 2.0 7.0 4.0	0.01 0.007 0.01 0.006 0.02 0.01	1960–2008 1970s 1980s 1990–1996 1997–2001 2002–2008	Cumulatively 158 km^3 for 1960–2008 (20% of pumpage of 807 km^3)	MODFLOW
	Feng et al. (2013)	8.3 (±1.1)	0.02 (±0.03)	2003–2010	2.5 km^3 $year^{-1}$ for shallow aquifers reported by Groundwater Bulletin of China Northern Plains	GRACE-derived total terrestrial water storage changes
	Huang et al. (2015)	2.5 (±0.4)-PP 1.5 (±0.2)-ECP	0.007 (±0.001) 0.004 (±0.0005)	2003–2012	Piedmont Plain (PP) East Central Plain (ECP)	GRACE-derived total terrestrial water storage changes
Indus	Cheema et al. (2014)	31	0.09	2007	68 km^3 of total groundwater abstraction	Remote sensing combined with a hydrological model and spatial information on canal water supplies
Northern India	Rodell et al. (2009)	17.7 (±4.5)	0.05 (±0.01)	2002–2008	Cumulative 109 km^3 for 2002–2008	GRACE-derived total terrestrial water storage changes
Northern India and surrounding regions	Tiwari et al. (2009)	54 (±9)	0.15 (±0.03)	2002–2008		GRACE-derived total terrestrial water storage changes
	Jacob et al. (2012)	35	0.1	2003–2010		GRACE-derived total terrestrial water storage changes

Table 1 continued

		GWD (km^3 $year^{-1}$)	SLR (mm $year^{-1}$)	Year	Notes	Sources
Bangladesh	Shamsudduha et al. (2012)	0.44(±1.24)–2.04(± 0.79)-wet seasons 0.52(± 0.5)–2.83(± 0.42)-annual	0.001(± 0.004)–0.006(± 0.002) 0.002(± 0.001)–0.008(± 0.001)	2003–2007	Depletion of 0.52 (± 0.30)–0.85 (± 0.17) km^3 $year^{-1}$ from borehole hydrographs	GRACE-derived total terrestrial water storage changes
California's Central Valley	Famiglietti et al. (2011)	3.1 (± 0.6)	0.009 (± 0.002)	2003–2010	Cumulative 20.3 km^3 for 2003–2010	GRACE-derived total terrestrial water storage changes
	Scanlon et al. (2012a)	2.0 6–8	0.006 0.017–0.022	1962–2003 2006–2010	Cumulative 24.6 km^3 for 1976–1977, 49.3 km^3 for 1987–1992, 140 km3 since the 1860s, and 80 km^3 since the 1960s	MODFLOW
	Scanlon et al. (2012b)	8.9 (± 0.9)	0.025 (± 0.0025)	2006–2010	Cumulative 31.0 (± 3.0) km^3 for 2006–2010	GRACE-derived total terrestrial water storage changes
High Plains Aquifer	Scanlon et al. (2012a)	5.7 7.0 12.5	0.016 0.02 0.035	1950–2007 1987–2007 2003–2013	Cumulative 330 km^3 after pre-development in the 1950s	MODFLOW
Canning Basin	Richey et al. (2015)	3.6	0.01	2003–2013	Australia	GRACE-derived total terrestrial water storage changes

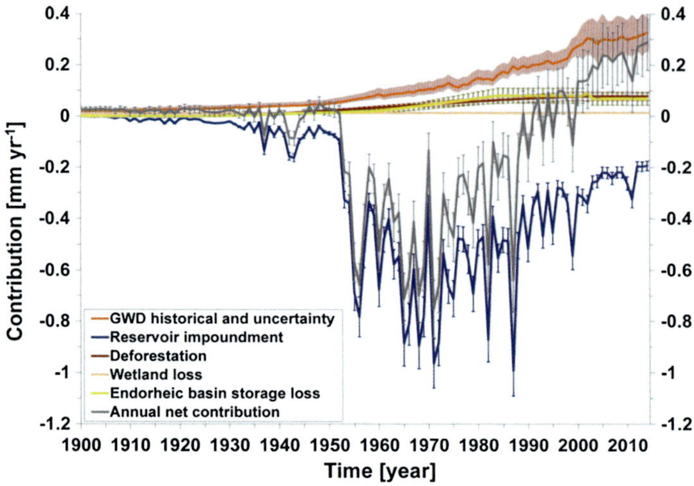

Fig. 1 Time series of the estimated annual contribution of terrestrial water storage change to global sea level over the period 1900–2014 (rates in mm $year^{-1}$) (modified from Wada et al. 2016)

availability without using groundwater pumping and recharge data. A volume-based study by Konikow (2011) estimated global GWD to be 145 (±39) km^3 $year^{-1}$ (0.41 ±0.1 mm $year^{-1}$) based on measurements of changes in groundwater storage from in situ observations, calibrated groundwater modeling, GRACE satellite data, and some extrapolations using the fixed ratio of depletion to abstraction observed in the USA (15.4%).

One critical assumption of most existing global estimates of GWD impacts on sea level change is that nearly 100% of the GWD ends up to the ocean, assuming all other stores (atmospheric moisture and surface waters) to remain constant. However, groundwater pumping can also perturb regional climate due to land–atmosphere interactions (Lo and Famiglietti 2013). Over the Ogallala Aquifer in the Great Plains (USA) groundwater-fed irrigation enhances regional precipitation by 15–30% during July from the easternmost part of the aquifer to as far downwind as Indiana (DeAngelis et al. 2010) and a downwind precipitation by 20–30% over the Midwest (Kustu et al. 2010, 2011). The latest study by Wada et al. (2016) used a coupled climate-hydrological model simulation to track the fate of water pumped from underground, and found that the fraction of GWD that ends up in the ocean is 80%, over which roughly two-thirds result from an increase in runoff to the ocean, while the remainder results from the enhanced net flux of precipitation minus evaporation over the ocean. They estimated that the contribution of GWD to global SLR amounts to 0.02 (±0.004) mm $year^{-1}$ in 1900 and increased to 0.27 (±0.04) mm $year^{-1}$ in 2000 (Fig. 1). This indicates that most studies likely overestimated the cumulative contribution of GWD to global SLR during the 20th century and early 21st century by 5–10 mm.

4 Water Impoundment Behind Dams

To acquire the history and the amount of anthropogenic water impoundment is a non-trivial task. An open depository of worldwide reservoir information is the ICOLD (International Commission On Large Dams; http://www.icold-cigb.org/) registry, which is "complete" to the extent of contributions from willing countries and water authorities. Based on the

ICOLD registry and augmented with various other partial sources, Chao et al. (2008) compiled a dataset of 29,484 named artificial dams down to numerous ones with nominal water capacity less than 0.01 km^3 the tally of which inevitably becomes grossly incomplete. They found a power law relationship between reservoir number count and reservoir capacity, to the power of −0.52, which assures that the smaller reservoirs contain negligible amount of water in global sea level contribution. Using this relationship, they estimated a total global impounded water volume of 8300 km^3. This estimate was updated slightly by Wada et al. (2012b). In parallel, Lehner et al. (2011) introduced the Global Reservoir and Dam database (GRanD) that contains information for 6862 dams and their associated reservoirs with a total storage capacity of 6197 km^3. By extrapolation, they estimated a global storage capacity of 8070 km^3.

We make yet another updated database by combining the available sources after applying careful quality control. The result is a list of 48,064 reservoirs that have a combined capacity of 7968 km^3. This new estimate included all reservoirs with a capacity larger than 1 km^3, plus all reservoirs down to smaller than 0.01 km^3 adopted from the ICOLD registry. The database also includes geographic locations for most of 144 reservoirs exceeding 10 km^3 that have a combined capacity of 4331 km^3 (see Supplementary Information). Our estimate is somewhat smaller than the estimate in Chao et al. (2008) because some reservoirs were found to be duplicated.

Figure 2 gives the time history of the growth of the total capacity according to the nominal years of completion for each reservoir, along with a continental breakdown. Also

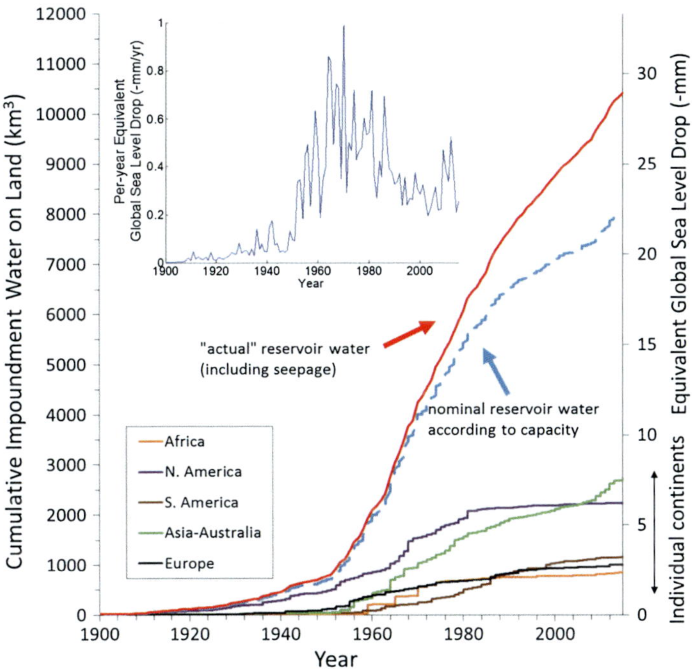

Fig. 2 Cumulative amount of water impoundment in artificial reservoirs as a function of time during the last century, with the equivalent sea level drop on the right-hand scale. The blue dashed curve is the nominal capacity, and the red curve is the "actual" impoundment (*see text*). Also shown is the continental breakdown. The *inset* is the per-year water impoundment

shown is the per-year capacity, which clearly reflects the history of the human activity in dam building. From the design capacity of the reservoir to the actual amount of water impoundment, we make assumptions following Chao et al. (2008) that: (1) No reservoir is kept at full capacity at all times. The actual percentage of water storage varies from one climatic regime to the next, most typically from season to season as well as interannually. Here we assume that on average 85% of the design capacity is used. (2) There is additional water impounded by the reservoirs that seeps underground to raise the local groundwater level. This additional water is estimated to be 5% of the capacity for the first year and continues to increase slowly according to the square-root of time since completion until reaching twice of the design capacity. Certain large reservoirs are excluded from this additional water seepage because they are not "new" reservoirs but rather dam-raised lake levels of existing lakes, including the Owen Falls/Lake Victoria (the largest in the list) and several large water projects (Table 2). Lake Victoria should no more be considered as an artificial added capacity, since this lake stored 2.5 m in the early sixties (around 170 km^3), but this was progressively lost in the following years. The actual water retention varies greatly depending on regional geology; our modeling may represent upper limits in some cases. (3) Reservoirs suffer from silt accumulation with time. That presents serious problems for water management, but has little consequence, at least to the first order, on anthropogenic contributions to sea level change as the impounded water is replaced by an equal volume of silt that would have otherwise been deposited in the ocean. In known cases, the disappearance of natural floods due to reservoirs may even encourage more silt deposit along river banks (Chao et al. 2008).

Using the new reservoir database and applying the stated assumption (presented in Fig. 2), we estimate that humans have impounded 10,416 km^3 of water behind dams, accounting for 28.9 mm drop in global mean sea level. During the second half of the 20th

Table 2 Large reservoirs that are excluded from this additional water seepage due to that they are not "new" reservoirs but rather dam-raised lake levels of existing lakes, including the Owen Falls/Lake Victoria (the largest in the list) and several large water projects

Country	Reservoir name
Canada	Manicouagan
	Jenpeg
	Smallwood Reservoir
	Missi Falls Control
	Ear Falls Dam
	Whitesand Rapids
	Pipmuacan
	Keenleyside
China	Sanhezha
Finland	TainionkoskI
Russia	Irkutsk
	Verkhne-Tulomskaya
	Onda
	Kumskaya
	Verkhne-Svirskaya
USA	Structure 308

century, when the dam-building activity was at its highest, the average rate of sea level change was −0.51 mm/year and is significant in comparison with other natural and anthropogenic sources of sea level change. It has been noted (Chao et al. 2008) that water impoundment can largely account for the slowing in the rate of SLR observed during the second half of the 20th century.

5 Deforestation, Wetland Degradation, and Lake Storage Changes

5.1 Deforestation/Afforestation

At present, global deforestation is a net consequence of tropical forest loss and temperate-boreal forest gain (FAO 2015; Keenan et al. 2015; MacDicken 2015; Sloan and Sayer 2015). Net deforestation releases carbon and water stored in both biotic tissues and soil, which leads to positive SLR through three primary processes. First, forest clearing eliminates evapotranspiration and thus increases total surface runoff in the hydrological budget (Meherhomji 1991, Milly et al. 2010). Gornitz et al. (1997) estimated an equivalent of 0.13 mm year^{-1} SLR attributed to deforestation-induced runoff increase in the 1980s, yet this estimate may overshoot the contribution of net global deforestation as it excludes synchronous temperate forest regrowth and restoration. The global deforestation rate has slowed in the past two decades (FAO 2015; Sloan and Sayer 2015). Based on the net deforestation rate of 3.3 million ha year^{-1} between 2010 and 2015 (FAO 2015), we here apply the same method proposed by Gornitz et al. (1997) and update a SLR equivalent of no more than 0.03 mm year^{-1} attributed to the present deforestation-induced runoff increase.

Oxidation of deforested biomass (i.e., carbohydrates in dry matter) produces carbon dioxide (CO_2) and water, along which additional water stored within plants (except dry matter) is also released. Based on a bulk forest dry-to-wet mass ratio of 0.25 (Rohrig 1991), Gornitz et al. (1997) calculated this combined contribution from net global deforestation in the 1980s to be ~0.02 mm year^{-1} SLR. Sahagian et al. (1994), Sahagian (2000) and Vörösmarty and Sahagian (2000) further took into account an equal amount of water stored in soil, fallen leaves and surrounding atmosphere and estimated 0.14 mm year^{-1} SLR resulting from tropical deforestation alone from 1940 to 1990. Given the net loss of 0.2 Gt year^{-1} in forest carbon stock from 2010 to 2015 (FAO 2015) and assuming that such carbon loss is completely emitted, we estimate that water release from recent global deforestation, combining the amounts from both oxidation and plant storage, likely dropped to 1.5 Gt year^{-1}, equivalent to less than 0.005 mm year^{-1} SLR.

Widespread deforestation also triggers complex climate feedbacks, as suggested by both modeling and observation-based studies (Butt et al. 2011; Chagnon and Bras 2005; Nobre et al. 2009; Shukla et al. 1990; Spracklen et al. 2012). Reduced evapotranspiration may suppress regional precipitation and thus groundwater recharge (Huntington 2008; Spracklen et al. 2012), counteracting the contribution of increased runoff to SLR. Deforestation-induced emission of CO_2 and water vapor, both major greenhouse gases, leads to additional global warming (Ciais et al. 2013), which accelerates glacier and ice cap melting and reinforces SLR. Nevertheless, contributions of these climate feedbacks are coupled with and attenuated by other carbon sinks. For example, about half of the elevated CO_2 concentration from anthropogenic emissions since 1980s, including deforestation, was taken up by ocean and terrestrial biosphere (Khatiwala et al. 2009; Sabine et al. 2004). If uncertainties from the land-atmospheric coupling are excluded, we suggest that the current net global deforestation, synthesizing the direct impacts of runoff increase and water release from oxidation and plant storage, leads to an upper-bound SLR of ~0.035 mm year^{-1}.

5.2 Wetland Degradation

Wetland degradation contributes to SLR primarily through (1) direct water drainage or removal from standing inundation, soil moisture, and plant storage and (2) water release from vegetation oxidation and peat combustion. The latter process is similar to that of deforestation (Sect. 5.1), except that additional water is released from the burning of peat harvested as an important fuel source in certain regions (van der Werf et al. 2010). The scale of wetland contribution to SLR remains poorly constrained, largely because of our incomplete knowledge of global wetland changes. There is still no universal consensus on the definition of wetland (Mitra et al. 2005) due to the diversity of wetland types. This introduces fundamental ambiguity in quantifying wetland extents and water/carbon storage. Thus, a major advancement of global wetland inventory and monitoring is needed to reduce uncertainties in estimates of their contributions to SLR.

Preliminary estimates were inferred by several studies from regional wetland records in the past two centuries. Sahagian et al. (1994) calculated that, in the USA alone, wetland drainage since 1780 had led to an average SLR of 0.006 mm year^{-1}. This estimation was based on a documented wetland loss of 0.22 million ha (mha) year^{-1} in the USA (Mitsch and Gosselink 1993) and an assumed ~1 m water column depth. Milly et al. (2010) assumed a global wetland area of 856 mha based on Mitsch and Gosselink (1993) and a 50% area loss of the original wetland. By spreading that drainage over the same 220 years, they obtained a global contribution of ~0.06 mm year^{-1} SLR, 10 times of that of the USA. Davidson (2014) recently argued that the presumed global wetland loss of 50% could be substantially underestimated. By synthesizing 189 reports of wetland area changes, this analysis shows that 71% of the original natural wetlands that fall under the Ramsar Convention for wetland types (Matthews 1993) had been lost by early 21st century, and 67% was lost during past two centuries, implying that the estimate of Milly et al. (2010) is likely conservative. Here we consider a recent wetland loss rate of 0.565% year^{-1} since 1990 (Davidson 2014) and a present global wetland area of 371 mha averaged from three databases: Matthews natural wetlands (Matthews and Fung 1987), ISLSCP (Darras 1999), and DISCover (Belward et al. 1999; Loveland and Belward 1997). If we also assume a uniform 1 m depth of water in wetlands (Milly et al. 2010), the contribution of recent global wetland drainage to SLR would be 0.067 mm year^{-1}.

The assessments above, however, exclude the impacts of biomass oxidation and peat combustion. According to Armentano and Menges (1986) and Gornitz et al. (1997), the net global carbon emission from 1795 to 1980, integrating carbon storage reduction by peat combustion and in temperate and tropical wetlands sums to 22–101 million t C year^{-1}. Water release associated with such carbon emission was converted by Gornitz et al. (1997) similarly to that of deforestation (Sect. 5.1), yielding a SLR of 0.001–0.002 mm year^{-1}. Inferred from multiple sources, Mitra et al. (2005) inferred from multiple sources that global mean carbon densities are 210–700 t C ha^{-1} in wetland soils (including peatland) and ~50 t C ha^{-1} in vegetation biomass. If we apply the same wetland area and loss rate as used for assessing wetland water drainage, the annual reduction of wetland carbon stock since 1990, if completely emitted, releases water equivalent to 0.003–0.007 mm year^{-1} SLR. Integrating the impacts of wetland drainage, oxidation and peat combustion, we here suggest that the recent global wetland degradation results in an upper-bound SLR of $\sim$0.075 mm year^{-1}.

5.3 Lake Storage Changes

Lakes store the greatest mass of liquid water on the terrestrial surface (Oki and Kanae 2006). Variation in lake water storage shares an intrinsic bond with that of the entire terrestrial water storage that links directly to SLR (Reager et al. 2016). Compared to some other storage forms such as glaciers and groundwater, lakes have more active interactions with surface and land–atmosphere fluxes, thus typically more dynamic in budget (Sheng et al. 2016; Song et al. 2014; Wang et al. 2012). In many regions of the world, lakes are monitored as "sentinels" of both climate change and anthropogenic impacts (Adrian et al. 2009; Smith et al. 2005; Song et al. 2016; Wang et al. 2014). However, also because of this "dynamic" nature, along with lakes' extensive distribution in various environments, changes in lake water storages on a global scale are poorly known, and their overall contribution to SLR remains unclear.

Existing studies of lake storage contributions in recent decadal and centennial periods focus on big endorheic (landlocked) lakes, whose growth and declines likely signal the opposite loss and gain in the exorheic (ocean connected) storage. In the past century, perhaps the greatest contributor in global lake storage was the largest endorheic lake, the Caspian Sea (Milly et al. 2010), where water level exhibits substantial oscillations attributed to meteorological, geological, and anthropogenic factors (Ozyavas et al. 2010). A centennial record of the Caspian Sea level, measured by both gauging stations and recent satellite altimetry (Cretaux et al. 2011; Golytsyn and Panin 1989; Klige and Myagkov 1992; Schwatke et al. 2015), reveals an enduring drop of $\sim$3 m during 1900–1977 inducing 0.05 mm year^{-1} SLR and a $\sim$2-m increase in the subsequent two decades or −0.12 mm year^{-1} SLR (Milly et al. 2010), followed by another drop of 0.722 (±0.026) m from 1995 to the near present (year 2014) or, as we calculate here, 0.047 (±0.002) mm year^{-1} SLR. If we assume that the lake level variation kept pace with groundwater changes which can be approximated using the method proposed by Sahagian et al. (1994), the overall contribution of the Caspian Sea, including both surface and groundwater storage variations, has been about 0.03 mm year^{-1} SLR since 1900, 0.075 (±0.002) mm year^{-1} since 1995, or 0.109 (±0.004) mm year^{-1} since 2002.

In contrast to Caspian Sea's level fluctuations, the Aral Sea has been falling constantly over the past half a century. Between 1960 and 1990, the water storage in the Aral Sea Basin declined at a striking rate of 64 km^3 year^{-1}, equivalent to 0.18 mm year^{-1} SLR (Sahagian 2000; Sahagian et al. 1994; Vörösmarty and Sahagian 2000). The main culprit was attributed to be upstream water diversion for irrigation (Perera 1993), which was modeled by Pokhrel et al. (2012) to be $\sim$500 km^3 during 1951–2000, equivalent to 0.03 mm year^{-1} SLR. This modeled irrigation-induced contribution is substantially smaller than the estimate of Sahagian et al. (1994) which accounts for basin-scale water storage changes inferred from historical documentation (Micklin 1992). Such a discrepancy likely implies the importance of other factors such as enhanced land-to-atmospheric flux and climate variations. Dramatic decline in the Aral Sea continued in the recent decade, with an annual rate of 6.043 (±0.082) km^3 year^{-1} measured from 2002 to 2014 (Schwatke et al. 2015). If we assume that groundwater drainage has kept pace with lake level reduction (Sahagian et al. 1994), the Aral Sea has contributed to the recent SLR by 0.0358 (±0.0003) mm year^{-1}.

In the recent couple of decades, advancement of altimetric and gravimetric satellites has enabled more extensive and frequent monitoring of lake water storage all over the world. Evident level and volume decrease after 2000 was revealed in major lakes along East Africa's Great Rift Valley, such as Lakes Victoria, Tanganyika, and Malawi (Ahmed et al. 2014; Awange et al. 2008; Becker et al. 2010; Ramillien et al. 2014; Swenson and Wahr 2009). Several large lakes in the Middle East, such as the Aral Sea and Urmia Lake, continued with striking water storage declines in early 21st century driven by climate change and human water diversion (Cretaux et al. 2011; Schwatke et al. 2015; Singh et al. 2012; Tourian et al. 2015). On the contrary, most endorheic lakes across High Asia exhibited rapid expansions due to wetting climate and increasing glacier meltwater supply (Song et al. 2013; Zhang et al. 2013), which resulted in a small negative contribution to SLR (Jacob et al. 2012). The endorheic Lake Chad in Africa, despite substantial shrinkage and desiccation in the 20th century, has experienced a steady budget recovery of 0.05–0.06 m year^{-1} since the 1990s (Schwatke et al. 2015), counteracting the recent SLR but by a trivial amount ($\sim$0.0002 mm year^{-1}). As Milly et al. (2010) suggested, lake water storage is dominated by regional interannual variation, and thus trends cannot be extrapolated beyond the study periods or areas. A holistic and continuous understanding of its overall contribution to global SLR requires real-time observations of detailed variations in surface water volume, which fortunately are being approached by some of the near-future satellites such as NASA's Surface Water and Ocean Topography (SWOT) mission.

6 Climate-Driven Water Storage Change

Interannual to decadal changes in the sea level budget are important in understanding long-term changes in rates of SLR. The rate of SLR is strongly influenced by the transfer of water between ocean and land, the rate of which changes in response to internal climate variability such as the Pacific Decadal Oscillation (PDO) and the El Niño Southern Oscillation (ENSO), mountain glacier changes, human management practices, and to climatic change in atmospheric transport and delivery of moisture to the continents. Understanding this short-term variability is important so that the longer-term observational sea level record can be correctly interpreted and that the detection and attribution of underlying trends can be improved. A comprehensive understanding of global SLC (sea level change) depends upon a strong understanding of land water storage and its variability.

The term "climate-driven land water storage" can be used to describe variability in global hydrology and water storage in several states, including global snow, surface water, soil moisture, and groundwater storage. These sources comprise what is acknowledged to be one of the most important components of decadal sea level budgets, but also one of the most difficult to observe and characterize globally (Church et al., 2013). The annual cycle of land water storage represents an amplitude of 17 $\pm$ 4 mm sea level equivalent (SLE) of water mass, moved through the seasonal distribution of water from ocean to land (e.g., Wouters et al. 2011). Because of this large-amplitude oscillation, natural changes in the interannual to decadal cycling of water can have a large effect on the apparent rate of SLC over decadal and shorter time periods (Milly et al. 2003).

From the years 2003–2011, the altimetry reported rate of SLR was $\sim$2.4 mm year^{-1} (Cazenave et al. 2014). However, increased mass loss from glaciers (Gardner et al. 2013) and ice sheets (Shepherd et al. 2012) during this time period made this lower rate difficult to reconcile with component-based SLC budgets. Recent research using observations from

NASA's GRACE satellite mission to detect changes in global land and ocean mass identified climate-driven changes in land water storage as the major contributor to the apparent decrease in the rate of SLR (i.e., large net transfer from ocean to land, Reager et al. 2016). Reager et al. (2016) estimated a total continental land mass change (including glaciers) over their 2002–2014 study period of 0.32 ± 0.13 mm $year^{-1}$ of SLR (i.e., ocean gaining). This result compares well with trends in land and ocean mass change in studies by Riva et al. (2010), Llovel et al. (2010), (2014), and Jensen et al. (2013) when differences in time periods are accounted for. To isolate a hydrology-only "land water storage" signal, Reager et al. (2016) then removed global land glacier mass loss trends. Estimating a glacier rate of 0.65 ± 0.09 mm $year^{-1}$ of SLR (ocean gaining), glacier-free land gained or stored water at a rate of 120 ± 60 Gt $year^{-1}$ equivalent to –0.33 ± 0.16 mm $year^{-1}$ of SLR (i.e., land gaining). Increases in precipitation over land caused positive storage trends in some regions: e.g., large flooding periods in the upper Missouri River basin (Reager et al. 2014); recovery from drought in the Amazon (Chen et al. 2010), the Zambezi and Niger basins in Africa (e.g., Ramillien et al. 2014), and weaker gains in Northern Australia associated with La Niña (Fasullo et al. 2013). Using the GRACE observations to represent the net land water storage mass gain (equivalent to –0.33 ± 0.16 mm $year^{-1}$ of SLR) over land (i.e., the sum of both the human- and climate-driven components), subtracting the IPCC estimate for the human-driven component of 0.38 ± 0.12 mm $year^{-1}$ of SLR (Church et al. 2013) provides an estimate of the climate-driven land water storage change. Applying this method, an estimated –0.71 ± 0.20 mm $year^{-1}$ SLE of land water storage uptake is required to close the observed land water storage balance. This number agrees well with the hypothesis and numbers presented by Cazenave et al. (2014) who provided much lower SLR estimate over a recent decade (Table 3) and supports the concept that ENSO-driven modulations of the global water cycle are of first-order importance in decadal-scale sea level budgets, roughly comparable to the magnitudes of ice mass losses from glaciers and ice sheets (Fig. 3).

Reager et al. (2016) use a direct observation of the global land mass change, including detailed measurements of glacier change. In their approach, they provide a thorough and rigorous quantification of the uncertainty for all mass change terms over land, including not only the measurement and leakage errors, but also uncertainties in the GRACE post-

Table 3 Global land water storage budget (2002–2014) in mm $year^{-1}$

		2002–2014 (mm $year^{-1}$)
Observed SLR	Church et al. (2013) (1993–2010)	3.2 (±0.4)
	Cazenave et al. (2014) (2003–2011)	2.4
Estimated Land water storage		
Groundwater	Wada et al. (2016)	0.30 (±0.1)
Reservoir impoundment	This study	−0.24 (±0.02)
Deforestation (after 2010)	This study	0.035
Wetland loss (after 1990)	This study	0.075
Endorheic basin storage loss		
Caspian	This study	0.109 (±0.004)
Aral Sea	This study	0.036(±0.0003)
Climate-driven land water storage	Reager et al. (2016)	−0.71 (±0.2)
Net land water storage	This study	−0.40 (±0.2)
	IPCC AR5 (1993–2010)	0.38 (±0.12)

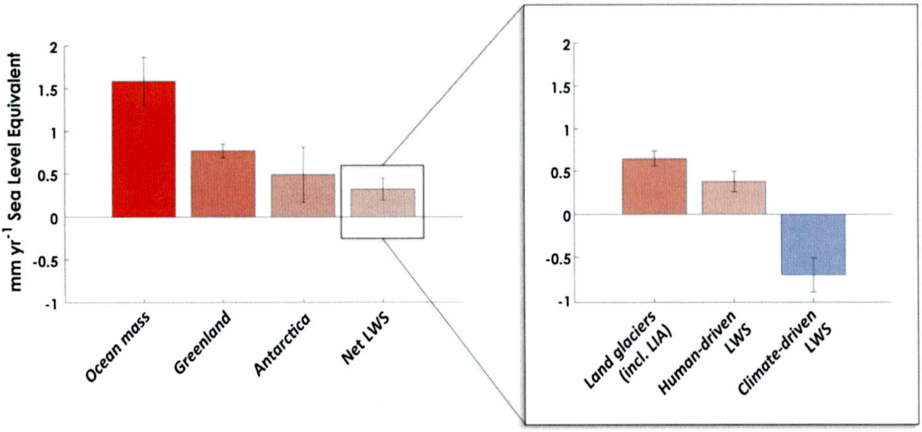

Fig. 3 Global mass budget estimate from Reager et al. (2016). This includes a disaggregation of the land water storage estimate into land glaciers, human-driven, and climate-driven water storage

processing corrections for geocenter, mean pole, and glacial isostatic adjustment (GIA), as well as the uncertainties in glacier change. Because of this, the confidence intervals presented in Reager et al. (2016) should be considered the most comprehensive assessment of uncertainties present to date. This study marks a new era in which changes in land water storage and the potential impact of climate-driven changes in hydrology can be measured from land water storage space and applied to improve the interpretation of the sea level record.

Sea level variability due to climate-driven hydrology will change with the length of the record and does not represent a long-term offset to global SLR (Dieng et al. 2015). However, over the last decade, climate-driven land water storage uptake is of opposite sign and of comparable magnitude to ice losses from glaciers and ice sheets and nearly twice as large as mass losses from direct human-driven changes in land water storage. Climate-driven changes in land water storage are now observable on a global scale and that these changes are large and necessary for closure of decadal-scale sea level budgets.

7 Updated Land Water Contribution and Discussion

Based on separate contribution from different land water storage components, we estimate the net land water storage contribution to SLR is −0.40 (±0.2) during the period of 2002–2014 (Table 3). When considering terrestrial water contribution to SLR including groundwater, reservoir impoundment, water release due to deforestation, marsh drainage or wetland loss, and storage loss from endorheic lakes, it is important to note that reservoir impoundment due to dam building is of the opposite sign in its sea level contribution, suggesting that the volume of water accumulated in reservoirs up to 2015 amounts to ~30 mm sea level equivalent. Lettenmaier and Milly (2009) indicated that the volume of silt accumulated in reservoirs should be removed, which is equal to ~4 mm sea level equivalent. Indeed, silting-up of existing reservoirs may already be, or in coming decades may become, a larger effect on impoundment than the construction of new reservoir capacity (Wisser et al. 2013). It is also important to note that climate-driven land water storage change has a large negative contribution to SLR over the period 2002–2014.

Together with reservoir impoundment, the overall negative contribution amounts to nearly 1 mm year^{-1} over the period 2002–2014, which cancels out all the positive contribution from groundwater, and the remaining land water storage change. However, climate-driven land water storage likely has large decadal variability over past and future periods, which are not yet accounted for in estimated sea level variations.

During the period 1900–1950, the net contribution of terrestrial water sources to global SLR is small with the average rate of +0.014 (±0.008) mm year^{-1}; however, as a result of increased dam building during 1950–1990 the net contribution became consistently negative with the average rate of −0.34 (±0.025) mm year^{-1}. Since the 1990s dam building has been tapering off and GWD has been steadily increasing, the net contribution became positive with the average rate of +0.12 (±0.08) mm year^{-1} over the period 1993–2010 (Wada et al. 2016). During the recent decade, GWD is the most important positive terrestrial water contribution (Wada et al. 2012). The increase is driven by growing water demand for population and agricultural production over intense irrigated regions including India, Pakistan, China, Iran, USA, Mexico, and Saudi Arabia.

For the contribution from deforestation, wetland loss, and endorheic basin storage loss, we estimate the overall contribution to be 0.22 mm year^{-1} over the period 2002–2014, which still gives a significant positive contribution (40% of the overall positive contribution). However, large uncertainty still remains and the contribution likely varies substantially due to human and climate influences over time.

Using the latest available estimates of land water contributions, we estimate the net land water contribution during the period 2002–2014 to be largely negative (−0.40 mm year^{-1}). Although the time periods are not exactly the same, this suggests a large discrepancy from the estimated net land water contribution reported in the IPCC AR5 (0.38 mm year^{-1}; 1993–2010).

Acknowledgements The International Space Science Institute (ISSI) in Bern, Switzerland, and specifically Anny Cazenave and Nicolas Champollion, are acknowledged for hosting the ISSI Workshop on Integrative study of the mean sea level and its components. Y. Wada is supported by Japan Society for the Promotion of Science (JSPS) Oversea Research Fellowship (grant no. JSPS-2014-878). A portion of this research was conducted at the Jet Propulsion Laboratory, California Institute of Technology, under contract with NASA. We acknowledge for the ICOLD Register Committee chairman Patrick Ledelliou, and Dam Surveillance past chairman Bernard Goguel, who removed duplicates from the ICOLD Register database. We wish to thank one anonymous referee and Anny Cazenave for the constructive comments and suggestions, which substantially improved the quality of the manuscript.

References

Abraham JP et al (2013) A review of global ocean temperature observations: implications for ocean heat content estimates and climate change. Rev Geophys 51(3):450–483. doi:10.1002/rog.20022

Adrian R et al (2009) Lakes as sentinels of climate change. Limnol Oceanogr 54:2283–2297. doi:10.4319/lo.2009.54.6_part_2.2283

Ahmed M, Sultan M, Wahr J, Yan E (2014) The use of GRACE data to monitor natural and anthropogenic induced variations in water availability across Africa. Earth Sci Rev 136:289–300

Armentano TV, Menges ES (1986) Patterns of change in the carbon balance of organic soil-wetlands of the temperate zone. J Ecol 74:755–774. doi:10.2307/2260396

Awange JL, Sharifi MA, Ogonda G, Wickert J, Grafarend EW, Omulo MA (2008) The falling Lake Victoria water level: GRACE, TRIMM and CHAMP satellite analysis of the lake basin. Water Resour Manag 22:775–796

Becker M, LLovel W, Cazenave A, Güntner A, Crétaux J-F (2010) Recent hydrological behavior of the East African great lakes region inferred from GRACE, satellite altimetry and rainfall observations. Cr Geosci 342:223–233

Belward AS, Estes JE, Kline KD (1999) The IGBP-DIS global 1-km land-cover data set DISCover: a project overview. Photogramm Eng Remote Sens 65:1013–1020

Boening C, Willis JK, Landerer FW, Nerem RS, Fasullo J (2012) The 2011 La Niña: so strong, the oceans fell. Geophys Res Lett 39(19):n/a-n/a. doi:10.1029/2012GL053055

Bredehoeft JD (2002) The water budget myth revisited: why hydrogeologists model? Ground Water 40:340–345. doi:10.1111/j.1745-6584.2002.tb02511.x

Butt N, de Oliveira PA, Costa MH (2011) Evidence that deforestation affects the onset of the rainy season in Rondonia. Braz J Geophys Res Atmos 116:D11120. doi:10.1029/2010jd015174

Cao G, Zheng C, Scanlon BR, Liu J, Li W (2013) Use of flow modeling to assess sustainability of groundwater resources in the North China Plain. Water Resour Res. doi:10.1029/2012WR011899

Cazenave A, Dieng HB, Meyssignac B, von Schuckmann K, Decharme B, Berthier E (2014) The rate of sea-level rise. Nat Clim Change 4(5):358–361

Chagnon FJF, Bras RL (2005) Contemporary climate change in the Amazon. Geophys Res Lett. doi:10.1029/2005gl022722

Chao BF, Wu YH, Li YS (2008) Impact of artificial reservoir water impoundment on global sea level. Science 320:212–214. doi:10.1126/science.1154580

Cheema MJ, Immerzeel WW, Bastiaanssen WG (2014) Spatial quantification of groundwater abstraction in the irrigated Indus basin. Ground Water 52:25–36. doi:10.1111/gwat.12027

Chen JL, Wilson CR, Tapley BD (2010) The 2009 exceptional Amazon flood and interannual terrestrial water storage change observed by GRACE. Water Resour Res. doi:10.1029/2010WR009383

Chen J, Famigliett JS, Scanlon BR, Rodell M (2016) Groundwater storage changes: present status from GRACE observations. Surv Geophys 37:397–417, doi:10.1007/s10712-015-9332-4, Special Issue: ISSI Workshop on Remote Sensing and Water Resources

Church JA, et al. (2013) Sea level change. In: Climate change 2013: the physical science basis. Contribution of working group I to the fifth assessment report of the intergovernmental panel on climate change (Stocker TF, Qin D, Plattner G-K, Tignor M, Allen SK, Boschung J, Nauels A, Xia Y, Bex V, Midgley PM (eds)). Cambridge University Press, Cambridge, United Kingdom and New York, NY, USA

Ciais P et al. (2013) Carbon and other biogeochemical cycles. In: Stocker TF et al. (eds) Climate change 2013: the physical science basis. Contribution of working group I to the fifth assessment report of the Intergovernmental Panel on Climate Change. Cambridge University Press, Cambridge, United Kingdom and New York, NY, USA, pp 465–570

Cretaux JF et al (2011) SOLS: a lake database to monitor in the Near Real Time water level and storage variations from remote sensing data. Adv Space Res 47:1497–1507. doi:10.1016/j.asr.2011.01.004

Darras S (1999) IGBP-DIS wetlands data initiative, a first step towards identifying a global delineation of wetalnds. IGBP-DIS Office, Toulouse

Davidson NC (2014) How much wetland has the world lost? Long-term and recent trends in global wetland area. Mar Freshwater Res 65:934–941. doi:10.1071/Mf14173

DeAngelis A, Dominguez F, Fan Y, Robock A, Kustu MD, Robinson D (2010) Evidence of enhanced precipitation due to irrigation over the Great Plains of the United States. J Geophys Res. doi:10.1029/2010JD013892

Dieng HB, Champollion N, Cazenave A, Wada Y, Schrama E, Meyssignac B (2015) Total land water storage change over 2003–2013 estimated from a global mass budget approach. Environ Res Lett. doi:10.1088/1748-9326/10/12/124010

Döll P, Hoffmann-Dobrev H, Portmann FT, Siebert S, Eicker A, Rodell M, Strassberg G (2012) Impact of water withdrawals from groundwater and surface water on continental water storage variations. J Geodyn 59–60:143–156. doi:10.1016/j.jog.2011.05.001

Döll P, Müller Schmied H, Schuh C, Portmann FT, Eicker A (2014) Global-scale assessment of groundwater depletion and related groundwater abstractions: combining hydrological modeling with information from well observations and GRACE satellites. Water Resour Res 50:5698–5720. doi:10.1002/2014WR015595

Döll P, Douville H, Güntner A, Müller Schmied H, Wada Y (2016) Modelling freshwater resources at the global scale: challenges and prospects. Surv Geophys 37:195–221, Special Issue: ISSI Workshop on Remote Sensing and Water Resources

FAO (2015) Global forest resources assessment 2015: how have the world's forests changed? Rome

Famiglietti JS, Lo M, Ho SL, Bethune J, Anderson KJ, Syed TH, Swenson SC, de Linage CR, Rodell M (2011) Satellites measure recent rates of groundwater depletion in California's Central Valley. Geophys Res Lett. doi:10.1029/2010GL046442

Famiglietti JS (2014) The global groundwater crisis. Nature Clim Change 4:945–948. doi:10.1038/nclimate2425

Fasullo JT, Boening C, Landerer FW, Nerem RS (2013) Australia's unique influence on global sea level in 2010–2011. Geophys Res Lett 40:4368–4373. doi:10.1002/grl.50834

Feng W, Zhong M, Lemoine J-M, Biancale R, Hsu H-T, Xia J (2013) Evaluation of groundwater depletion in North China using the Gravity Recovery and Climate Experiment (GRACE) data and ground-based measurements. Water Resour Res 49:2110–2118. doi:10.1002/wrcr.20192

Foster S, Loucks DP (eds) (2006) Non-renewable groundwater resources: a guidebook on socially-sustainable management for water-policy makers, IHP-VI, Series on Groundwater No. 10, UNESCO, Paris, France

Gardner AS, Moholdt G, Cogley JG, Wouters B, Arendt AA, Wahr J, Berthier E, Hock R, Pfeffer WT, Kaser G, Ligtenberg SRM, Bolch T, Sharp MJ, Hagen JO, van den Broeke MR, Paul F (2013) A reconciled estimate of glacier contributions to sea level rise: 2003 to 2009. Science 340:852–857. doi:10.1126/science.1234532

Gleick PH (2003) Global freshwater resources: soft-path solutions for the 21st century. Science 302:1524–1528. doi:10.1126/science.1089967

Golytsyn GS, Panin GN (1989) Once more on the water level changes of the Caspian Sea. Vestnik Akademii Nauk SSSR 9:59–63 **(in Russian)**

Gornitz V (1995) Sea-level rise: a review of recent past and near-future trends. Earth Surf Process Landforms 20:7–20. doi:10.1002/esp.3290200103

Gornitz V, Rosenzweig C, Hillel D (1997) Effects of anthropogenic intervention in the land hydrologic cycle on global sea level rise. Global Planet Change 14:147–161. doi:10.1016/s0921-8181(96)00008-2

Gornitz V (2001) In sea level rise: history and consequences. In: Douglas BC, Kearney MS, Leatherman SP (eds), Academic Press, San Diego, pp 97–119

Gregory JM, White NJ, Church JA, Bierkens MFP, Box JE, van den Broeke MR, Cogley JG, Fettweis X, Hanna E, Huybrechts P, Konikow LF, Leclercq PW, Marzeion B, Oerlemans J, Tamisiea ME, Wada Y, Wake LM, van de Wal RSW (2013) Twentieth-century global-mean sea level rise: is the whole greater than the sum of the parts? J Clim 26:4476–4499. doi:10.1175/JCLI-D-12-00319.1

Huang Z, Pan Y, Gong H, Yeh PJ, Li X, Zhou D, Zhao W (2015) Subregional-scale groundwater depletion detected by GRACE for both shallow and deep aquifers in North China Plain. Geophys Res Lett 42:1791–1799. doi:10.1002/2014GL062498

Huntington TG (2008) Can we dismiss the effect of changes in land-based water storage on sea-level rise? Hydrol Process 22:717–723. doi:10.1002/hyp.7001

Jacob T, Wahr J, Pfeffer WT, Swenson S (2012) Recent contributions of glaciers and ice caps to sea level rise. Nature 482:514–518. doi:10.1038/nature10847

Jensen L, Rietbroek R, Kusche J (2013) Land water contribution to sea level from GRACE and Jason-1 measurements. J Geophys Res Oceans 118:212–226. doi:10.1002/jgrc.20058

Keenan RJ, Reams GA, Achard F, de Freitas JV, Grainger A, Lindquist E (2015) Dynamics of global forest area: results from the FAO global forest resources assessment. Forest Ecol Manag 352:9–20. doi:10.1016/j.foreco.2015.06.014

Khatiwala S, Primeau F, Hall T (2009) Reconstruction of the history of anthropogenic CO_2 concentrations in the ocean. Nature 462:346–U110 doi:10.1038/nature08526

Klige RK, Myagkov MS (1992) Changes in the water regime of the Caspian Sea. Geo J 27:299–307

Konikow LF (2011) Contribution of global groundwater depletion since 1900 to sea-level rise. Geophys Res Lett 38:L17401. doi:10.1029/2011GL048604

Kustu M, Fan Y, Robock A (2010) Large-scale water cycle perturbation due to irrigation pumping in the US High Plains: a synthesis of observed streamflow changes. J Hydrol 390:222–244. doi:10.1016/j.jhydrol.2010.06.045

Kustu MD, Fan Y, Rodell M (2011) Possible link between irrigation in the U.S. High Plains and increased summer streamflow in the Midwest. Water Resour Res 47:W03522. doi:10.1029/2010WR010046

Lehner B, Reidy Liermann C, Revenga C, Vörösmarty C, Fekete B, Crouzet P, Döll P, Endejan M, Frenken K, Magome J, Nilsson C, Robertson JC, Rödel R, Sindorf N, Wisser D (2011) High-resolution mapping of the world's reservoirs and dams for sustainable river-flow management. Fron Ecol Environ 9:494–502. doi:10.1890/100125

Lettenmaier DP, Milly PCD (2009) Land waters and sea level. Nat Geosci 2:452–454. doi:10.1038/ngeo567

Llovel W, Becker M, Cazenave A, Crétaux J-F, Ramillien G (2010) C R Geosci 342:179–188. doi:10.1016/j.crte.2009.12.004

Llovel W, Willis JK, Landerer FW, Fukumori I (2014) Deep-ocean contribution to sea level and energy budget not detectable over the past decade. Nat Clim Change 4:1031–1035. doi:10.1038/nclimate2387

Lo M-H, Famiglietti JS (2013) Irrigation in California's Central Valley strengthens the southwestern U.S. water cycle. Geophys Res Lett. doi:10.1002/grl.50108

Loveland TR, Belward AS (1997) The IGBP-DIS global 1 km land cover data set, DISCover: first results. Int J Remote Sens 18:3291–3295

MacDicken KG (2015) Global Forest Resources Assessment 2015: what, why and how? Forest Ecol Manag 352:3–8. doi:10.1016/j.foreco.2015.02.006

Matthews E, Fung I (1987) Methane emission from natural wetlands: global distribution, area, and environmental characteristics of sources. Global Biogeochem Cycles 1:61–86

Matthews GVT (1993) The Ramsar Convention on wetlands: its history and development. Ramsar Convention Bureau, Gland

Meherhomji VM (1991) Probable impact of deforestation on hydrological processes. Clim Change 19:163–173. doi:10.1007/Bf00142223

Micklin PP (1992) The aral crisis—introduction to the special issue. Post-Sov Geogr 33:269–282

Milly PCD, Cazenave A, Gennero MC (2003) Contribution of climate-driven change in continental water storage to recent sea-level rise. Proc Natl Acad Sci USA 100:13158–13161

Milly PCD, Cazenave A, Famiglietti JS, Gornitz V, Laval K, Lettenmaier DP, Sahagian DL, Wahr JM, Wilson CR (2010) Terrestrial water-storage contributions to sea-level rise and variability. In: Church JA, Woodworth PL, Aarup T, Wilson WS (eds) Understanding sea-level rise and variability. Wiley-Blackwell, Oxford. doi:10.1002/9781444323276.ch8

Mitra S, Wassmann R, Vlek PLG (2005) An appraisal of global wetland area and its organic carbon stock. Curr Sci India 88:25–35

Mitsch WJ, Gosselink JG (1993) Wetlands, 2nd edn. Van Nostrand Reinhold, New York

Nobre P, Malagutti M, Urbano DF, de Almeida RAF, Giarolla E (2009) Amazon deforestation and climate change in a coupled model simulation. J Clim 22:5686–5697. doi:10.1175/2009jcli2757.1

Oki T, Kanae S (2006) Global hydrological cycles and world water resources. Science 313:1068–1072. doi:10.1126/science.1128845

Ozyavas A, Khan SD, Casey JF (2010) A possible connection of Caspian Sea level fluctuations with meteorological factors and seismicity. Earth Planet Sc Lett 299:150–158. doi:10.1016/j.epsl.2010.08.030

Pala C (2006) Once a terminal case, the North Aral Sea shows new signs of life. Science 312:183. doi:10.1126/science.312.5771.183

Pala C (2011) In Northern Aral Sea, rebound comes with a big catch. Science 334:303. doi:10.1126/science.334.6054.303

Perera J (1993) A sea turns to dust. New Sci 140:24–27

Pokhrel YN, Hanasaki N, Yeh PJ-F, Yamada T, Kanae S, Oki T (2012) Model estimates of sea level change due to anthropogenic impacts on terrestrial water storage. Nat Geosci 5:389–392. doi:10.1038/ngeo1476

Pokhrel YN, Koirala S, Yeh PJ-F, Hanasaki N, Longuevergne L, Kanae S, Oki T (2015) Incorporation of groundwater pumping in a global Land Surface Model with the representation of human impacts. Water Resour Res 51:78–96. doi:10.1002/2014WR015602

Postel SL (1999) Pillar of sand: can the irrigation miracle Last?. W.W. Norton, New York USA. ISBN 0-393-31937-7

Ramillien G, Frappart F, Seoane L (2014) Application of the regional water mass variations from GRACE Satellite Gravimetry to large-scale water management in Africa. Remote Sens 6:7379–7405

Reager JT, Thomas BF, Famiglietti JS (2014) River basin flood potential inferred using GRACE gravity observations at several months lead time. Nat Geosci 7:588–592. doi:10.1038/ngeo2203

Reager JT, Gardner AS, Famiglietti JS, Wiese DN, Eicker A, Lo MH (2016) A decade of sea level rise slowed by climate-driven hydrology. Science 351(6274):699–703. doi:10.1126/science.aad8386

Richey AS, Thomas BF, Lo M-H, Reager JT, Famiglietti JS, Voss K, Swenson S, Rodell M (2015) Quantifying renewable groundwater stress with GRACE. Water Resour Res 51:5217–5238. doi:10.1002/2015WR017349

Riva REM, Bamber JL, Lavallée DA, Wouters B (2010) Sea-level fingerprint of continental water and ice mass change from GRACE. Geophys Res Lett 37:L19605. doi:10.1029/2010GL044770

Rodell M, Velicogna I, Famiglietti JS (2009) Satellite-based estimates of groundwater depletion in India. Nature 460:999–1002. doi:10.1038/nature08238

Rohrig E (1991) Biomass and productivity. In: Rohrig E (ed) Ecosystems of the world. Elsevier, New York, pp 165–174

Sabine CL et al. (2004) The oceanic sink for anthropogenic CO_2 Science 305:367–371 doi:10.1126/science.1097403
Sahagian D (2000) Global physical effects of anthropogenic hydrological alterations: sea level and water redistribution. Global Planet Change 25:39–48. doi:10.1016/S0921-8181(00)00020-5
Sahagian DL, Schwartz FW, Jacobs DK (1994) Direct anthropogenic contributions to sea level rise in the twentieth century. Nature 367:54–57. doi:10.1038/367054a0
Scanlon BR, Faunt CC, Longuevergne L, Reedy RC, Alley WM, McGuire VL, McMahon PB (2012a) Groundwater depletion and sustainability of irrigation in the U.S. High Plains and Central Valley. Proc Natl Acad Sci USA 109:9320–9325 doi:10.1073/pnas.1200311109
Scanlon BR, Longuevergne L, Long D (2012b) Ground referencing GRACE satellite estimates of groundwater storage changes in the California Central Valley, USA. Water Resour Res 48:W04520. doi:10.1029/2011WR011312
Schwatke C, Dettmering D, Bosch W, Seitz F (2015) DAHITI—an innovative approach for estimating water level time series over inland waters using multi-mission satellite altimetry. Hydrol Earth Syst Sci 19:4345–4364
Shamsudduha M, Taylor RG, Longuevergne L (2012) Monitoring groundwater storage changes in the highly seasonal humid tropics: validation of GRACE measurements in the Bengal Basin. Water Resour Res 48:W02508. doi:10.1029/2011WR010993
Sheng Y, Song C, Wang J, Lyons EA, Knox BR, Cox JS, Gao F (2016) Representative lake water extent mapping at continental scales using multi-temporal Landsat-8 imagery. Remote Sens Environ 185:129–141
Shepherd A et al (2012) A reconciled estimate of ice-sheet mass balance. Science 338(6111):1183–1189. doi:10.1126/science.1228102
Shukla J, Nobre C, Sellers P (1990) Amazon deforestation and climate change. Science 247:1322–1325 doi:10.1126/science.247.4948.1322
Singh A, Seitz F, Schwatke C (2012) Inter-annual water storage changes in the Aral Sea from multi-mission satellite altimetry, optical remote sensing, and GRACE satellite gravimetry. Remote Sens Environ 123:187–195
Sloan S, Sayer JA (2015) Forest Resources Assessment of 2015 shows positive global trends but forest loss and degradation persist in poor tropical countries. Forest Ecol Manag 352:134–145. doi:10.1016/j.foreco.2015.06.013
Smith LC, Sheng Y, MacDonald GM, Hinzman LD (2005) Disappearing Arctic lakes. Science 308:1429. doi:10.1126/science.1108142
Solomon S et al. (eds.) (2007) Climate change 2007: the physical science basis. Contribution of Working Group I to the Fourth Assessment Report of the Intergovernmental Panel on Climate Change, Cambridge Univ. Press, Cambridge, UK
Song C, Huang B, Ke L (2013) Modeling and analysis of lake water storage changes on the Tibetan Plateau using multi-mission satellite data. Remote Sens Environ 135:25–35. doi:10.1016/j.rse.2013.03.013
Song C, Huang B, Richards K, Ke L, Phan VH (2014) Accelerated lakeexpansion on the Tibetan Plateau inthe 2000s: induced by glacial melting or other processes? Water Resour Res 50:3170–3186. doi:10.1002/2013WR014724
Song C, Sheng Y, Ke L, Yong N, Wang J (2016) Glacial lake evolution in the southeastern Tibetan Plateau and the cause of rapid expansion of proglacial lakes linked to glacial-hydrogeomorphic processes. J Hydrol 540:504–514
Spracklen DV, Arnold SR, Taylor CM (2012) Observations of increased tropical rainfall preceded by air passage over forests. Nature 489:U127–282. doi:10.1038/nature11390
Strassberg G, Scanlon BR, Rodell M (2007) Comparison of seasonal terrestrial water storage variations from GRACE with groundwater-level measurements from the High Plains Aquifer (USA). Geophys Res Lett 34:L14402. doi:10.1029/2007GL030139
Swenson S, Wahr J (2009) Monitoring the water balance of Lake Victoria, East Africa, from space. J Hydro 370:163–176
Syed TH, Famiglietti JS, Chambers DP, Willis JK, Hilburn K (2010) Satellite-based global ocean mass balance reveals water cycle acceleration and increasing continental freshwater discharge, 1994–2006. Proc Natl Acad Sci USA 107:17916–17921. doi:10.1073/pnas.1003292107
Tapley BD, Bettadpur S, Ries JC, Thompson PF, Watkins MM (2004) GRACE measurements of mass variability in the Earth system. Science 305:503–505. doi:10.1126/science.1099192
Taylor RG, Scanlon B, Döll P, Rodell M, van Beek R, Wada Y, Longuevergne L, LeBlanc M, Famiglietti JS, Edmunds M, Konikow L, Green TR, Chen J, Taniguchi M, Bierkens MFP, MacDonald A, Fan Y, Maxwell RM, Yechieli Y, Gurdak JJ, Allen DM, Shamsudduha M, Hiscock K, Yeh PJ-F, Holman I,

Treidel H (2013) Groundwater and climate change. Nature Clim Change 3:322–329. doi:10.1038/nclimate1744

Tiwari VM, Wahr J, Swenson S (2009) Dwindling groundwater resources in northern India, from satellite gravity observations. Geophys Res Lett 36:L18401. doi:10.1029/2009GL039401

Tourian M, Elmi O, Chen Q, Devaraju B, Roohi S, Sneeuw N (2015) A spaceborne multisensor approach to monitor the desiccation of Lake Urmia in Iran. Remote Sens Environ 156:349–360

van der Werf GR et al (2010) Global fire emissions and the contribution of deforestation, savanna, forest, agricultural, and peat fires (1997–2009). Atmos Chem Phys 10:11707–11735. doi:10.5194/acp-10-11707-2010

Van Dijk AIJM, Renzullo LJ, Wada Y, Tregoning P (2014) A global water cycle reanalysis (2003–2012) merging satellite gravimetry and altimetry observations with a hydrological multi-model ensemble. Hydrol Earth Syst Sci 18:2955–2973. doi:10.5194/hess-18-2955-2014

Vörösmarty CJ, Sahagian D (2000) Anthropogenic disturbance of the terrestrial water cycle. Bioscience 50:753–765. doi:10.1641/0006-3568(2000)050[0753:ADOTTW]2.0.CO;2

Voss KA, Famiglietti JS, Lo M, de Linage C, Rodell M, Swenson SC (2013) Groundwater depletion in the Middle East from GRACE with implications for transboundary water management in the Tigris-Euphrates-Western Iran region. Water Resour Res. doi:10.1002/wrcr.20078

Wada Y, van Beek LPH, van Kempen CM, Reckman JWTM, Vasak S, Bierkens MFP (2010) Global depletion of groundwater resources. Geophys Res Lett 37:L20402. doi:10.1029/2010GL044571

Wada Y, van Beek LPH, Bierkens MFP (2012a) Nonsustainable groundwater sustaining irrigation: a global assessment. Water Resour Res 48:W00L06, doi:10.1029/2011WR010562, Special Issue: Toward Sustainable Groundwater in Agriculture

Wada Y, van Beek LPH, Sperna Weiland FC, Chao BF, Wu Y-H, Bierkens MFP (2012b) Past and future contribution of global groundwater depletion to sea-level rise. Geophys Res Lett 39:L09402. doi:10.1029/2012GL051230

Wada Y (2016) Modelling groundwater depletion at regional and global scales: Present state and future prospects. Surv Geophys 37:419–451, doi:10.1007/s10712-015-9347-x, Special Issue: ISSI Workshop on Remote Sensing and Water Resources

Wada Y, Lo M-H, Yeh PJ-F, Reager PJ-F, Famiglietti JS, Wu R-J, Tseng Y-H (2016) Fate of water pumped from underground causing sea level rise. Nature Clim Change, doi:10.1038/nclimate3001, early online

Wang J, Sheng Y, Hinkel KM, Lyons EA (2012) Drained thaw lake basin recovery on the western Arctic Coastal Plain of Alaska using high-resolution digital elevation models and remote sensing imagery. Remote Sens Environ 119:325–336. doi:10.1016/j.rse.2011.10.027

Wang J, Sheng Y, Tong TSD (2014) Monitoring decadal lake dynamics across the Yangtze Basin downstream of Three Gorges Dam. Remote Sens Environ 152:251–269. doi:10.1016/j.rse.2014.06.004

Wisser D, Frolking S, Hagen S, Bierkens MFP (2013) Beyond peak reservoir storage? A global estimate of declining water storage capacity in large reservoirs. Water Resour Res 49:5732–5739. doi:10.1002/wrcr.20452

Wouters B, Riva REM, Lavallée DA, Bamber JL (2011) Seasonal variations in sea level induced by continental water mass: first results from GRACE. Geophys Res Lett. doi:10.1029/2010GL046128

Zhang G, Yao T, Xie H, Kang S, Lei Y (2013) Increased mass over the Tibetan Plateau: From lakes or glaciers? Geophys Res Lett 40:2125–2130. doi:10.1002/grl.50462

Surv Geophys (2017) 38:153–185
DOI 10.1007/s10712-016-9379-x

Glacial Isostatic Adjustment and Contemporary Sea Level Rise: An Overview

Giorgio Spada[1]

Received: 25 November 2015 / Accepted: 14 July 2016 / Published online: 22 August 2016

Abstract Glacial isostatic adjustment (GIA) encompasses a suite of geophysical phenomena accompanying the waxing and waning of continental-scale ice sheets. These involve the solid Earth, the oceans and the cryosphere both on short (decade to century) and on long (millennia) timescales. In the framework of contemporary sea-level change, the role of GIA is particular. In fact, among the processes significantly contributing to contemporary sea-level change, GIA is the only one for which deformational, gravitational and rotational effects are simultaneously operating, and for which the rheology of the solid Earth is essential. Here, I review the basic elements of the GIA theory, emphasizing the connections with current sea-level changes observed by tide gauges and altimetry. This purpose is met discussing the nature of the "sea-level equation" (SLE), which represents the basis for modeling the sea-level variations of glacial isostatic origin, also giving access to a full set of geodetic variations associated with GIA. Here, the SLE is employed to characterize the remarkable geographical variability of the GIA-induced sea-level variations, which are often expressed in terms of "fingerprints". Using harmonic analysis, the spatial variability of the GIA fingerprints is compared to that of other components of contemporary sea-level change. In closing, some attention is devoted to the importance of the "GIA corrections" in the context of modern sea-level observations, based on tide gauges or satellite altimeters.

Keywords Sea level rise · Glacial isostasy · Global change

✉ Giorgio Spada
giorgio.spada@gmail.com

[1] Dipartimento di Scienze Pure ed Applicate (DiSPeA), Università degli Studi di Urbino "Carlo Bo", Urbino, Italy

1 Introduction

During the "altimetry era" (1992-today), GIA modeling and its applications have seen a considerable development, stimulated by the recognized important role that GIA has in our understanding of current sea-level rise (Milne and Mitrovica 1996; Peltier 2001; Mitrovica et al. 2001; Mitrovica and Milne 2003; Mitrovica et al. 2011). However, the basic principles of GIA and the physics governing the process of loading and un-loading of the Earth's crust were established in the late seventies (Farrell 1972; Peltier and Andrews 1976; Farrell and Clark 1976; Clark et al. 1978), well before the awareness of climate change arose within the scientific community and in the society (Bindoff et al. 2007; Church et al. 2013). In a first stage, GIA modeling was motivated by the need for understanding the variations in relative sea level that occurred in the geological history of the Earth, whose geographical variability have long puzzled Earth scientists (Woodward 1888; Farrell and Clark 1976). Today, GIA is tightly integrated into the science of global change (Church et al. 2013), and it is subject to a continuous development motivated by the increasing amount of high-precision observations available from various fields of geophysics and geodesy (see, e.g., Wöppelmann et al. 2007; King et al. 2010; Spada et al. 2012; Mémin et al. 2014).

Sometimes, GIA is confused with post-glacial rebound (PGR), the movement of the Earth's crust in response to the melting of continental ice sheets.[1] As a matter of fact, PGR is only one of the aspects of GIA, which is particularly manifest in the regions that were covered by thick ice sheets at the last glacial maximum (LGM, $\approx 21,000$ years ago), like North America and the Baltic region in Europe, and in the immediately surrounding areas. Indeed, the direct observation of PGR in Fennoscandia and northern Europe (see Steffen and Wu 2011 and references therein) stimulated the whole development of the GIA theory (Haskell 1935, 1936). GIA is not only describing the ongoing viscous response to past ice sheet changes. Rather, it includes a wide range of phenomena associated with the isostatic disequilibrium induced by the ice melting, without any limitation of spatial and temporal scales. GIA is associated with global and regional temporal variations of the Earth's gravity field (Peltier 2004), with the three-dimensional displacements of the Earth's surface both in the near and in the far field of the former ice sheets (King et al. 2010; Serpelloni et al. 2013), with loading- and un-loading-induced stress variations in the crust and the mantle (Spada et al. 1991; Steffen et al. 2012; Brandes et al. 2015), and with fluctuations of the Earth's rotation axis, both involving lateral movements of the pole or true polar wander (TPW) (Spada et al. 1992; Ricard et al. 1993; Steinberger and O'Connell 1997; Mitrovica et al. 2005; Cambiotti et al. 2010; Mitrovica and Wahr 2011; Nakada 2009) and changes in the length of day (Munk and MacDonald 1960; Sabadini et al. 1982; Spada et al. 1992). Furthermore, current sea-level changes are affected by GIA (Douglas 1991, 1997; Spada et al. 2012), which has also significantly impacted the Earth's geography during the last millennia (Peltier 1994, 1996). Projections of future sea-level variations also depend on the continuing isostatic disequilibrium being maintained during the following centuries (Slangen 2012; Bamber and Riva 2010).

In this overview, the focus will be on the relationships between the contemporary sea-level variations, sampled by tide gauges and satellite altimetry, and GIA. For a long time, the main concern of sea-level studies was the determination of a globally averaged sea-level rise from instrumental observations (Spada and Galassi 2012). Although, in the

[1] https://en.wikipedia.org/wiki/Post-glacial_rebound.

1940s, it was already recognized that GIA can constitute a possible source of contamination for the sea-level data (Gutenberg 1941), it was only starting from the late 1980s that reliable GIA corrections to the instrumental observations could be determined. This was made possible by the development of a realistic model describing the interactions between the cryosphere, the solid Earth and the oceans (Peltier and Tushingham 1989). These are designed to fit the histories of past sea-level revealed from geological observations from specific locations (e.g., Tushingham and Peltier 1992), and useful for the interpretation of current sea-level variations. During the last few decades, the sea-level research has shifted from the determination of global trends to the recognition of regional, or even local, trends and of their causes (Cazenave and Nerem 2004; Cazenave et al. 2009; Milne et al. 2009; Cazenave and Llovel 2010). This has been stimulated by an enormous increase in the amount and in the quality of sea-level data, and by the improved understanding of some of the causes of sea-level variability (Milne et al. 2009; Meyssignac and Cazenave 2012; Church et al. 2013). In this new context, GIA research has regained interest and this geophysical process is now recognized as one of the most important sources of regional sea-level change (Milne et al. 2009; Kopp et al. 2015; Spada and Galassi 2016), caused by either the melting of the late-Pleistocene ice sheets or current ice sheets fluctuations driven by global warming (Church et al. 2013).

Many of the advances in our understanding of Earth's rheology have been possible by the study of the phenomena associated with GIA. For this reason, the analysis of the impact of this process on regional sea-level change requires some discussion of the underlying physical ideas. In the following, this is done by discussing the "Sea level equation" (or SLE) introduced by Farrell and Clark in 1976. This integral (implicit) equation describes the response of the Earth to surface loads characterized by any timescale, and it is therefore useful for predicting sea-level variations either associated with the melting of late-Pleistocene ice sheets or induced by the effects of global warming. Although the SLE has been primarily used to study relative sea-level change, from its solution a full set of geophysical quantities associated with GIA can be obtained, such as horizontal and vertical crustal rates of displacement, time variations of the gravitational potential of the Earth and absolute sea-level variations. Global maps of these fields, showing their spatial variability, define the GIA "fingerprints" (Plag and Jüettner 2001) which can be useful—in principle—to identify the ice sources that are responsible for these geodetic variations (Bamber and Riva 2010; Mitrovica et al. 2011; Spada et al. 2013). The GIA fingerprints also provide the means for evaluating "GIA corrections", which play an important role in the interpretation of current geodetic variations and have been of fundamental importance for the assessment of the secular global mean sea-level rise from tide gauge observations (Spada and Galassi 2012).

In this work, the GIA associated with the melting of past ice sheets will be discussed adopting one specific model, i.e., ICE-5G(VM2) of Peltier (2004), whose chronology since the LGM is available on line.[2] The SLE will be solved using an improved version of the open-source program SELEN[3] (Spada and Stocchi 2007), whose features and limitations are described in Spada et al. (2012). However, it is to be emphasized that the development of global GIA models has been considerable during the last few decades. The suite of ICE-X GIA models initiated by Peltier and collaborators (Peltier and Andrews 1976; Wu and Peltier 1983; Tushingham and Peltier 1991; Peltier 1994, 2004) has recently culminated with ICE-6G(VM5a) (Peltier et al. 2015). In parallel, Kurt Lambeck and co–workers at the

[2] http://www.atmosp.physics.utoronto.ca/~peltier/data.php.

[3] http://geodynamics.org/cig/software/selen/.

Research School of Earth Sciences of the National Australian University (ANU) have progressively developed, by successive refinements, another family of global GIA models that implement a detailed definition of water load in the SLE (Nakada and Lambeck 1987; Lambeck et al. 2003), now merged into the CALSEA package.[4] The ICE-X and the ANU models assume a spherically symmetrical Earth, characterized by layers having a Maxwell viscoelastic rheology. However, they differ in the assumptions about the a priori viscosity profile, they are constrained by different sets of relative sea level (RSL) observations since the LGM, and they adopt distinct strategies for solving the SLE (Mitrovica and Milne 2003). A description of some the differences between the ICE-X and the ANU models have been given in Schmidt et al. (2014), both in terms of ice sheets chronology and of geodetic responses in Northern Europe. For a global inter-comparison of GIA models, the reader is referred to Guo et al. (2012). The features of the global GIA models developed until year $\sim$2010 within the sea-level community have been comprehensively summarized in Whitehouse (2009).

The paper is organized as follows. In Sect. 2, some existing definitions of *sea level* are reviewed, and utilized in Sect. 3 to construct the SLE, the equation that governs GIA. The nature of the SLE is described in Sect. 4. Sections 5 and 6 deal with the fingerprints that describe patterns of sea-level change and of surface vertical displacements associated with GIA, for past and current ice sources, respectively. The regional variability of these fingerprints is illustrated in Sect. 7 and compared to that of other sources of sea-level change. The GIA corrections to relative and absolute sea-level variations are described in Sect. 8. In Sect. 9, the conclusions are drawn.

2 Definitions of Sea Level and Sea-Level Change

In GIA studies, several definitions involving *sea level* and *sea-level change* are commonly employed. From the perspective of constructing the SLE, and to avoid any ambiguity, in this section I provide some basic definitions about the concept of "sea level".

At a given location on the Earth's surface of spherical coordinates $\omega \equiv (\vartheta, \lambda)$, where ϑ is co–latitude and λ is longitude, and at time t, *sea level* is defined as

$$\mathrm{SL}(\omega, t) = R_{\mathrm{ss}}(\omega, t) - R_{\mathrm{se}}(\omega, t), \tag{1}$$

where R_{ss} is the radius of the sea surface, which is an equipotential surface of the Earth's gravity field (Heiskanen and Moritz 1981), and R_{se} is the radius of the solid surface of the Earth at the same location. Radii R_{ss} and R_{se} are measured in a geocentric reference frame with origin at the center of mass (CM) of the whole Earth.

The SLE does not directly involve SL; rather it involves its variation relative to a previous time $t_0 \leq t$ labeling an equilibrium reference state, a quantity defined as *sea-level change*:

$$S(\omega, t) = SL(\omega, t) - SL(\omega, t_0) \tag{2}$$

which can be directly observed at tide gauges (see Fig. 1). The average of S over the surface of the oceans, which will be introduced in Sect. 3, has an important role in GIA modeling since it defines the *"eustatic" sea-level change*.

[4] http://rses.anu.edu.au/highlights/view.php?article=188.

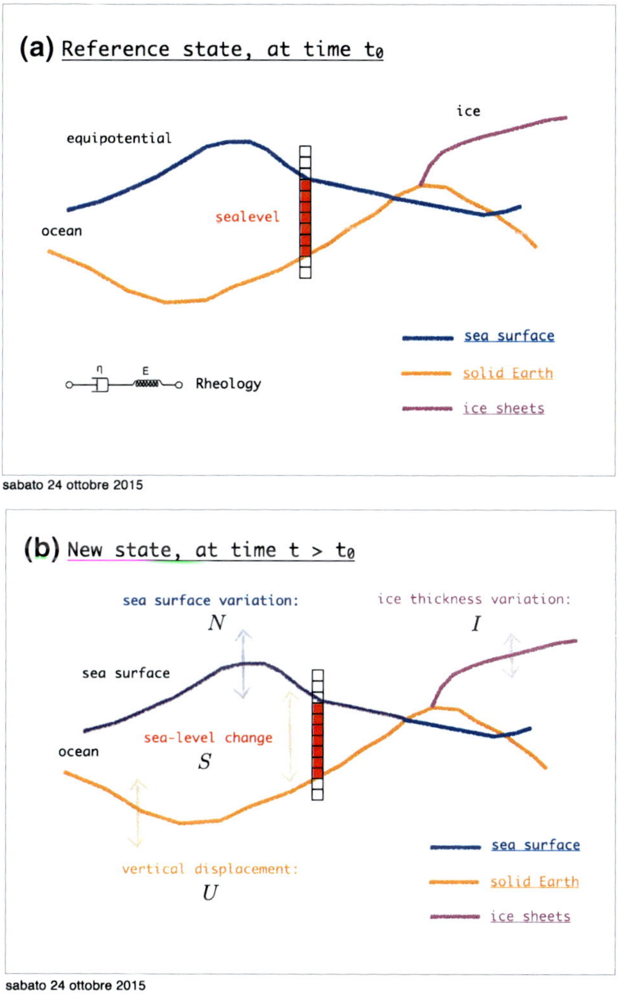

Fig. 1 Reference (**a**) and perturbed state (**b**), with an illustration of the geometrical meaning of the fundamental variables S (relative sea-level change), U (vertical displacement), N (absolute sea-level change), and I (ice thickness variation). In (**a**), the mechanical analogue of a Maxwell body (e.g., Mainardi and Spada 2011) is shown, to denote the sensitivity of S, U and N to mantle rheology. *The vertical bar* represents a tide gauge that directly samples S, while an altimeter samples N

Sometimes, a quantity known as *relative sea level change* (or RSL) is employed. RSL is commonly used in paleo-sea-level studies (see, e.g., Tushingham and Peltier 1992), where at one given location it is convenient to refer sea level to the present datum. At a given time t_{BP} before present (BP), *RSL* stems from the difference

$$\mathrm{RSL}(\omega, t_{\mathrm{BP}}) = SL(\omega, t_{\mathrm{BP}}) - SL(\omega, t_{\mathrm{p}}), \tag{3}$$

where SL is given by (1) and t_{p} is present time. By its own definition, RSL vanishes at present, i.e., $\mathrm{RSL} = 0$ for $t_{\mathrm{BP}} = t_{\mathrm{p}}$. Sea-level change and relative sea-level change are tightly related, since Eqs. (3) and (2) give

$$\mathrm{RSL}(\omega, t_{\mathrm{BP}}) = S(\omega, t_{\mathrm{BP}}) - S(\omega, t_{\mathrm{p}}). \tag{4}$$

Equation (2) can be written in a new form introducing the *sea surface variation*

$$N(\omega, t) = R_{\mathrm{ss}}(\omega, t) - R_{\mathrm{ss}}(\omega, t_0), \tag{5}$$

also referred to as *absolute sea-level change*, and the *vertical displacement* of the solid surface of the Earth

$$U(\omega, t) = R_{\mathrm{se}}(\omega, t) - R_{\mathrm{se}}(\omega, t_0), \tag{6}$$

which along with (1) give

$$S(\omega, t) = N(\omega, t) - U(\omega, t), \tag{7}$$

that represents the native form of the SLE. It states that S, directly observable using tide gauge instruments, stems from the difference between N, detectable from satellite altimetry, and U, measured from global positioning system (GPS) receivers (Wöppelmann et al. 2007; Wöppelmann and Marcos 2016; Pfeffer and Allemand 2016). Hence, it could be more appropriate to dub it "tide gauge equation". It is clear from (7) that knowledge of S does not suffice to determine N, and vice-versa, unless U is known.

Often, in GIA studies, the time derivatives (i.e., rates) of S, U and N evaluated at present time are considered, especially when dealing with GIA corrections to geodetic observations. Hereinafter, these quantities will be indicated as $\dot{S}$, $\dot{U}$ and $\dot{N}$, respectively, and the same notation will be employed to denote the rate of change of any other relevant variable.

3 The "Sea-Level Equation" for GIA Modeling

In this introduction to the SLE, I will largely follow the fundamental work of Farrell and Clark (1976). However, a different notation and sometimes different definitions will be adopted, more akin to the current geophysical literature on GIA. The reader is also referred to Peltier and Andrews (1976), Tushingham and Peltier (1991), Milne and Mitrovica (1996), Clark et al. (1978), Milne and Mitrovica (1998), Lambeck et al. (2003), Mitrovica and Milne (2003), Peltier (2004), Spada and Stocchi (2006), Tamisiea (2011) and Mitrovica et al. (2011) to gain more insight into the SLE, its physical meaning and the methods of solution. In some of our numerical simulations based on the SLE that will be presented in the following, shorelines migration and Earth rotation will be taken into account. However, for the sake of brevity, this theory Section illlustrates the SLE theory in the particular case of fixed shorelines and no-marine-based ice, following the "classical" GIA theory of Farrell and Clark (1976). Furthermore, details will not provided on how the SLE can be modified to account for Earth rotation. Nevertheless, it is important to note that during the last two decades there have been significant developments which have extended the classical theory in several aspects concerning the definition of the water load and the effects of Earth rotation on sea level.

The extension of the sea-level theory of Farrell and Clark (1976) to the case of shoreline migration and marine-based ice has been quite controversial. The subject has been reviewed by Mitrovica (2003) by scrutinizing some of the "sea-level theories" published in the literature to year 2003 (Johnston 1993; Yokoyama et al. 2000; Peltier and Drummond 2002). For an overview of the theory and for the details of the numerical implementation of the generalized SLE, the reader is referred to Mitrovica and Milne (2003) and Kendall et al. (2005), respectively, while for the description of the theory adopted by the ANU group and implemented in the SLE solver CALSEA, see Lambeck et al. (2003). Furthermore, the traditional theory widely used for studying the effects of glaciations on Earth rotation (Sabadini and Peltier 1981; Wu and Peltier 1984), often employed in GIA predictions (see, e.g., Milne and Mitrovica 1998), has also been deeply revised (Mitrovica et al. 2005; Nakada 2009; Cambiotti et al. 2010). The whole subject, which is relevant for the study of TPW on various timescales, has been reviewed in Mitrovica and Wahr (2011). At the heart of the revised rotational theory is the rejection of the assumptions that the Earth's lithosphere is perfectly elastic on long timescales and that the Earth's ellipticity can be suitably approximated by a model-dependent "tidal fluid Love number". The adoption of the effectively observed ellipticity in the linearized Liouville equations for Earth rotation (Ricard et al. 1993) has significant consequences. In particular, it leads to a systematic reduction of the rates of polar motion and, consequently, of the present-day effect of GIA on geodetic variations (Mitrovica and Wahr 2011).

Although elegant and appealingly simple, the symbolic form (7) of the SLE is unsuitable for applications. In what follows, taking inspiration from fundamental results about GIA (Farrell and Clark 1976; Clark et al. 1978; Tushingham and Peltier 1991; Mitrovica and Milne 2003), an explicit form of the SLE useful for numerical modeling of GIA is obtained. According to Farrell and Clark (1976) and Tamisiea (2011), the absolute sea-level change is

$$N(\omega, t) = G + c, \tag{8}$$

where c is a yet undetermined function of time, and by Brun's formula (Heiskanen and Moritz 1981) the *geoid height variation* is

$$G(\omega, t) = \frac{\Phi}{\gamma}, \tag{9}$$

where $\Phi(\omega, t)$ represents the *total variation of gravity potential*, γ being the reference surface gravity. Φ includes effects from mass redistribution at the surface and inside the Earth and from the deformations that these induce. It also accounts for variations in the centrifugal potential due to changes in Earth rotation (Milne and Mitrovica 1998). The constant term c in Eq. (8) is introduced since the sea surface does necessarily remain on the same equipotential surface as the volume of the oceans changes (Farrell and Clark 1976). Hence, $c(t)$ allows us to track the sea surface as a function of time (Tamisiea 2011). Substitution of Eq. (8) into (7) gives the SLE in the new form

$$S(\omega, t) = \frac{\Phi}{\gamma} - U + c. \tag{10}$$

The constant c is determined by imposing the constraint of mass conservation. Since the mass of the solid Earth is constant, the total mass of the system composed by the ice sheets and the oceans must be the same in the reference and in the current states, i.e.:

$$m_{\mathrm{i}}(t) + m_{\mathrm{o}}(t) = 0, \tag{11}$$

where

$$m_{\mathrm{i}}(t) = \int_i \rho_{\mathrm{i}} I \mathrm{d}A \tag{12}$$

is the mass variation of the ice sheets, where ρ_{i} is the ice density, dA is the area element, the integral is over the ice-covered regions, and

$$I(\omega,t) = T(\omega,t) - T_0(\omega,t_0), \tag{13}$$

is the variation of the thickness T of the continental ice sheets. Furthermore,

$$m_o(t) = \int_o \rho_w S \mathrm{d}A \tag{14}$$

is the mass variation of the oceans, over which the integration is performed, and ρ_w is water density. Substituting (12) and (14) into (11) using (10) gives:

$$c(t) = S_e - \left\langle \frac{\Phi}{\gamma} - U \right\rangle, \tag{15}$$

where

$$\langle \cdots \rangle = \frac{1}{A_o} \int_o (\cdots) \mathrm{d}A \tag{16}$$

indicates the average over the surface of the oceans with (constant) area A_o, and where

$$S_e(t) = -\frac{m_i}{\rho_w A_o} \tag{17}$$

represents a spatially uniform term often referred to as *"eustatic" sea-level change*, whose meaning in this context will be discussed in Sect. 4. From (10) and (15), the SLE takes the somewhat more explicit form

$$S(\omega,t) = \left(\frac{\Phi}{\gamma} - U \right) + S_e - \left\langle \frac{\Phi}{\gamma} - U \right\rangle. \tag{18}$$

Following Farrell and Clark (1976), U and Φ can be obtained by spatiotemporal convolution with the *surface load variation*

$$\mathcal{L}(\omega,t) = \rho_i I + \rho_w S \mathcal{O}, \tag{19}$$

where the two terms on the right-hand side are associated with the waxing and waning of the grounded ice sheets (*ice load*) and with the redistribution of meltwater in the ocean basins (*meltwater load*), respectively, and

$$\mathcal{O}(\omega) = \begin{cases} 1, & \text{if } \omega \in \text{oceans} \\ 0, & \text{if } \omega \in \text{land} \end{cases} \tag{20}$$

is the *ocean function*. From (19), vertical displacement stems from two terms, i.e.,

$$U(\omega,t) = \rho_i G_u \otimes_i I + \rho_w G_u \otimes_o S, \tag{21}$$

where G_u is the Green's function for vertical displacement, and $\otimes_i$ and $\otimes_o$ are spatiotemporal convolutions over the ice- and water-covered regions, respectively (Farrell and Clark 1976). Similarly, the total variation of the gravity potential is

$$\Phi(\omega,t) = \rho_i G_\phi \otimes_i I + \rho_w G_\phi \otimes_o S, \tag{22}$$

where G_ϕ is the corresponding Green's function. Explicit expressions for the elastic and viscous components of G_u and G_ϕ, involving the load-deformation coefficients (LDCs) $h'(t)$ and $k'(t)$ (Farrell 1972; Wu and Peltier 1982), are given in, e.g., Spada and Stocchi (2006). The quantity that enters directly into the SLE is the combination

$$\frac{G_{\mathrm{s}}}{\gamma}(\omega,t)=\frac{G_{\Phi}}{\gamma}-G_{\mathrm{u}}, \tag{23}$$

known as "sea-level Green's function".

Since the Earth is not expanding and its total mass is conserved

$$U_{00}(t)=\Phi_{00}(t)=0, \tag{24}$$

where $U_{00}(t)$ and $\Phi_{00}(t)$ are the degree $l=0$ and order $m=0$ harmonic components of the spherical harmonic expansions of U and Φ, respectively (Spada and Stocchi 2006). Denoting by

$$\overline{(\cdots)}=\frac{1}{A_{\mathrm{e}}}\int_{e}(\cdots)\mathrm{d}A \tag{25}$$

the average over the whole surface of the Earth, Eq. (24) implies

$$\overline{U}(\omega,t)=\overline{\Phi}(\omega,t)=0. \tag{26}$$

It is to be remarked, however, that the relative sea-level change and the absolute sea-level change do not vanish when averaged over the whole Earth. In fact, using (7) and (8) and the expression for c (15) yields

$$\overline{S}(\omega,t)=\overline{N}(\omega,t)=c \tag{27}$$

for both fields.

By substitution of Eqs. (21) and (22) into (18) one obtains

$$S(\omega,t)=+\frac{\rho_{\mathrm{i}}}{\gamma}G_{\mathrm{s}}\otimes_{\mathrm{i}}I+\frac{\rho_{\mathrm{w}}}{\gamma}G_{\mathrm{s}}\otimes_{\mathrm{o}}S+S_{\mathrm{e}}-\frac{\rho_{\mathrm{i}}}{\gamma}\langle G_{\mathrm{s}}\otimes_{\mathrm{i}}I\rangle-\frac{\rho_{\mathrm{w}}}{\gamma}\langle G_{\mathrm{s}}\otimes_{\mathrm{o}}S\rangle, \tag{28}$$

which represents the SLE in a form amenable for geophysical applications. This form of the SLE, first obtained by Farrell and Clark (1976), is sometimes referred to as *"gravitationally self-consistent"*, since the sea-level variations predicted by the SLE are consistent with the variations of the gravitational field induced by the time-evolving surface loads (Peltier 1989; Milne and Mitrovica 1996). A *"topographically self-consistent"* (Peltier 1994) version of the SLE can be obtained also allowing for the horizontal migration of shorelines during the whole period since the LGM, until they would fit their present shapes and the present topography (Lambeck et al. 2003; Mitrovica and Milne 2003; Kendall et al. 2005). It is worth noting that Eq. (28) can be applied to GIA problems of any characteristic timescale, ranging from millennia to decades. These two aspects of GIA will be addressed in Sects. 5 and 6 below.

4 Applications of the Sea-Level Equation

Assuming that the melting history of the ice sheets and the Earth's rheology are given, the basic unknown of the SLE (28) is sea-level change S. Once the SLE is solved for S, various geophysical quantities become accessible, such as, for example, the bedrock displacement U (from Eq. 21), the absolute sea-level change N (from Eqs. 8, 15 and 22), and the total variation of gravitational potential Φ induced by surface loading (from Eq. 22). Using an expression similar to (21), the horizontal displacements of the Earth's surface in response to GIA can be also retrieved (Spada and Stocchi 2006).

However, even in the most favorable practical geophysical contexts, the spatial and temporal details of I are seldom exactly known and Earth's rheology is not constrained a priori. Hence, the SLE is generally used iteratively as a tool for constraining the ice thickness variations, starting from observations of RSL across a given region, and from an a priori reasonable guess of the rheological profile of the mantle based for example on classical PGR studies (Haskell 1935, 1936). Since the seminal work of Farrell and Clark (1976), the iterative trial-and-error process characterizing the solution of the SLE has been employed in a number of GIA studies on various spatiotemporal scales (Nakada and Lambeck 1988; Mitrovica and Milne 2003; Peltier 2004; Spada and Stocchi 2007). Their final outcome is often a set of admissible rheological profiles and melting scenarios for the ice sheets that are consistent with a set of observations, which normally involve sea-level variations.

The SLE (28) has the form of an implicit (or integral) equation, since the unknown function S appears explicitly on the left-hand side, but it is also embedded in the spatiotemporal convolution integrals on the right-hand side. For this reason, the SLE cannot be solved explicitly unless some drastic simplifying assumptions are made (Spada and Stocchi 2006), as discussed below. The SLE is a linear equation as long as shorelines are not allowed to migrate horizontally, i.e., if the ocean function $\mathcal{O}$ (20) and consequently the area of the surface oceans A_o are not time-dependent Mitrovica and Milne (2003). Sometimes, the terms only containing function I on the right-hand side of Eq. (28) are collectively referred to as the *glacial-isostatic* (gi) component of S, while those depending on S would define the *hydro-isostatic* (hi) component. This separation, which would allow us to write the SLE in the symbolic (and appealing simple) form

$$S(\omega, t) = S_{\mathrm{gi}} + S_{\mathrm{e}} + S_{\mathrm{hi}} \tag{29}$$

is, however, illusory, since the SLE is implicit in Mitrovica and Milne (2002), i.e., S_{hi} depends on the whole S.

A closed-form solution of the SLE is possible in a very special, but important, case. Assuming for one moment that the four terms containing the sea-level Green's function G_s are simply omitted in (28), the SLE reduces to:

$$S(\omega, t) = S_{\mathrm{e}}, \tag{30}$$

showing that, in this case, sea-level change would match the spatially uniform *eustatic* value (17) everywhere. Indeed, following Spada and Stocchi (2006), condition $G_s = 0$ is equivalent to assuming *i)* a perfectly rigid Earth ($G_u = 0$, or $U = 0$), and *ii)* that the mass redistribution of the ice sheets and of the ocean masses are not perturbing the Earth's gravity field ($G_\phi = 0$, or $\Phi = 0$). I note that setting $I = 0$ in the SLE (28) (i.e., assuming for one moment stationary ice sources as shown in Farrell and Clark (1976)), would not ensure eustatic sea-level variations. Indeed, the self-attraction of the ocean masses and the yielding of the ocean floor described by the remaining terms in the SLE would produce geographically variable sea-level patterns, even in the absence of ice load. Following Farrell and Clark (1976), it is worth mentioning that the term *eustatic* has received many qualifications (Dott 1992) since the word was introduced by Suess in 1906. For this reason, some authors prefer to avoid using this term. In Chapter 13, the Fifth Assessment Report (AR5) of the Intergovernmental Panel on Climate Change (IPCC) (Church et al. 2013), devoted to Sea Level Change, the term *eustatic* is not employed. Here, the term *eustatic* denotes the sea-level variations that one would observe for $G_s = 0$ (rigid, non-gravitating Earth), assuming fixed shorelines. In such ideal conditions, the SLE predicts an eustatic sea-level change. This straightforward observation clarifies the fundamental reasons for which GIA is responsible for geographically variable sea-level variations, namely the deformations of the Earth induced by surface loads (governed by G_u) and the gravity field variations induced by the same deformations and from the direct gravitational effects of the loads (associated with G_ϕ) (Farrell and Clark 1976). These effects, acting simultaneously, define several regions over the Earth's surface that are characterized by Holocene RSL curves of comparable shape (sometimes referred to as "Clark's zones", see Clark et al. 1978, Clark and Lingle 1979, Lambeck and Chappell 2001, Stocchi and Spada 2007, Khan et al. 2015). At present time, they also define the geometry of the so-called sea-level fingerprints (Plag and Jüettner 2001; Bamber and Riva 2010; Mitrovica et al. 2011; Spada et al. 2013), which characterize the patterns of the present-day effects of GIA, and that will be described in detail below.

The eustatic solution to the SLE (30) generally constitutes a very poor approximation to the actual solution. However, it is worth noting that a general and elegant relationship exists between S_e and the solution of the SLE, which has been explicitly illustrated in Farrell and Clark (1976). Taking the ocean-average of both sides of (28) and assuming constant densities gives

$$\langle S(\omega,t)\rangle = +\frac{\rho_i}{\gamma}\langle G_s \otimes_i I\rangle + \frac{\rho_w}{\gamma}\langle G_s \otimes_o S\rangle + \langle S_e\rangle - \frac{\rho_i}{\gamma}\Big\langle\langle G_s \otimes_i I\rangle\Big\rangle - \frac{\rho_w}{\gamma}\Big\langle\langle G_s \otimes_o S\rangle\Big\rangle, \tag{31}$$

which, since $\langle\langle f\rangle\rangle = \langle f\rangle$ for any function $f = f(\omega)$ and $\langle k(t)\rangle = k(t)$ for any spatially constant function $k(t)$, provides the result

$$\langle S(\omega,t)\rangle = S_e, \tag{32}$$

showing that the spatially averaged sea-level variation in general dynamic conditions (i.e., deformable and gravitating Earth), matches the eustatic value that one would observe in static conditions (i.e., rigid and non-gravitating Earth). Equation (32) is a simple consequence of mass conservation. It is independent of the ice sheets chronology, of the rheology of the Earth's mantle, of the shape of the coastlines and their possible time evolution, and of the existence of a rotational feedback on sea level. Furthermore, it holds regardless of the time scale of the GIA process.

Recalling the definition of eustatic sea-level change (17), result (32) implies that, from the knowledge of the time-dependent spatial average $\langle S\rangle$, one would immediately obtain the time history of the mass balance of the ice sheets (i.e., $m_i(t)$) , without invoking any other geophysical observation (this argument assumes that the area of the oceans did not change significantly since the LGM). Unfortunately, the geographically sparse sea-level observations from the Holocene (see, e.g., Tushingham and Peltier 1991) are far from permitting a reliable estimate of $\langle S\rangle$, which prevents a straightforward assessment of the time evolution of the major continental ice sheets during the last millennia. This is inferred, instead, using evidence from locations that are in the remote far field with respect to the regions that were formerly deeply covered by ice at the LGM $\sim$21,000 years BP, where it can be assumed that the effects of isostatic disequilibrium, governed by the terms containing the sea-level Green's function G_s in Eq. (28), are small compared to the eustatic component of sea-level rise (Peltier 2004).

Because of the integral nature of the SLE, a successive substitution approach to its solution is generally adopted. This holds for both the traditional sea-level theory (Farrell and Clark 1976) and for the more recent theories that account for migration of shorelines,

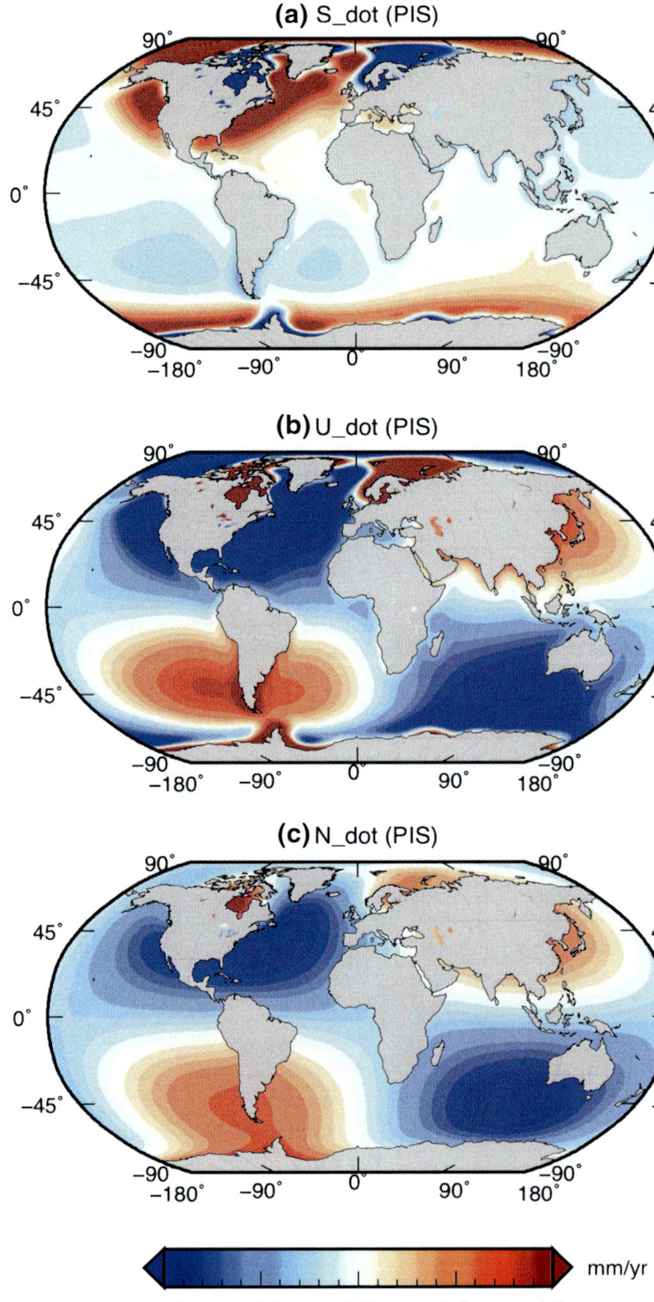

Fig. 2 Past ice sheet (PIS) fingerprints for rates of relative sea-level (**a**), of vertical displacement (**b**), and of absolute sea-level change (**c**), computed by means of SELEN (Spada et al. 2012), for model ICE-5G(VM2) (Peltier 2004). The equal-area grid (Tegmark 1996) used for these maps and for all the following GIA maps have a spacing of ~ 50 km

Earth rotation, and marine-based ice (Mitrovica and Milne 2003; Lambeck et al. 2003; Kendall et al. 2005). The method is borrowed from the theory of linear integral equations in one variable, and it is usually referred to as Neumann series method (Jerri 1999). Indeed, the structure of the SLE (28) is analogous to that of an integral, non-homogeneous Fredholm equation of the second kind:

$$u(x) = f(x) + \lambda \int_a^b K(x,x')u(x')\mathrm{d}x', \tag{33}$$

where function u is the unknown (which here plays the role of S), λ is a constant, the integration interval (a, b) is fixed by analogy with the fixed shoreline approximation in (28), $f(x)$ represents all the terms that do not depend on S and includes the source term I, and the kernel K, which is of convolution type, i.e., $K(x,x') = K(x - x')$, represents the sea-level Green's function G_s (note that the variable I would depend upon S when the interactions between the ice sheet and oceans are taken into account; see Boer et al. 2014 and Konrad 2015). Although the SLE is actually three-dimensional (S is a function of position on the Earth's surface and of time), while (33) is one-dimensional, the approach to the solution is similar in the two cases. In a zeroth order approximation, a reasonable guess $u^{(0)}$ is assumed and the integral are evaluated numerically, giving a first-order approximation $u^{(1)}$. The process is iterated until it converges, obtaining in this way various approximations to the real solution u. In view of Eq. (32), the obvious first guess will be $u^{(0)} = S_e$, which is not—in general—the right solution of the SLE, but it is certainly a valid solution *on the average*. The convergence of the iterative approach to the true solution is rather fast; normally three to five iterations are required (Spada and Stocchi 2007; Spada et al. 2012).

5 GIA Fingerprints for the Melting of Late-Pleistocene Ice Sheets

Figure 2 shows the present-day rates of sea-level change $\dot{S}$ (a), of vertical displacement $\dot{U}$ (b) and of absolute sea-level change $\dot{N}$ (c) for model ICE-5G(VM2) (Peltier 2004), obtained using the SLE solver SELEN (Spada et al. 2012). SELEN implements the pseudo-spectral approach to solving the SLE (Mitrovica and Peltier 1991), which has now superseded the "finite element" method employed in, e.g., Clark et al. (1978), Peltier et al. (1978) and Wu and Peltier (1983). Here, the finely layered viscosity profile VM2 has been averaged over three layers (shallow upper mantle, transition zone and lower mantle). An incompressible rheology is assumed, and the effects of Earth rotation are taken into account adopting the classical theory described in Milne and Mitrovica (1998). Note that the three maps, which represent the three fundamental "fingerprints" for the past ice sheets (PIS) are not linearly independent, because of the SLE (7). In the context of GIA, the term "sea-level fingerprint" was first coined in Plag and Jüettner (2001) and adopted in numerous studies since then (Mitrovica et al. 2001; Clark et al. 2002; Bamber and Riva 2010; Mitrovica et al. 2011; Spada et al. 2013). Basically, the fingerprints represent spatial patterns showing S, U or N (or their rates of change, as in Fig. 2) in response to the melting of continental ice sheets (Tamisiea 2011). Potentially, their geometry can be useful to identify the location and the magnitude of the ice sources that are responsible for their shapes (Douglas 2008), although the problem of their visibility in maps of sea-level change is far from being completely solved at present (Kopp et al. 2010; Cazenave and Llovel 2010; Spada and Galassi 2016). The problem of the effective "strength" of the GIA fingerprints in comparison with other sources of contemporary sea-level change will be addressed in Sect. 7.

Since in the GIA model ICE-5G(VM2), the melting of the PIS is assumed to end ~4000 years BP, the fields in Fig. 2 only manifest delayed viscoelastic deformations, gravity changes and rotational effects caused by the slow isostatic readjustment of the Earth (Spada 1992). The melting of present-day ice sheets will produce geometrically different fingerprints, which will be discussed in Sect. 6. Some of the features shown on the maps of Fig. 2 have a straightforward interpretation. For instance, the signature of the PGR associated with the un-loading of the former ice sheets (Cathles 1975) is manifest by broad uplifting areas across the polar regions in both hemispheres and by the surrounding collapsing forebulge regions. Some other features are, however, of more difficult interpretation, being associated with the phenomena of "ocean siphoning" and "continental levering", whose mechanisms have been explained invoking hydro-isostatic effects associated with the meltwater loading (Mitrovica and Milne 2002). The very large scale lobes shown in particular by $\dot{U}$ and $\dot{N}$, characterized by a harmonic degree $\ell = 2$ and order $m = 1$ tesseral symmetry, are caused by the rotational response of the Earth to the isostatic disequilibrium forced by the melting of the continental ice sheets (Milne and Mitrovica 1998; Peltier 2004). Adopting a purely eustatic theory and assuming a non-rotating Earth, all the sea-level fingerprints shown in Fig. 2 would simply vanish.

For the PIS sea-level fingerprints in Fig. 2a, one numerically obtains $\langle \dot{S} \rangle \approx 3 \times 10^{-6}$ mmyear^{-1}. In Tamisiea (2011), similar results are found using the deglaciation model ICE-3G (Tushingham and Peltier 1991, 1992), with values $\langle \dot{S} \rangle < 2 \times 10^{-3}$ mmyear^{-1} for a broad range of viscosity profiles. These findings strongly suggest a vanishing $\langle \dot{S} \rangle$. This is confirmed by the SLE theory outlined above. In fact, from (17):

$$\dot{S}_e = -\frac{\dot{m}_i}{\rho_w A_o} = 0, \tag{34}$$

where it has been assumed that A_o and ρ_w are constant and I have taken into account that in model ICE-5G(VM2) the rate of present-day mass loss is $\dot{m}_i = 0$. Hence, (32) gives

$$\langle \dot{S} \rangle = \dot{S}_e = 0. \tag{35}$$

This result shows that the melting of PIS currently generates *only* regional variations of relative sea level that average to zero globally over the oceans (in a sense, the PIS fingerprint for $\dot{S}$ is "massless"). Hence, within the assumption of fixed shorelines (i.e., $\dot{A}_o = 0$), GIA from the melting of the PIS is not contributing to present-day eustatic sea-level variations, a result that is independent of assumptions about the Earth's rheological profile (however, allowing for the horizontal migration of shorelines would imply $\dot{S}_e = m_i \dot{A}_o / \rho_w A_o^2 \neq 0$ even for $\dot{m}_i = 0$). I note that, due to the marked variability of the sea-level fingerprint for S, averaging the rate of relative sea-level change over the coastlines would produce significantly different results. I find a value $\langle \dot{S} \rangle = -0.7$ mmyear^{-1} for the global coastlines, and $\langle \dot{S} \rangle = -0.8$ mmyear^{-1} at the locations of all the revised local reference tide gauges of the permanent service for mean sea level (PSMSL)[5] with a record length ≥ 50 years.

[5] http://www.psmsl.org.

Table 1 PIS ocean-averaged values of S, U and N obtained for model ICE-5G(VM2), adopting different assumptions in GIA modeling

RF MS	No No	Yes No	No Yes	Yes Yes
$\langle S \rangle$	+0.00	+0.00	−0.17	−0.14
$\langle U \rangle$	−0.23	−0.27	−0.21	−0.24
$\langle N \rangle$	−0.23	−0.27	−0.38	−0.39

Here RF and MS stand for "rotational feedback" and "moving shorelines", respectively. Units are mmyear^{-1} throughout

A vanishing $\langle \dot{S} \rangle$, however, does not imply a vanishing $\langle \dot{N} \rangle$, which highlights the importance of discerning among the various existing definitions of sea level (see Sect. 2 above). In this respect, the only theoretical conclusion one can reach stems directly from the native form of the SLE (7), which gives

$$\langle \dot{N} \rangle = \langle \dot{U} \rangle, \tag{36}$$

where (35) has been explicitly used. By numerically solving the SLE adopting model ICE-5G(VM2), and using (22) and (15) in (8), I obtain the estimate:

$$\langle \dot{N} \rangle = -0.27 \quad \text{mmyear}^{-1} \tag{37}$$

showing that the average rate of *absolute* sea-level change from past ice melting is small, but not negligible, across the oceans. The same holds for the rate of crustal uplift in Fig. 2b. The estimated rate $\langle \dot{N} \rangle$ is not strongly dependent on assumptions about the rheology of the Earth (Tamisiea 2011). Furthermore, including the horizontal migration of shorelines in modeling would not drastically change the results. The numerical results obtained from a few simulations are shown in Table 1.

6 GIA Fingerprints for Contemporary Glacial Melting

Figure 3 shows maps of rate of relative sea-level change $\dot{S}$ (left) associated with the melting of the major contemporary ice sources (hereafter referred to as CIS), namely Greenland (GRE, a), Antarctica (ANT, b), and Glaciers and Ice Caps (GIC, c). The geographical distribution of the latter is taken from Meier (1984). In all the three cases, following Spada and Galassi (2016), it has been assumed that ice melting occurs at a constant rate and uniformly across the CIS. I note that this is a poor approximation that may be valid only as long as the fingerprints are observed from a relatively large distance, since in this case the effective distribution of the ice sources plays a minor role (Bamber and Riva 2010; Spada et al. 2012).

Similarly to the GIA fingerprints for the PIS, shown in Fig. 2, those for the CIS considered here are obtained as solutions of the SLE, so they are based on the same physics. However, there are some important differences. First, since for the CIS the characteristic timescale of melting is relatively short (from a few decades to one century), the rheological behavior of the Earth can be approximated to that of an elastic body (Slangen 2012; Spada et al. 2013, 2014). Therefore, the fingerprints in Fig. 3 are not sensitive to mantle viscosity,

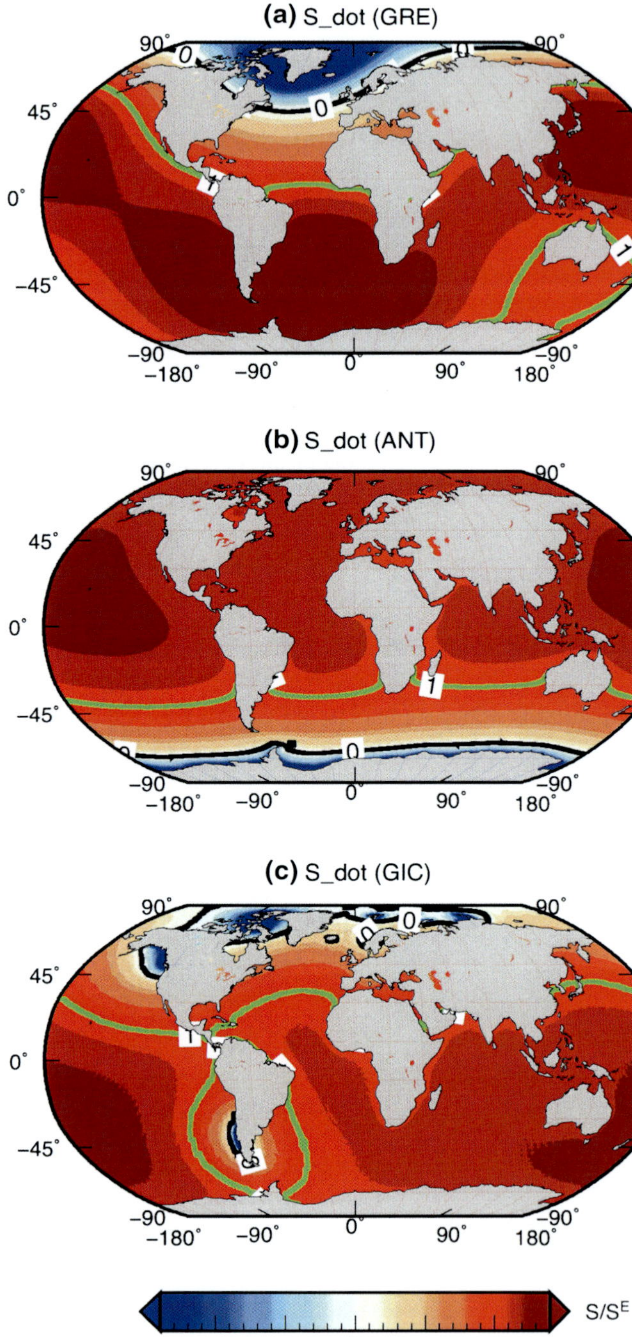

since the Green's functions only depend on the elastic moduli and on the density profile of the Earth. These are consistent with the seismological preliminary reference Earth model (Dziewonski and Anderson 1981). Second, contrary to the PIS fingerprints shown in

◀**Fig. 3** CIS fingerprints for relative sea-level obtained using SELEN Spada et al. (2012) and assuming a uniform mass loss from Greenland (**a**), Antarctica (**b**) and Glaciers and Ice Caps (**c**). The values are normalized dividing by the eustatic (uniform) rate of sea-level change, $\dot{S}_e$. The *green contour* denotes the eustatic value ($\dot{S}/\dot{S}_e = 1$); a *black contour* marks the neutral regions where relative sea-level is not changing ($\dot{S}/\dot{S}_e = 0$)

Fig. 2, those for CIS are not "massless" (i.e., $\dot{m}_i \neq 0$). As a consequence, from (32), the ocean-average of the fingerprint for the rate of present-day relative sea-level change is

$$\langle \dot{S} \rangle = \dot{S}_e = -\frac{\dot{m}_i}{\rho_w A_o} \neq 0. \tag{38}$$

Since the ocean-averages of the CIS fingerprints in Fig. 3 do not vanish, they have been normalized by $\dot{S}_e$ to better appreciate deviations from the predictions of the eustatic theory. Despite the different spatial patterns of sea-level change associated with the three CIS sources, all the fingerprints share a common feature, i.e., a marked sub-eustatic value in the vicinity of the sources (blue hues) and super-eustatic amplitudes at larger distances (red). Assuming a rigid Earth, uniform non-gravitating oceans (i.e., the mutual attraction between different portions of the ocean mass is neglected) and a localized (point-like) ice source, a global pattern of sea-level change similar to that suggested by the numerical results shown was predicted already by Woodward in 1888, in his pioneering work on the form and position of mean sea-level. In particular, his theory predicts a rate of relative sea-level change

$$\dot{S}_w(\psi) = \frac{a}{m_e}\left(\frac{1}{2\sin(\psi/2)} - \frac{\rho_e}{3\rho_w} - 1\right)\dot{m}_i, \tag{39}$$

where the first term in brackets on the right-hand side represents the effect of gravitational attraction between the ice source and the ocean masses, the second describes the eustatic component of sea-level change and the last ensures mass conservation. In (39), a is the radius of the Earth, m_e is its mass, ψ is the angular separation between a given location and the ice source, ρ_e is the average Earth's density and ρ_w is the density of water. Of course, since the Earth is assumed to be rigid ($\dot{U}_w = 0$), in Woodward's theory, absolute and relative sea-level change coincide

$$\dot{N}_w = \dot{S}_w. \tag{40}$$

Due to the point-like source and to the spherical symmetry of the Earth and of the oceans, in Woodward's problem $\dot{S}_w$ is only ψ-dependent and there are no longitudinal effects on sea level. As shown in detail in Spada and Stocchi (2006), for $\dot{m}_i \leq 0$, corresponding to an instantaneous melting of the point mass, formula (39) would predict a sea-level fall ($\dot{S}_w \leq 0$) in the vicinity of the source (for angular distances $\psi \lesssim 10°$), caused by the decreased gravitational pull acting between the ice source and the surrounding oceans. To conserve mass, sea-level would rise elsewhere, with values below the eustatic level approximately for $\psi \lesssim 60°$, and above for $\psi \gtrsim 60°$. Although in the numerical results of Fig. 3 the elastic uplift enhances the sea-level fall in the near field of the load and rotational effects emphasize the rise in the far field, the pattern predicted by Woodward's theory can be easily identified globally, which clearly confirms the importance of gravitational effects in GIA modeling (Mitrovica and Milne 2003).

Since for the CIS the rate of mass loss is $\dot{m}_{\rm i} \neq 0$, the relationship between the ocean-averaged absolute sea-level change and vertical displacement is

$$\langle \dot{N} \rangle - \langle \dot{U} \rangle = \dot{S}_{\rm e} \neq 0, \tag{41}$$

which substantially differs from its PIS counterpart (36), and that obviously cannot be solved for $\langle \dot{N} \rangle$ and $\langle \dot{U} \rangle$ simultaneously. To estimate these average values, one must rely upon the numerical solution of the SLE. The CIS fingerprints for $\dot{N}$ and $\dot{U}$ are shown in the left and right columns of Fig. 4, respectively, for the same three spatially homogeneous sources considered in Fig. 3 above (GRE, ANT and GIC). The fingerprints are computed for $\dot{m}_{\rm i} = -100$ Gt year^{-1}, corresponding to a spatially uniform sea-level rise of ~ 0.27 mmyear^{-1} (the numerical equivalence with Eq. 37 is fortuitous). By linearity of the SLE, the fingerprints corresponding to other values of $\dot{m}_{\rm i}$ can be obtained by rescaling these solutions. From Table 2, it can be noted that the average values of the rate of vertical

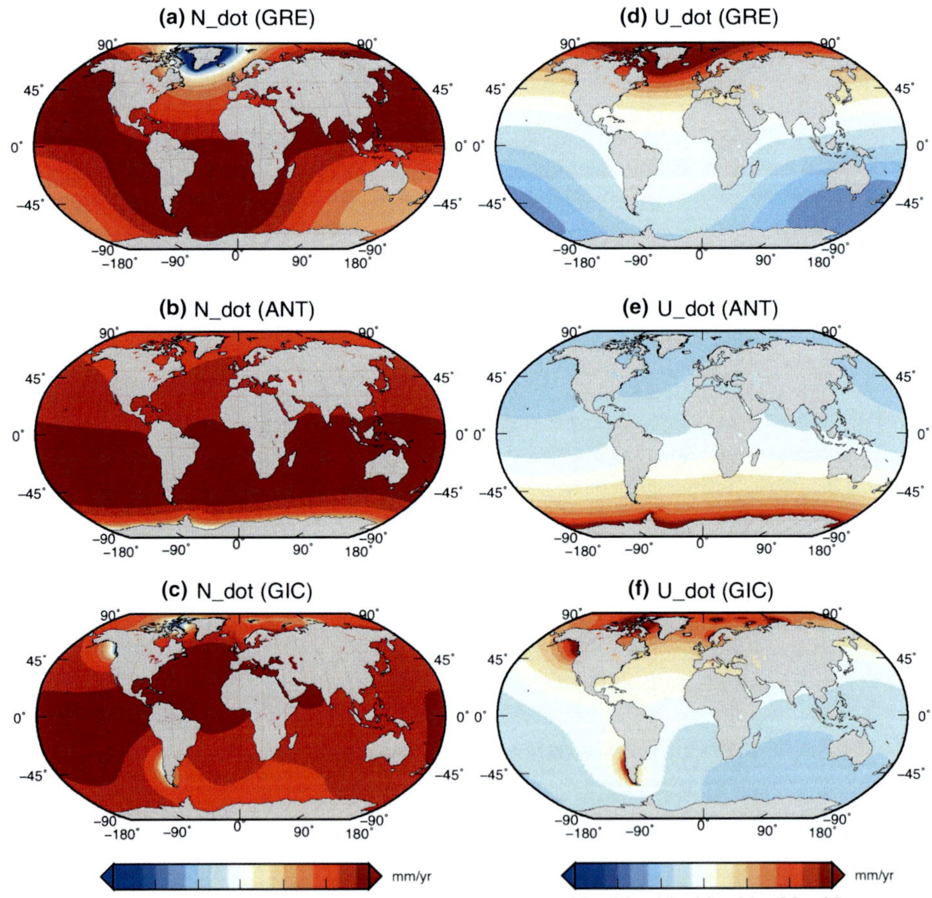

Fig. 4 CIS fingerprints for the rate of absolute sea-level change ($\dot{N}$, *left*) and for the rate of crustal uplift ($\dot{U}$, *right*) associated to the present-day ice melting. The ice sources are the same as in Fig. 3. In all frames, $\langle \dot{S} \rangle = \dot{S}_{\rm e} = 0.27$ mmyear^{-1}

displacement is small compared to the averaged rate of sea-level change. Indeed, to a very high level of precision (≈0.02 mmyear^{-1}), the results for the CIS indicate that

$$\langle \dot{U} \rangle \approx 0, \tag{42}$$

i.e., very close to the whole-Earth average

$$\overline{\dot{U}} = 0, \tag{43}$$

that is expected from the general law (27), valid for both the CIS and the PIS. This finding provides a remarkable rule of thumb

$$\langle \dot{N} \rangle \approx \langle \dot{S} \rangle, \tag{44}$$

valid for all the three CIS sources considered, which indicates that for the elastic fingerprints in Fig. 4 relative and absolute sea-level change are essentially the same on the average (rescaling the mass balance of the ice sources by a factor as large as 5 would only alter this result at the 0.1 mmyear^{-1} level). For Woodward's model, Eq. (44) would hold rigorously, and according to (40) it would be also valid locally.

7 Regional Variability of the GIA Fingerprints

The simplest way to characterize the PIS and CIS fingerprints is by the evaluation of spatial averages over the surface of the oceans. In Sects. 5 and 6, these averages have been shown to provide clues to the meaning of the GIA fingerprints. Furthermore, they have an important role in the definition of the so-called GIA corrections that will be discussed in Sect. 8. Here, to better describe the global fingerprints variability, I first analyze their spatial frequency histograms. Figure 5a, relative to the PIS fingerprints, shows that while $\dot{S}$ takes most of its values in the narrow interval between −0.5 and +0.5 mmyear^{-1}, for $\dot{N}$ and $\dot{U}$ the histograms show a more marked symmetry and a flatter distribution, with some values exceeding ±1 mmyear^{-1}. This difference is caused by the rotational feedback on sea level, associated with a signal of degree $l = 2$ and order $m = 1$ separately affecting $\dot{N}$ and $\dot{U}$ (using the revised rotation theory of Mitrovica and Wahr (2011) would reduce the spread of $\dot{N}$ and $\dot{U}$, however). The effect is less evident on $\dot{S} = \dot{N} - \dot{U}$, since a partial cancelation of the rotational lobes shown by the PIS fingerprints occurs (see Fig. 2). For the CIS fingerprints, shown in Fig. 5b, the variability of $\dot{N}$ is strongly reduced relative to the PIS regardless of the source of ice melt considered. The reason is the less pronounced

Table 2 CIS ocean-averaged values of the fingerprints associated with the melting of present-day sources (GRE, ANT and GIC)

Ice source	GRE	ANT	GIC
$\langle S \rangle$	+0.27	+0.27	+0.27
$\langle U \rangle$	−0.02	−0.01	−0.02
$\langle N \rangle$	+0.25	+0.26	+0.25

A mass balance of −100 Gt year^{-1} for all the sources is assumed, and melting occurs uniformly across the ice masses. The rotational feedback on sea-level is included; units are mmyear^{-1}

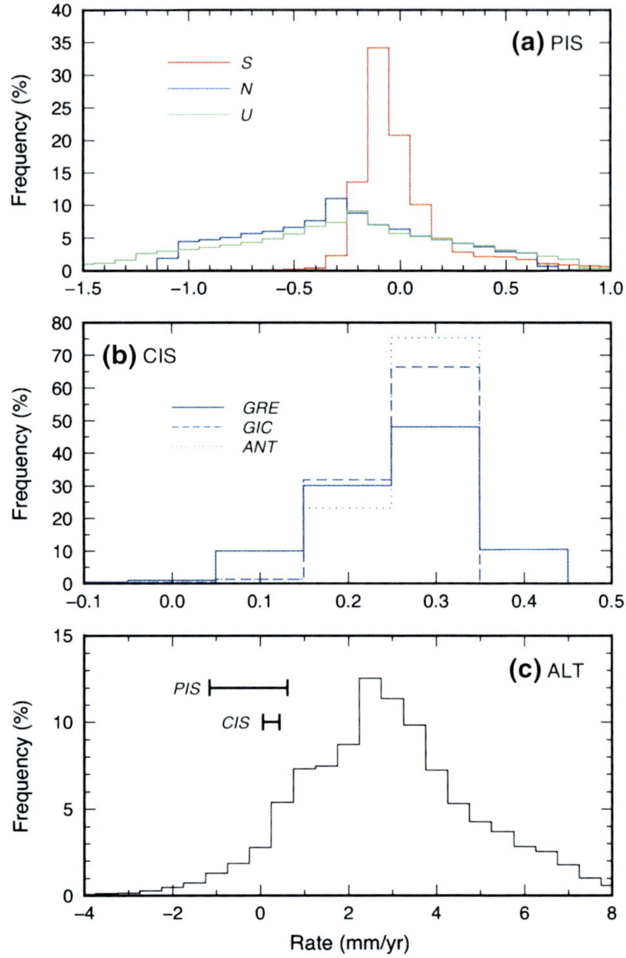

Fig. 5 Frequency histograms of the PIS fingerprints (**a**) and of the CIS fingerprints for the rate of absolute sea-level change $\dot{N}$ (**b**). The ALT frequency histogram for the observed rate of absolute sea-level change is shown in (**c**), where the *horizontal bars* show the 2σ ranges of the distributions of the PIS $\dot{N}$ in **a** and of the CIS $\dot{N}$ for GRE in **b**. Since the grid used in SELEN is equal-area (Tegmark 1996; Spada et al. 2012), the frequency distributions do not change with resolution from the poles to the equator

effects associated with Earth rotational feedback, qualitatively visible in Fig. 4, which for the CIS only acts through the elastic rotational deformations.

Figure 5c shows the spatial frequency histogram for the observed altimetric rate of sea-level change from the AVISO (Archiving, Validation and Interpretation of Satellites Oceanographic data) webpage,[6] for the time period 1993–2012. The frequency histogram for altimetry (ALT) is evidently asymmetrical, something that so far has not received attention but that would probably merit some investigation. The ALT map from which the histogram has been obtained, reproduced in Fig. 6, has the same pixel size of those showing the PIS and the CIS fingerprints in Figs. 3 and 4, respectively. The ALT average rate is ~ 3 mmyear^{-1}, substantially exceeding the $\dot{N}$ rate expected for PIS (~ -0.3 mmyear^{-1}, see last line in Table 1), and for the CIS ($\sim +0.25$ mmyear^{-1} assuming a melting rate of 100 Gt year^{-1} from each of the sources, see Table 2). It is apparent that the PIS and the CIS contribution to $\dot{N}$ is limited to the relatively narrow

[6] http://www.aviso.oceanobs.com.

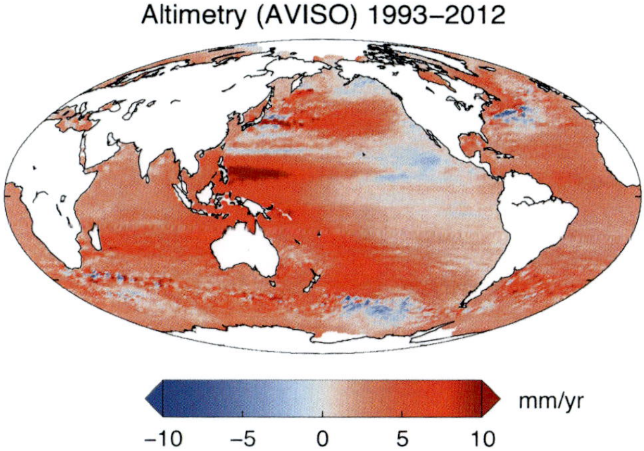

Fig. 6 Time-average rate of absolute sea-level change during 1993–2012 from altimeter data. The source is: Archiving, Validation and Interpretation of Satellites Oceanographic data (AVISO), obtained from http://www.aviso.oceanobs.com/en/data/products/ocean-indicators-products/mean-sea-level.html. The AVISO data, which are not corrected for the effects of GIA, are mapped on the same equal-area grid used for the GIA fingerprints (see caption of Fig. 2). The spatial variability of sea level is essentially associated with the non-uniform changes in ocean thermal expansion and salinity variations (Cazenave and Llovel 2010; Spada and Galassi 2016)

window between -1 and $+1$ mmyear^{-1} (see horizontal segments in Fig. 5c), in the same range of values in which a kink in the ALT histogram is visible. However, it can be verified that the PIS and the CIS sources are not the cause of this anomaly in the histogram. Figure 6 confirms previous observations suggesting a minor role of the PIS and the CIS in the regional variability of the observed rate of sea-level change (Cazenave et al. 2009; Meyssignac et al. 2012; Spada and Galassi 2016), which is dominated by thermosteric and halosteric effects (Levitus 2005; Lombard et al. 2005a, b; Ishii and Kimoto 2009; Levitus et al. 2009; Meyssignac and Cazenave 2012). It has been proposed (Cazenave et al. 2009; Meyssignac 2012; Meyssignac et al. 2012) that the CIS signatures could become visible in the ALT sea-level maps in the future, when the rate of melting of glaciers will further increase in response to global warming (Church et al. 2013).

To quantitatively analyze the spatial pattern of the sea-level fingerprints so far described, I employ the "orthonormal functions" (ON) method (Hwang 1991, 1993), which has been recently implemented in Spada and Galassi (2016). The ON method is an alternative approach to using empirical orthogonal functions (EOFs) (Navarra and Simoncini 2010) and generalizes the traditional spherical harmonic (SH) decomposition method to functions that are not defined over the whole sphere (Wunsch and Stammer 1995; Spada and Galassi 2016). The technique is therefore particularly suitable for an analysis of the altimetric record shown in Fig. 5c, which is undefined across the continents and part of the polar regions. For an arbitrary 2-D function f, the ON expansion is

$$f(\omega) = \sum_{l=0}^{l_{\max}} \sum_{m=0}^{l} \left(\alpha_{\mathrm{f,lm}}\, O_{\mathrm{lm}}(\omega) + \beta_{\mathrm{f,lm}}\, Q_{\mathrm{lm}}(\omega) \right), \tag{45}$$

where l and m are the degree and order , $l_{\max}$ is the truncation degree, O_{lm} and Q_{lm} are the "cosine" and "sine" OF functions, respectively. By the orthonormality of the ONs, the (real) coefficients $(\alpha_{f,lm}, \beta_{f,lm})$ in Eq. (45) can be obtained by numerical quadrature across the domain over which the field f is defined (see Spada and Galassi 2016 for details).

Figure 7 shows the power spectra of the ALT field (7a) and of the sea-level fingerprints considered in this study (7b). In analogy with the SH method (see, e.g., Wieczorek 2007), the spectra have been computed according to

$$S_f(l) = \sum_{m=0}^{l} \left(\alpha_{f,lm}^2 + \beta_{f,lm}^2 \right). \tag{46}$$

The ALT spectrum is clearly red and for $l \geq 6$ it decays with l following a power law $S_f(l) \approx l^{-1.8}$. The power spectra for the PIS and the CIS fingerprints show a less regular behavior with varying l. Their power is concentrated in the low-degree harmonics $(0 \leq l \leq 2)$, corresponding to their ocean-average $(l = 0)$, to the east–west and north–south variance in their patterns $(l = 1)$ and to effects from Earth rotation $(l = 2)$. For $l \geq 3$, they decay very quickly with $\approx l^{-5}$. In general, the PIS fingerprints show a more enhanced power compared to the CIS fingerprints. The GRE and the GIC components exhibit a larger spatial variability because of the more asymmetrical distribution of the corresponding ice sources compared to the nearly zonal ANT. From Fig. 6b, it clearly appears that the power contained in the PIS and CIS fingerprints is in general is several orders of magnitude lower than the one shown by the ALT (gray line) and that this gap increases with l. This demonstrates the difficult (if not impossible) detectability of the PIS and CIS fingerprints in the ALT maps (Spada and Galassi 2016).

8 GIA Corrections to Relative and Absolute Sea-Level Observations

Since virtually all current geodetic observations are potentially affected by GIA, it is very often necessary to perform suitable corrections. This is true, in particular, when one wants to decontaminate the data from the effects of the melting of the PIS to enlighten the contribution of present-day climatic variations. Following the seminal work of Peltier and Tushingham (1989), the decontamination has been usually performed using "1-D" GIA models, i.e., assuming that the Earth's rheology only varies with depth (see list of references in Table 1 of Spada and Galassi 2012). However, there are notable exceptions (Kendal et al. 2006). Here, I briefly review some of the key aspects of the "GIA correction" problem, which, however, would certainly deserve more attention because of its relevance in the climate change sciences. In particular, GIA corrections will be illustrated in some detail in the context of sea-level observations by tide gauges and satellite altimetry. A discussion of GIA corrections of data recovered by the gravity recovery and climate experiment (GRACE) would merit a separate section. For this topic, which has been recently debated (Chambers et al. 2012; Peltier et al. 2015), the reader is referred to Tamisiea (2011) and to references therein. A recent example showing the importance of GIA corrections to GPS observations is given by Serpelloni et al. (2013).

From a historical perspective, the importance of GIA corrections was first recognized in the context of secular sea-level rise. Table 1 in Spada and Galassi (2012), subsequently revised and updated in Spada et al. (2015), reports all previous estimates of global mean sea-level rise (hereinafter GMSLR) from tide gauge observations, published in the literature since the seminal work of Gutenberg (1941). Here, the rate of GMSLR is defined as

Fig. 7 ON power spectra for the ALT field (**a**) and for the PIS and CIS fingerprints considered in this study (**b**), shown as a function of the angular degree l. According to Jean's relation (Jeans 1923; Dahlen and Tromp 1998), the *horizontal wavelength* corresponding to degree l is $\lambda \approx 40,000\,\mathrm{km}/(l + {}^1/_2)$. The *gray curve* in **b** reproduces the ALT spectrum from **a**. *Source* Spada and Galassi (2016)

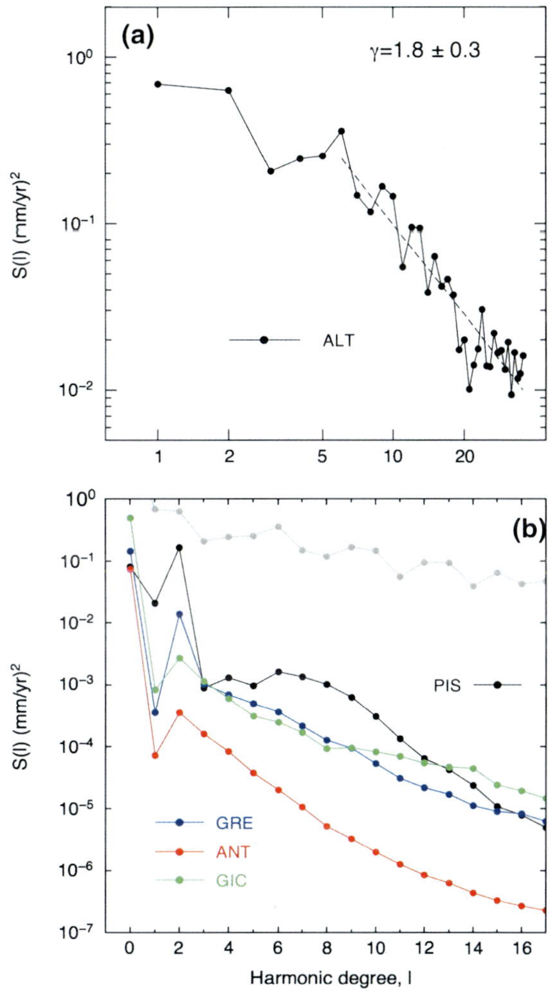

$$\rho = \frac{\dot{V}_o}{A_o}, \tag{47}$$

where $\dot{V}_o$ is the rate of change of the ocean volume (including mass and steric contributions) and A_o is the present-day area of the surface of the oceans, respectively. Estimates of ρ for the twentieth century, reported since the year 2000 (Church et al. 2001; Nakada and Inoue 2005; Church and White 2006; Bindoff et al. 2007; Hagedoorn et al. 2007; Church and White 2011; Spada and Galassi 2012; Church et al. 2013; Hay et al. 2015) point to values in the range between 1.4 and 1.8 mmyear^{-1}, where the uncertainty is associated to a number of factors (some of them are listed below). The mean value of long-term sea-level rise obtained in all the studies in the literature so far is 1.61 mmyear^{-1} and the weighted average of values for which an uncertainty estimate is known, is $1.63 \pm 0.06\,\mathrm{mmyear}^{-1}$

(see Spada et al. 2015 for details on how the average is evaluated). According to the IPCC AR5 (Church et al. 2013), during 1901–1990 the observed global mean sea-level rise has been in the range of 1.3–1.7 $\mathrm{mm year}^{-1}$. It is worth noting that for some authors (Pirazzoli 1986, 1993; Gröger and Plag 1993) ρ has been considered undeterminable, mainly because of the sparse distribution of tide gauges and the marked regional variability shown by contemporary sea-level rise (Spada and Galassi 2012). GIA modeling has been employed to perform corrections to tide gauge observations only since 1989, when global solutions of the SLE adopting realistic assumptions regarding the Earth's rheology became available Peltier and Tushingham (1989). GIA corrections help to obtain estimates of the rate of GMSLR, provided that the set of N_{tg} tide gauges considered has a sufficient global coverage and proper selection criteria have been applied (Douglas 1991, 1997; Spada and Galassi 2012). The application of the GIA correction is of great importance, since a large number of sufficiently long tide gauge records are from regions that were covered by ice at the LGM (Jevrejeva et al. 2014). However, due to the global nature of GIA, no tide gauge record can be assumed to be totally unaffected (Spada and Galassi 2012).

To illustrate the meaning of the GIA correction at tide gauges, following Spada and Galassi (2012), I consider the design equation

$$\dot{S}^k_{\mathrm{obs}} = \dot{S}^k_{\mathrm{pis}} + \dot{S}^k_{\mathrm{cis}} + \dot{S}^k_{\mathrm{ste}} + \dot{S}^k_{\mathrm{oth}}, \tag{48}$$

where $\dot{S}^k_{\mathrm{obs}}$ is the rate of relative sea-level observed at the k-th tide gauge station, $\dot{S}^k_{\mathrm{pis}}$ is the PIS contribution to sea-level change obtained from a specific GIA model (e.g., from the sea-level fingerprint of Fig. 2c), $\dot{S}^k_{\mathrm{cis}}$ denotes the effect from the melting of the CIS, $\dot{S}^k_{\mathrm{ste}}$ is the contribution of steric effects including the thermosteric and halosteric components, and $\dot{S}^k_{\mathrm{oth}}$ represents cumulatively any other causes, including variations in the land water storage, tectonic and seismic contributions, sediments compaction, *et cetera*. In general, the GIA term $\dot{S}^k_{\mathrm{pis}}$ in Eq. (48) depends quite significantly on the model employed to compute it (Guo et al. 2012; Jevrejeva et al. 2014); a discussion about this sensitivity has been given by Spada and Galassi (2012). Since the isostatic disequilibrium induced by the melting of the PIS has a characteristic timescale of a few millennia (see, e.g., Turcotte and Schubert 2014), $\dot{S}^k_{\mathrm{pis}}$ can be effectively considered constant with the possible exception of the tide gauges in the Baltic Sea (Spada et al. 2014). For this reason, GIA corrections are not applied in studies concerned with the contemporary global sea-level acceleration (Woodworth 1990; Douglas 1992; Church and White 2006; Jevrejeva et al. 2008, 2014; Olivieri and Spada 2013; Hogarth 2014; Spada et al. 2015).

Since tide gauges are localized along the global coastlines (see Fig. 8) and the selection criteria applied in global studies impose relatively small N_{tg} values (e.g., $N_{\mathrm{tg}} = 23$ in the study of Douglas (1997) and, similarly, $N_{\mathrm{tg}} = 22$ in Spada and Galassi 2012), averaging Eq. (48) over the oceans is unfortunately unfeasible in a straightforward way. In Spada and Galassi (2012), it is shown that the GIA-corrected observed rates, averaged over the tide gauges

$$\rho'_{\mathrm{tg}} = \frac{1}{N_{\mathrm{tg}}} \sum_{k=1}^{N_{\mathrm{tg}}} (\dot{S}^k_{\mathrm{obs}} - \dot{S}^k_{\mathrm{pis}}) \tag{49}$$

would directly provide the rate of GMSLR (ρ) only under a number of assumptions: (1) $\dot{S}^k_{\mathrm{pis}}$ is evaluated correctly and it is not strongly sensitive to the particular GIA model adopted,

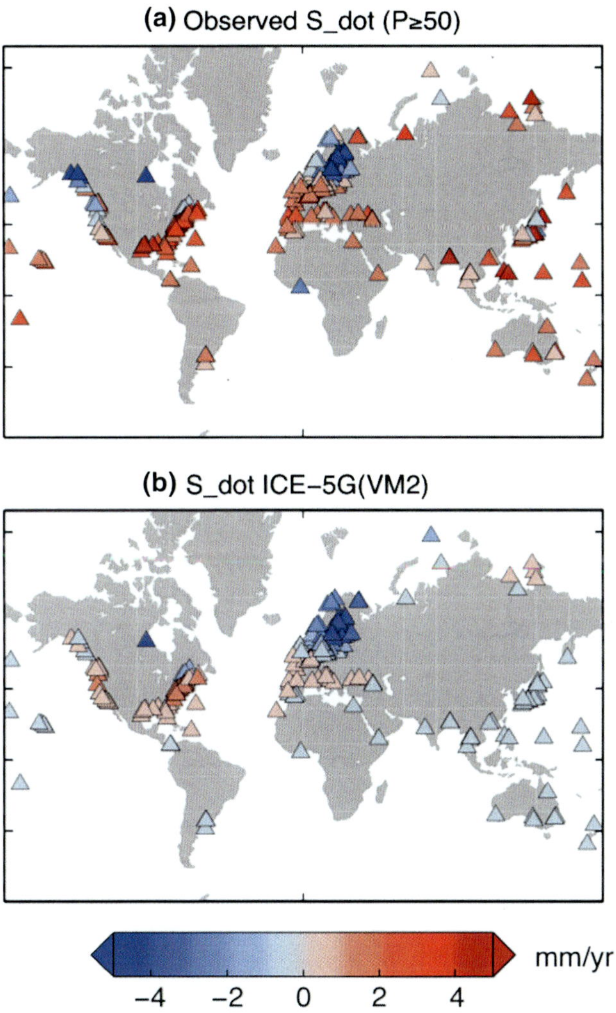

Fig. 8 **a** Rate of relative sea-level change $\dot{S}^{k}_{\text{obs}}$ where k labels the revised local reference PSMSL sites with a record that covers a period of at least 50 years, **b** the GIA correction $\dot{S}^{k}_{\text{psi}}$ at the same locations

(2) the tide gauges records are sufficiently long so that they can be assumed to be representative of global sea-level change (Douglas 2001), (3) the spatial coverage of tide gauges is sufficiently dense to capture the global average sea-level change associated with the melting of the CIS, (4) the steric component of sea-level rise can be considered spatially uniform over the secular timescale and (5) the selected tide gauge sites are tectonically stable or, at least, tectonic deformations average out to zero. These assumptions were certainly not met in the past, and it is also likely that they will never be rigorously met in the future. In view of this, one can tentatively write

$$\rho'_{\text{tg}} = \rho + \dot{\epsilon}_{\text{tg}} \tag{50}$$

where ρ is given by (47) and $\dot{\epsilon}_{\rm tg}$ is a cumulative uncertainty associated with the process of extracting the rate of GMSLR from a specific set of tide gauges. It should be noted that $\dot{\epsilon}_{\rm tg}$ will strongly depend on the choice of the tide gauges, and its amplitude cannot be estimated a priori.

Covering the oceans (see Fig. 6) and involving absolute sea-level variations, the altimetric data have a substantially different nature compared to tide gauge data (see, e.g., Cazenave and Nerem 2004; Cazenave and Llovel 2010). By analogy with (48), the design equation for altimetry reads

$$\dot{N}_{\rm obs}(\omega) = \dot{N}_{\rm pis} + \dot{N}_{\rm cis} + \dot{N}_{\rm ste} + \dot{N}_{\rm oth}, \tag{51}$$

which, contrary to Eq. (48), can be safely averaged over the oceans to obtain

$$\rho'_{\rm alt} = <\dot{N}_{\rm cis}> \; + <\dot{N}_{\rm ste}> \; + <\dot{N}_{\rm oth}> \, , \tag{52}$$

where

$$\rho'_{\rm alt} = <\dot{N}_{\rm obs}> \; - <\dot{N}_{\rm pis}> \, , \tag{53}$$

represents the GIA-corrected, average rate of GMSLR observed by altimetry. According to the rule of thumb valid for the CIS (44), the first term in Eq. (52) is $\approx <\dot{S}_{\rm cis}> \; = -\dot{m}_{\rm i,cis}/(\rho_{\rm w} A_{\rm o}) = \dot{V}_{\rm o,cis}/A_{\rm o}$, where I have used (38) and the mass conservation constraint, and $V_{\rm o,cis}$ is the variation of the ocean volume due to the melting of the CIS. The second term in Eq. (52) is $<\dot{N}_{\rm ste}> \; = \; <\dot{S}_{\rm ste}> = \dot{V}_{\rm o,ste}/A_{\rm o}$, since to a first approximation the thermal expansion of the oceans is not changing their mass and, consequently, does not load the Earth ($\dot{U}_{\rm ste} = 0$). However, a more refined estimate should account for vertical displacements induced by loading across the continental shelves, in response to horizontal pressure gradients associated with the oceans warming (Richter et al. 2013; Gregory et al. 2013). Hence, with $\dot{V}_{\rm o} = \dot{V}_{\rm o,cis} + \dot{V}_{\rm o,ste}$, (47) gives

$$\rho'_{\rm alt} = \rho + \dot{\epsilon}_{\rm alt}, \tag{54}$$

where the term $\dot{\epsilon}_{\rm alt}$ includes $<\dot{N}_{\rm oth}>$ and other possible sources of uncertainty, including those arising from the incomplete coverage of the altimetric observations in the polar regions (see, e.g., Cazenave et al. 2009). I note that, since no obvious relationship exists between $\dot{\epsilon}_{\rm alt}$ and $\dot{\epsilon}_{\rm tg}$, a match between $\rho'_{\rm alt}$ and $\rho'_{\rm tg}$ should not be expected a priori. This is not due to errors inherent in the experimental assimilation of tide gauge and altimetric data—which have not been taken into account here—rather it reflects the different nature of the two types of sea-level observations.

Adopting the GIA model ICE-5G(VM2) (Peltier 2004), the altimetric corrections to absolute sea-level change $<\dot{N}_{\rm pis}>$ have been found in the range between ~ -0.2 and -0.4 mmyear^{-1}(see Table 1). These values correspond to different setups of the GIA models, which are, however, characterized by the same rheological profile (i.e., a volume-averaged VM2). In a more thorough study (Tamisiea 2011), the viscosity profile in ICE-5G(VM2) has been changed by $\sim$1 order of magnitude across the upper mantle and the lower mantle in order to test the sensitivity of $<\dot{N}_{\rm pis}>$ to mantle rheology. The range of values so obtained for the GIA correction spans from ~ -0.15 to -0.50 mmyear^{-1}. According to Tamisiea (2011), using the previous GIA model ICE-3G (Tushingham and Peltier 1991) instead of ICE-5G(VM2) would change these values at $\sim$20 % at most. These results clearly indicate that the GIA correction to absolute sea-level change is sensitive to the GIA model parameters. However, it is also clear that its amplitude is relatively small, approximately 10 times smaller than the altimetry-derived average rate of absolute sea-level change (~ 3 mmyear^{-1}) and comparable with its experimental uncertainty (~ 0.5 mmyear^{-1}) (Cazenave and Llovel 2010). This motivates the application of the common rule of thumb of adding 0.3 mmyear^{-1} in order to correct the time-averaged averaged altimetric rate for GIA (Church et al. 2013).

9 Conclusions

I have reviewed some of the key elements of GIA, with a particular attention to their links with the contemporary sea-level variations, observed either by tide gauge instruments or by satellite altimetry.

The various geophysical processes that accompany GIA have been described by means of the SLE, which establishes the relationship between relative sea-level, absolute (geocentric) sea-level and the vertical displacement of the Earth's solid surface (Farrell and Clark 1976). The properties of this integral equation have been qualitatively illustrated, with the aim of enlightening the origin of the non-uniform sea-level variations that the SLE implies. Spatial averages of relative sea-level change over the surface of the oceans define the eustatic sea-level variations, which one would observe ideally in the very special conditions of a rigid and non-gravitating Earth. Although the term *eustatic* has been used in the past with many qualifications (Suess 1906; Farrell and Clark 1976; Dott 1992) and tends now to be abandoned, it has played an important role in the framework of GIA, which has been highlighted in the discussion.

Among the mechanisms that significantly contribute to contemporary sea-level rise (Cazenave and Llovel 2010), GIA is the only one involving the cryosphere and simultaneously causing global deformations of the solid Earth, variations of the gravitational potential and changes in the Earth's angular velocity. These processes affect both relative and absolute sea-level change. Current GIA models that describe the time evolution of the late-Pleistocene ice sheets, as for example ICE-5G(VM2) (Peltier 2004), assume that their melting lasted until 4000 years ago. Hence, the contribution of current global sea-level rise is relatively small and does not involve ocean mass variations at present. On the contrary, the current melting of the ice sheets driven by global warming is varying the mass of the oceans and it is responsible for regional sea-level variations as well.

The patterns of the GIA-induced local and regional sea-level changes have been studied carefully in the last decades until very recently (Kopp et al. 2015; Khan et al. 2015), but probably their spatial variability has been somewhat overemphasized (Spada and Galassi 2016). Here, following Spada and Galassi (2016), the observed altimetric record and the fingerprints for absolute sea-level have been decomposed on the basis of the so-called ON functions (Hwang 1991, 1993). This allows for a quantitative evaluation of the spectral properties of these fields and for a comparison, degree-by-degree, with the regional patterns observed by altimetry. The power held in the GIA fingerprints is only a very modest fraction of the total power contained in the contemporary sea-level variations. The largest GIA contribution is associated with the harmonic component of harmonic degree 2, which is associated with the rotational variations induced by the global isostatic disequilibrium.

In the last part of the overview, I have discussed the GIA corrections to tide gauge and altimetric observations. For tide gauges, the importance of the correction has been recognized very early in the development of the sea-level change science, but it was only in the late 1980s that reliable models to perform such correction became available (Spada and Galassi 2012). On the one hand, it is clear that reliable GIA corrections are a fundamental requisite for a correct estimate of the rate of GMSLR, which enlightens the importance of the solid Earth influence on sea-level (Conrad 2013). On the other hand, it should be remarked that GIA models have evolved considerably during the last three decades and will certainly evolve in the future, stimulated by with the growing amount of the proxy data that constrain the history of sea-level in the last few thousand years. This will require GIA models including increasing complexities, like 3-D lateral viscosity variations or nonlinear or composite rheologies, along the lines of significant achievements during the last two decades (Giunchi et al. 1997; D'Agostino et al. 1997; Giunchi and Spada 2000; Zhong et al. 2003; Gasperini and Dal 2004; Latychev et al. 2005; Kendal et al. 2006; Spada et al. 2006; Whitehouse et al. 2006; Dal Forno and Gasperini 2007; Wal et al. 2010; Dal Forno et al. 2012; Geruo et al. 2013; Wal et al. 2013, 2015). Hence, the realization of GIA corrections is not established once and for all, and the refinement of their possible range of variation will be certainly a challenge for future studies (Tamisiea 2011).

Acknowledgments This paper is an outcome of the ISSI Workshop "Integrative Study of Sea Level Budget", held in Bern in February 2015. Two anonymous reviewers are acknowledged for their very constructive comments. I have greatly benefited from stimulating discussions on GIA and the SLE with students and colleagues during the POLENET/SERCE GIA training school held in Gibraltar Island (Ohio, USA) in September 2015. Daniele Melini is thanked for providing support with the numerical simulations. Gaia Galassi and Marco Olivieri are acknowledged for very insightful suggestions during the various stages of preparation of this manuscript. The open-source program SELEN (a *SE*a *L*evel *E*quatio*N* solver) is available from the Computational Infrastructure for Geodynamics (CIG), at the address http://geodynamics.org/cig/software/selen/), or from the author. All figures have been drawn using the Generic Mapping Tools of Wessel and Smith (1998). The PSMSL is acknowledged for making available the tide gauge data from the web page http://www.psmsl.org. This work is funded by a DiSPeA (Dipartimento di Scienze Pure e Applicate) research Grant (CUP H32I16000000005) and by Programma Nazionale di Ricerche in Antartide (PNRA 2013/B2.06, CUP D32I14000230005). I benefited from the warm and relaxing atmosphere of the Naturalistic Annex of the Museum of Bagnacavallo (RA), Italy, where the paper was drafted.

References

Bamber J, Riva R (2010) The sea level fingerprint of recent ice mass fluxes. Cryosphere 4(4):621

Bindoff NL, Willebrand J, Artale V, Cazenave A, Gregory JM, Gulev S, Hanawa K, Le Quéré C, Levitus S, Nojiri Y, et al (2007) In: Solomon S (ed) Climate Change 2007: the physical science basis. Contribution of working group I to the fourth assessment report of the intergovernmental panel on climate change (Cambridge University Press, Cambridge)

Brandes C, Steffen H, Steffen R, Wu P (2015) Intraplate seismicity in northern Central Europe is induced by the last glaciation. Geology G36:710–711

Cambiotti G, Ricard Y, Sabadini R (2010) Ice age true polar wander in a compressible and non-hydrostatic Earth. Geophys J Int 183(3):1248

Cathles L (1975) The viscosity of the Earth's mantle. Princeton University Press, Princeton

Cazenave A, Chambers D, Cipollini P, Fu L, Hurell J, Merrifield M, Nerem S, Plag H, Shum C, Willis J (2009) The challenge for measuring sea level rise and regional and global trends. Proc OceanObs 9:135

Cazenave A, Dominh K, Guinehut S, Berthier E, Llovel W, Ramillien G, Ablain M, Larnicol G (2009) Sea level budget over 2003–2008: a reevaluation from GRACE space gravimetry, satellite altimetry and Argo. Glob Planet Change 65(1):83

Cazenave A, Llovel W (2010) Contemporary sea level rise. Annu Rev Mar Sci 2:145

Cazenave A, Nerem RS (2004) Present-day sea level change: observations and causes. Rev Geophys 42(3):1–20. doi:10.1029/2003RG000139

Chambers DP, Wahr J, Tamisiea ME, Nerem RS (2012) Reply to comment by WR Peltier et al. on Ocean mass from GRACE and glacial isostatic adjustment. J Geophys Res Solid Earth 117(11):B11404

Church J, Gregory J, Huybrechts P, Kuhn M, Lambeck K, Nhuan M, Qin D, Woodworth P (2001) Changes. In: Houghton J, Ding Y, Griggs D, Noguer M, Van der Linden P, Dai X, Maskell K (eds) Sea level. Cambridge University Press, Cambridge, pp 639–694

Church J, Clark P, Cazenave A, Gregory J, Jevrejeva S, Levermann A, Merrifield M, Milne G, Nerem R, Nunn P, Payne A, Pfeffer W, Stammer D (2013) Contribution of working group I to the fifth assessment report of the intergovernmental panel on climate change. In: Stocker T, Qin D, Plattner GK, Tignor M, Allen S, Boschung J, Nauels A, Xia Y, Bex V, Midgley P (eds) Climate change 2013: the physical science basis. Cambridge University Press, Cambridge, pp 1138–1191

Church JA, White NJ (2006) A 20th century acceleration in global sea-level rise. Geophys Res Lett 33(1):L01602

Church JA, White NJ (2011) Sea-level rise from the late 19th to the early 21st century. Surv Geophys 32(4–5):585

Clark JA, Farrell WE, Peltier WR (1978) Global changes in postglacial sea level: a numerical calculation. Quat Res 9(3):265

Clark PU, Mitrovica J, Milne G, Tamisiea M (2002) Sea-level fingerprinting as a direct test for the source of global meltwater pulse IA. Science 295(5564):2438

Clark JA, Lingle CS (1979) Predicted relative sea-level changes (18,000 years BP to present) caused by late-glacial retreat of the Antarctic ice sheet. Quat Res 11(3):279

Conrad CP (2013) The solid Earth's influence on sea level. Geol Soc Am Bull 125(7–8):1027

D'Agostino G, Spada G, Sabadini R (1997) Postglacial rebound and lateral viscosity variations: a semi-analytical approach based on a spherical model with Maxwell rheology. Geophys J Int 129(3):F9

Dahlen F, Tromp J (1998) Theoretical global seismology. Princeton University Press, Princeton

Dal Forno G, Gasperini P (2007) Modelling of mantle postglacial relaxation in axisymmetric geometry with a composite rheology and a glacial load interpolated by adjusted spherical harmonics analysis. Geophys J Int 169(3):1301

Dal Forno G, Gasperini P, Spada G (2012) In: VII Hotine-Marussi symposium on mathematical geodesy. Springer, pp 393–397

De Boer B, Stocchi P, Van De Wal R et al (2014) A fully coupled 3-D ice-sheet-sea-level model: algorithm and applications. Geosci Model Dev 7(5):2141

Dott RH (1992) An introduction to the ups and downs of eustasy. Geol Soc Am Mem 180:1

Douglas B (1991) Global sea level rise. J Geophys Res 96(C4):6981

Douglas B (1992) Global sea level acceleration. J Geophys Res Oceans 97(C8):12699

Douglas B (1997) Global sea rise: a redetermination. Surv Geophys 18(2–3):279

Douglas B (2001) Sea level change in the era of the recording tide gauge. Int Geophys Ser 75:37

Douglas B (2008) Concerning evidence for fingerprints of glacial melting. J Coast Res 24(sp2):218

Dziewonski AM, Anderson DL (1981) Preliminary reference Earth model. Phys Earth Planet Inter 25(4):297

Farrell W (1972) Deformation of the Earth by surface loads. Rev Geophys 10(3):761

Farrell W, Clark J (1976) On postglacial sea-level. Geophys J Roy Astron Soc 46:647

Gasperini P, Dal Forno G, Boschi E (2004) Linear or non-linear rheology in the Earth's mantle: the prevalence of power-law creep in the postglacial isostatic readjustment of Laurentia. Geophys J Int 157(3):1297

Geruo A, Wahr J, Zhong S (2013) Computations of the viscoelastic response of a 3-D compressible Earth to surface loading: an application to Glacial Isostatic Adjustment in Antarctica and Canada. Geophys J Int 192(2):557

Giunchi C, Spada G, Sabadini R (1997) Lateral viscosity variations and post-glacial rebound: effects on present-day VLBI baseline deformations. Geophys Res Lett 24(1):13

Giunchi C, Spada G (2000) Postglacial rebound in a non-Newtonian spherical Earth. Geophys Res Lett 27(14):2065

Gregory JM, White N, Church J, Bierkens M, Box J, Van den Broeke M, Cogley J, Fettweis X, Hanna E, Huybrechts P et al (2013) Twentieth-century global-mean sea level rise: is the whole greater than the sum of the parts? J Clim 26(13):4476

Gröger M, Plag HP (1993) Estimations of a global sea level trend: limitations from the structure of the PSMSL global sea level data set. Glob Planet Change 8(3):161

Guo J, Huang Z, Shum C, van der Wal W (2012) Comparisons among contemporary glacial isostatic adjustment models. J Geodyn 61:129

Gutenberg B (1941) Changes in sea level, postglacial uplift, and mobility of the Earth's interior. Geol Soc Am Bull 52(5):721

Hagedoorn JM, Wolf D, Martinec Z (2007) An estimate of 8 mean sea-level rise inferred from tide-gauge measurements using glacial-isostatic models consistent with the relative sea-level record. Pure Appl Geophys 164:791

Haskell N (1935) The motion of a viscous fluid under a surface load. J Appl Phys 6(8):265

Haskell N (1936) The motion of a viscous fluid under a surface load part II. J Appl Phys 7(2):56

Hay CC, Morrow E, Kopp RE, Mitrovica JX (2015) Probabilistic reanalysis of twentieth-century sea-level rise. Nature 517(7535):481

Heiskanen WA, Moritz H (1981) Physical Geodesy. Technical University, Physical Geodesy (Institute of Physical Geodesy

Hogarth P (2014) Preliminary analysis of acceleration of sea level rise through the twentieth century using extended tide gauge data sets. Oceans 119(11):7645

Hwang C (1991) Orthogonal functions over the oceans and applications to the determination of orbit error, geoid and sea surface topography from satellite altimetry. Rep./Dep. of Geodetic Science A. Surveying. The Ohio State Univ.; N414

Hwang C (1993) Spectral analysis using orthonormal functions with a case study on the sea surface topography. Geophys J Int 115(3):1148

Ishii M, Kimoto M (2009) Reevaluation of historical ocean heat content variations with time-varying XBT and MBT depth bias corrections. J Oceanogr 65:287

Jeans J (1923) The propagation of earthquake waves. Proc R Soc Lond Ser A 102:554

Jerri A (1999) Introduction to integral equations with applications. Wiley, Hoboken

Jevrejeva S, Moore J, Grinsted A, Matthews A, Spada G (2014) Trends and acceleration in global and regional sea levels since 1807. Glob Planet Change 113:11

Jevrejeva S, Moore J, Grinsted A, Woodworth P (2008) Recent global sea level acceleration started over 200 years ago?. Geophys Res Lett 35(8):L08715

Johnston P (1993) The effect of spatially non-uniform water loads on prediction of sea-level change. Geophys J Int 114(3):615

Kendall RA, Mitrovica JX, Milne GA (2005) On post-glacial sea level-II. Numerical formulation and comparative results on spherically symmetric models. Geophys J Int 161(3):679

Kendall RA, Latychev K, Mitrovica JX, J.E. Davis, M.E. Tamisiea (2006) Decontaminating tide gauge records for the influence of glacial isostatic adjustment: the potential impact of 3-D Earth structure. Geophys Res Lett 33(24):L24318

Khan NS, Ashe E, Shaw TA, Vacchi M, Walker J, Peltier W, Kopp RE, Horton BP (2015) Holocene relative sea-Level changes from near-, intermediate-, and far-field locations. Curr Clim Change Rep 1(4):247–262

King MA, Altamimi Z, Boehm J, Bos M, Dach R, Elosegui P, Fund F, Hernández-Pajares M, Lavallee D, Cerveira PJM et al (2010) Improved constraints on models of glacial isostatic adjustment: a review of the contribution of ground-based geodetic observations. Surv Geophys 31(5):465

Konrad H (2015) Sea-level and solid-earth feedbacks on ice-sheet dynamics. Ph.D. thesis, Freie Universität Berlin

Kopp R, Mitrovica J, Griffies S, Yin J, Hay C, Stouffer R (2010) The impact of Greenland melt on local sea levels: a partially coupled analysis of dynamic and static equilibrium effects in idealized water-hosing experiments. Clim Change 103:619

Kopp RE, Hay CC, Little CM, Mitrovica JX (2015) Geographic variability of sea-level change. Curr Clim Change Rep 1(3):192

Lambeck K, Purcell A, Johnston P, Nakada M, Yokoyama Y (2003) Water-load definition in the glacio-hydro-isostatic sea-level equation. Quat Sci Rev 22(2):309

Lambeck K, Chappell J (2001) Sea level change through the last glacial cycle. Science 292(5517):679

Latychev K, Mitrovica JX, Tromp J, Tamisiea ME, Komatitsch D, Christara CC (2005) Glacial isostatic adjustment on 3-D Earth models: a finite-volume formulation. Geophys J Int 161(2):421

Levitus S (2005) Warming of the world ocean. Geophys Res Lett 32:L02604

Levitus S, Antonov JI, Boyer TP, Locarnini RA, Garcia HE, Mishonov AV (2009) Global ocean heat content 1955–2008 in light of recently revealed instrumentation problems. Geophys Res Lett 36:L07608

Lombard A, Cazenave A, Le Traon P, Ishii M (2005a) Contribution of thermal expansion to present-day sea-level change revisited. Glob Planet Change 47:1

Lombard A, Cazenave A, DoMinh K, Cabanes C, Nerem R (2005b) Thermosteric sea level rise for the past 50 years; comparison with tide gauges and inference on water mass contribution. Glob Planet Change 48:303

Mainardi F, Spada G (2011) Creep, relaxation and viscosity properties for basic fractional models in rheology. Eur Phys J Spec Top 193:133

Meier MF (1984) Contribution of small glaciers to global sea level. Science 226(4681):1418

Mémin A, Spada G, Rogister Y, Hinderer J, et al (2014) Decadal geodetic variations in Ny-Ålesund (Svalbard): role of past and present ice-mass changes. Geophys J Int 198(1):285–297

Meyssignac B (2012) La variabilité régionale du niveau de la mer. Ph.D. thesis, Université Paul Sabatier-Toulouse III
Meyssignac B, Melia DS y, Becker M, Llovel W, Cazenave A (2012) Tropical Pacific spatial trend patterns in observed sea level: internal variability and/or anthropogenic signature? Clim Past 8(2):787
Meyssignac B, Cazenave A (2012) Sea level: a review of present-day and recent-past changes and variability. J Geodyn 58:96
Milne GA, Gehrels WR, Hughes CW, Tamisiea ME (2009) Identifying the causes of sea-level change. Nat Geosci 2(7):471
Milne G, Mitrovica J (1996) Postglacial sea level change on a rotating Earth: first results from a gravitationally self-consistent sea-level equation. Geophy J Int 126(3):F13
Milne GA, Mitrovica JX (1998) Postglacial sea-level change on a rotating Earth. Geophys J Int 133(1):1
Mitrovica JX, Tamisiea ME, Davis JL, Milne GA (2001) Recent mass balance of polar ice sheets inferred from patterns of global sea-level change. Nature 409(6823):1026
Mitrovica JX (2003) Recent controversies in predicting post-glacial sea-level change. Quat Sci Rev 22(2):127
Mitrovica JX, Wahr J, Matsuyama I, Paulson A (2005) The rotational stability of an ice-age earth. Geophys J Int 161(2):491
Mitrovica J, Gomez N, Morrow E, Hay C, Latychev K, Tamisiea M (2011) On the robustness of predictions of sea level fingerprints. Geophys J Int 187(2):729
Mitrovica J, Milne G (2002) On the origin of late Holocene sea-level highstands within equatorial ocean basins. Quat Sci Rev 21(20):2179
Mitrovica JX, Milne GA (2003) On post-glacial sea level: I. General theory. Geophys J Int 154(2):253
Mitrovica JX, Peltier W (1991) On postglacial geoid subsidence over the equatorial oceans. J Geophys Res Solid Earth 96(B12):20053
Mitrovica JX, Wahr J (2011) Ice age Earth rotation. Annu Rev Earth Planet Sci 39:577
Munk WH, MacDonald GJ (1960) The rotation of the Earth: a geophysical discussion. Cambridge University Press, New York
Nakada M (2009) Polar wander of the Earth associated with the quaternary glacial cycle on a convecting mantle. Geophys J Int 179(1):569
Nakada M, Inoue H (2005) Rates and causes of recent global sea-level rise inferred from long tide gauge data records. Quat Sci Rev 24(10):1217
Nakada M, Lambeck K (1987) Glacial rebound and relative sea-level variations: a new appraisal. Geophys J Int 90(1):171
Nakada M, Lambeck K (1988) The melting history of the late Pleistocene Antarctic ice sheet. Nature 333(6168):36
Navarra A, Simoncini V (2010) A guide to empirical orthogonal functions for climate data analysis. Springer, Berlin
Olivieri M, Spada G (2013) Intermittent sea-level acceleration. Glob Planet Change 109:64. doi:10.1016/j.gloplacha.2013.08.004
Peltier W, Farrell W, Clark J (1978) Glacial isostasy and relative sea level: a global finite element model. Tectonophysics 50(2):81
Peltier W (1989) Geophysics, encyclopedia of Earth science. Springer, Berlin
Peltier WR (1994) Ice age paleotopography. Science 265(5169):195
Peltier WR (1996) Mantle viscosity and ice-age ice sheet topography. Science 273(5280):1359
Peltier W (2001) Global glacial isostatic adjustment and modern instrumental records of relative sea level history. Int Geophys 75:65
Peltier W (2004) Global glacial isostasy and the surface of the ice-age Earth: the ICE-5G (VM2) model and GRACE. Annu Rev Earth Planet Sci 32:111
Peltier W, Argus D, Drummond R, Moore A (2012) Postglacial rebound and current ice loss estimates from space geodesy: the new ICE-6G (VM5a) global model. AGU Fall Meet Abstr 1:02
Peltier W, Andrews J (1976) Glacial-isostatic adjustment— I. The forward problem. Geophys J Int 46(3):605
Peltier WR, Argus DF, Drummond R (2015) Space geodesy constrains ice age terminal deglaciation: the global ICE-6G-C (VM5a) model. J Geophys Res B: Solid Earth 120(1):450–487
Peltier W, Drummond R (2002) A broad-shelf effect upon postglacial relative sea level history. Geophys Res Lett 29(8):10-1–10-4
Peltier W, Tushingham A (1989) Global sea level rise and the greenhouse effect: might they be connected? Science 244(4906):806

Pfeffer J, Allemand P (2016) The key role of vertical land motions in coastal sea level variations: a global synthesis of multisatellite altimetry, tide gauge data and GPS measurements. Earth Planet Sci Lett 439:39

Pirazzoli P (1986) Secular trends of relative sea-level (RSL) changes indicated by tide-gauge records. Tech. rep., CNRS-INTERGEO, 191 Rue Saint Jacques, 75005 Paris, France

Pirazzoli P (1993) Global sea-level changes and their measurement. Glob Planet Change 8(3):135

Plag HP, Jüettner HU (2001) Inversion of global tide gauge data for present-day ice load changes (scientific paper). Mem Natl Inst Polar Res 54:301 special issue

Ricard Y, Spada G, Sabadini R (1993) Polar wandering of a dynamic Earth. Geophys J Int 113(2):284

Richter K, Riva R, Drange H (2013) Impact of self-attraction and loading effects induced by shelf mass loading on projected regional sea level rise. Geophys Res Lett 40(6):1144

Sabadini R, Yuen DA, Boschi E (1982) Polar wandering and the forced responses of a rotating, multilayered, viscoelastic planet. J Geophys Res 87(B4):2885

Sabadini R, Peltier W (1981) Pleistocene deglaciation and the Earth's rotation: implications for mantle viscosity. Geophys J Int 66(3):553

Schmidt P, Lund B, Näslund JO, Fastook J (2014) Comparing a thermo-mechanical Weichselian Ice Sheet reconstruction to reconstructions based on the sea level equation: aspects of ice configurations and glacial isostatic adjustment. Solid Earth 5(1):371

Serpelloni E, Faccenna C, Spada G, Dong D, Williams SD (2013) Vertical GPS ground motion rates in the Euro-Mediterranean region: new evidence of velocity gradients at different spatial scales along the Nubia-Eurasia plate boundary. J Geophys Res Solid Earth 118(11):6003

Slangen A (2012) Modelling regional sea–level changes in recent past and future. Ph.D. thesis, Utrecht University, the Netherlands

Spada G (1992) Rebound Post-glaciale e Dinamica Rotazionale di un Pianeta Viscoelastico Stratificato. Ph.D. thesis, University of Bologna, Bologna, Italy

Spada G, Yuen D, Sabadini R, Boschi E (1991) Lower-mantle viscosity constrained by seismicity around deglaciated regions. Nature 351(6321):53

Spada G, Ricard Y, Sabadini R (1992) Excitation of true polar wander by subduction. Nature 360(6403):452

Spada G, Sabadini R, Yuen DA, Ricard Y (1992) Effects on post-glacial rebound from the hard rheology in the transition zone. Geophys J Int 109(3):683

Spada G, Antonioli A, Cianetti S, Giunchi C (2006) Glacial isostatic adjustment and relative sea-level changes: the role of lithospheric and upper mantle heterogeneities in a 3-D spherical Earth. Geophys J Int 165(2):692

Spada G, Ruggieri G, Sørensen LS, Nielsen K, Melini D, Colleoni F (2012) Greenland uplift and regional sea level changes from ICESat observations and GIA modelling. Geophys J Int 189(3):1457

Spada G, Bamber J, Hurkmans R (2013) The gravitationally consistent sea-level fingerprint of future terrestrial ice loss. Geophys Res Lett 40(3):482

Spada G, Olivieri M, Galassi G (2014) Anomalous secular sea-level acceleration in the Baltic Sea caused by isostatic adjustment. Ann Geophys 57(4):S0432

Spada G, Olivieri M, Galassi G (2015) A heuristic evaluation of long-term global sea level acceleration. Geophys Res Lett 42:4166

Spada G, Galassi G (2012) New estimates of secular sea level rise from tide gauge data and GIA modelling. Geophys J Int 191(3):1067

Spada G, Galassi G (2016) Spectral analysis of sea level during the altimetry era, and evidence for GIA and glacial melting fingerprints. Glob Planet Chang 143:34–49. doi:10.1016/j.gloplacha.2016.05.006

Spada G, Melini D, Galassi G, Colleoni F (2012) Modeling sea level changes and geodetic variations by glacial isostasy: the improved SELEN code. arXiv:1212.5061

Spada G, Stocchi P (2006) The sea level equation, theory and numerical examples. Aracne, Roma

Spada G, Stocchi P (2007) SELEN: a Fortran 90 program for solving the "Sea level equation". Comput Geosci 33(4):538

Steffen R, Eaton DW, Wu P (2012) Moment tensors, state of stress and their relation to post-glacial rebound in northeastern Canada. Geophys J Int 189(3):1741

Steffen H, Wu P (2011) Glacial isostatic adjustment in Fennoscandia review of data and modeling. J Geodyn 52(3):169

Steinberger B, O'Connell RJ (1997) Changes of the Earth's rotation axis owing to advection of mantle density heterogeneities. Nature 387(6629):169

Stocchi P, Spada G (2007) Glacio and hydro-isostasy in the Mediterranean Sea: Clark's zones and role of remote ice sheets. Ann Geophys 50:6

Suess E (1906) Face of the Earth. Clarendon Press, Oxford

Tamisiea ME (2011) Ongoing glacial isostatic contributions to observations of sea level change. Geophys J Int 186(3):1036

Tegmark M (1996) An icosahedron-based method for pixelizing the celestial sphere. Astrophys J 470:L81

Turcotte DL, Schubert G (2014) Geodynamics. Cambridge University Press, Cambridge

Tushingham A, Peltier W (1991) ICE-3G—a new global model of late Pleistocene deglaciation based upon geophysical predictions of post-glacial relative sea level change. J Geophys Res 96(B3):4497

Tushingham A, Peltier W (1992) Validation of the ICE-3G model of Wuerm–Wisconsin deglaciation using a global data base of relative sea level histories. J Geophys Res 97(B3):3285

van der Wal W, Wu P, Wang H, Sideris MG (2010) Sea levels and uplift rate from composite rheology in glacial isostatic adjustment modeling. J Geodyn 50(1):38

van der Wal W, Barnhoorn A, Stocchi P, Gradmann S, Wu P, Drury M, Vermeersen B (2013) Glacial isostatic adjustment model with composite 3-D Earth rheology for Fennoscandia. Geophys J Int 194(1):61

van der Wal W, Whitehouse PL, Schrama EJ (2015) Effect of GIA models with 3D composite mantle viscosity on GRACE mass balance estimates for Antarctica. Earth Planet Sci Lett 414:134

Wessel P, Smith WHF (1998) New, improved version of generic mapping tools released. Eos Trans Am Geophys Un 79(47):579

Whitehouse P. (2009) State of the art report. Svensk Kärnbränslehantering AB. In: Swedish Nuclear Fuel and Waste Management Co., Stockholm, p 105. http://skb.se/upload/publications/pdf/TR-09-11.pdf

Whitehouse P, Latychev K, Milne GA, Mitrovica JX, Kendall R (2006) Impact of 3-D Earth structure on Fennoscandian glacial isostatic adjustment: Implications for space-geodetic estimates of present-day crustal deformations. Geophys Res Lett 33(13):L13502

Wieczorek MA (2007) The gravity and topography of the terrestrial planets. Treat Geophys 10:165

Woodward R (1888) On the form and position of mean sea level. USGS Bull 48:87

Woodworth P (1990) A search for accelerations in records of European mean sea level. Int J Clim 10(2):129

Wöppelmann G, Miguez BM, Bouin MN, Altamimi Z (2007) Geocentric sea-level trend estimates from GPS analyses at relevant tide gauges world-wide. Glob Planet Change 57(3):396

Wöppelmann G, Marcos M (2016) Vertical land motion as a key to understanding sea level change and variability. Rev Geophys 54(1):64–92

Wunsch C, Stammer D (1995) The global frequency-wavenumber spectrum of oceanic variability estimated from TOPEX/POSEIDON altimetric measurements. J Geophys Res Oceans 100(C12):24895

Wu P, Peltier W (1982) Viscous gravitational relaxation. Geophys J Int 70(2):435

Wu P, Peltier W (1983) Glacial isostatic adjustment and the free air gravity anomaly as a constraint on deep mantle viscosity. Geophys J Int 74(2):377

Wu P, Peltier W (1984) Pleistocene deglaciation and the Earth's rotation: a new analysis. Geophys J Int 76(3):753

Yokoyama Y, Lambeck K, De Deckker P, Johnston P, Fifield LK (2000) Timing of the last glacial maximum from observed sea-level minima. Nature 406(6797):713

Zhong S, Paulson A, Wahr J (2003) Three-dimensional finite-element modelling of Earth's viscoelastic deformation: effects of lateral variations in lithospheric thickness. Geophys J Int 155(2):679

Part II
Sea Level Processes at Regional Scale

Surv Geophys (2017) 38:187–215
DOI 10.1007/s10712-016-9383-1

Causes of the Regional Variability in Observed Sea Level, Sea Surface Temperature and Ocean Colour Over the Period 1993–2011

B. Meyssignac[1] · C. G. Piecuch[2] · C. J. Merchant[3] · M.-F. Racault[4,5] · H. Palanisamy[1] · C. MacIntosh[3] · S. Sathyendranath[4,5] · R. Brewin[4,5]

Received: 22 March 2016 / Accepted: 11 August 2016 / Published online: 9 September 2016

Abstract We analyse the regional variability in observed sea surface height (SSH), sea surface temperature (SST) and ocean colour (OC) from the ESA Climate Change Initiative datasets over the period 1993–2011. The analysis focuses on the signature of the ocean large-scale climate fluctuations driven by the atmospheric forcing and do not address the mesoscale variability. We use the ECCO version 4 ocean reanalysis to unravel the role of ocean transport and surface buoyancy fluxes in the observed SSH, SST and OC variability. We show that the SSH regional variability is dominated by the steric effect (except at high latitude) and is mainly shaped by ocean heat transport divergences with some contributions from the surface heat fluxes forcing that can be significant regionally (confirming earlier results). This is in contrast with the SST regional variability, which is the result of the compensation of surface heat fluxes by ocean heat transport in the mixed layer and arises from small departures around this background balance. Bringing together the results of SSH and SST analyses, we show that SSH and SST bear some common variability. This is because both SSH and SST variability show significant contributions from the surface heat fluxes forcing. It is evidenced by the high correlation between SST and buoyancy-forced SSH almost everywhere in the ocean except at high latitude. OC, which is determined by phytoplankton biomass, is governed by the availability of light and nutrients that essentially depend on climate fluctuations. For this reason, OC shows significant correlation with

Electronic supplementary material The online version of this article (doi:10.1007/s10712-016-9383-1) contains supplementary material, which is available to authorized users.

✉ B. Meyssignac
benoit.meyssignac@legos.obs-mip.fr

1 LEGOS, CNES, CNRS, IRD, UPS, Université de Toulouse, 31000 Toulouse, France

2 Atmospheric and Environmental Research, Inc., 131 Hartwell Avenue, Lexington, MA 02421, USA

3 University of Reading, Reading RG6 6AH, UK

4 Plymouth Marine Laboratory (PML), Prospect Place, The Hoe, Plymouth PL1 3DH, UK

5 National Centre for Earth Observation (NCEO), PML, Plymouth PL1 3DH, UK

SST and SSH. We show that the correlation with SST displays the same pattern as the correlation with SSH with a negative correlation in the tropics and subtropics and a positive correlation at high latitude. We discuss the reasons for this pattern.

Keywords Sea level · Sea surface temperature · Ocean colour · Ocean heat content · Climate · ESA Climate Change Initiative · Wind forcing · Buoyancy forcing · ECCO reanalysis

1 Introduction

Oceans have been routinely monitored from space for more than 30 years now. In 1978, NASA launched Seasat, TIROS-N and Nimbus-7, the first three satellites dedicated to ocean observations. Seasat carried five complementary sensors and provided the first estimates from space of sea surface height (SSH), surface wind stress, sea surface temperature (SST), surface wave field and polar ice extent. Unfortunately, these estimates covered only a short period because Seasat failed after 105 days in space. TIROS-N lasted more than 2 years in orbit and produced the first really useful maps of SST. The Nimbus-7 carried the Coastal Zone Colour Scanner (CZCS), which remained operational for seven and a half years, from October 1978 until June 1986. CZCS was the first satellite sensor specifically developed to study ocean colour properties. Since then, about a dozen ocean-observing satellite missions have been launched by several space agencies to monitor continuously more than 10 essential ocean variables including SSH, ocean currents, tides, wave height (with radar altimeters), wind speed and direction (with microwave scatterometers), SST (with infrared and microwave radiometers), sea surface salinity (with microwave imaging radiometers with aperture synthesis), ocean colour (with multispectral imagers) and ocean bottom pressure (with space gravimetry).

In >30 years of measurements, satellite missions have revolutionized our understanding of the oceans. By providing a global mapping repeated with high temporal resolution, they have revealed the intense spatio-temporal variability, which characterizes the cycles of the oceans. This new picture challenged earlier views based on previous sparse measurements collected from ships and buoys. It spurred oceanographers to make considerable progress in the understanding of the role of the ocean in the physical (e.g., climate and weather), chemical (e.g., carbon cycle) and biological (e.g., primary production) processes of the Earth.

In this paper, we are interested in the role played by the ocean in the climate system. We revisit the 30-year-long satellite record of ocean observations and summarize what we have learned from it about the ocean variability at climatic timescales (i.e. interannual to multidecadal timescales). We take advantage of the satellite archive, which gives an almost global view of the oceans, to explore the regional ocean variability.

Among all ocean variables remotely sensed from space, SSH, SST and ocean colour (OC) have long enough continuous records ($\sim$20 years or longer) to address the interannual to multidecadal variability in the ocean. For this reason, we focus on these three variables. We also analyse the hydrographic data (temperature and salinity) obtained from in situ measurements because these data are covering a large part of the ocean since the 1970s (Rhein et al. 2013; Abraham et al. 2013) and provide highly valuable information to interpret the signal showed by the other variables. Our objectives are to (1) describe the dominant spatio-temporal patterns of nonseasonal SSH, SST and OC variability, (2) discuss the current state of knowledge regarding the physical mechanisms responsible for these patterns and (3) explain the covariance at climatic timescales between these spatio-temporal patterns. The outline of the paper is as follows: Sect. 2 recalls the physical background governing the SSH, SST and OC variability and explains how this variability is linked to the atmospheric forcing and the ocean circulation. Section 3 shows the spatio-temporal patterns of SSH, SST and OC obtained from the satellite archive and uses the outputs of an ocean reanalysis to unravel the role of the atmospheric forcing and the ocean circulation in these patterns. In Sect. 3, we also take the opportunity to show how the patterns in SSH, SST and OC relate to each other and we discuss to what extent the information in SSH, SST and OC provided by satellites enables us to monitor the mechanisms responsible for the ocean variability at climatic timescales. Section 4 summarizes the paper, reviews the future issues and draws some conclusions.

2 Physical Background

2.1 SSH Variations Deduced from the Pressure Budget of the Water Column

SSHs indicate the level of the top of ocean water columns relative to a defined reference, which is constant with time; we use in general the mean sea surface as a reference. SSH variations indicate variations in the volume of water columns and are governed by two processes: the changes in mass and the changes in density of those water columns. Both processes can be tracked through vertical pressure changes in the pressure budget. Mathematically, the pressure budget under the hydrostatic and Boussinesq approximations can be written as follows (Gill and Niller 1973):

$$\eta = \frac{1}{\rho_0 g}(p_{\mathrm{b}} - p_{\mathrm{a}}) + \eta_{\mathrm{st}} \tag{1}$$

where η is the SSH, p_{a} is the sea surface atmospheric pressure, p_{b} is the ocean bottom pressure, ρ_0 is a constant density, g is the acceleration due to gravity and η_{st} is the steric sea level. η_{st} is given by $\eta_{\mathrm{st}} = \frac{1}{\rho_0}\int_{-H}^{0}\rho \mathrm{d}z$ in which H is the ocean bottom depth and ρ the water density deviation from ρ_0.

The pressure budget breaks down SSH into three signals: (1) a signal which depends on the atmospheric pressure (i.e. the inverted barometer response, Wunsch and Stammer 1997), (2) a signal which depends on bottom pressure and gravity and (3) a signal which depends on the water density along the water column. The first signal is isostatic (i.e. dynamically irrelevant) on climate timescales and thus will not be analysed here. The second signal represents the mass component in sea level. It arises essentially from changes in bottom pressure caused by the water mass redistribution by the ocean circulation in response to the atmospheric forcing. This signal is the dominant component in sea level changes at high frequencies (for periods <1 months, Forget and Ponte 2015). It is much smaller at interannual and higher timescales, but it remains sizeable in particular at high latitudes (see Sect. 3). The mass signal in sea level can arise also from changes in the gravity field of the Earth (i.e. changes in local g). These changes can be due to ongoing ice loss in ice sheets, for example, or to the current solid earth response to ice loss in ice sheets that occurred during the last deglaciation (Tamiseia 2011). However, this component in sea level change is significant only at timescales longer than multidecadal timescales and will not be considered here. The third sea level signal in the pressure budget represents the steric component in sea level. This signal arises from changes in the sea water density due to changes in the water column temperature (thermosteric sea level) or salinity (halosteric sea level). These changes in temperature and salinity are caused by buoyancy forcing at the surface or redistribution of heat and freshwater within the ocean by the ocean circulation in response to the atmospheric forcing. The redistribution of heat and freshwater can occur through advection or diffusion (including diapycnal, isopycnal and convective mixing) such that steric sea level can be written as follows:

$$\eta_{\mathrm{st}} = \eta_{\mathrm{T}} + \eta_S = \eta_{\mathrm{T}}^{\mathrm{a}} + \eta_{\mathrm{T}}^{\mathrm{d}} + \eta_{\mathrm{T}}^{\mathrm{BF}} + \eta_S^{\mathrm{a}} + \eta_S^{\mathrm{d}} + \eta_S^{\mathrm{BF}} \quad (2)$$

where subscripts T and S refer to thermosteric and halosteric sea level and a, d and BF refer to advection, diffusion and buoyancy forcing (a more detailed explanation of the η_{st} budget can be found in Piecuch and Ponte 2011). Thermosteric effects (η_{T}) dominate steric sea level variability over most of the ocean (see Sect. 3; Köhl 2014; Forget and Ponte 2015). At interannual and longer timescales, heat advection in the oceans ($\eta_{\mathrm{T}}^{\mathrm{a}}$) plays the leading role in thermosteric sea level variability nearly everywhere, suggesting that the heat redistribution in the ocean is nearly adiabatic. However, in many locations at all latitudes, the air-sea heat flux ($\eta_{\mathrm{T}}^{\mathrm{BF}}$) role is sizeable and can reach the same order of magnitude as the advection term. This is unlike the heat diffusion term which is negligible almost everywhere (except locally in the Arctic and along the margin of Antarctica close to deep water formation regions, Forget and Ponte 2015) and thus will be neglected here.

The interannual and longer time-scale variability in steric sea level is predominantly the result of internal reorganization of water masses in the ocean forced by anomalies in surface wind stress (Stammer et al. 2013). When this contribution is removed, the remaining variability is explained by buoyancy forcing anomalies (i.e. surface heat and freshwater exchanges) and the intrinsic oceanic variability spontaneously generated by the ocean circulation under the repeated seasonal atmospheric forcing (Penduff et al. 2011; Sérazin et al. 2015). It is essentially surface heat fluxes anomalies, which explain most of this residual variability (see Sect. 3). The surface heat fluxes can be written as follows:

$$\eta_{\mathrm{T}}^{\mathrm{BF}} \approx \varepsilon \frac{Q_{\mathrm{net}}}{\rho_{\mathrm{ml}} C_p} = \varepsilon \frac{Q_{\mathrm{sh}} + Q_{\mathrm{lh}} + Q_{\mathrm{sw}} + Q_{\mathrm{lw}}}{\rho_{\mathrm{ml}} C_p} \quad (3)$$

where ε is the thermal expansion coefficient, ρ_{ml} is the density of the upper ocean mixed layer and C_p is its heat capacity. Surface heat fluxes are caused by the turbulent energy fluxes (sum of the sensible heat flux—Q_{sh}—and latent heat flux—Q_{lh}–), which are broadly proportional to the wind speed, the air-sea temperature and humidity differences. Radiative fluxes (sum of the downward solar radiative flux—Q_{sw}—and the longwave radiative flux—Q_{lw}—) are functions of air temperature, humidity and cloudiness.

2.2 SST Variations Deduced from the Heat Budget of the Upper Ocean Mixed Layer

SST from space closely reflects upper ocean mixed layer temperatures, being within 0.2 °C of upper ocean mixed layer temperatures, other than where dynamic processes drive near-surface temperature gradients (Grodsky et al. 2008). It is governed by the processes regulating the exchange of energy at the sea surface and at the bottom of the mixed layer.

These processes are both atmospheric (wind stress and buoyancy fluxes) and oceanic (heat transport by currents, vertical mixing and boundary layer depth influence). Mathematically, the heat budget can be written as follows (Deser et al. 2010).

$$\frac{\partial T}{\partial t} = \frac{Q_{\mathrm{net,ml}}}{\rho C_p h} + \left(\overrightarrow{U_{\mathrm{geo}}} + \overrightarrow{U_{\mathrm{ek}}}\right) \cdot \vec{\nabla} T + \frac{W_{\mathrm{e}} - W_{\mathrm{ek}}}{h} \cdot (T - T_b) \quad (4)$$

where T is the SST, h is the mixed layer depth, $\overrightarrow{U_{\mathrm{geo}}}$ is the geostrophic current velocity, $\overrightarrow{U_{\mathrm{ek}}}$ is the Ekman current velocity, W_{e} is the vertical entrainment rate, W_{ek} is the Ekman pumping velocity and T_b is the temperature at depth that is entrained in the mixed layer. Ekman and geostrophic currents contribute to the heat budget of the mixed layer through horizontal advection, while entrainment velocity and Ekman pumping change the SST through vertical advection. $Q_{\mathrm{net,ml}}$ is the net surface energy flux which enters the mixed layer. It is the sum of the turbulent energy fluxes, the component of downward solar radiative flux, which is absorbed within the mixed layer and the longwave radiative flux. In general, $Q_{\mathrm{net,ml}} \approx Q_{\mathrm{net}}$ and both terms depend on the same variables: wind speed, air-sea temperature and humidity difference, cloudiness. However, in regions of clear water and shallow mixed layer, they can be different because a non-negligible portion of the downward solar radiative flux penetrates below the mixed layer (see Sect. 3).

Thus, both SSH (through the steric sea level η_{st}) and SST have a dependence on the surface heat fluxes $\frac{Q_{\mathrm{net}}}{\rho_{\mathrm{ml}} C_p}$. This common dependence is expected to give rise to correlated variability at interannual and longer timescales, in particular in response to the atmospheric forcing. In contrast, SST has a different dependency on ocean circulation from SSH: SST depends on the circulation of the upper mixed layer, while SSH depends on the circulation of the ocean from the surface down to the bottom (see, for example, Fig. 1d in Piecuch and Ponte 2011). This will result in different responses of the SST and SSH to the oceanic circulation and variability at interannual and longer timescales that are not correlated. In Sect. 3, we will explore with the use of an ocean reanalysis the complex relation which exists between SSH and SST variability through their common response to surface heat fluxes.

2.3 Bio-Optics of Ocean Colour

In the global ocean, phytoplankton biomass is essentially governed by the availability of light and nutrients (locally temperature and concentration of predators, i.e. zooplankton, also can play a role). A common approach to estimate phytoplankton biomass is to measure the concentration of chlorophyll (the main pigment in phytoplankton cells), because of the central role this pigment plays in photosynthesis, because it is produced uniquely by plants, and because it is easily measured, and can be estimated from satellite ocean colour observations.

Phytoplankton cells are viable in the upper layer of the ocean, where sufficient light is available and where recycled nutrients are available, and additional nutrients can be brought up from the deep oceans through upwelling, wind mixing, advection and other physical processes. The solar irradiance penetrating the ocean will be attenuated with depth due to the optical properties of pure seawater itself, and also due to the presence of particles, in particular phytoplankton cells and the chlorophyll pigment they contain.

Ocean colour is determined by the spectral variance of reflectance, defined as the ratio of upwelling irradiance (radiant flux per unit surface area, W m^{-2}) at the surface of the ocean to the downwelling irradiance at the same depth. In satellite applications, it is also customary to use the term remote-sensing reflectance for the ratio of upwelling radiance (radiant flux per unit surface area and unit steradian, W m^{-2} str^{-1}) to downwelling irradiance at the surface. Both irradiance reflectance and remote-sensing reflectance, which vary with the wavelength of light considered, are functions of absorption (a) and back-scattering (b_b) coefficients of light, which in turn, are affected by the absorption and back-scattering properties of phytoplankton, which vary with the concentration of phytoplankton in the water, and also with the type of phytoplankton present. Though other substances, such as detritus and coloured dissolved organic material contribute to the variability in ocean colour, it is often assumed, as a first approximation, that phytoplankton may be treated as the single, independent variable that determines ocean colour in the open ocean. Remote-sensing reflectance at wavelength λ can be written as:

$$R_{\mathrm{rs}}(\lambda) = f(a(\lambda), b_b(\lambda)) \tag{5}$$

where f is a function that increases with back-scattering b_b and decreases with absorption a. The function f also incorporates the effect of the angular structure in the light field on R_{rs}. As noted earlier, the absorption and back-scattering coefficients are both functions of the concentration of phytoplankton in the water, typically measured as the biomass B in chlorophyll units:

$$a(\lambda) = a_w(\lambda) + a_B(\lambda) + a_Y(\lambda) + a_X(\lambda) \tag{6}$$

and

$$b_b(\lambda) = b_{bw}(\lambda) + b_{bB}(\lambda) + b_{bX}(\lambda) \tag{7}$$

where the subscripts w, B, Y and X stand for pure seawater, chlorophyll concentration, concentration of coloured dissolved organic matter (sometimes called yellow substance) and particles in suspension other than phytoplankton, respectively. In the visible domain of the electromagnetic spectrum, reflectance is a small part of the solar flux that reaches just below the surface, of the order of 5 %. The rest penetrates into the ocean. The same optical properties that determine reflectance at the sea surface also dictate the rate of light penetration into the ocean. Inside the water column, the decrease in irradiance level with depth can be described using an exponential function as

$$I(z, \lambda) = I_0(\lambda)e^{-K(\lambda)z} \tag{8}$$

where $I(z, \lambda)$ is irradiance at depth z and wavelength λ, $I_0(\lambda)$ is incident irradiance (just below the surface) and $K(\lambda)$ is the diffuse vertical attenuation coefficient in m^{-1} (Kirk 1994). The attenuation coefficient can be expressed as

$$K(\lambda) = g(a(\lambda), b_b(\lambda)) \tag{9}$$

where g is an increasing function of both a and b_b.

Because K increases with chlorophyll, phytoplankton-rich waters display a high attenuation coefficient, and the irradiance will not penetrate as deep as in low chlorophyll waters. The photic depth (defined as the depth at which irradiance reaches 1 % of I_0) is deeper in clear waters and shallower in waters characterized by a high chlorophyll biomass (e.g., Edwards et al. 2001).

The energy absorbed at a particular depth yields a local temperature increase with time t given by (e.g., Lewis et al. 1983)

$$\frac{\partial T}{\partial t} = \frac{1}{\rho_0 c_p} \frac{\partial I}{\partial z} \tag{10}$$

where $T(z)$ is the temperature change (in °K) of the water due to the heating, c_p is the specific heat capacity (in J kg^{-1} K^{-1}) and ρ_0 is the density of the water (kg m^{-3}). This equation does not account for the decrease in downwelling light with depth due to back-scattering at depth z (which does not contribute to local heating), or for heat gains by attenuation of upwelling irradiance at that depth (Edwards et al. 2001; Zaneveld et al. 1981).

Since I tends to zero below the photic depth, the heating will be confined to a layer near the surface in chlorophyll-rich waters whereas, in low chlorophyll waters, the heat energy will penetrate farther down the water column. In other words, in an oceanic region where surface chlorophyll concentration is high, the upper ocean layer will be warmer (and the deeper layer will be cooler) compared with a region where chlorophyll is absent (e.g., Sathyendranath et al. 1991; Edwards et al. 2001; Wu et al. 2007; Zhai et al. 2011).

These considerations highlight the relationship between SST and ocean colour: the optical properties have a modulating influence on the distribution of solar heating with depth, and hence on SST. Within the mixed layer, it would be reasonable to assume that the chlorophyll concentration would be uniform and equal to the value at the surface determined from ocean colour remote sensing. Where the photic depth is significantly deeper than the mixed layer, the vertical structure in chlorophyll concentration would have to be taken into account (Lewis et al. 1983).

3 Spatio-Temporal Patterns in SSH, SST and OC and Their Relation

In this section, we analyse the spatio-temporal patterns of SSH, SST and OC obtained from the satellite archive. To estimate these patterns, we use the ESA Climate Change Initiative (CCI) project datasets because they have been developed to be the most homogeneous and stable satellite records at interannual to decadal timescales as possible (Ablain et al. 2015; Merchant et al. 2014; Sathyendranath et al. 2016). Among SSH, SST and OC-CCI records, the SSH and SST records cover the same period: 1993–2014. In the following, we focus our analysis on this period 1993–2014 because it is the longest period covered by the CCI datasets. In addition to satellite datasets, we use an update of the analysis of ocean subsurface temperature and salinity by Ishii and Kimoto (2009) to estimate the steric effect in SSH variability. This analysis is based on temperature and salinity data from the World Ocean Database and Atlas, the Global Temperature-Salinity in the tropical Pacific from IRD (l'Institut de Recherche pour le Development, France) the Centennial in situ Observation-Based Estimates sea surface temperature and Argo profiling floats data. This in situ measurements record has become almost global since 2006 for the upper 2000 m (Argo array). Before 2006, the historical in situ measurements are sparse (time, space and depth), and they are sparser for salinity than for temperature profiles. In particular, a large fraction of the deep/abyssal ocean (below 700 m depth before 2006 and below 2000 m depth after 2006) still lacks in in situ measurements. We also make use of an ocean reanalysis to unravel the role of the atmospheric forcing and the ocean circulation in the patterns of SSH, SST and OC. The reanalysis used here is the ECCO (Estimating the Circulation and Climate of the Ocean)—Production version 4 Release 1 solution (Forget et al. 2015; Forget and Ponte

2015), hereafter referred to simply as ECCO. This product represents a model solution that has been constrained to observations (e.g., satellite altimetry, Argo floats and historical hydrography) based on the method of Lagrange multipliers. The optimization is achieved by making iterative adjustments to the initial conditions, boundary conditions and internal model parameters (consult Wunsch and Heimbach 2007, for more details on the general procedure). This solution covers the period 1992–2011. The ocean model setup is global, including the Arctic, and is fully coupled to an interactive sea ice and snow model. The spatial grid has a nominal 1-degree horizontal resolution, telescoping to 1/3-degree in the tropics and effectively 40 km in the Arctic, and uses 50 vertical levels. Initial-guess bulk-formula surface forcing is taken from the Interim European Centre for Medium-Range Weather Forecasts Reanalysis (ERA-Interim) of Dee et al. (2011) and iteratively adjusted as outlined above. The model also uses parameterization schemes to incorporate the effects of geostrophic eddies, vertical mixing and salt plumes (see Forget et al. 2015, for more details on this solution). We have chosen ECCO v4 because it has been shown to fit well to observations of SSH (Forget and Ponte 2015), in situ temperature and salinity (Forget et al. 2015), SST (Buckley et al. 2014) as well as other ocean circulation and climate variables (e.g., Piecuch et al. 2015).

In this paper, we focus on the climate fluctuations of the ocean and we analyse their signature on the SSH, SST and OC variables. In several regions, such as western boundary current regions, individual meso-scale eddies can generate substantial variability in ocean variables and mask the underlying climate fluctuations. To remove this noise from eddies, we perform a spatio-temporal smoothing on the SST and SSH data (see Sect. 3.3.2 for OC analysis) which filters out the smaller spatial scales and the shorter timescales consistently in observations and model. The spatio-temporal smoothing consists in applying first a 30-day boxcar window and then a spatial Gaussian filter which removes the variability on scales smaller than 3°. As in Forget and Ponte (2015), we have chosen 30 days for the temporal smoothing scale because it corresponds to the longest repeat cycle among observation datasets, and thus it enables us to get full maps of each observation dataset before performing the spatial smoothing. For the spatial smoothing scale, we have chosen 3° because it enables us to separate properly the ocean mesoscale, which relates to the baroclinic Rossby radius of deformation that is of the order of hundreds of km, from the ocean large-scale climate fluctuations driven by the atmospheric forcing, which are of the order of thousands of km (Forget and Ponte 2015).

In the following sections, the spatio-temporal patterns of SSH, SST and OC at interannual and decadal timescales are analysed in terms of standard deviation and trend of the time series. For SSH and SST data, the linear trend over 1993–2011 is estimated simultaneously with the seasonal cycle from monthly time series. After removing the trend and the seasonal cycle, we apply a 13-month Hanning window to remove remaining intraseasonal signals and then we estimate the standard deviation of the residual time series. For OC data, the linear trend over 1998–2010 is estimated based on monthly chlorophyll anomalies

3.1 Sea Level and How it Relates to Ocean Mass, Temperature and Salinity

3.1.1 Regional Variations

SSH observations from the CCI sea level dataset show considerable interannual variability (Fig. 1a). This variability is maximum within the tropics. Temperature and salinity

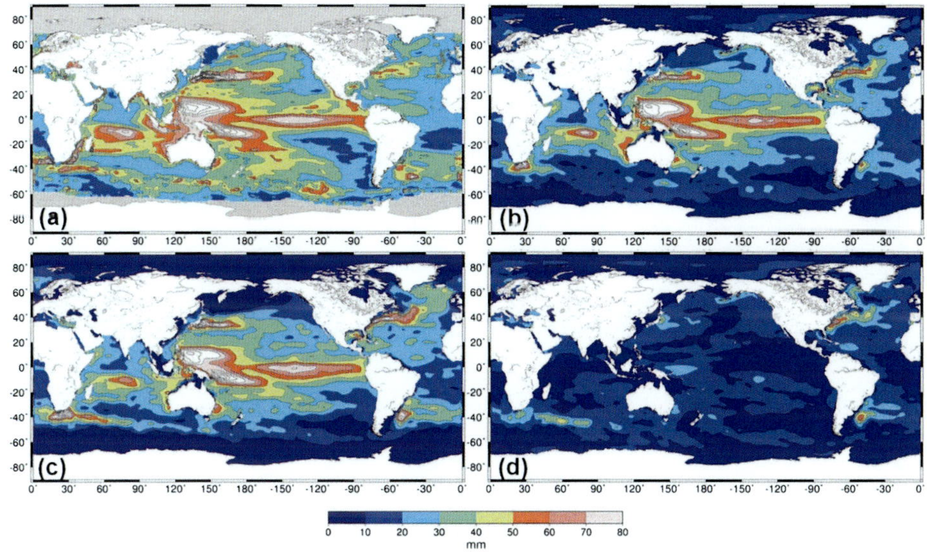

Fig. 1 **a** Standard deviation in SSH (η) from the CCI sea level dataset. **b–d** Standard deviation in, respectively, steric height (η_{Steric}), thermosteric height (η_T) and halosteric height (η_S) from an update of Ishii and Kimoto (2009). Heights are in mm. Note that thermosteric and halosteric data have large uncertainties in regions where the in situ temperature and salinity observations are poor like in the southern ocean before 2006

observations have a very sparse coverage in the Arctic and do not allow evaluation of the steric effect in this region. But, for the rest of the ocean, they show that most of the interannual variability in sea level is dominated by the interannual variability in steric sea level (Fig. 1b). This result is confirmed by independent data from space gravimetry over the period 2004–2014. Indeed, the Gravity Recovery and Climate Experiment (GRACE) mission, which measures the time varying gravity field of the Earth, provides continuous estimates of the ocean bottom pressure since 2004. These estimates confirm that ocean bottom pressure interannual variability is smaller than the steric variability by an order of magnitude except at high latitudes and in shallow shelf seas (>60°N and <55°S, Piecuch et al. 2013; Ponte and Piecuch 2014).

At mid and low latitudes, the steric sea level signal is essentially due to temperature changes (Fig. 1c). Salinity changes play only a local role (Fig. 1d), but this role can be sizeable in several regions like the Eastern Indian Ocean (Llovel and Lee 2015) or the North Atlantic (Wunsch et al. 2007; Köhl 2014; Forget and Ponte 2015).

The picture is similar for the trends in sea level as for the standard deviation. They are largely dominated by the steric effect (Fig. 2a, b). The halosteric effect is much smaller than the thermosteric effect, but it is sizeable in many regions and should not be neglected (Fig. 2c, d). Interestingly, in the few regions where the halosteric signal is sizeable, like in the southern tropical Pacific, its effect tends to compensate the thermosteric effect. Such compensation suggests nearly adiabatic transport of the water masses in these regions (an example of such adiabatic transport is heaving of the water column) as suggested by previous authors (e.g., Wunsch et al. 2007; Durack et al. 2014).

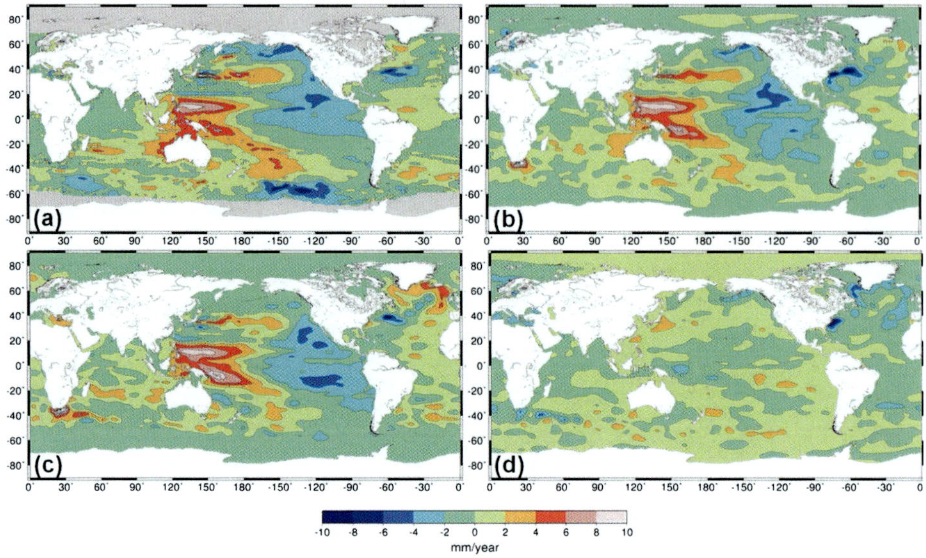

Fig. 2 **a** Trends over 1993–2011 in SSH (η) from the CCI sea level dataset (Ablain et al. 2015). **b–d** Trends over 1993–2011 in, respectively, steric height (η_{Steric}), thermosteric height (η_{T}) and halosteric height (η_S) from an update of Ishii and Kimoto (2009). Trends are in mm/year. *Note* that thermosteric and halosteric data have large uncertainties in regions where the in situ temperature and salinity observations are poor like in the Southern Ocean before 2006

3.1.2 The Effects of Wind Stress and Surface Buoyancy Fluxes

The ECCO model estimate shows similar patterns in SSH variability and trends as those obtained from the CCI observations (Figs. 3a, 4a). The ECCO estimate of the ocean bottom pressure (Figs. 3c, 4c) and of the steric effect (Figs. 3b, 4b) corroborates the notion that SSH variability and trends are dominated by the steric effect in general. At high latitudes, in the polar oceans where altimetry data are not available, ECCO also provides SSH estimates. In these regions, the SSH variability and trends appear to be substantial and actually dominated by the mass signal (Figs. 3c, 4c). This mass signal in the Arctic and in the Southern Ocean has been confirmed by GRACE observations since 2004 (Purkey et al. 2014; Makowski et al. 2015). It is the result of the barotropic circulation caused by wind forcing (Frankcombe et al. 2013; Volkov and Landerer 2013; Volkov 2014; Peralta-Ferriz et al. 2014; Fukumori et al. 2015; Makowski et al. 2015). In the Arctic, ECCO shows that the steric effect is actually sizeable and should not be neglected in comparison with the mass signal. This steric effect has a significant halosteric component coming from the variability of the freshwater inputs in the Arctic and the sea ice (see Fig. 4 in Köhl 2014, which suggests that both mixed layer processes and heaving of isopycnals contribute to interannual halosteric variability in the Arctic). At low and mid latitudes, ECCO confirms that the sea level variability and trends are almost entirely of thermosteric origin with some local halosteric effect, which tends to compensate the thermosteric effect as in observations.

An advantage of reanalyses over observations is that they allow unambiguous identification of the anomalous forcings, which are responsible for the interannual variability and the trends in SSH. In this subsection, we use perturbation experiments based on the ECCO model setup to distinguish the influences of wind stress and buoyancy exchanges on sea

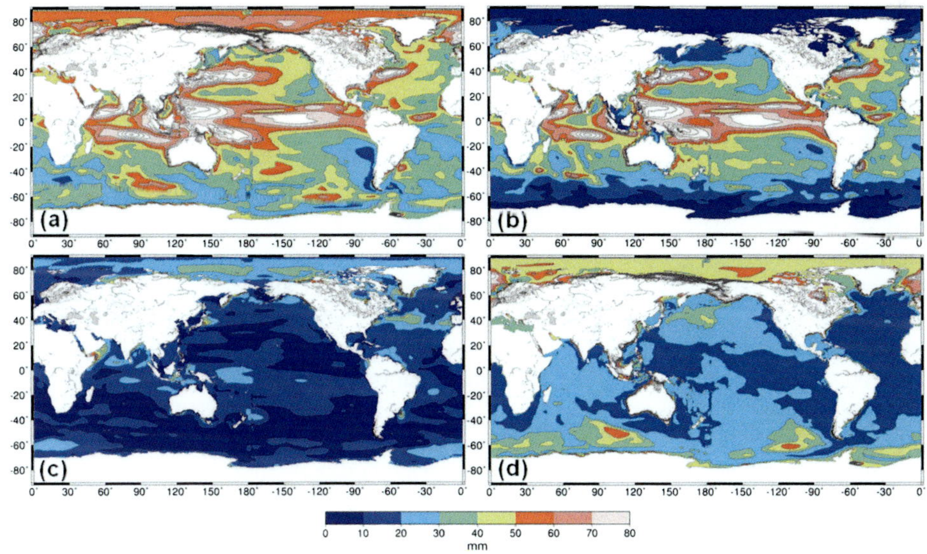

Fig. 3 **a–d** Standard deviation in, respectively, SSH (η), thermosteric height (η_{T}), halosteric height (η_S) and bottom pressure (p_{b}) from ECCO version 4. Heights are in mm

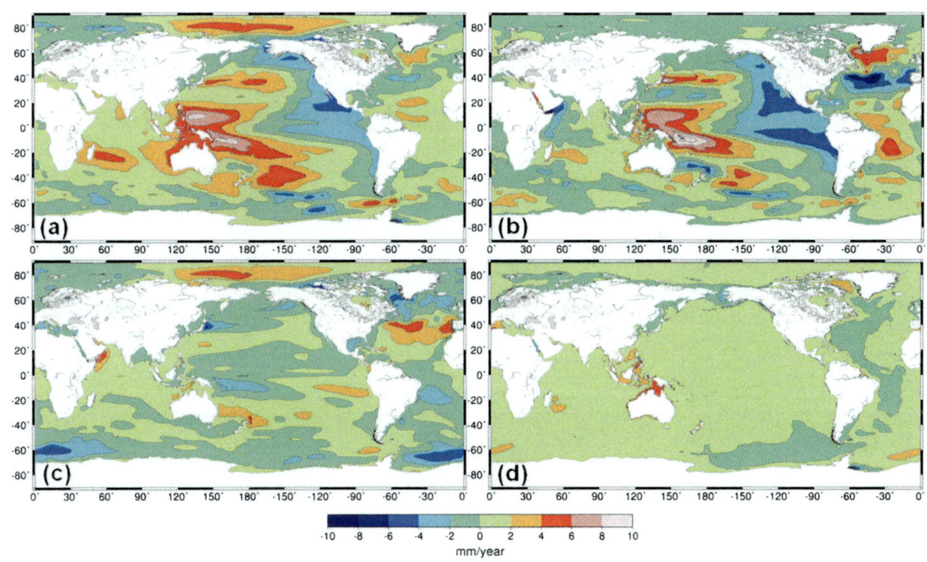

Fig. 4 **a–d** Trends over 1993–2011 in, respectively, SSH (η), thermosteric height (η_{T}), halosteric height (η_S) and bottom pressure (p_{b}) from ECCO version 4. Trends are in mm/year

level. This set of experiments is described more extensively by Forget and Ponte (2015). In more detail, each member of that set employs different surface boundary conditions. In each run, a common forcing component was prescribed, which comprised both fully variable surface buoyancy exchanges and a climatological mean seasonal cycle in wind stress. What differed between the simulations was that interannual changes in the wind stress were or were not turned off over the global ocean. The difference between these two experiments (shown in Figs. 5a, 6a) represents the oceanic response to interannual and decadal wind stress changes; the experiment without interannual or decadal wind stress changes (but retaining the common forcing component; shown in Figs. 5b, 6b) reflects the ocean's adjustment to interannual and decadal changes in buoyancy exchanges as well as any nonlinear intrinsic changes (cf. Penduff et al. 2011; Piecuch and Ponte 2012)

The perturbation experiments reveal that SSH variability is essentially forced by surface wind stress anomalies (see Fig. 5a, c and also Stammer et al. 2013; Forget and Ponte 2015). The remainder variability, which is forced by anomalous surface fluxes of buoyancy along with any nonlinear intrinsic variability (cf. Penduff et al. 2011), is in general significantly smaller except in some regions (Fig. 5b, d) such as the region of the Antarctic circumpolar current (ACC), in the Kuroshio extension, in the Arctic, in the North Atlantic, in the North Pacific and in the tropical Pacific, confirming earlier results from Thompson and Ladd (2004), Cabanes et al. (2006) and Piecuch and Ponte (2012, 2013).

In the case of SSH trends over 1993–2011, the perturbation experiments reveal that both wind stress forcing and buoyancy forcing play a leading role in SSH trends but in different regions. In the tropics, it is the wind stress anomalies which are responsible for the SSH trends (Fig. 6a). Indeed, the large positive pattern in the western tropical Pacific has been associated with a deepening of the thermocline in response to trade wind intensifications, probably linked to the negative phase of the Pacific decadal oscillation in recent decades (Timmermann et al. 2010; Merrifield 2011; McGregor et al. 2012; Qiu and Chen 2012; Meyssignac et al. 2012; Moon and Song 2013; Meehl et al. 2013; Palanisamy et al. 2015). In the extra tropics, the SSH trends are essentially driven by the buoyancy forcing (Fig. 6b), while the wind stress forcing contribution is smaller but remains significant in

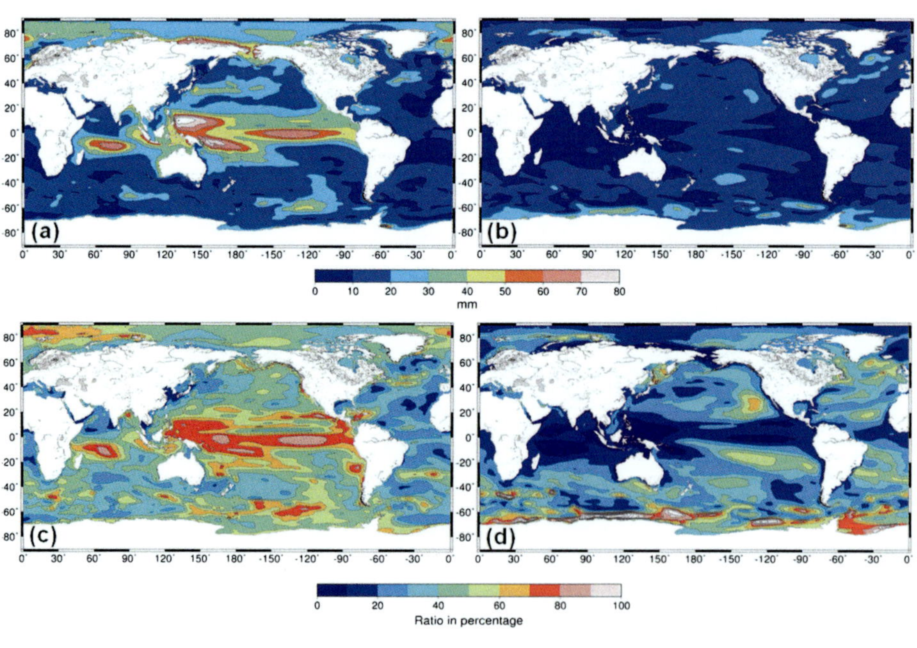

Fig. 5 Interannual variability in η due to wind stress (**a**) and to buoyancy forcing (**b**) from ECCO version 4. (See the text for the definition of η due to wind stress and to buoyancy forcing). Ratio of the interannual variability in η due to wind stress (**c**) and buoyancy forcing (**d**) over the interannual variability in total η

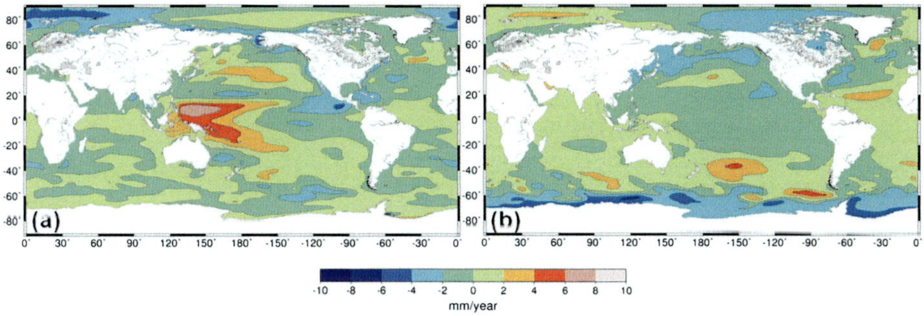

Fig. 6 Trends in η between 1993 and 2011 (in mm/year) from ECCO version 4. **a** Contribution due to wind stress; **b** contribution due to buoyancy forcing

several regions (such as the North and South Pacific and the Arctic, see Fig. 6a, b). Note that the trends in SSH driven by the buoyancy forcing are positive all over the ocean reflecting the general warming of the ocean. Interestingly, they are maximum in the subtropical gyres and in the North Atlantic subpolar gyre.

3.2 SST and How it Relates to Sea Level and Ocean Temperature and Salinity

3.2.1 Regional Variations

Interannual variability in SST is presented in Fig. 7 as the standard deviation of de-seasonalised, de-trended monthly SST anomalies. Very low SST variability at high latitudes in areas of perennial sea ice reflects the damping of variability in SST by the freezing and melting of that sea ice: in the SST data set, SST is set to the freezing temperature of sea water where the ocean is ice covered. Variability is greater in areas of seasonal sea ice, where variations in sea ice extent are reflected also in SST.

In the extra tropics poleward of sea ice, the variability in monthly SST is seen to be relatively high, typically of order 1 K. Much of the large-scale development of SST anomalies in the extra tropics is driven by large-scale reorganization of atmospheric circulation anomalies. Atmospheric variability with timescales longer than ~ 10 days is effective at driving SST anomalies that reflect the temperature of the upper mixed layer, because of the large thermal inertia of the upper ocean mixed layer (Frankignoul and

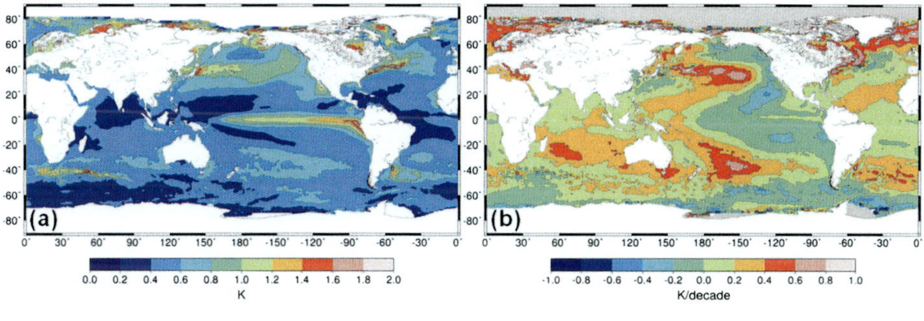

Fig. 7 Variability and trends in SST from the CCI dataset (Merchant et al. 2014). **a** Standard deviation in K. **b** Linear trend in over 1993–2011 in K per decade

Hasselmann 1977; Deser et al. 2003). (Note that, however, SST can respond strongly to higher frequency atmospheric variability when the depth of the response is shallow, as in diurnal variability.) The large-scale nature of atmospheric teleconnection patterns is imprinted upon the SST anomaly field essentially via the surface energy fluxes and Ekman currents (e.g., Cayan 1992; Marshall et al. 2001; Visbeck et al. 2003). Spatial variations in mixed layer depth (which themselves reflect the recent history of wind stress) modify the effective thermal inertial and also play a role in determining the magnitude of the SST anomalies.

Across much of the tropics, variability in monthly SST is low, <0.5 K. The clear exception is the equatorial Pacific Ocean, where SST variability >1.5 K is present eastwards of 180°W, associated with the El Nino Southern Oscillation (ENSO). ENSO is a coupled mode of variability, in which large-scale atmospheric circulation anomalies develop in close interaction with the SST variability, in contrast to extratropical anomalies (e.g., Deser et al. 2010).

Local processes such as upwelling, entrainment and lateral advection also contribute to SST variability. For example, vertical advection plays a prominent role along the coastal and equatorial upwelling zones, with variability being wind-driven. Horizontal advection is important along the western boundary current regions (e.g., the Gulf Stream and Kuroshio Current). Oceanic processes also play an indirect role in SST variability by affecting the depth of the upper ocean mixed layer.

The trends in SST over the period are relatively uniform around 0.3 K decade^{-1} across much of the Atlantic and Indian Oceans. The North Atlantic, including the Barents Sea, the sub-Greenland gyre and Labrador Sea, shows a warming trend exceeding 0.5 K decade^{-1}. Across the Pacific Ocean, the SST trend largely reflects the change in the tendency of the Pacific Decadal Oscillation over the period (Fig. 7a), as previously noted with respect to SSH.

3.2.2 The Effect of Surface Heat Fluxes and Ocean Transport

The ECCO model estimate shows generally similar patterns in SST variability as those obtained from the CCI observations (Fig. 8a), except at mid latitudes and in the Southern Ocean where it tends to underestimate the SST variability notably in eddy active regions (in the Kuroshio and Gulf Stream extensions, the Malvinas current and the Agulhas current). It shows also generally similar patterns in SST trends (Fig. 9a) except in the North Atlantic (from the Gulf Stream extension to the Barents sea) and in the eastern part of the north Pacific where trends are underestimated by a few tenths of K/decade and in the north-east Indian Ocean and China sea where trends are overestimated by a few tenths of K/decade.

As for SSH, we use ECCO output to infer the role of surface heat fluxes and ocean transport divergences in SST variations. We integrate vertically the upper ocean temperature budget (Eq. 7 in Piecuch and Ponte 2012) over a constant climatological mixed layer depth computed for each grid cell. The role of surface heat fluxes on SST is diagnosed by calculating the changes in SST due to the surface heat forcing term $\frac{Q_{\text{net,ml}}}{\rho C_p h}$. The ocean transport divergences effect is diagnosed by calculating the changes in SST due to the sum of the advection and diffusive terms. This form of the upper ocean temperature budget does not distinguish between Ekman and geostrophic transport contributions as in Eq. (4). However, it allows the same separation of SST in terms of surface heat forcing and ocean transport divergences as for steric sea level (Sect. 3.1). In that sense, it allows us to

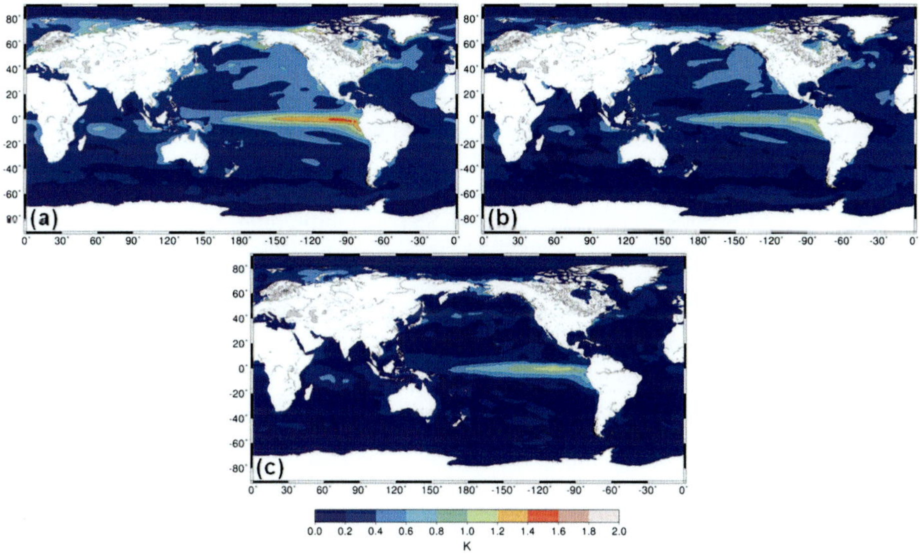

Fig. 8 **a** Standard deviation in SST from ECCO. **b** standard deviation in SST due to the surface heat fluxes (after removing the cancellation part with ocean transport divergences). **c** Standard deviation in SST due to the ocean transport divergences (after removing the cancellation part with surface heat fluxes)

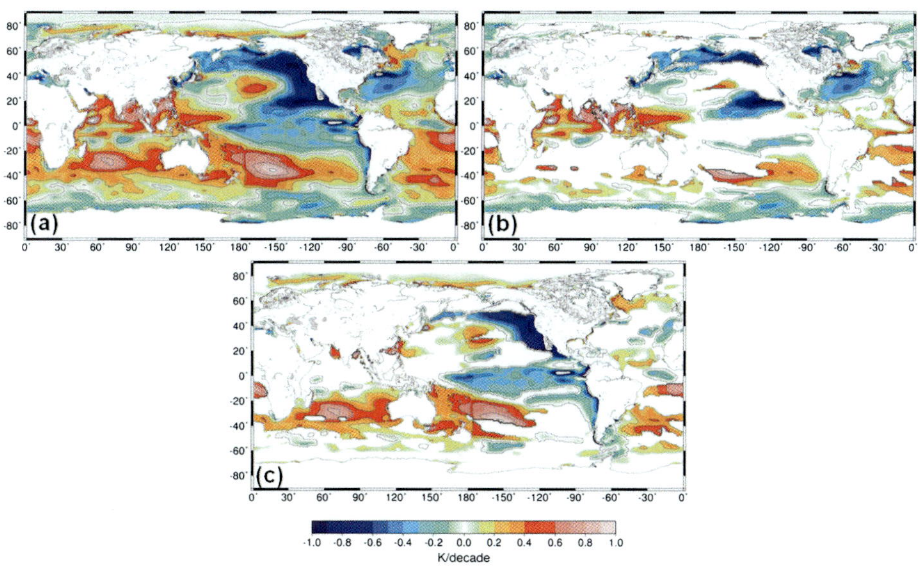

Fig. 9 Same as Fig. 8 but for trends

compare easily between upper ocean heat and steric sea level budgets, being able to infer whether surface heat fluxes or ocean transport divergences are more or less important for upper ocean heat or steric sea level changes.

Both standard deviation patterns in SST due to surface heat flux and ocean transport divergences are very similar (see Fig S1a, b in the supplementary material). They show high variability in the tropics and upwelling regions and low variability at higher latitudes with a minimum in the Arctic Ocean north of 80°N and in the Southern Ocean. The amplitudes of both standard deviation maps are also significantly higher than the standard deviation of the SST map almost everywhere: this is a result of integrating the ocean temperature budget over time. The effect of surface heat fluxes on SST is opposite to the effect of ocean transport divergences almost everywhere and they tend to cancel out (see Fig S1a, b). However, the cancellation of surface heat fluxes effect with ocean transport divergences effect is not exact and gives rise to SST variability. Figure 8b and c shows the standard deviation of the residual in SST due to surface heat fluxes and ocean transport, respectively, after removing the cancellation part (see also Fukumori and Wang 2013). Figure 8b shows that the interannual variability in SST due to the surface heat flux forcing is larger almost everywhere except in the tropical Pacific where ocean transport divergences play a similar role to surface heat fluxes in the local SST variability.

The same compensation process between surface heat fluxes and ocean transport divergences effects on SST occurs at longer timescales. The trend pattern in SST due to surface heat flux and ocean transport is very similar and tends to cancel out each other (see Fig. S2a, b in the supplementary material). Trends in SST due to surface heat fluxes are very large in the tropics and very low in boundary current regions and at high latitudes (Fig. S2a), while trends in SST due to ocean transport divergences are opposite in the same regions reflecting that the ocean gains most of the heat in the tropics and transports it to higher latitudes where it is released to the atmosphere or buried in the deeper layers of the ocean. However, the effect of surface heat fluxes and ocean transport divergences on SST trends does not fully cancel out everywhere. Figure 9b, c shows the residual in SST trends due to excess surface heat fluxes or ocean transport divergences, respectively, after removing the cancellation part. Figure 9b shows that surface heat fluxes dominate over ocean transport divergences and explain the SST trends in the tropical Indian Ocean, in the western tropical Pacific Ocean, in the North Pacific and North Atlantic subtropical gyres. It also dominates in the North Atlantic subpolar gyre and on the southern edge of the South Pacific subtropical gyre. In all upwelling regions (California current, Humbolt current and to a lesser extent in the Benguela current), it is the ocean transport divergences effect which dominates and explain the SST trends. In the eastern tropical Pacific, in the Indian and South Pacific subtropical gyres and in the North Atlantic subpolar gyre, it also dominates. Interestingly enough, the dominance of the ocean transport divergences over the surface heat fluxes, which explain the negative trends in SST in the California current, the Humbolt current and the eastern tropical Pacific, is consistent with the increase in cold deep water entrainment in these regions in response to increasing trade winds associated with the decreasing Pacific Decadal Oscillation (PDO) over the period 1993–2011.

The case of the western tropical Pacific region where the SST trends are dominated by surface heat fluxes warrants further discussion. At first sight, this result seems inconsistent with Fig. 6, which indicates positive SSH trends in this same region due to wind forcing (see Sect. 3.1.2). But there is a possible interpretation. The time-mean SST budget in the western tropical Pacific Ocean is a balance between strong increase due to surface heat fluxes and strong decrease due to ocean transport divergences (see Fig. S2). Over the study period, well-reported-on wind stress changes act to reduce the magnitude of that background ocean transport divergence contribution, making it less negative (and resulting in the steric sea level rise in this area). This results in their being an excess of surface heat forcing contribution (or deficit of ocean transport divergence contribution), which leads to the result in Fig. 9 that the western tropical Pacific SST trends are attributed to surface heat flux forcing.

3.2.3 Comparison with SSH Variability

At interannual timescales, SSH and SST show a fairly large common variability in the tropical band and in upwelling regions. Figure 10a shows the correlation map of SST and SSH detrended interannual time series computed from the CCI observational datasets. The correlation is higher than 0.7 in most of the tropics, in the California current, the Humbolt current and in the Canary current. In the Benguela current, the correlation is not significant, but it is potentially because this narrow coastal current is not well sampled by satellite altimetry and its variability in SSH is not captured properly by the CCI product (Note that the correlation is significant and high in the Benguela current in the ECCO estimate which support this hypothesis—see Fig. 10b). In the extra tropics, the correlation is in general non-significant except south-east of Greenland and in some local eddies in the Antarctic circumpolar current, in the Kuroshio extension and in the Gulf Stream. The ECCO estimate confirms this global picture and shows a similar pattern in the correlation map for SST and SSH as in the CCI correlation map (see Fig. 10b). The one difference between the ECCO correlation map and the CCI correlation map is that the ECCO estimate shows actually significant correlations in large regions of the extra tropics, but these correlations are low and below 0.6 in general.

To get insights into the cause of the common variability in SSH and SST, we correlate the detrended interannual time series of SST with the wind-driven SSH response (see Fig. 10c) and with the buoyancy-forced SSH response (see Fig. 10d). Figure 10c and d shows that, in general, SST is more closely related to the buoyancy-forced sea level response than it is to either the total (wind + buoyancy) sea level or the wind-driven sea level. In particular, for buoyancy-driven sea level, the average correlation coefficient between sea level and sea surface temperature is 0.55 and the correlation coefficient is

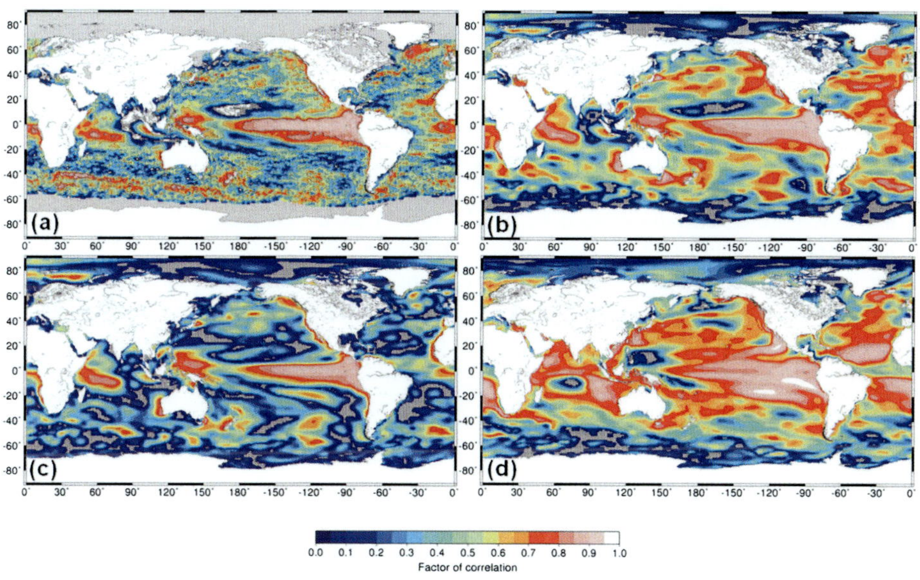

Fig. 10 Correlation coefficient between SST and SSH for CCI (**a**), between SST and total SSH for ECCO (**b**), between SST and the wind forcing driven SSH (**c**), between SST and the buoyancy-forced SSH (**d**). Only positive correlations are shown because actually no significant negative correlation between SST and SSH was found

significant over 68 % of the global ocean. These same numbers for the wind-driven sea level are 0.25 and 29 %, respectively. And these same numbers are 0.49 and 63 % for the fully forced (wind + buoyancy-driven) sea level, respectively. The reason for the high correlation between SST and buoyancy-forced SSH is that the SST variability is dominated in most regions of the ocean by the SST response to surface heat fluxes (see Sect. 3.2.2; Fig. 8b) and that in many regions also surface heat fluxes effects dominate over freshwater fluxes effects in the buoyancy-forced sea level response. As a result, surface heat fluxes appear as the main cause for the common variability in SSH and SST. In a few regions, this statement does not hold. In regions where the freshwater effect on the buoyancy forcing is dominant over the heat flux effect, like in the Pacific warm pool or in the Arctic region, the correlation of buoyancy-forced sea level with SST is not significant (see Fig. 10c). In the tropics and in the upwelling regions, the wind stress is also a cause of sizeable common variability in SST and SSH along with the surface heat flux (see Fig. 10d). This is because wind stress variability in the tropics generates at the same time a zonal pressure gradient and a zonal tilt in the thermocline, which make, respectively, sea level and SST vary in phase in this region.

The picture is different for trends in SSH and SST. The relationship between SST and buoyancy-forced sea level is somewhat less clear for trends than it is for interannual to decadal timescales. For example, the correlation coefficient between the spatial patterns in SST (Fig. 9a) and SSH (Fig. 2b) is 0.32, between SST and wind-driven SSH (Fig. 6a) is 0.15, and between SST and buoyancy-forced sea level (Fig. 6b) is 0.28. This means that at long timescales, while SST is more correlated with buoyancy-driven sea level than it is with wind-driven sea level, SST is even more correlated (but still modestly so) with the total sea level. This indicates that probably the relationship between SST and buoyancy-forced sea level is quite complex in general, and probably depends critically on timescale, among other factors. One striking feature when looking at Figs. 9a and 6b is that, out of the tropics, the SST pattern seems quite similar to the buoyancy-forced sea level except around Greenland. Around Greenland, sea level trends are more subdued, while surface temperature trends are more pronounced. This decoupling is potentially due at least in part to freshwater fluxes and halosteric sea level changes in this region. Indeed, models suggest that such salinity effects on sea level are important in this region and act to compensate and offset sea level changes due to temperature effects (i.e. thermosteric height, Köhl 2014). In the eastern tropical Pacific, the negative SST trends are in phase with the negative SSH trends which are essentially caused by wind stress. This is consistent with the increasing trade winds associated with the decreasing PDO since the late 1970s. Another interesting region is the Arctic (and to a lesser extent the south of the Southern Ocean) where the SST pattern does not correlate with any sea level pattern. The reason is probably that mass (bottom pressure) trends probably play an important role in sea level trends as well as the halosteric effects and hence make the SSH trends independent of the SST trends.

3.3 Ocean Colour and How it Relates to SST and SSH

3.3.1 General Principles

To investigate how ocean colour relates to SST and SSH, we must first understand the dominant processes driving the biophysical interactions, which include the constraints on phytoplankton dynamics imposed by the physical environment, and also the feedback mechanisms by which phytoplankton could modify their environment.

In the open ocean waters of the tropics and subtropics, light is plentiful all-year round and phytoplankton biomass increases when nutrients become available in the mixed layer through various vertical processes, such as wind mixing and upwelling. The transport of cold waters from depths to the surface, carrying nutrients with them, reduces SST. In such cases, a decrease in SST can be considered a proxy for nutrient supply. These regions are therefore characterized by a negative correlation between SST and chlorophyll concentration.

In contrast, in high latitudes, light availability is the dominant driver: phytoplankton biomass shows a clear seasonality (tied to the annual solar cycle), and surface nutrients are generally replenished by deep-mixing in winter. In these regions, the seasonal changes in light at the sea surface are accompanied by seasonal warming of the waters, such that an increase in SST may be taken here to be a proxy for increase in available light.

Sea level variations can reflect alterations in ocean circulation and ocean density, occurring in response to changes in wind forcing, ocean warming and changes in water and ice mass exchange between the land and the oceans (Church et al. 2013). Some of these processes form zones of convergence and divergence in the ocean, which are characterized by enhanced water column stratification and upwelling, respectively. In the tropics and subtropics, zonal wind stress patterns cause convergence zones and increase sea level (Palanisamy et al. 2015). Such zones are also characterized by deep thermoclines and low nutrient availability, which are unfavourable for phytoplankton production (Kahru et al. 2010); a negative correlation between SSH and chlorophyll concentration can be observed (Turk et al. 2001). In high latitudes, oceanic convergence zones also show an increase in SSH, but the conditions may still favour phytoplankton production due to enhanced upper ocean light availability and supply of nutrients from winter mixing. In these regions, positive correlations can be found between SSH and chlorophyll (e.g., Wilson and Coles 2005; Brewin et al. 2014).

3.3.2 Correlation Analyses Based on CCI Datasets

Ocean colour observations from the OC-CCI dataset show high inter-annual variability (Fig. 11a) at high latitudes and in coastal upwelling highly productive regions, whereas low variability is observed in the oligotrophic gyres (where phytoplankton production is minimum). Trends based on monthly chlorophyll anomalies are shown in Fig. 11b. Correlation coefficients calculated using monthly means and monthly anomalies over the period 1998–2010 between chlorophyll (OC-CCI) and sea level (SL-CCI), and between chlorophyll (OC-CCI) and SST (SST-CCI) are presented in Fig. 12. Although the correlation between chlorophyll and SSH tends to be weaker when compared with that between chlorophyll and SST, they display similar regional patterns throughout most of the global oceans. The tropics and subtropics typically show negative correlations (i.e. chlorophyll is low when SST and SSH are high), whereas the high latitudes typically show positive correlations (i.e. chlorophyll is high when SST and sea level are high), as expected from the rationale presented above. In high latitudes, positive correlations between chlorophyll and SST are weaker when monthly anomalies (representative of inter-annual variability) are used compared with monthly means (when the seasonality has not been removed), but they show similar patterns (Fig. 12). The correlation patterns displayed in the tropics, subtropics and high latitudes are consistent with previous studies, which have been carried out for different time periods and different satellite sensors (e.g., Wilson and Coles 2005; Brewin et al. 2012, 2014; Siegel et al. 2013). This consistency suggests that the patterns are independent of the time period selected for the

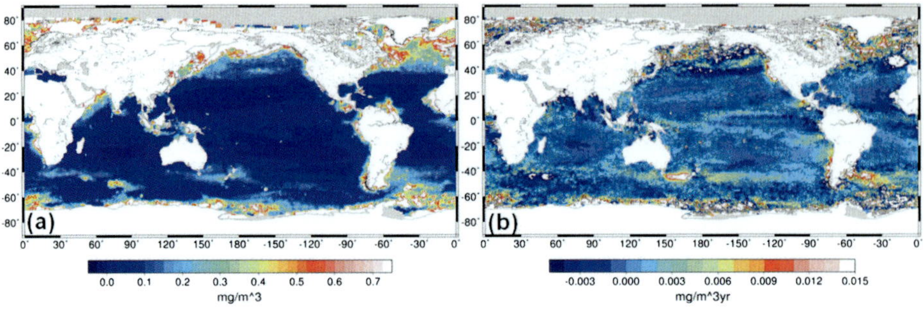

Fig. 11 **a** Standard deviation in chlorophyll concentration. **b** Linear trend in chlorophyll concentration anomalies calculated over the period 1998–2010. Only the linear regression coefficients, which are significant at the 95 % confidence level, are shown in colour. Non-significant linear regression coefficients are in *grey*

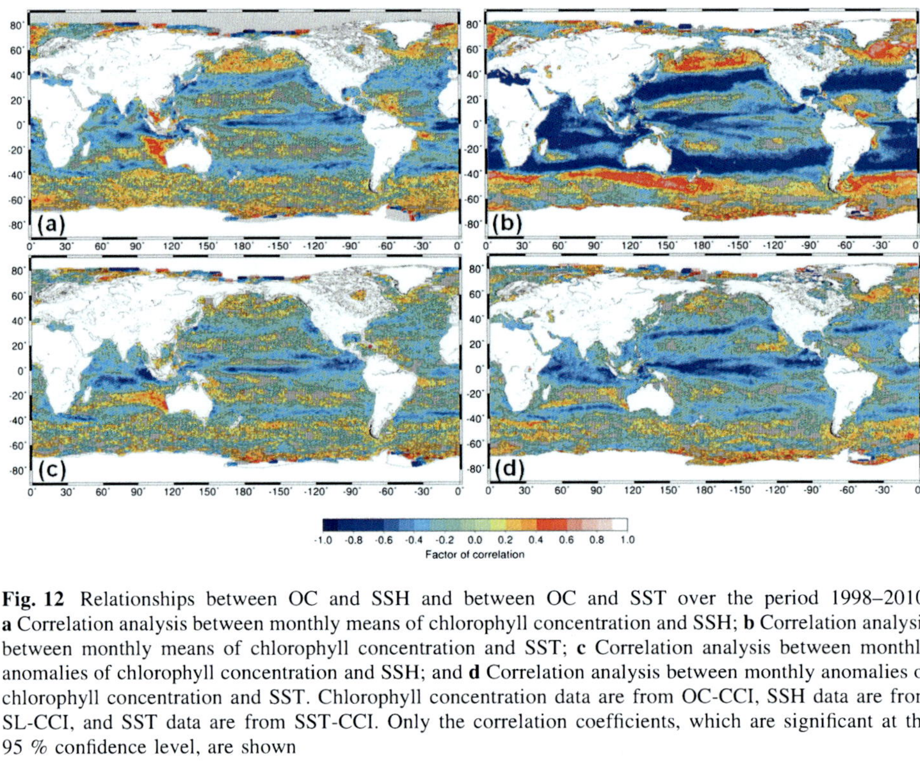

Fig. 12 Relationships between OC and SSH and between OC and SST over the period 1998–2010. **a** Correlation analysis between monthly means of chlorophyll concentration and SSH; **b** Correlation analysis between monthly means of chlorophyll concentration and SST; **c** Correlation analysis between monthly anomalies of chlorophyll concentration and SSH; and **d** Correlation analysis between monthly anomalies of chlorophyll concentration and SST. Chlorophyll concentration data are from OC-CCI, SSH data are from SL-CCI, and SST data are from SST-CCI. Only the correlation coefficients, which are significant at the 95 % confidence level, are shown

analysis and of differences in chlorophyll-retrieval and atmospheric correction algorithms used (Brewin et al. 2014; note also that OC-CCI product includes data from three sensors—SeaWiFS, MODIS and MERIS—which have been processed using SeaDAS—Fu et al. 1998—and POLYMER atmospheric correction algorithms—Steinmetz et al. 2011—and show a significant increase in data coverage (Sathyendranath and Krasemann 2014; Racault et al. 2015; Sathyendranath et al. 2016).

3.3.3 *Regional Variations*

Other considerations have to be invoked to explain variations superimposed on these large-scale, latitudinal patterns. Some of these regions are examined more closely below.

3.3.3.1 High Latitude HNLC Regions In certain regions that are known as high nutrient-low chlorophyll (HNLC) regions (Boyd et al. 2007), some trace nutrients (notably iron) may remain low, even when other nutrients such as nitrogen, phosphorus and silica are available, limiting phytoplankton production. In HNLC regions in high latitudes (including parts of the Pacific and the Southern Ocean), the correlation between chlorophyll and SST may become weak or negative, indicating that phytoplankton growth is predominantly limited by nutrient availability rather than light availability. Based on observed relationships between chlorophyll and SST and chlorophyll and SSH, Le Quéré et al. (2002) estimated that in the Southern Ocean between 40°S and 65°S, 44 % of the total primary production was predominantly light limited, and 10 % was driven by changes in stratification and nitrogen supply, which in turn was associated with changes in SST. The SST-related changes were higher in the south Atlantic part of the Southern Ocean, accounting for 30 % of changes in the total primary production. To explain the remaining variability in biological productivity, the authors also considered changes in atmospheric dust deposition. However, in this region, atmospheric dust supply tends to remain low, originating from small dust sources in Argentina, Australia and South Africa (Prospero et al. 2002) and a weak positive correlation between dust deposition and chlorophyll was found only in the west Pacific region of the Southern Ocean (Le Quéré et al. 2002). Their study demonstrates that simple assumptions about light limitation and nutrient (nitrogen and iron) limitation may not be sufficient to explain fully the observed variations in chlorophyll distribution.

3.3.3.2 Continental Shelves and Coastal Regions In continental shelves and coastal regions, additional physical processes may also control the supply of nutrient in the euphotic zone, including coastal upwelling, Ekman pumping and river outflow. In the eastern subtropical North Atlantic, between 10°N and 25°N, the negative correlation found between chlorophyll and SSH has been shown to result from an inverse relationship between SSH and nutricline depth (Pastor et al. 2014). In this region, the changes in nutricline depth and nutrient supply to the surface are governed primarily by changes in Ekman pumping (driving vertical advection processes), rather than by changes in stratification.

Atmospheric dust supply from the North African and Asian deserts, as well as coastal river discharge can also provide significant sources of nutrient supply for phytoplankton growth and modify the general latitudinal relationships between chlorophyll and SSH and between chlorophyll and SST. For instance, in the vicinity of the Amazon and Orinoco River plumes, positive relationships between chlorophyll and SST and chlorophyll and SSH are observed (Fig. 12a, b). The chlorophyll concentration and the concentration of coloured dissolved organic matter are higher in these river plumes than in surrounding waters (Smith and Demaster 1996; Hu et al. 2004). The location and direction of these river plumes are influenced by anticyclonic eddies, characteristic of higher SST and SSH, which are capable of transporting the plumes hundreds of miles offshore (Johns et al. 1990; Corredor et al. 2004).

3.3.3.3 Effect of Eddies A prominent feature is shown off Western Australia where there is a clear positive relationship between chlorophyll and SSH (Fig. 12a, c). In this region, anticyclonic eddies are thought to entrain shelf waters with regionally high chlorophyll concentration (Pearce and Griffiths 1991; Moore et al. 2007). These anticyclonic eddies then propagate westward and higher chlorophyll values are maintained through sub-mesoscale injection of nutrients into the anticyclonic-eddy boundary (Moore et al. 2007).

3.4 Bio-Optical Heating Influence on SST, Mixed Layer Depth (MLD), Air-Sea Flux and How it May Affect SSH

The absorption of solar irradiance by phytoplankton cells (described in Sect. 2.3) has been shown to influence considerably the heating rate of the upper layer of the world oceans. Using a mixed layer model that incorporates ocean colour observations, Sathyendranath et al. (1991) examined the effect of light attenuation on surface temperature by comparing model results forced by a pure-water case, and by a chlorophyll-dependent case. The authors estimated biologically induced heating of surface temperature could reach a maximum of up to +4 °C from August to September in the Arabian Sea, using satellite data for 1979. Subsequently, several investigators have used three-dimensional ocean models to study the influence of phytoplankton on SST and MLD at the regional and global scales (e.g., Nakamoto et al. 2000; Manizza et al. 2005; Wu et al. 2007; Zhai et al. 2011). As phytoplankton absorb a significant fraction of the incident solar radiation and increase the temperature in the surface layer, less radiation penetrates to greater depths. The increased temperature contrast between surface and deep waters can enhance water column stability and reduce the depth of the mixed layer. For instance, Nakamoto et al. (2000) reported, in the Arabian Sea, a bio-optical heating of SST up to +0.6 °C and mixed layer depth decrease by 20 m. During the Summer season, in the Labrador Sea, Wu et al. (2007) showed SST changes between −1.0 and +2.0 °C, and mean MLD difference of 10 m (i.e. ~20–50 % shallower than it would be if phytoplankton were absent). Using a global ocean model, Manizza et al. (2005) observed temperature differences caused by the absorption of heat by phytoplankton between −0.2 and +0.6 °C at mid and high latitudes, with minimum values in winter and maximum values during the spring bloom. In their model results, the amplitude of the seasonal changes in the water column thermal structure and the MLD increased from low to high latitudes, with maximum MLD decrease by up to 20 m observed at 60° in both hemispheres.

Solar radiation and phytoplankton are not the only factor controlling SST and MLD in the oceans. Additional physical processes such as air-sea heat exchange, wind mixing and horizontal advection play important roles (see Sect. 3.2). Zhai et al. (2011) demonstrated increased heat loss from the ocean to the air associated with bio-optical heating of the upper ocean caused by the presence of phytoplankton in the Gulf of St. Lawrence. Atmospheric heat gain induced by the presence of phytoplankton has been further examined in coupled ocean–atmosphere models, which show an amplification of the seasonal cycle of temperature in the troposphere, modulating the tropical convection patterns and atmospheric circulation (Shell et al. 2003) and affecting large patterns of climate variability such as ENSO (Zhang 2015).

The strong influence of the absorption of solar irradiance by phytoplankton cells on the heating rate of the upper layer of the world oceans suggests that bio-optical heating should have an influence not only on SST but also on SSH. This hypothesis is further supported by the fact that SST and buoyancy-forced SSH show large common variability, which is due to the surface heat forcing (see Sect. 3.2.3).

4 Conclusions

In this paper, we have analysed the regional variability of observed SSH, SST and OC from the CCI datasets over the period 1993–2011. We have focused our analysis on the signature of the ocean large-scale climate fluctuations driven by the atmospheric forcing, and we did not consider the mesoscale activity, which has been filtered out. We used the ECCO version 4 ocean reanalysis to unravel the role of ocean transport and surface buoyancy fluxes in the ocean climate fluctuations and analyse the associated signatures on the observed SSH, SST and OC. Our analyses corroborate the findings from past studies on the role of ocean transport and surface buoyancy fluxes in the SSH variations at interannual to decadal timescales (Piecuch and Ponte 2011; Piecuch et al. 2013; Fukumori and Wang 2013; Forget and Ponte 2015). They also provide new insights on the role of ocean transport and surface buoyancy fluxes in the SST variations and on the relation between the SSH, SST and OC variations at interannual to decadal timescales.

In agreement with previous studies, we show that the variability and trends in observed SSH over the last two decades are largely dominated by the steric effect except in shallow shelf seas where the mass effect is of the same order of magnitude as the steric effect and at high latitudes (>60°N and <55°S) where it dominates over the steric effect. The steric sea level signal is essentially due to temperature changes. Salinity changes play only a local role, but this role can be sizeable in several regions, in particular in the North Atlantic, in the Arctic and in the Southern Ocean. The ECCO reanalysis reveals that the observed steric sea level variability and trends are essentially forced by surface wind stress anomalies, in particular in the tropics where the SSH variability and trends are the most intense over the last two decades. The buoyancy forcing plays also a sizeable role but of smaller amplitude and more uniformly distributed. In the extra tropics, where the wind stress forcing effect on SSH is smaller, the buoyancy forcing effect becomes significant and explains a sizeable part of the SSH variability and trends. In general, on average over the ocean, the buoyancy forcing effect on SSH trends is positive which reflects the penetration of heat into the ocean and the global warming of the ocean (Gregory et al. 2001; Suzuki and Ishii 2011). This result confirms that different reasons explain the global sea level rise from ocean warming and the regional sea level rise. While the buoyancy fluxes only are responsible for the total heat that enters the ocean and the associated global mean sea level rise, both ocean transport divergences caused by wind stress anomalies and the non-uniform buoyancy forcing (essentially at mid to high latitudes) are responsible for the regional distribution of the heat within the ocean and thus for the regional sea level departures around the global mean.

Concerning the SST variability and trends over the last two decades, we show that the effect of local surface heat fluxes and ocean transport divergences locally is large and opposite to each other almost everywhere in the ocean. This finding primarily reflects the nature of the local steady-state SST budget (in our case, integrated over time), wherein local heat forcing is compensated for and entirely cancelled out by the action of ocean transport processes. However, the cancellation between local forcing and ocean transports is not exact, and the residual tendencies are what give rise to the resulting SST variability and trends. This result gives a unique perspective on the observed SST changes that, to our knowledge, has not been strongly emphasized elsewhere: almost everywhere, the mixed layer heat content is approximately in a local steady-state balance with the ocean transport divergences effect almost entirely compensating the heat fluxes effect; in this context, observed changes in SST are interpreted as relatively small departures from this background steady-state. In terms of interannual variability in SST, the effect of the surface heat flux forcing is larger than the effect of the ocean transport almost everywhere except in the tropical Pacific where both play a similar role. In terms of trends in SST, the picture is not so clear: both effects play a dominant role but in different regions.

Thus, we have provided complementary descriptions of mechanisms responsible for SSH and SST variability and trends. On the one hand, our exploration of SSH was more dynamical, focusing on the relevant forcing mechanisms (namely winds or buoyancy), without pinpointing particular processes (e.g., circulation versus mixing). On the other hand, our investigation of SST was more kinematical, targeting the underlying physical processes (e.g., advection or diffusion), without identifying the responsible drivers in all cases (i.e. wind or buoyancy). These separate analyses put on display the capabilities of the ocean reanalysis machinery, and how such frameworks can help us understand, in a very detailed fashion, the nature of oceanic variability and change. In any case, bringing together the results of our SSH and SST analyses, we can see that SSH and SST bear some common variability. The main reason is that both SSH and SST variability show significant contributions from the surface heat fluxes forcing. This is evidenced by the high correlation between SST and buoyancy-forced SSH almost everywhere except in a few regions where the freshwater effect on buoyancy-forced SSH is dominant over the heat flux effect, as in the Pacific warm pool or in the Arctic region. In the tropics and in the upwelling regions, the wind stress forcing is also a cause of sizeable common variability in SST and SSH along with the surface heat flux forcing because wind stress variability in the tropics generate at the same time a zonal pressure gradient and a zonal tilt in the thermocline, which make, respectively, sea level and SST vary in phase in this region. SSH and SST bear also some common features in their trend patterns over the last two decades. However, the picture is more complicated than for the interannual to decadal variability. This indicates that probably the relationship between SST and buoyancy-forced sea level or wind forced sea level is quite complex in general, and likely depends critically on time-scale, among other factors.

OC is a variable that is fundamentally different from SST and SSH because it depends on biological processes, which are forced by underlying physical processes. The concentration of chlorophyll, the main photosynthetic pigment present in all phytoplankton cells, can be used as key measure of the phytoplankton population and is a central variable in models used to estimate primary production (the rate at which phytoplankton produces organic matter from dissolved inorganic CO_2), and export production (approximately 20 % of net photosynthesis at global scale is exported to the deep ocean, indicating a strong coupling between biological activity and the oceanic carbon sink, Laws et al. 2000). The growth of phytoplankton is primarily governed by the availability of light and nutrients, which in turn depend mostly on climate forcing conditions. Thus, OC variability and trends bear a strong dependence on climate fluctuations, with implications for the oceanic primary production and export production, and can show significant correlations with SSH and SST. Although the correlation between OC and SSH tends to be weaker than between OC and SST, they display similar regional patterns throughout most of the global oceans. The tropics and subtropics show negative correlations because in these regions phytoplankton growth and export production are driven by nutrient availability in the mixed layer. Here, increases in nutrient availability are mainly driven by vertical entrainment (deeper mixed layers) and upwelling which influence SST (cooler waters entrained into the mixed layer) and SSH (surface divergence). At high latitudes, phytoplankton growth and export production are primarily driven by light availability. At these latitudes, changes in light at the sea surface are accompanied by warming of the waters which causes an increase in SST and buoyancy-forced sea level and explain the positive correlation between OC, SSH and SST.

As discussed earlier, phytoplankton can have a feedback effect on climate fluctuations. The absorption of solar irradiance by phytoplankton pigments tends to increase the heating rate of surface layers, stabilize the water column and reduce the depth of the mixed layer. As a result it changes the distribution of the heat and the stratification in the upper ocean. Furthermore, it has been suggested that temperature might affect primary production and remineralization differently, and hence modulate biologically mediated air-sea carbon exchange (Matsumoto 2007). The regional effect of bio-optical feedbacks on SST has been reported by several studies as noted in Sect. 3.4, but no apparent effects on SSH have been reported yet. Given the significant influence of the absorption of solar irradiance by phytoplankton on the heating rate of the upper layer, and the importance of the surface heat fluxes in the SSH variability, we may speculate that bio-optical heating could influence SSH variability indirectly. However, it is not a trivial task to evaluate the influence of bio-optical heating on SSH and, to our knowledge, it has not been undertaken yet. Noting that localization of solar heating close to the surface would alter air-sea exchange of heat, a possible starting point for future studies could be to examine the impact of phytoplankton on the total heat content of the water column, which would impact SSH. The question remains whether such an impact could be a significant one.

Acknowledgments This work was supported by the CNES. It is based on observations from Topex/Poseidon, Jason 1/2, ENVISAT, ERS1/2 and Altikaa. The authors want to thank Gaël Forget (Massachusetts Institute of Technology) for providing the output of the ECCO perturbation forcing experiments and Ou Wang (Jet Propulsion Laboratory) for providing the fields necessary for computing the SST budgets. C.G. Piecuch participation was supported by NASA grant NNX14AJ51G. This paper is an outcome of the ISSI Workshop "Integrative Study of Sea Level Budget", held in Bern in February 2015. M-F.R., S.S., and R.J.W.B.would like to acknowledge funding from the ESA Living Planet Fellowship programme, the ESA OC-CCI project and the NERC's UK National Centre for Earth Observation.

References

Ablain M, Cazenave A, Larnicol G, Balmaseda M, Cipollini P, Faugère Y, Fernandes MJ, Henry O, Johannessen JA, Knudsen P, Andersen O, Legeais J, Meyssignac B, Picot N, Roca M, Rudenko S, Scharffenberg MG, Stammer D, Timms G, Benveniste J (2015) Improved sea level record over the satellite altimetry era (1993–2010) from the Climate Change Initiative project. Ocean Sci 11:67–82. doi:10.5194/os-11-67-2015

Abraham JP, Baringer M, Bindoff NL, Boyer T, Cheng LJ, Church JA, Conroy JL, Domingues CM, Fasullo JT, Gilson J, Goni G, Good SA, Gorman JM, Gouretski V, Ishii M, Johnson GC, Kizu S, Lyman JM, Macdonald AM, Minkowycz WJ, Moffitt SE, Palmer MD, Piola AR, Reseghetti F, Schuckmann K, Trenberth KE, Velicogna I, Willis JK (2013) A review of global ocean temperature observations: implications for ocean heat content estimates and climate change. Rev Geophys 51:450–483. doi:10.1002/rog.20022

Boyd PW, Jickells T, Law CS, Blain S, Boyle EA, Buesseler KO, Coale KH, Cullen JJ, de Baar HJW, Follows M, Harvey M, Lancelot C, Levasseur M, Owens NPJ, Pollard R, Rivkin RB, Sarmiento J, Schoemann V, Smetacek V, Takeda S, Tsuda A, Turner S, Watson AJ (2007) Mesoscale iron enrichment experiments 1993–2005: synthesis and future directions. Science 315:612–617. doi:10.1126/science.1131669

Brewin RJW, Hirata T, Hardman-Mountford NJ, Lavender SJ, Sathyendranath S, Barlow R (2012) The Influence of the Indian Ocean Dipole on interannual variations in phytoplankton size structure as revealed by earth observation. Deep Sea Res II 77–80:117–127. doi:10.1016/j.dsr2.2012.04.009

Brewin RJW, Mélin F, Sathyendranath S, Steinmetz F, Chuprin A, Grant M (2014) On the temporal consistency of chlorophyll products derived from three ocean-colour sensors. ISPRS J Photogramm Remote Sens 97:171–184. doi:10.1016/j.isprsjprs.2014.08.013

Buckley MW, Ponte RM, Forget G, Heimbach P (2014) Low-frequency SST and upper-ocean heat content variability in the North Atlantic. J Clim 27:4996–5018. doi:10.1175/JCLI-D-13-00316.1

Cabanes C, Huck T, Colin de Verdière A (2006) Contributions of wind forcing and surface heating to interannual sea level variations in the Atlantic Ocean. J Phys Oceanogr 36(9):1739–1750

Cayan DR (1992) Latent and sensible heat flux anomalies over the northern oceans: driving the sea surface temperature. J Phys Oceanogr 22:859–881

Church JA, Clark PU, Cazenave A, Gregory JM, Jevrejeva S, Levermann L, Merrifield MA, Milne GA, Nerem RS, Nunn PD, Payne AJ, Pfeffer WT, Stammer D, Unnikrishnan AS (2013) Sea level change. In: Stocker TF, Qin D, Plattner GK, Tignor M, Allen SK, Boschung J, Nauels A, Xia Y, Bex V, Midgley PM (eds) Climate change 2013: the physical science basis. Contribution of Working Group I to the fifth assessment report of the Intergovernmental Panel on Climate Change. Cambridge University Press, Cambridge

Corredor JE, Morell JM, Lopez JM, Capella JE, Armstrong RA (2004) Cyclonic eddy entrains Orinoco River Plume in eastern Caribbean. EOS Trans Am Geophys Union 85(20):197–202

Dee D, Uppala S, Simmons A, Berrisford P, Poli P, Kobayashi S, Andrae U, Balmaseda M, Balsamo G, Bauer P, Bechtold P, Beljaars ACM, van de Berg L, Bidlot J, Bormann N, Delsol C, Dragani R, Fuentes M, Geer AJ, Haimberger L, Healy SB, Hersbach H, Hólm EV, Isaksen L, Kållberg P, Köhler M, Matricardi M, McNally AP, Monge-Sanz BM, Morcrette J-J, Park B-K, Peubey C, de Rosnay P, Tavolato C, Thépaut J-N, Vitart F (2011) The ERA-interim reanalysis: configuration and performance of the data assimilation system. Q J R Meteorol Soc 137(656):553–597. doi:10.1002/qj.828

Deser C, Alexander MA, Timlin MS (2003) Understanding the persistence of sea surface temperature anomalies in midlatitudes. J. Clim 16:57–72. doi:10.1175/1520-0442(2003)016<0057:UTPOSS>2.0.CO;2

Deser C, Alexander MA, Xie SP, Phillips AS (2010) Sea surface temperature variability: patterns and mechanisms. Ann Rev Mar Sci 2:115–143. doi:10.1146/annurev-marine-120408-151453

Durack PJ, Wijffels SE, Gleckler PJ (2014) Long-term sea-level change revisited: the role of salinity. Environ Res Lett 9:114017. doi:10.1088/1748-9326/9/11/114017

Edwards AM, Platt T, Wright DG (2001) Biologically induced circulation at fronts. J Geophys Res 106:7081–7095. doi:10.1029/2000JC000332

Forget G, Ponte RM (2015) The partition of regional sea level variability. Prog Oceanogr 137:173–195. doi:10.1016/j.pocean.2015.06.002

Forget G, Campin JM, Heimbach P, Hill CN, Ponte RM, Wunsch C (2015) ECCO version 4: an integrated framework for non-linear inverse modeling and global ocean state estimation. Geosci Model Dev 8:3071–3104. doi:10.5194/gmd-8-3071-2015

Frankcombe LM, Spence P, Hogg AM, England MH, Griffies SM (2013) Sea level changes forced by Southern Ocean winds. Geophys Res Lett 40:5710–5715. doi:10.1002/2013GL058104

Frankignoul C, Hasselmann K (1977) Stochastic climate models. Part 2. Application to sea-surface temperature variability and thermocline variability. Tellus 29:284–305. doi:10.1111/j.2153-3490.1977.tb00740.x

Fu G, Baith KS, McClain CR (1998) The SeaWiFS data analysis system. In: Proceedings of the 4th Pacific Ocean remote sensing conference, Qingdao, July 1998, pp 73–79

Fukumori I, Wang O (2013) Origins of heat and freshwater anomalies underlying regional decadal sea level trends. Geophys Res Lett 40:563–567. doi:10.1002/grl.50164

Fukumori I, Wang O, Llovel W, Fenty I, Forget G (2015) A near-uniform fluctuation of ocean bottom pressure and sea level across the deep ocean basins of the Arctic Ocean and the Nordic Seas. Prog Oceanogr 134:152–172. doi:10.1016/j.pocean.2015.01.013

Gill A, Niller P (1973) The theory of the seasonal variability in the ocean. Deep Sea Res 20(2):141–177. doi:10.1016/0011-7471(73)90049-1

Gregory JM, Church JA, Boer GJ, Dixon KW, Flato GM, Jackett DR, Lowe JA, O'Farrell SP, Roeckner E, Russell GL, Stouffer RJ, Winton M (2001) Comparison of results from several AOGCMs for global and regional sea-level change 1900–2100. Clim Dyn 18:225–240. doi:10.1007/s003820100180

Grodsky SA, Carton JA, Liu H (2008) Comparison of bulk sea surface and mixed layer temperatures. J Geophys Res 113:C10026. doi:10.1029/2008JC004871

Hu C, Montgomery ET, Schmitt RW, Muller-Karger FE (2004) The dispersal of the Amazon and Orinoco River water in the tropical Atlantic and Caribbean Sea: observation from space and S-PALACE floats. Deep Sea Res II 51:1151–1171. doi:10.1016/j.dsr2.2004.04.001

Ishii M, Kimoto M (2009) Reevaluation of historical ocean heat content variations with time-varying XBT and MBT depth bias corrections. J Oceanogr 65:287–299. doi:10.1007/s10872-009-0027-7

Johns WE, Lee TN, Schott F, Zantopp RJ, Evans RH (1990) The North Brazil Current retroflection: seasonal structure and eddy variability. J Geophys Res 95:22103–22120. doi:10.1029/JC095iC12p22103

Kahru M, Gille ST, Murtugudde R, Strutton PG, Manzano-Sarabia M, Wang H, Mitchell BG (2010) Global correlations between winds and ocean chlorophyll. J Geophys Res. doi:10.1029/2010JC006500

Kirk JTO (1994) Light and photosynthesis in aquatic ecosystems, 2nd edn. Cambridge University Press, New York

Köhl A (2014) Detecting processes contributing to interannual halosteric and thermosteric sea level variability. J Clim 27:2417–2426. doi:10.1175/JCLI-D-13-00412.1

Laws EA, Falkowski PG, Smith WOS Jr, Ducklow H, McCarthy JJ (2000) Temperature effects on export production in the open ocean. Global Biogeochem Cycles 14:1231–1246

Le Quéré C, Bopp L, Tegen I (2002) Antarctic circumpolar wave impact on marine biology: a natural laboratory for climate change study. Geophys Res Lett. doi:10.1029/2001GL014585

Lewis MR, Cullen JJ, Platt T (1983) Phytoplankton and thermal structure in the upper ocean: consequences of nonuniformity in chlorophyll profile. J Geophys Res 88:2565–2570. doi:10.1029/JC088iC04p02565

Llovel W, Lee T (2015) Importance and origin of halosteric contribution to sea level change in the southeast Indian Ocean during 2005–2013. Geophys Res Lett 42:1148–1157. doi:10.1002/2014GL062611

Makowski JK, Chambers DP, Bonin JA (2015) Using ocean bottom pressure from the gravity recovery and climate experiment (GRACE) to estimate transport variability in the southern Indian Ocean. J Geophys Res Oceans 120:4245–4259. doi:10.1002/2014JC010575

Manizza M, Le Quere C, Watson AJ, Buitenhuis ET (2005) Bio-optical feedbacks among phytoplankton, upper ocean physics and sea-ice in a global model. Geophys Res Lett. doi:10.1029/2004GL020778

Marshall J, Johnson H, Goodman J (2001) A study of the interaction of the North Atlantic Oscillation with ocean circulation. J Clim 14:1399–1421. doi:10.1175/1520-0442(2001)014<1399:ASOTIO>2.0.CO;2

Matsumoto K (2007) Biology-mediated temperature control on atmospheric pCO_2 and ocean biogeochemistry. Geophys Res Lett. doi:10.1029/2007GL031301

McGregor S, Gupta AS, England MH (2012) Constraining wind stress products with sea surface height observations and implications for Pacific ocean sea level trend attribution. J Clim 25(23):8164–8176. doi:10.1175/JCLI-D-12-00105.1

Meehl GA, Hu A, Arblaster JM, Fasullo J, Trenberth KE (2013) Externally forced and internally generated decadal climate variability associated with the Interdecadal Pacific Oscillation. J Clim 26:7298–7310. doi:10.1175/JCLI-D-12-00548.1

Merchant CJ, Embury O, Roberts-Jones J, Fiedler E, Bulgin CE, Corlett GK, Good S, McLaren A, Rayner N, Morak-Bozzo S, Donlon C (2014) Sea surface temperature datasets for climate applications from phase 1 of the European Space Agency Climate Change Initiative (SST CCI). Geosci Data J 1:179–191. doi:10.1002/gdj3.20

Merrifield MA (2011) A shift in western tropical Pacific sea level trends during the 1990s. J Clim 24:4126–4138. doi:10.1175/2011JCLI3932.1

Meyssignac B, Salas y Melia D, Becker M, Llovel W, Cazenave A (2012) Tropical Pacific spatial trend patterns in observed sea level: internal variability and/or anthropogenic signature? Clim Past 8:787–802. doi:10.5194/cp-8-787-2012

Moon JH, Song YT (2013) Sea level and heat content changes in the western North Pacific. J Geophys Res Oceans 118:2014–2022. doi:10.1002/jgrc.20096

Moore TS, Matear RJ, Marra J, Clementson L (2007) Phytoplankton variability off the Western Australian Coast: mesoscale eddies and their role in cross-shelf exchange. Deep Sea Res II 54:943–960. doi:10.1016/j.dsr2.2007.02.006

Nakamoto S, Kumar SP, Oberhuber JM, Muneyama K, Frouin R (2000) Chlorophyll modulation of sea surface temperature in the Arabian Sea in a mixed-layer isopycnal general circulation model. Geophys Res Lett 27:747–750. doi:10.1029/1999GL002371

Palanisamy H, Cazenave A, Delcroix T, Meyssignac B (2015) Spatial trend patterns in the Pacific Ocean sea level during the altimetry era: the contribution of thermocline depth change and internal climate variability. Ocean Dyn. doi:10.1007/s10236-014-0805-7

Pastor MV, Palter JB, Pelegri JL, Dunne JP (2014) Physical drivers of interannual chlorophyll variability in the eastern subtropical North Atlantic. J Geophys Res 118:3871–3886. doi:10.1002/jgrc.20254

Pearce AF, Griffiths RW (1991) The mesoscale structure of the Leeuwin current—a comparison of laboratory models and satellite imagery. J Geophys Res 96:16739–16757. doi:10.1029/91JC01712

Penduff T, Juza M, Barnier B, Zika J, Dewar WK, Treguier AM, Molines JM, Audiren N (2011) Sea level expression of intrinsic and forced ocean variabilities at interannual time scales. J Clim 24:5652–5670. doi:10.1175/JCLI-D-11-00077.1

Peralta-Ferriz C, Morison JH, Wallace JM, Bonin JA, Zhang JL (2014) Arctic Ocean circulation patterns revealed by GRACE. J Clim 27:1445–1468. doi:10.1175/JCLI-D-13-00013.1

Piecuch CG, Ponte RM (2011) Mechanisms of interannual steric sea level variability. Geophys Res Lett. doi:10.1029/2011GL048440

Piecuch CG, Ponte RM (2012) Importance of circulation changes to Atlantic heat storage rates on seasonal and interannual time scales. J Clim 25:350–362. doi:10.1175/JCLI-D-11-00123.1
Piecuch CG, Ponte RM (2013) Buoyancy-driven interannual sea level changes in the tropical South Atlantic. J Phys Oceanogr 43:533–547. doi:10.1175/JPO-D-12-093.1
Piecuch CG, Quinn KJ, Ponte RM (2013) Satellite-derived interannual ocean bottom pressure variability and its relation to sea level. Geophys Res Lett 40:3016–3110. doi:10.1002/grl.50549
Piecuch CG, Fukumori I, Ponte RM, Wang O (2015) Vertical structure of ocean pressure variations with application to satellite-gravimetric observations. J Atmos Ocean Tech 32:603–613. doi:10.1175/JTECH-D-14-00156.1
Ponte RM, Piecuch CG (2014) Interannual bottom pressure signals in the Australian-Antarctic and Bellingshausen basins. J Phys Oceanogr 44(5):1456–1465. doi:10.1175/JPO-D-13-0223.1
Prospero JM, Ginoux P, Torres O, Nicholson SE, Gill TE (2002) Environmental characterization of global sources of atmospheric soil dust identified with the Nimbus 7 total ozone mapping spectrometer (TOMS) absorbing aerosol product. Rev Geophys 40:1002. doi:10.1029/2000RG000095
Purkey SG, Johnson GC, Chambers DP (2014) Relative contributions of ocean mass and deep steric changes to sea level rise between 1993 and 2013. J Geophys Res Oceans 119:7509–7522. doi:10.1002/2014jc010180
Qiu B, Chen S (2012) Multidecadal sea level and gyre circulation variability in the northwestern tropical Pacific Ocean. J Phys Oceanogr 42:193–206. doi:10.1175/JPO-D-11-061.1
Racault MF, Raitsos DE, Berumen ML, Brewin RJW, Platt T, Sathyendranath S, Hoteit I (2015) Phytoplankton phenology indices in coral reef ecosystems: application to ocean-color observations in the Red Sea. Remote Sens Environ 160:222–234. doi:10.1016/j.rse.2015.01.019
Rhein M, Rintoul SR, Aoki S, Campos E, Chambers D, Feely RA, Gulev S, Johnson GC, Josey SA, Kostianoy A, Mauritzen C, Roemmich D, Talley LD, Wang F (2013) Observations: ocean. In: Stocker TF, Qin D, Plattner GK, Tignor M, Allen SK, Boschung J, Nauels A, Xia Y, Bex V, Midgley PM (eds) Climate change 2013: the physical science basis. Contribution of Working Group I to the fifth assessment report of the Intergovernmental Panel on Climate Change. Cambridge University Press, Cambridge, pp 255–316. doi:10.1017/CBO9781107415324.010
Sathyendranath S, Krasemann H (2014) Climate assessment report: Ocean Colour Climate Change Initiative (OC-CCI)—phase one (ESA OC-CCI, http://www.esa-oceancolour-cci.org/?q=documents)
Sathyendranath S, Gouveia AD, Shetye SR, Ravindran P, Platt T (1991) Biological control of surface temperature in the Arabian Sea. Nature 349:54–56. doi:10.1038/349054a0
Sathyendranath S, Brewin RJW, Brockmann C, Brotas V, Ciavatta S, Chuprin A, Couto AB, Doerffer R, Dowell M, Grant M, Groom S, Horseman A, Jackson T, Krasemann H, Lavender S, Martinez Vicente V, Mélin Moore TS, Müller D, Regner P, Roy S, Steinmetz F, Swinton J, Taberner M, Thompson A, Valente A, Zühlke M, Brando VE, Feldman G, Franz B, Frouin R, Gould Jr. RW, Hooker S, Kahru M, Mitchell MG, Muller-Karger F, Sosik HM, Voss KJ, Werdell J, Platt T (2016) Creating an ocean-colour time series for use in climate studies: the experience of the Ocean-Colour Climate Change Initiative. Remote Sens Environ. Under review
Sérazin G, Penduff T, Grégorio S, Barnier B, Molines JM, Terray L (2015) Intrinsic variability of sea level from global ocean simulations: spatiotemporal scales. J Clim 28:4279–4292. doi:10.1175/JCLI-D-14-00554.1
Shell KM, Frouim R, Nakamoto S, Somerville RCJ (2003) Atmospheric response to solar radiation absorbed by phytoplankton. J Geophys Res. doi:10.1029/2003JD003440
Siegel DA, Behrenfeld MJ, Maritorena S, McClain CR, Antoine D, Bailey SW, Bontempi PS, Boss ES, Dierssen HM, Doney SC, Eplee RE Jr, Evans RH, Feldman GC, Fields E, Franz BA, Kuring NA, Mengelt C, Nelson NB, Patt FS, Robinson WD, Sarmiento JL, Swan CM, Werdell PJ, Westberry TK, Wilding JG, Yoder JA (2013) Regional to global assessments of phytoplankton dynamics from the SeaWiFS mission. Remote Sens Environ 135:77–91. doi:10.1016/j.rse.2013.03.025
Smith WO, Demaster DJ (1996) Phytoplankton biomass and productivity in the Amazon River plume: correlation with seasonal river discharge. Cont Shelf Res 16:291–319. doi:10.1016/0278-4343(95)00007-N
Stammer D, Cazenave A, Ponte RM, Tamisiea ME (2013) Causes for contemporary regional sea level changes. Ann Rev Mar Sci 5:21–46. doi:10.1146/annurev-marine-121211-172406
Steinmetz F, Deschamps PY, Ramon D (2011) Atmospheric correction in presence of sun glint: application to MERIS. Opt Express 19:9783–9800. doi:10.1364/OE.19.009783
Suzuki T, Ishii M (2011) Regional distribution of sea level changes resulting from enhanced greenhouse warming in the model for interdisciplinary research on climate version 3.2. Geophys Res Lett 38:L02601. doi:10.1029/2010GL045693

Tamisiea ME (2011) Ongoing glacial isostatic contributions to observations of sea level change. Geophys J Int 186:1036–1044. doi:10.1111/j.1365-246X.2011.05116.x

Thompson LA, Ladd CA (2004) The response of the North Pacific Ocean to decadal variability in atmospheric forcing: wind versus buoyancy forcing. J Phys Oceanogr 34:1373–1386. doi:10.1175/1520-0485(2004)034<1373:TROTNP>2.0.CO;2

Timmermann A, McGregor S, Jin FF (2010) Wind effects on past and future regional Sea level trends in the Southern Indo-Pacific*. J Clim 23(16):4429–4437

Turk D, McPhaden MJ, Busalacchi AJ, Lewis MR (2001) Remotely Sensed biological production in the equatorial Pacific. Science 293:471–474. doi:10.1126/science.1056449

Visbeck M, Chassignet EP, Curry RG, Delworth TL, Dickson RR, Krahmann K (2003) The ocean's response to North Atlantic Oscillation variability. In: Hurrell JW, Kushnir Y, Ottersen G, Visbeck M (eds) The North Atlantic Oscillation: climatic significance and environmental impact. American Geophysical Union, Washington, DC. doi:10.1029/134GM06

Volkov DL (2014) Do the North Atlantic winds drive the nonseasonal variability of the Arctic Ocean sea level? Geophys Res Lett 41:2041–2047. doi:10.1002/2013GL059065

Volkov DL, Landerer FW (2013) Non-seasonal fluctuations of the Arctic Ocean mass observed by the GRACE satellites. J Geophys Res Oceans 118:6451–6460. doi:10.1002/2013JC009341

Wilson C, Coles VJ (2005) Global climatological relationships between satellite biological and physical observations and upper ocean properties. J Geophys Res 110:C10001. doi:10.1029/2004JC002724

Wu Y, Tang CCL, Sathyendranath S, Platt T (2007) The impact of bio-optical heating on the properties of the upper ocean: a sensitivity study using a 3-D circulation model for the Labrador Sea. Deep Sea Res II 54:2630–2642. doi:10.1016/j.dsr2.2007.08.019

Wunsch C, Heimbach P (2007) Practical global oceanic state estimation. Phys D 230:197–208. doi:10.1016/j.physd.2006.09.040

Wunsch C, Stammer D (1997) Atmospheric loading and the oceanic "inverted barometer" effect. Rev Geophys 35:79–107. doi:10.1029/96RG03037

Wunsch C, Ponte RM, Heimbach P (2007) Decadal trends in sea level patterns: 1993–2004. J Clim 20:5889–5911. doi:10.1175/2007JCLI1840.1

Zaneveld JRV, Kitchen JC, Pak H (1981) The influence of optical water types on the heating rate of a constant depth mixed layer. J Geophys Res 86:6426–6428

Zhai L, Tang CCL, Platt T, Sathyendranath S (2011) Ocean response to attenuation of visible light by phytoplankton in the Gulf of St. Lawrence. J Mar Syst 88:285–297. doi:10.1016/j.jmarsys.2011.05.005

Zhang RH (2015) An ocean-biology-induced negative feedback on ENSO as derived from a hybrid coupled model of the tropical Pacific. J Geophys Res Oceans. doi:10.1002/2015JC011305

Surv Geophys (2017) 38:217–250
DOI 10.1007/s10712-016-9386-y

Spatial Patterns of Sea Level Variability Associated with Natural Internal Climate Modes

Weiqing Han[1] · Gerald A. Meehl[2] · Detlef Stammer[3] · Aixue Hu[2] · Benjamin Hamlington[4] · Jessica Kenigson[1] · Hindumathi Palanisamy[5] · Philip Thompson[6]

Received: 3 April 2016 / Accepted: 17 September 2016 / Published online: 4 October 2016

Abstract Sea level rise (SLR) can exert significant stress on highly populated coastal societies and low-lying island countries around the world. Because of this, there is huge societal demand for improved decadal predictions and future projections of SLR, particularly on a local scale along coastlines. Regionally, sea level variations can deviate considerably from the global mean due to various geophysical processes. These include changes of ocean circulations, which partially can be attributed to natural, internal modes of variability in the complex Earth's climate system. Anthropogenic influence may also contribute to regional sea level variations. Separating the effects of natural climate modes and anthropogenic forcing, however, remains a challenge and requires identification of the imprint of specific climate modes in observed sea level change patterns. In this paper, we review our current state of knowledge about spatial patterns of sea level variability associated with natural climate modes on interannual-to-multidecadal timescales, with particular focus on decadal-to-multidecadal variability. Relevant climate modes and our current state of understanding their associated sea level patterns and driving mechanisms are elaborated separately for the Pacific, the Indian, the Atlantic, and the Arctic and Southern Oceans. We also discuss the issues, challenges and future outlooks for understanding the regional sea level patterns associated with climate modes. Effects of these

✉ Weiqing Han
weiqing.han@colorado.edu

[1] Department of Atmospheric and Oceanic Sciences, University of Colorado, UCB 311, Boulder, CO 80309, USA

[2] Climate and Global Division, National Center for Atmospheric Research, Boulder, CO 80307, USA

[3] Oceanography, Remote Sensing of the Earth System and Coupled Climate Assimilation, Institut für Meereskunde, Universität Hamburg, Bundesstr. 53, 20146 Hamburg, Germany

[4] Department of Ocean, Earth and Atmospheric Science, Old Dominion University, Norfolk, VA 23529, USA

[5] LEGOS – OMP – CNES, UMR5566, 31401 Toulouse, CEDEX 9, France

[6] University of Hawaii Sea Level Center, Honolulu, HI 96822, USA

internal modes have to be taken into account in order to achieve more reliable near-term predictions and future projections of regional SLR.

Keywords Spatial patterns of sea level · Climate modes · Decadal sea level variability · Regional sea level change

1 Introduction

Sea level rise (SLR) is an important indicator for climate change, with direct impacts on coastal society and island countries and far-reaching effects on global population and economy. For this reason, there is huge societal demand for improved projections of future sea level change, particularly at local scale along coastlines (e.g., Milne et al. 2009; Church et al. 2011, 2013; National Research Council (NRC) Report 2012). In situ and satellite observations show that during the past few decades regional changes of sea level can deviate considerably from the global mean. For instance, since the early 1990s the rate of SLR trend in the western tropical Pacific was about three times the global mean value, whereas in the eastern tropical Pacific sea level varied very little (e.g., Merrifield 2011; McGregor et al. 2012; Han et al. 2014a; Thompson et al. 2014). These regional changes are shown to be associated with basin-wide spatial patterns, which exhibit distinct decadal variations (e.g., Lee and McPhaden 2008). For simplicity, in this paper "decadal variability" is collectively referred to as variations from one to a few decades (including multidecadal trend).

Various factors can cause sea level to change at regional or local scales (e.g., Stammer et al. 2013): changes in atmospheric and oceanic circulations (often referred to as dynamic change), large-scale deformation of ocean basins, variation in Earth's gravity field and local land movement (e.g., Church et al. 2013; Stammer et al. 2013; Kopp et al. 2015). Dynamic sea level change induced by changes in atmospheric and oceanic circulations is a major cause for contemporary decadal sea level variability as opposed to long-term anthropogenic changes, and a large fraction of the dynamic sea level change can be associated with natural internal climate modes in the Earth's coupled climate system (Stammer et al. 2013).

In this review, we summarize our current state of knowledge regarding the spatial patterns of sea level variability associated with natural climate modes, with particular emphasis on decadal timescales. We will also identify major science issues and challenges for understanding and extracting the imprints of internal climate modes in observed sea level change patterns, with a hope of contributing to decadal sea level predictions, which emerge as pressing priorities in climate research today (e.g., Goddard et al. 2009; Hurrell et al. 2009; Meehl et al. 2009, 2014; Pohlmann et al. 2009; Doblas-Reyes et al. 2011; Polkova et al. 2015). In Sects. 2–5 below, we review our current understanding of spatial patterns of sea level variability associated with climate modes in the Pacific, Indian, Atlantic, and Arctic and Southern Oceans, respectively, delineating the related oceanic processes whenever possible. In Sect. 6, we first provide a summary and then discuss remaining issues and challenges for future research on sea level variability associated with climate modes.

2 The Pacific

2.1 PDO-Related Sea Level Patterns

In the western tropical Pacific Ocean, intensified SLR has been observed since the early 1990s compared to the preceding decades (e.g., Merrifield 2011). Modeling studies suggest that warming of the tropical Indian and Atlantic Oceans enhances surface easterly trade winds and thus contributes to the intensified SLR (e.g., Luo et al. 2012; Han et al. 2014a; Hamlington et al. 2014; England et al. 2014; McGregor et al. 2014); however, a large portion of this rapid SLR—together with weak falls in the eastern basin—is part of the basin-scale sea level pattern associated with the Pacific Decadal Oscillation (PDO) or decadal variability of ENSO (Bromirski et al. 2011; Merrifield et al. 2012; Meyssignac et al. 2012; Zhang and Church 2012; Hamlington et al. 2013, 2014; Moon et al. 2013; Han et al. 2014a; Thompson et al. 2014; Palanisamy et al. 2015). The PDO is defined as the leading empirical orthogonal function (EOF) of sea surface temperature (SST) anomaly over the North Pacific (>20°N), and the leading principal component (PC1) is referred to as the PDO index (e.g., Mantua et al. 1997; Minobe 1997; Zhang et al. 1997; Garreaud and Battisti 1999; see review papers by Alexander 2010; Liu 2012). It is significantly correlated with ENSO (e.g., Alexander et al. 2002; Newman et al. 2003; Deser et al. 2004; Schneider and Cornuelle 2005; Vimont 2005). On decadal timescales, the PDO is highly correlated with ENSO (Zhang and Church 2012) and the Interdecadal Pacific Oscillation (IPO), a basin-wide decadal climate mode associated with decadal SST variability in the Pacific (e.g., Power et al. 1999; Folland et al. 2002; Meehl and Hu 2006). The correlation coefficients for PDO-IPO and IPO-NINO3.4 indices (8-year low-passed) are both 0.88 over the period of 1900–2008 (Han et al. 2014a; Zhang and Church 2012). Given their high correlations on decadal timescales, it has been suggested that the IPO may not be confidently treated as an independent climate mode to ENSO decadal variability (e.g., Trenberth et al. 2007) or that the PDO is a statistic mode rather than a physical mode with a single mechanism (see review by Newman et al. 2016). For consistency, unless specified otherwise, we will use the term PDO hereafter to represent decadal ENSO variability, PDO and IPO.

Both EOF analysis and multiple linear regression have been used to obtain the basin-scale spatial patterns of sea level variations associated with the PDO. Hamlington et al. (2013) performed EOF analysis on 20-year sliding trend maps of the annual mean reconstructed sea level data (e.g., Hamlington et al. 2011) for the 1950–2010 period, and showed that the leading EOF (EOF1) of global sea level change exhibits distinct spatial patterns (Fig. 1a), with SLR in the western tropical Pacific and Subtropical Gyre regions corresponding to sea level fall in the eastern basin in both hemispheres during a negative phase. Its temporal variability (PC1) is highly correlated with the PDO index, with a correlation coefficient of 0.96 (Fig. 1b) (see also Di Lorenzo et al. 2008). Consequently, EOF1 was defined as the PDO-related sea level variability, which explains 41 % variance (Hamlington et al. 2013) and dominates the satellite-observed, basin-wide sea level trends from 1993 to 2010 (Hamlington et al. 2014; Fig. 2a–c). The EOF1 patterns shown in Fig. 1a resemble the multiple linear regression patterns over the Pacific Ocean (e.g., Zhang and Church 2012; Si and Xu 2014; Frankcombe et al. 2015) (second row of Fig. 3) by regressing the observed, ocean model-simulated and reanalysis sea level data onto the PDO index. Regionally, they are similar to the North Pacific EOF1 of the upper 500 m thermosteric sea level from 1950 to 1998 (Lombard et al. 2005) and tropical Pacific EOF1 of

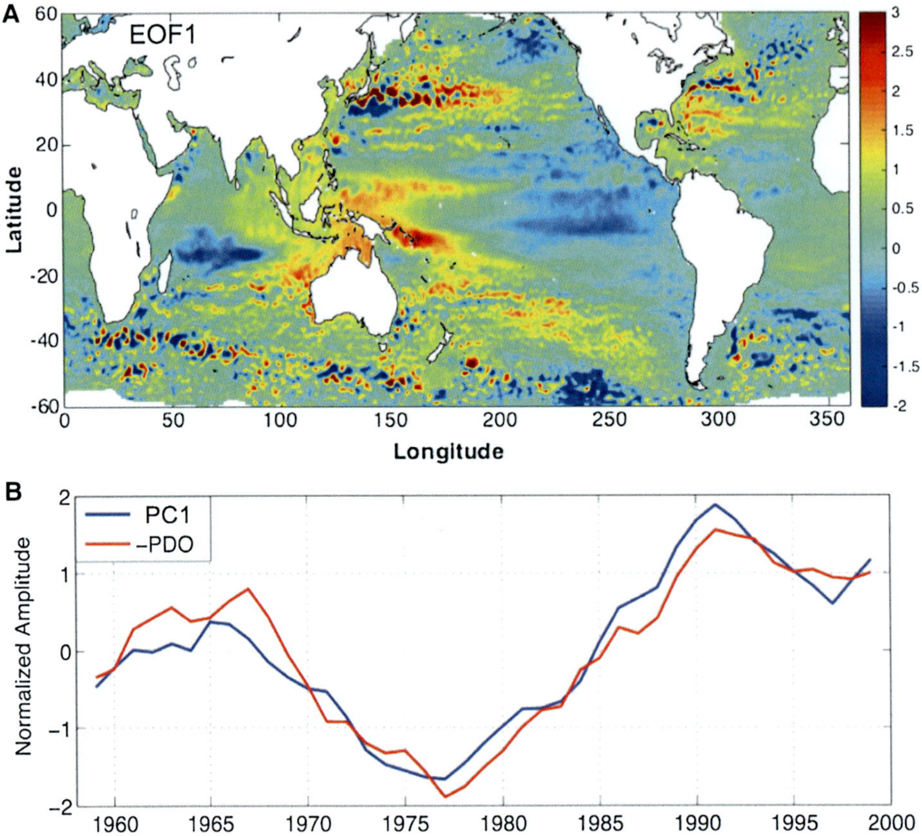

Fig. 1 **a** Spatial pattern of the leading EOF (EOF1) of 20-year sliding trends of reconstructed sea level dataset (Hamlington et al. 2011) for the 1958–1999 period, **b** principal component (PC) for the leading EOF (*blue*) and reversed PDO index (*red*). Customized from Hamlington et al. (2013)

reconstructed sea level from 1950 to 2009 (Meyssignac et al. 2012). Many studies have shown that decadal sea level variability in different regions of the Pacific including the marginal seas and US west coast is significantly correlated with the PDO index (e.g., Bromirski et al. 2011; Merrifield 2011; Marcos et al. 2012; Zhang and Church 2012).

For the period of 1998–2007, the decadal trend of sea level shows evidently different spatial patterns: Sea level is higher than normal in the central Pacific flanked by lower than normal sea level on either side of the basin (Fig. 4a). Compared to the PDO-related sea level patterns since the 1950s (Figs. 1, 3), this abnormal condition is due to the frequent occurrence of El Niño Modoki (or central Pacific El Niño) events during 2000–2004, which are associated with wind convergence to the dateline (Behera and Yamagata 2010; Fig. 4b). The sea level rise in the central Pacific succeeded a phase of lower than normal sea level associated with La Niña Modoki events toward the end of the 1990s (Behera and Yamagata 2010). This result suggests that decadal changes in ENSO behavior and its associated winds will induce changes in the spatial patterns of decadal sea level variations.

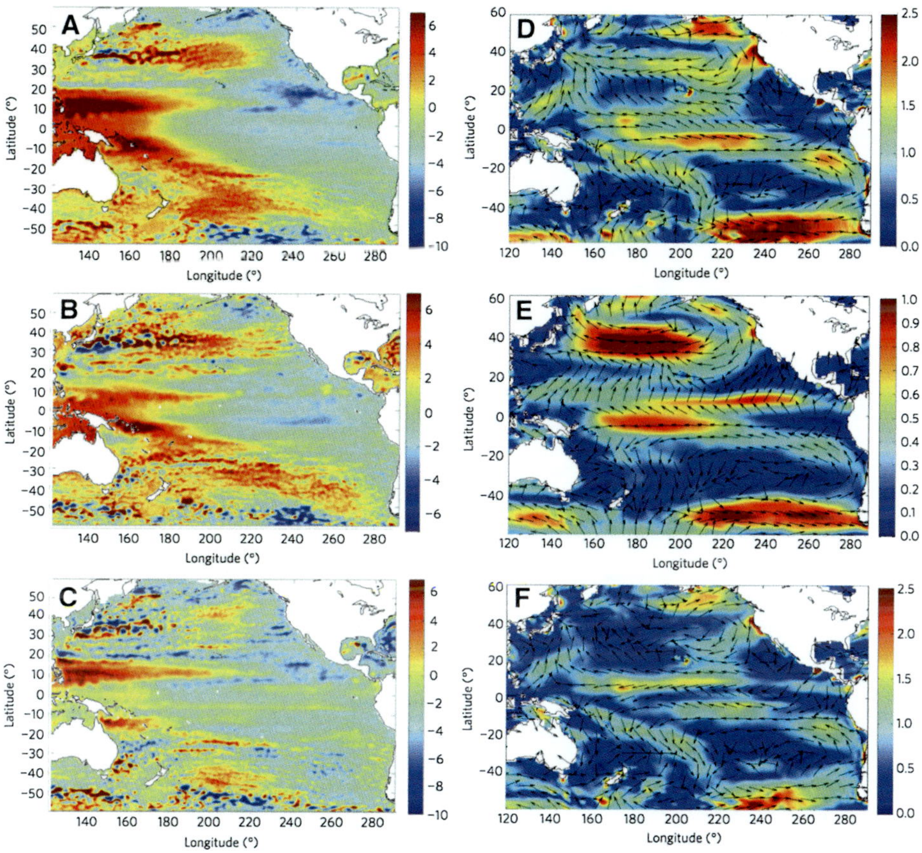

Fig. 2 **a** Satellite-observed sea level trends (mm year^{-1}) from 1993 to 2010 for Archiving, Validation and Interpretation of Satellite Oceanographic (AVISO) data, **b** PDO contribution estimated by the leading EOF of reconstructed sea level (see Fig. 1), **c** AVISO minus the PDO contribution, which is panel **a** minus panel **b**. The global mean sea level trend has been removed from the AVISO data. **d** Wind stress trends (mPa year^{-1}) from 1993 to 2010 from Operational Ocean Re-Analysis Series 3 (ORA-S3) data; **e** PDO contribution, **f** ORA-S3 trends minus PDO contributed trends, which is panel **d** minus panel **e**. Customized from Hamlington et al. (2014)

2.2 NPGO-Related Sea Level Patterns

In addition to the PDO, the North Pacific Gyre Oscillation (NPGO) is also associated with distinct spatial patterns of sea level change. The NPGO is defined as the second EOF of sea surface height anomaly (SSHa) over the Northeast Pacific region (180°W–110°W; 25°N–62°N) (Di Lorenzo et al. 2008). Its spatial patterns reflect a pair of counter-rotating gyres, with a positive NPGO corresponding to low sea level of the Alaskan Gyre in the north and high sea level in the Subtropical Gyre to the south (Fig. 5). It is argued that the NPGO is not limited to the Northeast Pacific, but exhibits global signatures in both SSH (Fig. 5, bottom panel) and SST fields (Di Lorenzo et al. 2008). Indeed, Merrifield (2011) showed that decadal sea level variations in some regions of the western tropical Pacific have higher correlations with the NPGO than with the PDO.

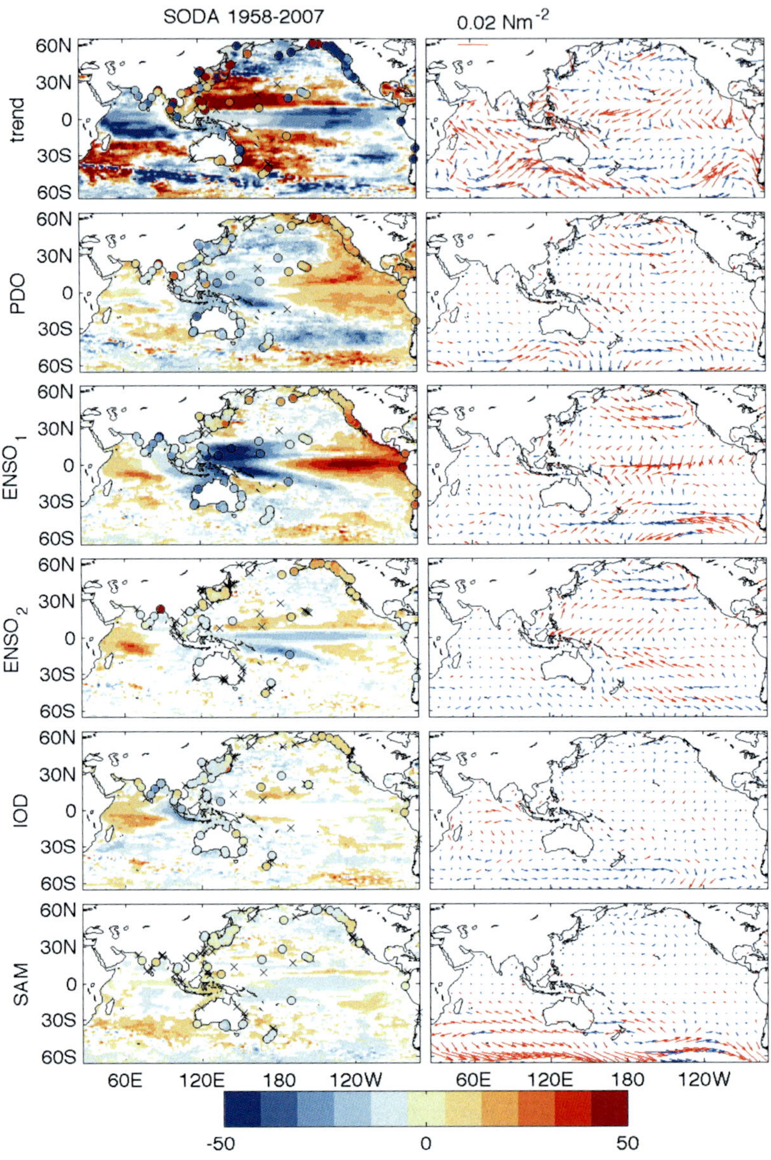

Fig. 3 (*Left column*) Regression coefficients between climate indices and sea level from 50-year (1958–2007) SODA product over 65°S–65°N of the Indo-Pacific basin (25°E–70°W), (*Right column*) Regression coefficients between climate indices and surface wind stress from ERA40 wind before 2001 and ERA-Interim wind after 2001. Tide gauges are shown as *colored circles*, and *crosses* indicate where the tide gauge regressions are not significant. *Red vectors* indicate significance at 95 % level. All indices were smoothed using a 5-month running mean with long-term mean seasonal cycle removed. The PDO index has the high-frequency component removed by smoothing the monthly PDO index with successive 25- and 37-month running mean. Using IPO index yields similar results. The ENSO indices represent interannual variability, having the low-frequency component removed. The first ENSO index (ENSO1) is the commonly used Multivariate ENSO Index, and ENSO2 describes the nonlinear atmospheric response to SST anomalies associated with the combination of ENSO and the annual cycle (see Frankcombe et al. 2015 for details). From Frankcombe et al. (2015)

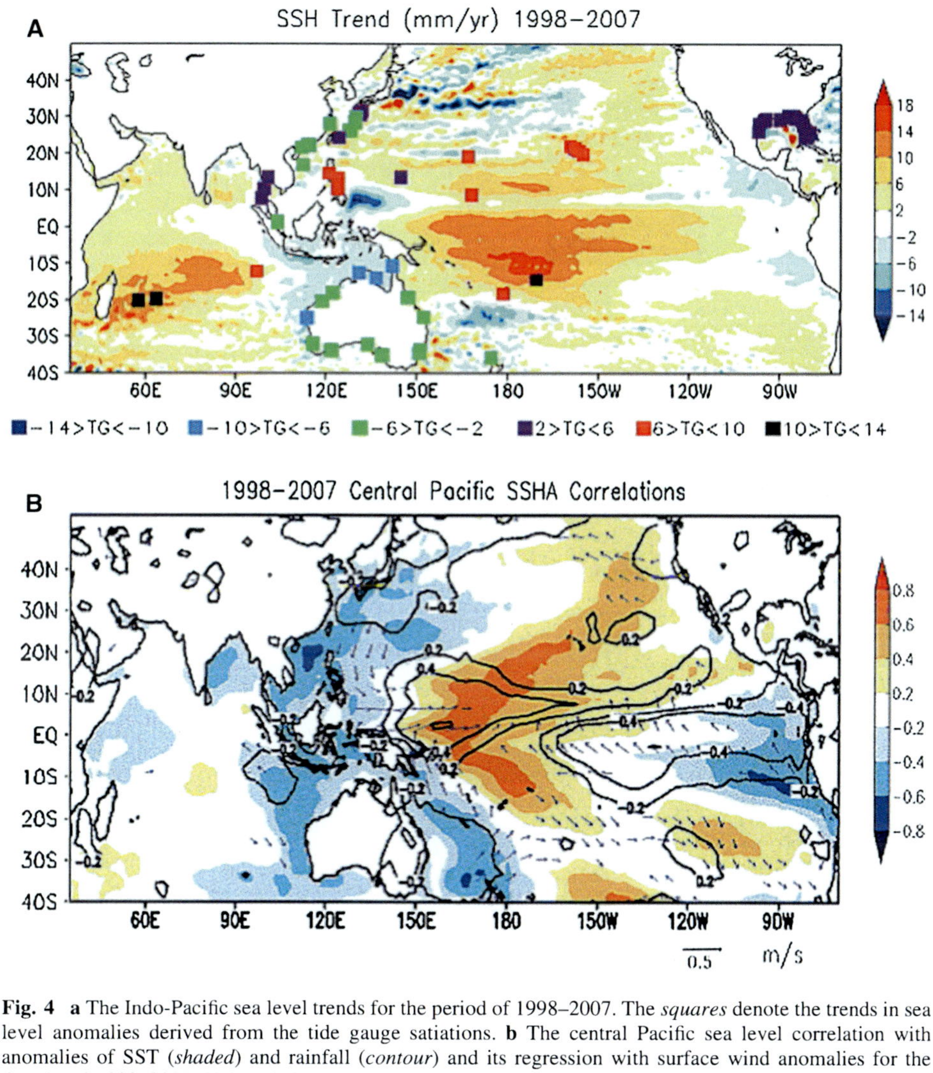

Fig. 4 **a** The Indo-Pacific sea level trends for the period of 1998–2007. The *squares* denote the trends in sea level anomalies derived from the tide gauge satiations. **b** The central Pacific sea level correlation with anomalies of SST (*shaded*) and rainfall (*contour*) and its regression with surface wind anomalies for the decade of 1998–2007. Values below the 95 % confidence level based on a 2-tailed t test are not shown. From Behera and Yamagata (2010)

2.3 Forcing and Processes

Previous studies suggest that over the past few decades change in wind forcing is the main cause for decadal sea level variability in the Pacific, including the intensified SLR in the western tropical Pacific since the early 1990s (e.g., Qiu 2002; Carton et al. 2005; Bindoff et al. 2007; Köhl et al. 2007; Köhl and Stammer 2008; Roemmich et al. 2007; Timmermann et al. 2010; Merrifield and Maltrud 2011; McGregor et al. 2012; Nidheesh et al. 2013; Qiu and Chen 2012; Han et al. 2014a). Given that a significant portion of the basin-scale coherent sea level pattern is associated with the PDO and NPGO, one can conclude that it is the surface wind changes associated with the PDO and NPGO that are the major cause for the basin-wide sea level patterns. Along the same line of arguments, regression

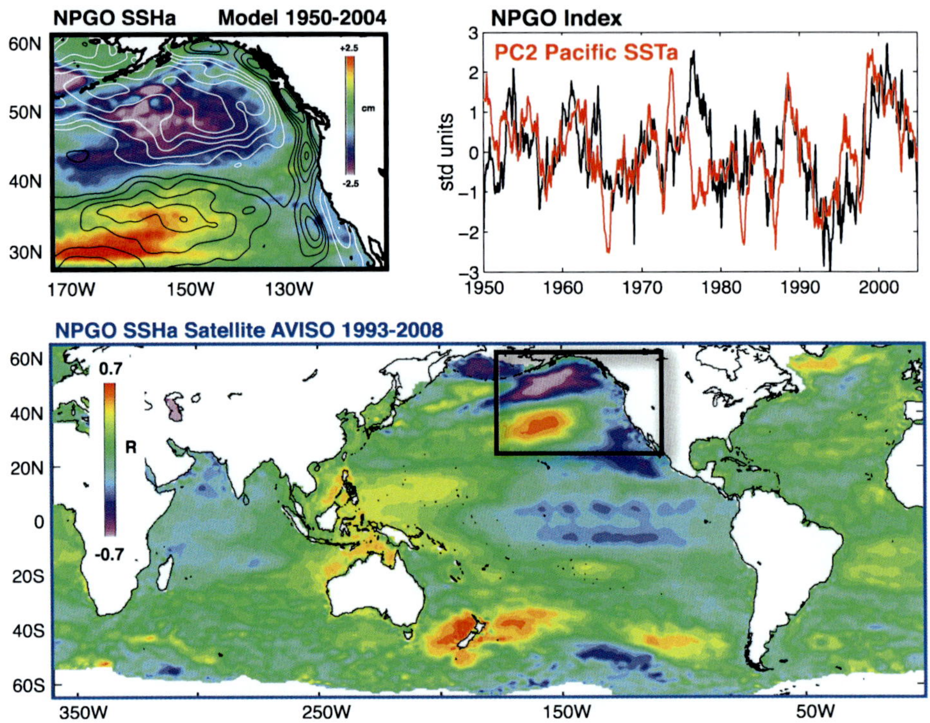

Fig. 5 (*Top left*) The 2nd EOF pattern of monthly SSH anomalies from the output of a regional ocean general circulation model (OGCM) for the 1950–2004 period; (*top right*) The NPGO index (*black curve*) defined as PC2 of SSH anomalies (SSHa) from the OGCM output, and PC2 of Pacific SST anomalies from NOAA ERSST v3 from 1950 to 2004 (*red curve*); (*Bottom*) Correlation map between NPGO index and monthly mean AVISO SSHa or the 1993–2008 period. Figure provided by Dr. Emanuele Di Lorenzo

analyses indeed show that PDO- and NPGO-related wind stress curl and alongshore wind (Fig. 6) can cause decadal sea level variations in the ocean interior and along the coasts by inducing Rossby waves (e.g., Qiu 2002), coastal Kelvin waves (e.g., Clarke and Lebedev 1997; Thompson et al. 2014), upwelling, and horizontal advection (Di Lorenzo et al. 2008; Bromirski et al. 2011). Hamlington et al. (2014) extracted the basin-wide surface winds associated with the PDO by regressing surface winds available from the European Centre for Medium-Range Weather Forecasts (ECMWF) operational ocean analysis/reanalysis system (ORA-S3) (Balmaseda et al. 2008) onto sea level PC1 from 1950 to 2010. The authors found that the linear trend of PDO-related wind from 1993 to 2010 (Fig. 2e) can explain a large portion of the observed trend (Fig. 2d). Note that while the PDO-winds have a single gyre structure over the North Pacific, the residual trend shows a double-gyre pattern over the Northeast Pacific region, which resembles the NPGO-winds in this region (compare Figs. 2e, f with 6a, b).

An ocean general circulation model (OGCM) experiment with surface wind and heat flux forcing (freshwater flux being fixed to climatology) reasonably reproduced the observed sea level variability over the Northeast Pacific region for the 1950–2004 period, and salinity variability in this experiment results from advection rather than from forcing by freshwater fluxes (Di Lorenzo et al. 2008). Independently, Bromirski et al. (2011) also examined the effects of surface wind and heat flux using OGCM experiments, showing that

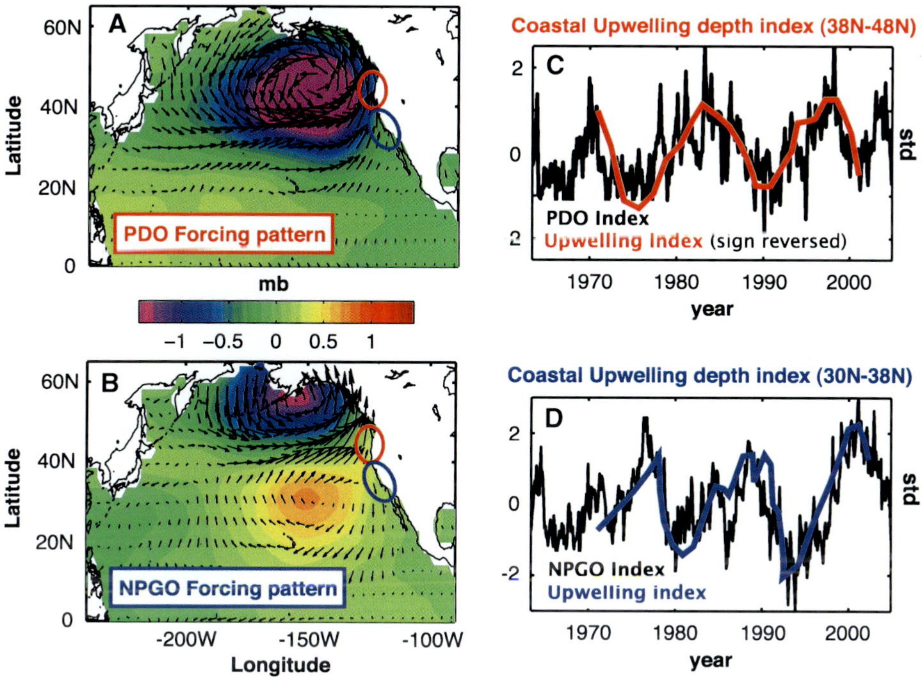

Fig. 6 Regression maps of **a** PDO and **b** NPGO indices with NCEP wind stress vectors and sea level pressure (*color scale*), **c** Coastal upwelling depth index from inverse model calculations averaged from 38°N to 48°N (area denoted by *red circle* in Fig. 6a, b) compared to PDO index, **d** Coastal upwelling depth index averaged from 30°N to 38°N (area denoted by *blue circle* in Fig. 6a, b) compared to NPGO index. A positive upwelling index implies a deeper upwelling cell. Adapted from Di Lorenzo et al. (2008)

local wind stress curl dominates heat flux in causing decadal sea level variability along the US west coast. Thompson et al. (2014) suggested the dominance of remote equatorial wind stress, rather than the local wind stress curl, in driving decadal sea level variability along the US west coast during recent decades. These results suggest that the decadal sea level patterns associated with the PDO and NPGO, which are dominated by the upper-ocean thermosteric sea level (e.g., Lombard et al. 2005), result primarily from changes in wind-driven ocean circulation, with surface heat and freshwater fluxes playing a minor role in the North Pacific Ocean. Indeed, the dominance of wind-driven ocean circulation in causing regional distributions of decadal sea level variability has been demonstrated, and the thermosteric and halosteric sea level components often have compensating effects due to heat and salt redistribution by advection [see reviews by Stammer et al. (2013) and Kopp et al. (2015)].

Diabatic fluxes, however, are shown to have significant contributions to the 1993–2004 trends of decadal sea level—particularly thermosteric sea level—in the western tropical Pacific Ocean (Fukumori and Wang 2013). Forcing by surface buoyancy flux is also shown to be non-negligible in causing interannual sea level variability in the tropical south Pacific and subtropical North Pacific, and the effect can be non-local due to advection and Rossby wave propagation (Piecuch and Ponte 2012; Forget and Ponte 2015). These results suggest that heat and freshwater fluxes associated with the PDO and NPGO may also induce significant decadal sea level variability in some regions. This aspect requires further study.

Note that even though the importance of winds has been emphasized, and PDO- and NPGO-related surface wind patterns have been extracted through regression analyses (Figs. 2, 3, 6), model experiments that use the extracted winds as surface forcing to quantify the roles played by the PDO and NPGO have not yet been performed. Moreover, the regression method may not cleanly isolate the wind signals associated with the climate modes (Palanisamy et al. 2015), and reanalysis winds that were used in the regression analyses have significant uncertainties (e.g., Wittenberg 2004), with apparent discrepancies or even opposite signs in their multidecadal trends since the 1960s (Nidheesh et al. 2013). Yet, reanalysis winds have been widely applied to investigate decadal climate variability due to their longer data records than satellite winds, and decadal trends of sea level over tropical oceans are very sensitive to the wind trends (McGregor et al. 2012). These issues will be further discussed in Sect. 6.

3 The Indian Ocean

3.1 Changes in Walker and Hadley Circulations and Related Sea Level Trend Patterns

The Indian Ocean sea level trends since the 1960s (from 1961 to 2001) exhibit a basin-wide pattern, with sea level falling in the southwest tropical basin and rising elsewhere (Fig. 7b) (e.g., Han et al. 2010; Timmermann et al. 2010; Dunne et al. 2012; NRC 2012).

Experiments using an OGCM and an atmospheric GCM (AGCM) show that surface winds associated with enhanced Indian Ocean Walker and Hadley circulations are the

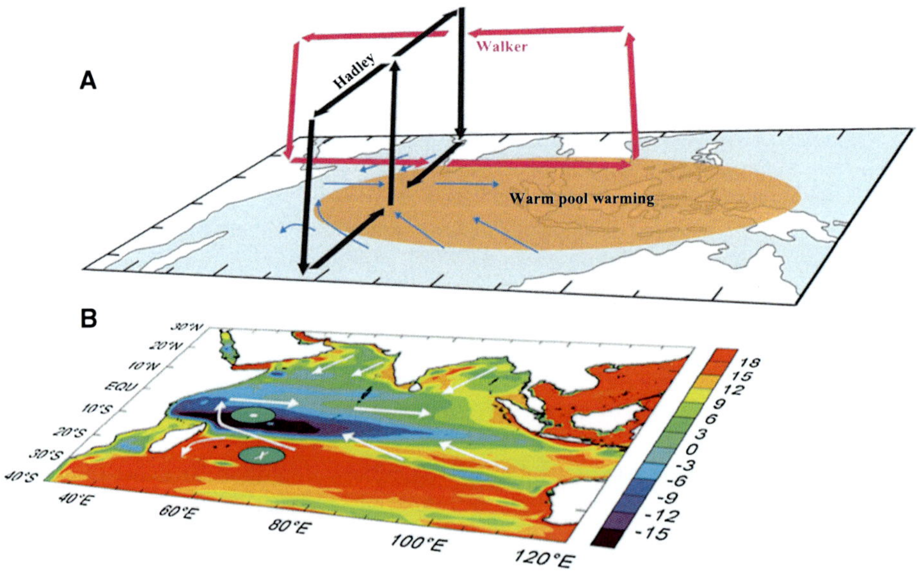

Fig. 7 A schematic diagram showing the mechanisms for the Indo-Pacific warm pool warming to cause the Indian Ocean (IO) sea level change. Warming enhances the IO regional Hadley and Walker cells (**a**); the two enhanced cells combine to form a specific pattern of surface wind change (surface *arrows* in **a** and **b**) together with the Ekman pumping velocity (positive—*circle* with *dot*; negative—*circle* with *x*), which drive the distinct sea level pattern (*color* contours in **b**). From Han et al. (2010)

major force for this basin-wide pattern (Fig. 7) (Han et al. 2010). Variation of the Indonesian Throughflow may have a significant contribution to the thermocline cooling and thus sea level fall in the southwest tropical Indian Ocean (Schwarzkopf and Böning 2011), as shown by OGCM experiments forced with different reanalysis winds from that of [Han et al. (2010); see Han et al. (2014b) for a review]. The "basin-wide patterns" of surface warming and thermocline cooling, however, are still mainly caused by Indian Ocean winds in the Schwarzkopf and Böning (2011) study (see their Fig. 2).

While the multidecadal trends of surface winds associated with the changing Walker and Hadley circulations are shown to be partly forced by the Indian Ocean warming (Han et al. 2010), which is attributed primarily to anthropogenic forcing since the 1950s (Du and Xie 2008; Dong et al. 2014), they are suggested to have a large contribution from natural climate variability (Timmermann et al. 2010). What natural climate modes account for the changes of Indian Ocean Walker and Hadley circulations, and what role does the changing Indian monsoon play? These important science issues remain to be explored.

3.2 Climate Modes, Sea Level Patterns and Processes

Overlying the multidecadal trend, satellite observations show large decadal variations of the basin-scale sea level patterns, with reversing trends from 1993 to 2000 and 2000 to 2006 (Lee and McPhaden 2008). Over the North Indian Ocean (north of 5°S), sea level experienced basin-wide falls from 1993 to 2003 but sharp rises from 2004 to 2013 (Srinivasu et al. 2016, Climate Dynamics, revised; Thompson et al. 2016). However, similar reversals did not occur over the south Indian Ocean (south of 5°S), the Pacific and the Atlantic Oceans. Both observational analyses and OGCM experiments suggest that winds over the Indian Ocean are the primary forcing for the basin-wide decadal sea level patterns (e.g., Trenary and Han 2013; Nidheesh et al. 2013; Zhuang et al. 2013), with the Indonesian Throughflow having a significant contribution primarily in the eastern basin (e.g., Trenary and Han 2013). A similar conclusion also holds for intraseasonal-to-interannual sea level variability over the south Indian Ocean (Trenary and Han 2012). The observed North Indian Ocean basin-wide decadal reversal of sea level results from the combined effect of changing surface turbulent heat flux and the cross-equatorial heat transport, with both being associated with decadal changes of surface winds (Srinivasu et al. 2016, Climate Dynamics, revised; Thompson et al. 2016). Thermosteric sea level is the primary contributor for the spatial patterns of decadal sea level variability (e.g., Fukumori and Wang 2013; Nidheesh et al. 2013), with halosteric sea level having apparent contributions in some regions (Shankar and Shetye 1999; Nidheesh et al. 2013) particularly in the southeast tropical Indian Ocean and near the west Australian coast in the past decade (Llovel and Lee 2015).

3.2.1 Effects of Climate Modes and Processes

To what extent are the observed patterns of decadal sea level induced by climate modes? Over the tropical Indian Ocean, two climate modes have been identified in SST anomaly (SSTa): the Indian Ocean basin mode and Indian Ocean Dipole (IOD). The decadal SSTa is dominated by the decadal Indian Ocean basin mode (DIOB), defined as the leading EOF of decadal SSTa with a basin-wide warming/cooling pattern and explaining 54 % of the variance (Fig. 8a) (Han et al. 2014b). The DIOB index, defined as PC1 of SSTa, is positively correlated with the PDO before 1985 ($r = 0.82$; Fig. 8b), similar to the ENSO influence on Indian Ocean interannual SSTa (Klein et al. 1999; Xie et al. 2009). The correlation, however, reverses to become negative after 1985 ($r = -0.69$; Fig. 8b) (Han

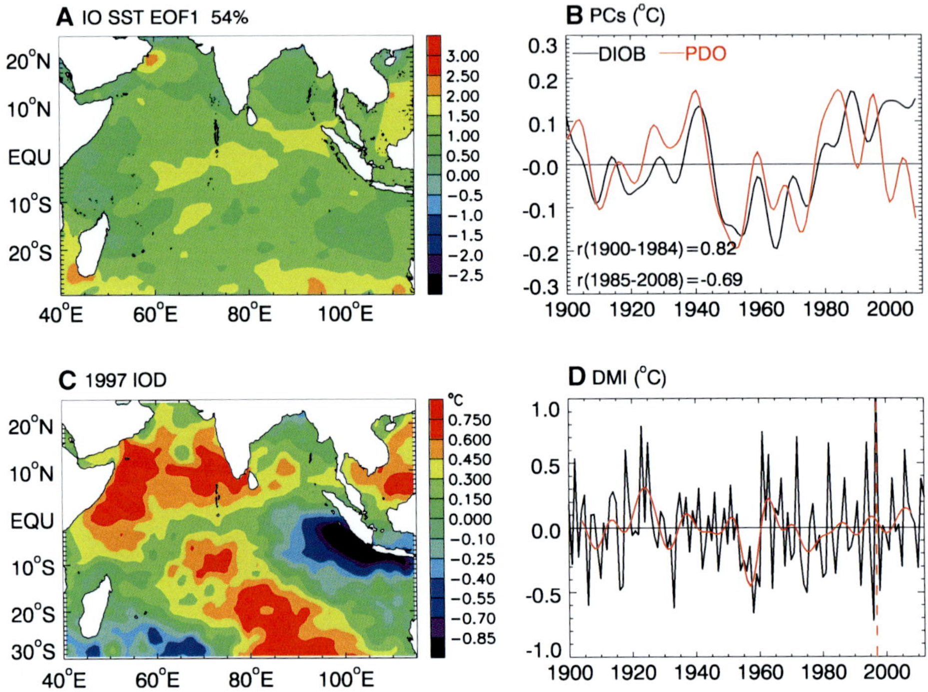

Fig. 8 **a** The leading EOF of SST anomaly (SSTa) for the Indian Ocean, based on 8-year low-pass-filtered monthly HadISST from 1900 to 2008, which explains 54 % variance and is referred to as the Decadal Indian Ocean Basin Mode (DIOB), **b** The PC1 of 8-year low-passed SSTa (*black*; DIOB index) and North Pacific SST PC1 (*red*; PDO index), **c** September, October and November mean SSTa for the 1997 IOD event, based on the detrended and demeaned SST from 1900 to 2012, **d** Dipole Mode Index (DMI; *black*) for each year, which is defined as the September, October and November mean SSTa difference between the western pole (50°E–70°E, 10°S–10°N) and eastern pole (90°E–110°E, 10°S–0°N); the *red curve* is the 8-year low-passed DMI, which shows decadal variability. *Dashed vertical line* marks the 1997 IOD

et al. 2014a). Causes for this change in character remain unclear, but the negative correlation indicates that Indian Ocean internal processes—likely related to monsoon variability—may be important for generating the DIOB. The IOD is an interannual coupled ocean–atmosphere climate mode (e.g., Saji et al. 1999; Webster et al. 1999), and its positive phase is associated with cold SSTa in the eastern tropical Indian Ocean and warm SSTa in the western tropical basin (Fig. 8c), reaching peak amplitudes during September–November. Its temporal variability is measured by the Dipole Mode Index (DMI; Fig. 8d), which exhibits large interannual variability with decadal modulation (black and red curves of Fig. 8d) (see also Ashok et al. 2004; Song et al. 2007; Tozuka et al. 2007 for IOD decadal variability).

Li and Han (2015) performed a suite of experiments using an eddy-permitting (0.25° × 0.25°) OGCM over the Indian Ocean basin for the 1948–2012 period, to assess the roles played by surface wind stress associated with climate modes, stochastic wind stress forcing and surface heat and freshwater fluxes. To extract the wind stress signals associated with the DIOB and IOD, Li and Han (2015) first performed multiple linear regression onto the observed Multivariate ENSO Index (MEI) and DMI, using National Center for Environmental Prediction (NCEP) reanalysis winds and SST data from HadISST (Hadley Centre

Sea Ice and Sea Surface Temperature data set). The wind stress and SST anomalies regressed onto MEI account for part of the DIOB because it is highly correlated with ENSO (or PDO) before 1985. Then, they employed a singular value decomposition (SVD) technique to obtain the covariance of the residual wind stress and SST with ENSO and IOD signals removed. The leading 7 SVD modes were obtained, which include the part of the DIOB that is independent of PDO. The sum of wind stress anomalies related to ENSO, IOD and the SVD modes measure the total effects of Indian Ocean climate modes. Experiment results of Li and Han (2015) show that the basin-wide decadal sea level patterns over the tropical Indian Ocean (north of 20°S) are forced mainly by wind stress associated with climate modes, with the maximum amplitude occurring in the southwest tropical Indian ocean thermocline ridge region (their Figs. 4, 6). Surface heat flux has a significant contribution in the subtropical basin between 20°S and 28°S, consistent with Fukumori and Wang (2013) for the effect of diabatic fluxes on 1993–2004 sea level trend in this region. Stochastic wind stress forcing has a large contribution in the southwest Indian Ocean (south of 30°S).

3.2.2 DIOB-Related Sea Level Patterns

Even though Li and Han (2015) demonstrated the dominant influence of climate modes on sea level in the tropical Indian Ocean, basin-wide sea level patterns associated with the DIOB alone, including both PDO influence and internal coupled processes related to monsoon variability, have not yet been assessed. Intuitively, the basin-wide warming and cooling associated with the DIOB will produce basin-wide sea level rise or fall due to thermal expansion. Changes in ocean circulation associated with the DIOB, however, remain unknown. Given that DIOB is significantly correlated with PDO before 1985, Indian Ocean sea level patterns associated with the PDO based on reconstructed sea level for 1950–2010 (Fig. 1a) and Simple Ocean Data Assimilation (SODA) product for 1958–2007 (second row in left column of Fig. 3) reflect a portion of sea level patterns associated with DIOB. The patterns from the two datasets over the Indian Ocean, however, show significant differences, indicating uncertainties involved with different datasets.

3.2.3 IOD-Related Sea Level Patterns

The spatial patterns of sea level associated with a negative IOD show SLR in the eastern basin, with the maximum amplitude being located at the equatorial basin and extending northward into the Bay of Bengal and southward to the west coast of Australia, which accompanies sea level fall in the western tropical basin with the maximum amplitude occurring in the Seychelles-Chagos thermocline ridge region (Figure 2 of Han and Webster 2002). These spatial patterns resemble that of SSHa EOF1 (Fig. 9a) (Rao et al. 2002), which represents the SSHa patterns associated with the IOD because its PC1 (Fig. 9b) is highly correlated with the DMI (Rao et al. 2002). The EOF1 patterns shown in Fig. 9 are similar to the regression patterns with DMI (fifth row of Fig. 3). The large SSHa amplitudes in the eastern basin and in the thermocline ridge region, together with the minimum amplitudes in the central equatorial basin and eastern Arabian Sea, agree with the patterns of interannual SSHa variance shown by Shankar et al. (2010). These results suggest that, under periodical wind forcing with periods longer than 17 months, sea level variability over the tropical Indian Ocean (north of 10°S) is in quasi-steady balance with wind variability, and the sea level maxima (minima) result from the constructive (destructive) interference between directly forced response and Rossby waves reflected from the eastern boundary (Shankar et al. 2010).

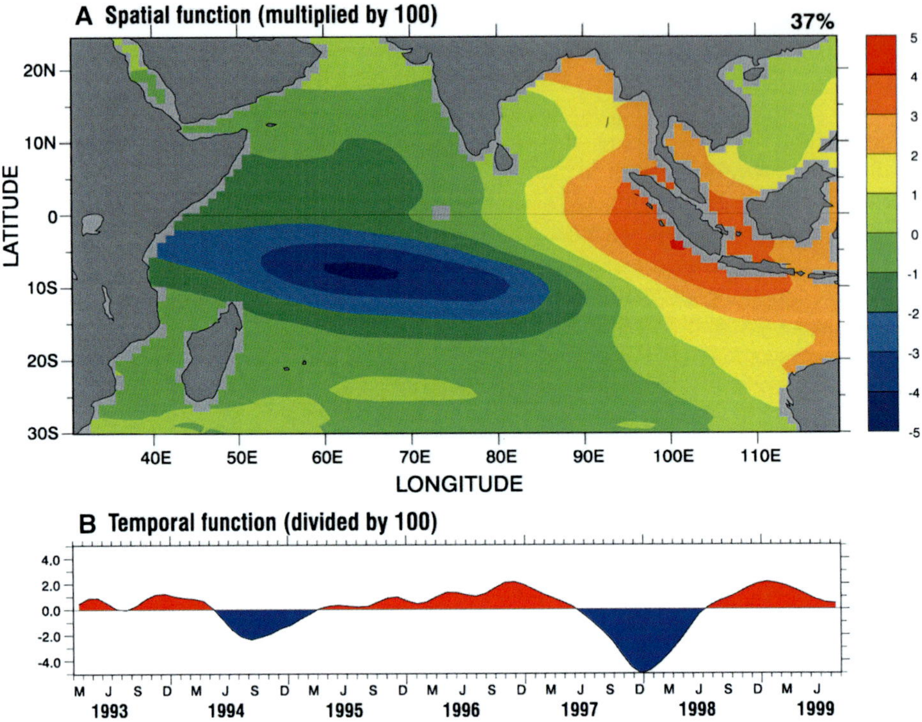

Fig. 9 The leading Complex EOF of SSH (**a**) and PC1 (**b**), based on Topex/Poseidon SSH anomaly from 1993 to 1999 (downloaded from: http://www.jamstec.go.jp/frcgc/research/d1/iod/fig_1.jpg)

Frankcombe et al. (2015) showed significant decadal variations in the relationship between IOD and sea level patterns by analyzing OGCM results and SODA data: the regression coefficients are enhanced over the Indian Ocean since the 1990s compared to the preceding decades (their Figs. 6, 7). By analyzing historical SST and reanalysis data, it has been shown that IOD events exhibit an increasing trend in frequency and strength during the twentieth century (Abram et al. 2008), and positive IOD events prevail after 1950, which corresponds to an upward trend of the IOD index (e.g., Kripalani and Kumar 2004; Ihara et al. 2008; Yuan et al. 2008; Cai et al. 2009; Fig. 8d). This upward trend, however, appears in some SST datasets but not in others, highlighting the uncertainties in these trends (Cai et al. 2013). With frequently occurring intensified positive IOD events, we expect enhanced and more frequent sea level rise in the thermocline ridge region and sea level fall in the eastern basin during boreal fall (Fig. 9), which may counteract the multidecadal sea level trend patterns (Fig. 7) in many regions.

3.2.4 Subtropical Dipole and Sea Level Patterns

In the south Indian Ocean, a subtropical dipole has also been identified in interannual SSTa with peak development in austral summer (Behera and Yamagata 2001; Suzuki et al. 2004). A positive phase is characterized by warm SSTa in the southwestern Indian Ocean south of Madagascar and cold SSTa in the eastern Indian Ocean off Australia. It also undergoes decadal variations, showing weakened amplitudes after 1979–1980 (Yan et al.

2013). Compared to other climate modes, few studies have investigated the influence of the subtropical dipole on sea level. Thompson et al. (2016) showed that periods of enhanced cross-equatorial overturning circulation in the Indian Ocean tend to occur during the positive subtropical dipole phase when the Mascarene High sea level pressure is strengthened and shifted to the north. The shift of this prominent atmospheric circulation alters wind stress curl, which drives changes in the overturning and affects heat content and sea level north of the equator.

4 The Atlantic

4.1 Sea level Variability, Forcing and Relation to Changes of Atlantic Meridional Overturning Circulation (AMOC) and the Gulf Stream

Tide gauge and satellite observations show significant interannual-to-decadal sea level variability over the Atlantic during the past several decades and century (e.g., Figure 1 of Kenigson and Han 2014). Observational analyses combined with model experiments have been carried out to understand their causes. In the western North Atlantic and along the US east coast, existing studies suggest the dominance of wind stress curl over the basin interior in driving westward-propagating Rossby waves, affecting interannual and decadal (periods >3 years) sea level variability from 18°N to 38°N (e.g., Sturges and Hong 1995; Hong et al. 2000; Thompson and Mitchum 2014). The regional along-shelf wind stress is shown to be important for interannual sea level variability from Nova Scotia to North Carolina (e.g., Andres et al. 2013; Woodworth et al. 2014). Near the eastern boundary of the North Atlantic, coherent decadal sea level variability has been observed during the past century (e.g., Kolker and Hameed 2007; Miller and Douglas 2007; Woodworth et al. 2010; Sturges and Douglas 2011; Calafat et al. 2012), and it has been attributed to forcing by local longshore winds and coastal wave propagation (e.g., Sturges and Douglas 2011; Calafat et al. 2012), with mass redistribution having a small contribution (Calafat et al. 2012). This result is in contrast to Woodworth et al. (2010), who suggested the importance of mass redistribution in affecting east Atlantic coastal sea level.

In the North Atlantic basin interior, OGCM experiments spanning 1951–2000 demonstrate that decadal variability in the leading EOF modes of SSHa and gyre circulation originate from the basin-scale thermal forcing, rather than from wind stress driving, and that low-frequency variations of SSH along the Gulf Stream reflect predominantly the Atlantic Meridional Overturning Circulation (AMOC) changes (Hakkinen 2001). The importance of buoyancy and mass forcing in affecting interannual variability of sea level through advection and Rossby wave propagation in the tropical South Atlantic and subtropical to subpolar North Atlantic has also been suggested (Piecuch and Ponte 2013; Forget and Ponte 2015). The leading EOF mode of satellite-observed or model-simulated SSHa exhibits a dipole pattern (Fig. 10): with low SSHa occurring in the Subpolar Gyre region (extending southwestward to the US northeast coast) being accompanied by high SSHa in the Subtropical Gyre region. Its PC1 shows large-amplitude decadal variability (Figs. 2, 3, 4 of Hakkinen 2001). This SSH dipole resembles the distinctive fingerprint of AMOC variability (Figure 1 of Zhang 2008), except that the extension to the US Northeast coast from the Subpolar Gyre disappears in the AMOC fingerprint. This difference may indicate the importance of local winds in driving coastal sea level particularly north of Cape Hatteras, as suggested by recent studies (Andres et al. 2013; Woodworth et al. 2014).

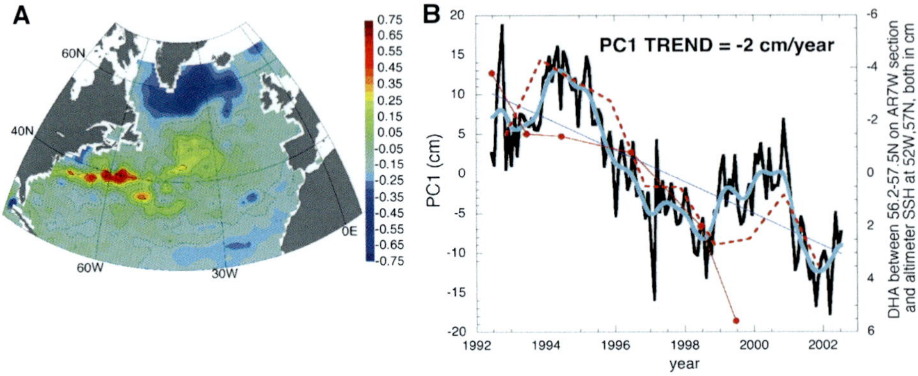

Fig. 10 (*Left*) EOF1 of altimeter SSH and (*right*) its PC1 (*black curve*). The *blue curve* is the binomially smoothed PC1. The dynamic height anomaly (DHA; in *solid red*; *dots* denote data points of the time series) computed in the central Labrador Sea (average from 56.2° to 57.5°N along the WOCE AR7/W section across the Labrador Sea from Newfoundland to Greenland) is shown in the *right panel* with its axis on *right*. The altimeter SSHa at 52°W, 57°N (12-month May-to-April average) is shown in *dashed red curve*. Adapted from Hakkinen and Rhines (2004)

Recently, accelerated SLR along the US northeast coast, particularly in the "hotspot" from Cape Hatteras to Cape Cod since the 1950s, has been detected using tide gauge observations (Sallenger et al. 2012). The weakened transport of the Gulf Stream, the upper branch of the AMOC, and the northward shift of the Gulf Stream during recent decades (Ezer 2013; Ezer et al. 2013; Yin and Goddard 2013) have been suggested to be the cause. The Gulf Stream sustains a sharp sea level gradient associated with geostrophic balance, suppressing sea level along the Atlantic coast by 1–1.5 m relative to the open ocean east of the Stream. A weakened Gulf Stream, therefore, weakens offshore sea level gradients and causes SLR along the US east coast. Furthermore, a weakened AMOC (e.g., due to surface warming and Greenland Ice Sheet melting) can induce rapid SLR along the US northeast coast and in the Subpolar Gyre region, based on OGCM and climate model simulations (e.g., Bingham and Hughes 2009; Yin et al. 2009; Hu et al. 2009). Andres et al. (2013), however, argued that the changes observed along the coast and over the shelf appear to influence the Gulf Stream path downstream of Cape Hatteras, rather than the changes in Gulf Stream transport affecting coastal sea level. Woodworth et al. (2014) analyzed the results from data-assimilative OGCM experiments (also see Williams et al. 2014) with and without wind stress forcing for the 1950–2009 period and suggested that although a relationship between US northeast coast sea level change and the AMOC variation can be identified (an increase of ~ 1.5 cm in sea level for a decrease of 1 Sv in MOC transport), it is the surface wind stress particularly the regional wind over the shelf that dominates the sea level variability along the US northeast coast.

4.2 NAO-Related Sea Level Patterns

Are the interannual and decadal sea level variations over the Atlantic Ocean associated with the known climate modes? Indeed, many studies have linked regional or basin-scale sea level variability, surface wind and heat flux to the North Atlantic Oscillation (NAO). These studies use EOF analysis by comparing the PCs with the NAO index or calculate correlation coefficients between the NAO index and regional sea level variability. The

NAO is a major atmospheric circulation anomaly pattern corresponding to fluctuations in Icelandic low sea level pressure (SLP) in the north and Azores high SLP to the south (e.g., Barnston and Livezey 1987; Hurrell 1995 and references therein). The winter (December–March) station-based NAO index is measured by the difference of normalized SLP between Azores high and Icelandic low (for NAO index and spatial patterns, see https://climatedataguide.ucar.edu/climate-data/hurrell-north-atlantic-oscillation-nao-index-station-based). A positive NAO phase corresponds to a positive SLP anomaly in Azores high and a negative SLP anomaly in Icelandic low. Note that variations in both buoyancy flux and surface wind stress, which in part are associated with the NAO, can induce changes in AMOC (e.g., Danabasoglu 2008; Schloesser et al. 2012, 2014; Yeager and Danabasoglu 2014) and thus affect SLR in the Subpolar Gyre, Subtropical Gyre, and likely along the US northeast coast.

A distinct 12–14-year spectral peak appears in all tide gauge stations along the US east coast from Charleston to Eastport with record lengths at least from 1930 to 2012 (Figure 1 of Kenigson and Han 2014). Hakkinen (2000) suggested that the 12–14-year cycles of US east coast sea level and basin-scale SST result mainly from surface heat flux forcing, whose leading EOF is associated with the NAO; but this does not exclude forcing of freshwater flux from enhancing or weakening the cycle. The author argued that the 12–14-year cycle is potentially a coupled ocean–atmosphere mode: Starting from the positive NAO phase, positive SST and oceanic heat content anomalies in the subtropics are advected to the Subpolar Gyre, where they are amplified by local heat flux, a positive feedback between the atmosphere and ocean. Meanwhile, this warm advection causes a negative feedback of the AMOC on itself, which is amplified by the positive feedback between the atmosphere and ocean in the Subpolar Gyre. As a result, the AMOC slows down and the opposite cycle starts (Hakkinen 2000). Andres et al. (2013) argued that the NAO has a strong influence on interannual variability of sea level along the US northeast coast from 1987 to 2012 but not during 1970–1987 (their Fig. 3a). Ezer (2013) and Ezer et al. (2013) suggested that the weakening Gulf Stream and AMOC, which affect the hotspot SLR acceleration during recent decades, are linked to the Atlantic Multidecadal Oscillation (AMO) but are also affected by the SLP and winds associated with the NAO. The effects of NAO-related wind stress curl and surface heat flux in the basin interior on interannual and decadal variability of the Gulf Stream have also been suggested by other studies (e.g., Curry and McCartney 2001; Di Nezio et al. 2009; Chaudhuri et al. 2011). This effect is shown to be important from 1986 to 1998 compared to the periods before and after (Meinen et al. 2010). Indeed, the 2009–2010 extreme SLR along the US northeast coast has been attributed to a strong negative NAO superimposed on a 30 % AMOC reduction (Goddard et al. 2015). Note that the inverted barometer (IB) effect accounts for approximately 50 % of the extreme SLR event (Piecuch and Ponte 2015).

Hakkinen and Rhines (2004) observed a downward trend of the Subpolar Gyre during the 1990s, which corresponds to SLR in the Subpolar Gyre and along the US northeast coast and sea level fall in the Gulf Stream and Subtropical Gyre (Fig. 10). They suggested that buoyancy forcing associated with the NAO is the major cause. The SSH patterns of Fig. 10 resemble the correlation map between the NAO index and satellite-observed SSHa from 1993 to 2013 (Fig. 11a) and the EOF1 patterns based on the observed upper 500 m thermosteric sea level from 1950 to 1998 (Lombard et al. 2005). The PC1 is significantly correlated with the NAO index with a correlation coefficient of 0.55, and the two agree well particularly on decadal timescales (Lombard et al. 2005). The EOF1 patterns of Lombard et al. (2005) are similar to the correlation map between the NAO index and the WOA13 upper 700 m thermosteric sea level from 1955 to 2013 (Fig. 11c), and to the

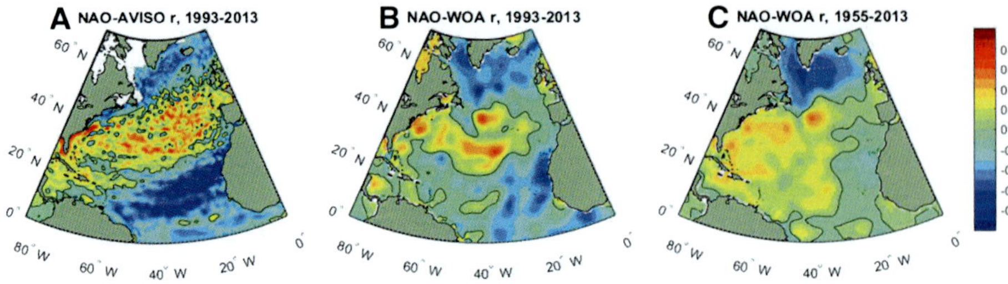

Fig. 11 **a** Correlation map between winter (DJFM mean) NAO index and annual mean AVISO satellite SSHa over the North Atlantic Ocean for the 1993–2013 period; zero correlation values are shown by the *black line contours*, **b** Same as **a** but for NAO and WOA13 upper 700 m thermosteric sea level (Levitus et al. 2012) from 1993 to 2013; **c** Same as **b** but for the 1955–2013 period. The global mean SLR is removed from each panel before the correlation calculation (The NAO index is downloaded from: www.climatedataguide.ucar.edu/climate-data/hurrell-north-atlantic-oscillation-nao-index-station-based)

spatial structure of upper 500 m temperature anomalies during a period of rapid change in the sea level circulation index of McCarthy et al. (2015). Note that there is a major difference between Fig. 11a, b: the correlation coefficients along the US northeast coast have opposite signs. This difference likely results from the lack of in situ observations near the coasts and the coarser $1° \times 1°$ resolution of thermosteric sea level data compared to the $0.25° \times 0.25°$ resolution AVISO data, which can better resolve sea level variability near the coast (compare Fig. 11a, b). Lozier et al. (2010) showed that changes in surface winds associated with the upward trend of NAO from 1950 to 2000 (Hurrell 1995) enhanced the mean AMOC strength over the Subpolar Gyre latitudinal band but weakened the AMOC in the Subtropical Gyre latitudinal band, suggesting that the AMOC changes can be gyre specific instead of a single, basin-scale overturning cell.

Over the Northeast Atlantic and west coast of Europe, decadal sea level variability is also significantly correlated with the NAO. Variations of SLP via IB effect and large-scale winds associated with the NAO are probably the cause, with wind stress playing a more important role (Miller and Douglas 2007; Calafat et al. 2012). Kolker and Hameed (2007) showed that changes in the center of action of the Icelandic low and Azores High play a crucial role in affecting decadal sea level variability and multidecadal trend along both the west and east coasts of the North Atlantic Ocean. The shift in the NAO center of action can affect surface winds, SLP and SST, thereby influencing sea level via a suite of coastal oceanographic processes. This result is important, indicating that the effects of NAO can be reflected in both the NAO index—which measures its strength variations—and in the location shift of its center of action. The relationship between sea level in the eastern North Atlantic and the NAO has also been demonstrated by earlier studies (e.g., Wakelin et al. 2003; Woolf et al. 2003; Yan et al. 2004; Tsimplis et al. 2006; Tsimplis and Shaw 2008). They suggested that the relationship is spatially and temporally variable, with stronger NAO influence during the second half of the twentieth century than the first half.

4.3 Atlantic Multidecadal Oscillation (AMO)-Related Sea Level Patterns

The AMO is defined as a mode of natural variability with coherent warming or cooling in the North Atlantic Ocean and an estimated period of 60–80 years (Schlesinger and Ramankutty, 1994; Delworth and Mann 2000; Kerr 2000; Enfield et al. 2001). The AMO index is based upon the detrended average SST anomalies in the North Atlantic, typically over 0–80°N (Trenberth and Shea 2006). Since 1940, the AMO index shows a negative transition (cooling trend) from the 1940s to 1970s and positive transition (warming trend) afterward, forming a ~60-year cycle with negative phase from the 1960s to 1990s. Surface warming over the North Atlantic during recent decades has been suggested to reflect partly anthropogenic forcing and partly positive AMO transition (Ting et al. 2009). By contrast, several recent studies suggest that the ~60-year oscillation in the AMO index in the past few decades can be externally forced by aerosols (natural and anthropogenic), volcanic eruptions, and the solar cycle (e.g., Booth et al. 2012; Knudsen et al. 2014). However, Zhang et al. (2013) argued that external forcing might not be the dominant causative factor. Mann et al. (2014) suggested that the "detrending method" failed to isolate the true internal variability, and the true AMO signal is likely to have been in a cooling phase during recent decades. Due to this controversy, in our discussion below we refer to the AMO index of Trenberth and Shea (2006) as "the ~60-year cycle of SST index."

Tide gauge observations detected robust accelerations of SLR along the highly populated US northeast coast since 1950 and especially since 1970 (Sallenger et al. 2012; Boon 2012; Ezer and Corlett 2012). Existing studies suggest that the ~60-year cycle in sea level, which is present in most tide gauge stations along the US northeast coast (Kenigson and Han 2014) with rapid SLR since 1970 coinciding with the positive transition of the ~60-year cycle of SST index, has a significant contribution to the observed SLR acceleration (Kopp 2013; Ezer 2013; Scafetta 2014; Kenigson and Han 2014). Using both Empirical Mode Decomposition (EMD) and ensemble EMD methods, Kenigson and Han (2014) constructed synthetic tide gauge data by extracting the leading oscillations at interannual-to-multidecadal timescales from tide gauge data, and extended the data back to 1813 by superimposing the oscillations on prescribed trends with known acceleration rates. Experiments with and without the ~60-year cycle demonstrated that the ~60-year cycle has indeed contributed a significant portion to the hotspot acceleration since 1970, and a record length of approximately twice the ~60-year period (see also Scafetta 2014) is required in order to adequately detect the long-term, nonlinear acceleration rate with errors <25 %.

By examining long tide gauge records during the twentieth century, Chambers et al. (2012) showed that a ~60-year cycle appears in the majority of tide gauge data over the Atlantic, Indian and Pacific Oceans. They suggested that there is a possibility that the ~60-year oscillation is present in global mean sea level, even though the tide gauge data are still too limited in time and space to make a definitive conclusion. In fact, Jevrejeva et al. (2008) found a ~60-year cycle in the global mean sea level of their reconstructed data since 1700 and speculated that it might be associated with the AMO.

What are the spatial patterns of decadal sea level variability induced by the natural internal AMO mode of the climate system? This issue has not yet been addressed. Note that the AMO can affect the North Atlantic SLR in two ways: The basin-wide warming during a positive AMO phase can increase sea level by thermal expansion, and the enhanced AMOC associated with a positive AMO (see review by Liu 2012 for MOC mechanisms and references therein) tends to weaken the US northeast coast and Subpolar Gyre SLR (e.g., Hu et al. 2009; Yin et al. 2009). The two effects combine to yield the spatial patterns of sea level change associated with the AMO. Kopp (2013) argued that the combination of the two effects produces a positive correlation between hotspot inter-decadal SLR and AMO index. As discussed above, however, the oscillation of SST index in the past few decades may be externally forced and thus its positive correlation with the hotspot sea level may not reflect the true internal AMO mode.

Several studies have shown prominent bidecadal (20–30-year) variability in SST, SLP and sea level over both the North and South Atlantic (e.g., Venegas et al. 1998; Danabasoglu 2008; Frankcombe and Dijkstra 2009; Chylek et al. 2011; Vianna and Menezes 2011, 2013). While most observational and modeling studies suggest that the bidecadal oscillations can be largely explained by internal ocean dynamics related to the AMOC and AMO (Huck and Vallis 2001; Von der Heydt and Dijkstra 2007; Danabasoglu 2008; Frankcombe and Dijkstra 2009; Frankcombe et al. 2010), others argue that they involve westward-propagating baroclinic Rossby waves of large-scale temperature and sea level anomalies (e.g., Sevellec and Fedorov 2013; Vianna and Menezes 2013). Sevellec and Fedorov (2013) demonstrated that the zonal structures of temperature anomalies alternate between a dipole (corresponding to AMOC variability) and one sign pattern (no AMOC variability). Consistent with this result, Vianna and Menezes (2013) showed that the leading CEOF mode of the bidecadal sea level signals is associated with the AMOC variability, whereas the second CEOF mode has distinguishable westward-propagating thermal Rossby waves and is not apparently related to AMOC change. The leading CEOF mode is characterized by in phase North and South Subtropical Gyres with an opposite sign in the tropical and subpolar regions (Fig. 12a), and it dominates sea level variability from 1915 to 1965 (Fig. 12). For the CEOF2 mode, the North and South Subtropical Gyres are not in phase (Fig. 12b) and it dominates the bidecadal sea level after 1970 (compare Fig. 12a–c) (also see Figure 1 of Vianna and Menezes 2013). These results indicate that patterns of sea level can vary on multidecadal timescales, corresponding to the time-varying effects of different oceanic processes (e.g., AMOC vs. Rossby waves).

5 The Arctic and Southern Oceans

5.1 The Arctic Ocean

The Arctic Oscillation (AO) is a large-scale pattern of climate mode characterized by SLP over the polar region varying in opposition to that over middle latitudes (about 45°N) on time scales ranging from weeks to decades (e.g., Thompson and Wallace 1998). The AO is also referred to as "Northern Annular Mode," and the NAO is its North Atlantic expression (http://nsidc.org/arcticseaicenews/tag/arctic-oscillation/). The AO index is constructed by projecting the daily 1000 mb height anomalies poleward of 20°N onto the loading pattern of the AO, which is the leading EOF of monthly mean 1000 mb height anomaly (National Weather Service Climate Prediction Center, http://www.cpc.noaa.gov).

In the past decade, several studies have focused on exploring the interannual-to-decadal sea level variability near the Norwegian coast and over the Arctic Ocean using tide gauge data, satellite observations and numerical models (e.g., Proshutinsky et al. 2001, 2004, 2007; Henry et al. 2012; Richter et al. 2012a, b; Volkov and Pujol 2012; Calafat et al. 2013). They found that both the IB effect and wind forcing are important in causing sea level variability, and the relative importance of the two has regional variations (e.g., Proshutinsky et al. 2001, 2004, 2007). They also found significant correlations between the AO index (red curve of Fig. 13) and 2- or 5-year running mean tide gauge records (black and blue curves of Fig. 13) but with distinct regional variations (Proshutinsky et al. 2007; Henry et al. 2012; Calafat et al. 2013). The correlations between AO index and regional mean sea level variability with IB effect (black curve) and without IB effect (blue curve) are 0.60 (0.56) for the Norwegian Sea, 0.79 (0.66) for the Barents

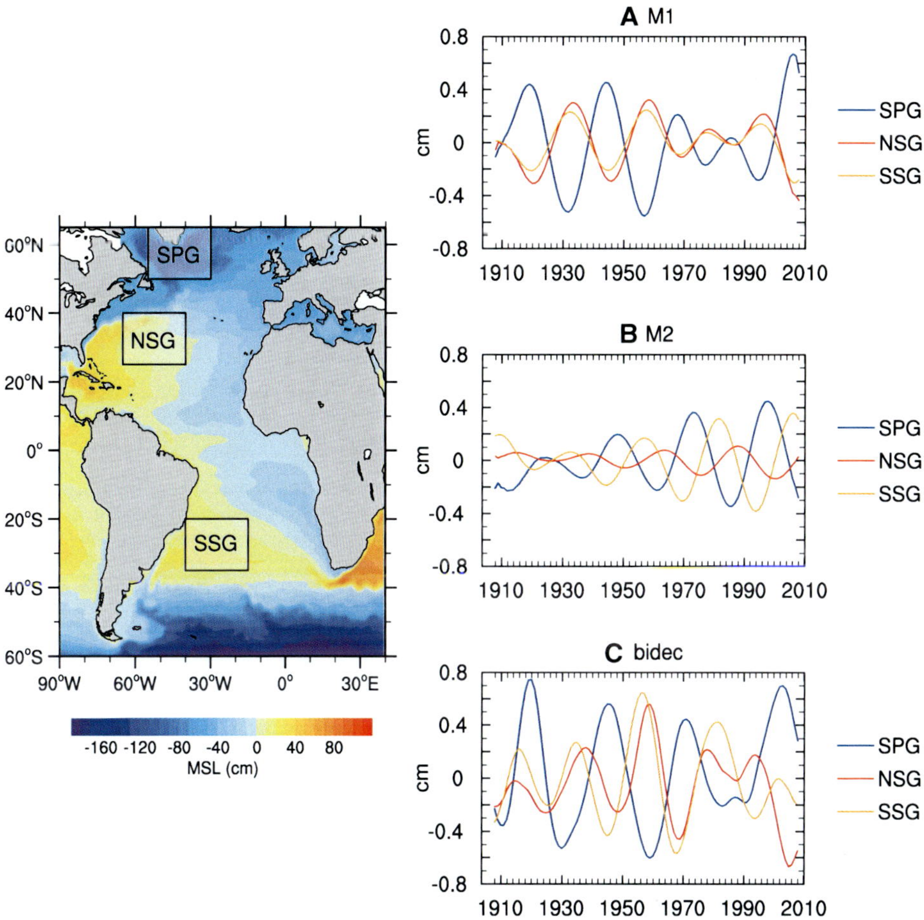

Fig. 12 Three space-averaged SSHa indices for the **a** EOF1 (M1), **b** EOF2 (M2) and **c** the bidecadal band, computed for each of the three boxes (SPG, NSG and SSG) shown on the map. SPG stands for Subpolar Gyre, NSG for North Subtropical Gyre, and SSG for South Subtropical Gyre. *Color contours* in the *map* show the 1908–2008 mean sea level from SODA data. The *map* is only used to illustrate the locations of the *boxes*. From Vianna and Menezes (2013)

Sea, 0.60 (0.46) for the Kara Sea, and 0.46 (0.37) for the Laptev Sea with all correlations exceeding the 95 % significance level. These correlations are based on the longest record available at each tide gauge station, with the longest record length being 1948–2010 (Table 1 of Calafat et al. 2013). Evidently, the AO can affect interannual and decadal sea level via the IB effect, as seen from the higher correlations with the IB influence (see above correlations outside the parentheses). The IB effect, however, may not be the dominant factor, because the correlations between the AO and sea level variability do not have dramatic decreases excluding the IB influence (correlations inside the parentheses), indicating that other factors (e.g., winds) associated with the AO are important in affecting Arctic sea level. In the East Siberian Sea and Chukchi Sea, the correlations are below 0.3 and not statistically significant above 95 % level.

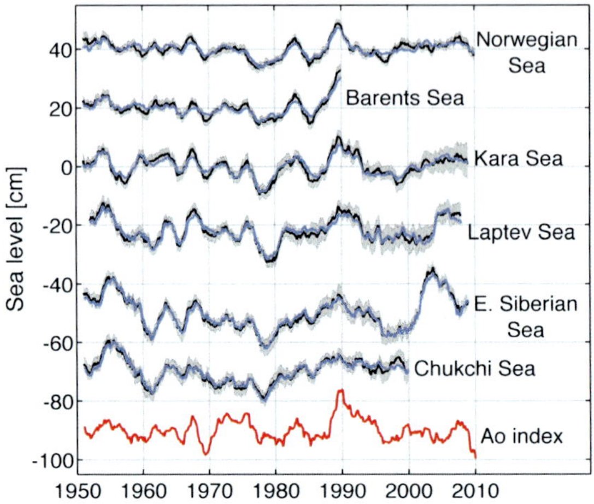

Fig. 13 Low-pass-filtered (2-year running mean) time series of total (*black line*) and IB-corrected (*blue line*) sea level in six regions along the continental boundaries of the Arctic Ocean: in a counterclockwise order they are the Norwegian coast, Barents Sea, Kara Sea, Laptev Sea, East Siberian Sea, and Chukchi Sea. The AO index (*red line*) is also shown (scaled to have the same standard deviation as the average of the sea level time series). The *gray-shaded area* represents the uncertainty of the time series of total sea level. Adapted from Calafat et al. (2013)

The AO effects exhibit significant decadal changes, with stronger impacts (higher correlations) on sea level from 1950 to 1995 and almost vanishing correlations after 1995 in all regions (Poshutinsky et al. 2007; Henry et al. 2012; Calafat et al. 2013). Near the Norwegian coast, in the Barents Sea and Kara Sea regions where AO influences are stronger, local wind forcing and sea level signals propagating from the eastern boundary of the North Atlantic are the primarily causes for sea level variations (Richter et al. 2012b; Calafat et al. 2013). In the East Siberian, Chukchi, and also Laptev Seas where the AO is not a good indicator of sea level variations throughout the 1950–2010 period, sea level variability is very sensitive to the change of regional wind and SLP patterns, particularly the longshore wind, which can reproduce the major features of the observed sea level variability from 1950 to 1999. After 1999, sea level variability in these regions, especially in the East Siberian Sea, is associated with both the longshore wind and strengthening Beaufort Gyre (Calafat et al. 2013). A detailed study that quantifies the basin-wide spatial patterns of decadal sea level variability associated with the AO and the related governing processes in the Arctic Ocean is needed, including the local forcing over the Arctic and remote forcing from the Atlantic Ocean.

5.2 The Southern Ocean

The Southern Annular Mode (SAM) is also referred to as the Antarctic Oscillation (Gong and Wang 1999; Limpasuvan and Hartmann 1999). As its northern hemisphere counterpart, the SAM is characterized by a deep, zonally symmetric structure with geopotential height anomalies of opposing signs in the polar cap region and in the surrounding zonal ring centered near 45°S (e.g., Thompson and Wallace 2000). The station-based SAM index is the pressure difference between the latitudes of 40°S and 65°S (Marshall 2003; See National Center for Atmospheric Research Climate Data Guide, https://climatedataguide.ucar.edu/, for SAM index since 1957). As such, the SAM index measures a "see-saw" of atmospheric mass between the middle and high latitudes of the Southern Hemisphere. Positive SAM index corresponds with stronger-than-average westerlies over the mid-high latitudes (50°S–70°S) and weaker westerlies in the midlatitudes (30°S–50°S).

Frankcombe et al. (2015) performed a multiple linear regression analysis by regressing the station-based SAM index onto 50-year SODA sea level and surface winds over the Indian and Pacific Oceans from 1958 to 2007 (bottom panels of Fig. 3). Their results suggest that the SAM is associated with significant sea level variability in the Southern Ocean including the southwest and southeast coasts of Australia, with visible signals in the tropical Pacific and Indian Oceans, but its influence is weak compared to the PDO, ENSO and IOD (Fig. 3). Similar conclusions can be drawn from satellite SSH data for the 1993–2007 period (Figure 3 of Frankcombe et al. 2015), even though this short period may alias the long-term trend with variability. The authors argued that the sea level signature in the equatorial Pacific associated with the SAM may partly reflect ENSO influence on the SAM, as suggested by existing studies (e.g., Ding et al. 2012).

Roemmich et al. (2007) suggested that the decadal spin up of the Subtropical Gyre over the South Pacific, which is associated with a 12-cm increase in satellite-observed SSH between 1993 and 2004 on large spatial scales centered at about (40°S, 170°W), results from the increased surface wind stress curl associated with an enhanced SAM. Regression analyses using a longer record of 1958–2007, however, do not show a strong influence of SAM in the South Pacific Subtropical Gyre region (Fig. 3, bottom row), suggesting that the influence of the SAM may be subject to significant decadal variability. On the other hand, the SAM can have apparent impacts on sea level and surface winds over the southern Indian Ocean (south of 30°S) and around Australia (bottom row of Fig. 3).

In addition to the zonally symmetric circumpolar forcing associated with SAM, there is evidence that zonal asymmetry in SAM may impact interbasin patterns of sea level change in the Southern Hemisphere. Thompson and Merrifield (2014) showed that a proxy for zonal differences in westerly wind strength over the Southern Ocean derived from SLP covaries with out-of-phase multidecadal sea level variations between the Indian-South Pacific and South Atlantic basins during most of the twentieth century. The authors hypothesized that this relationship is due to convergent and divergent transport in the Antarctic Circumpolar Current (ACC) forced by the zonal asymmetry in westerly wind strength. Zonal differences in recent westerly wind trends over the Southern Ocean are related to zonal differences in ocean bottom pressure trends in the ACC (Makowski et al. 2015).

6 Summary, Issues and Challenges

This paper reviews the state of our knowledge about spatial patterns of sea level variability and underlying mechanisms for each ocean basin. It focuses particularly on the patterns associated with natural internal climate modes on decadal-to-multidecadal timescales. Over the Pacific Ocean, the PDO (IPO or decadal variability of ENSO) and NPGO are the two dominant climate modes and they are associated with distinct spatial patterns of sea level change, with both having global signatures (Sect. 2; Figs. 1, 2, 3, 5). Decadal change in ENSO behavior, however, can significantly alter its associated sea level patterns (Fig. 4). Winds associated with these modes (Fig. 6) are the primary causes for the overall sea level patterns, which are dominated by the thermosteric component, with the halosteric component often compensating the thermosteric sea level due to heat and salt redistribution by advection. Heat and freshwater fluxes, however, may also have significant contributions in some regions during some decades. Further studies are needed to quantify the effects of heat versus freshwater fluxes associated with the climate modes. Over the Indian Ocean, sea level trend patterns since the 1960s are driven primarily by surface winds (Fig. 7; Sect. 3), but the nature of the climate modes that contribute to the changing winds remains elusive. On decadal timescales, wind stress associated with ocean–atmosphere coupled modes—ENSO (which causes DIOB before the mid-1980s) and IOD (Fig. 8)—is the major cause for decadal sea level variability north of 20°S. The sea level pattern associated with the IOD alone shows an east–west dipole (Fig. 9), and frequently occurring intensified positive IOD in a warming climate could induce frequent SLRs in the western basin but falls in the eastern Indian Ocean during boreal fall.

Over the Atlantic, the NAO-associated sea level patterns exhibit a dipole structure in the North Atlantic basin (Figs. 10, 11). In the basin interior, surface heat fluxes are suggested to be the major force for the decadal sea level patterns due to AMOC variations (Sect. 4). Effects of freshwater flux and sea ice transport associated with the NAO remain to be explored. Along the eastern boundary, longshore winds and coastal Kelvin waves are the primary causes for the coherent sea level changes. Along the west boundary (US east coast), some studies demonstrate the importance of interior wind stress curl and local wind over the shelf in driving decadal sea level variability, whereas others argue for the importance of AMOC and Gulf Stream variations. The 20–30-year decadal sea level signals observed in both the North and South Atlantic (Fig. 12) are associated with AMOC variations and oceanic Rossby waves. The upward trend of the AMO SST index during recent decades coincides with the observed accelerated SLR along the US northeast coast. The variation of the AMO index, however, might be partly externally forced and thus may not represent the true internal AMO. Over the Arctic, significant correlations between the AO index and tide gauge records have been found, but with distinct regional variations (Fig. 13; Sect. 5). The correlation coefficients are high in the Barents Sea, Norwegian Sea and Kara Sea, weaker for the Laptev Sea and not significant for the East Siberian Sea and the Chukchi Sea. Winds and to a lesser degree IB effects are important for driving the sea level variability, and in the Norwegian Sea coastal signals propagating from the eastern boundary of the North Atlantic also contribute. Finally, the SAM can have a significant influence on sea level in the Southern Indian and Pacific Oceans, with weak influence over the tropics compared to the PDO, ENSO and IOD. Zonal asymmetry in SAM-associated winds might have contributed to the asymmetry of decadal sea level variations in the southern ocean during most of the twentieth century.

While progress has been made toward understanding the spatial patterns of sea level variability associated with internal climate modes, immediate issues and challenges for advancing our understanding remain. These issues must be resolved in order to achieve more accurate depictions of climate modes' impacts on regional SLR and therefore contribute to improved decadal predictions and future projections of regional SLR. Relevant issues, challenges and future outlooks are discussed in each of the following aspects below. Understanding the impact of natural climate variability (coupled, forced or free internal ocean) on sea level is a prerequisite before anthropogenic finger prints can be identified.

(a) Representation of climate modes and associated fields Existing studies often use empirical analysis—particularly EOF—to identify climate modes. For instance, the PDO is defined as the leading EOF mode of SST in the North Pacific and its temporal variability is represented by PC1 (Sect. 2.1). While it is the leading empirical mode, the PDO is not a single physical mode of climate variability because it results from the combination of three groups of processes: (1) variability associated with the Aleutian low due to stochastic

whether noise and remote forcing from the tropics (largely ENSO), (2) ocean memory and (3) decadal variability of Kuroshio–Oyashio currents induced by westward-propagating oceanic Rossby waves (see review paper by Newman et al. 2016). Given that different processes may dominate the PDO during different decades, the PDO-associated SST, wind, and thus sea level patterns may have different manifestations. Using a fixed SST pattern (EOF1) to represent the PDO may not depict these pattern changes, resulting in limitations for assessing its associated wind and sea level patterns. Improved PDO indices that are able to represent this temporal evolution need to be sought.

To extract the surface winds/sea level associated with climate modes, existing studies generally use multiple linear regression by regressing the observed or reanalysis winds/sea level onto climate mode indices. Note that the regression coefficients remain constant (static) in time, which measure their overall correlations for the examined temporal period. In reality, however, their correlations may exhibit strong decadal variability within the period of interest, due to behavior changes of climate modes; but the static linear regression method cannot depict this decadal variability. For instance, Andres et al. (2013) showed that the NAO is strongly correlated with interannual sea level variability along the US northeast coast since 1987, but they are almost uncorrelated before 1987 (from 1970 to 1987). Advanced statistical techniques that can extract the time-evolving surface wind/sea level patterns associated with climate modes will be very helpful.

(b) Interactions among climate modes and effects of anthropogenic forcing The climate modes over each ocean basin may not be fully independent. For example, the PDO and DIOB are highly correlated with positive correlation before the mid-1980s and negative correlation afterward (Fig. 8b); the PDO can affect Indian Ocean winds partly through modulating the Asian-Australian monsoon (Meehl and Arblaster 2011, 2012); decadal variability of tropical Indian Ocean SST since the mid-1980s and multidecadal trend since the 1960s (e.g., Han et al. 2014a) as well as tropical Atlantic warming since the 1990s (e.g., McGregor et al. 2014) might have affected the tropical Pacific winds and sea level; and changes in ENSO-PDO phase relationship may enhance decadal sea level variability in the western tropical Pacific (Moon et al. 2015).

Furthermore, SST indices of decadal climate modes (e.g., PDO and AMO) can be affected by anthropogenic warming and other external forcing (e.g., Dong et al. 2014; Booth et al. 2012; Knudsen et al. 2014). Consequently, winds and sea level patterns regressed onto these indices may not represent the effects of pure natural, internal modes. For these reasons, separating the effects of internal climate modes on sea level from that of external forcing (both anthropogenic and natural) using observational analysis remains a challenge. To this end, large ensemble experiments of climate models with long integrations using anthropogenic, natural, and all forcings, respectively, will be helpful for the separation, even though different climate models may suffer from different biases. Efforts are being made in this line. For example, the National Center for Atmospheric Research (NCAR) has completed a 42-member ensemble experiment using the Community Earth System Model (CESM) from 1920 to 2100, with historical all forcing from 1920 to 2005 and RCP8.5 trajectory since 2006 (Kay et al. 2015). The 42-member ensemble mean will damp internal climate modes and thus measure the effect of external forcing.

(c) Effect of oceanic internal variability Several recent studies examined the effects of oceanic internal variability (instabilities) on sea level variability, by performing experiments using standalone, eddy-permitting ($1/3° \times 1/3°$ and $1/4° \times 1/4°$ grids) (e.g., Trenary and Han 2013; Li and Han 2015; Penduff et al. 2011) and eddy-resolving ($1/12° \times 1/12°$) (Sérazin et al. 2015) OGCMs. Over the Indian Ocean, Trenary and Han (2013) and Li and Han (2015) suggested that oceanic internal variability has a significant contribution to

decadal sea level variability in the subtropical south Indian Ocean, near the Somali coast and western Bay of Bengal. Penduff et al. (2011) demonstrated a significant increase of interannual SSHa variance by increasing OGCM resolution from 2° to ¼°. Sérazin et al. (2015) showed that the small-scale ($L < 6°$) SSHa variance is almost entirely of intrinsic origin at all timescales; the large-scale ($L > 12°$) low-frequency variability ($T > 18$ months), however, results largely from atmospheric forcing over most of the global ocean, but oceanic internal variability still has large amplitudes over the Gulf Stream, Kuroshio, and Antarctic Circumpolar Current regions. The intrinsic SSHa, which is independent of climate modes and unpredictable, affects the decadal predictability for regional sea level variability. Further research is required using high-resolution models to quantify the effects of oceanic intrinsic variability on decadal sea level variability and predictability.

(d) Limited observational records The short records of available datasets limit our ability to detect the full character of decadal climate variability. Modeling studies suggest that we need ~500 years of observations to sample the full range of ENSO decadal variability (e.g., Wittenberg 2009). Our reliable global-scale SST datasets based on in situ observations, however, are only ~one century long, and the records of in situ observed surface winds and sea level are even shorter. Yet, surface winds associated with climate modes are the major driver for the spatial patterns of sea level variability. Satellite observations (e.g., winds, sea level) have revolutionized our understanding of intraseasonal-to-interannual variability, but their records are too short for studying decadal variability. Therefore, these observations must be continued into the future to lengthen the record for studies of decadal climate variability.

Efforts have been made to generate reliable longer data records for climate studies (e.g., temperature, winds, sea level) by quality control existing historic datasets, assimilating observations into numerical models (reanalysis products), and applying statistical techniques to reconstruct basin- and global-scale datasets during earlier periods when observations are sparse in space. Apparent differences, however, exist among different reanalysis products, particularly with regard to their multidecadal trends (e.g., winds, Sect. 2.3). There are also significant differences among reconstructed sea level datasets before the satellite era regarding both global mean SLR (e.g., Christiansen et al. 2010; Calafat et al. 2014) and regional multidecadal trends over the Pacific (e.g., Moon et al. 2013). The discrepancies among different reconstructions are due primarily to different statistical approaches used before the satellite era (e.g., Christiansen et al. 2010; Calafat et al. 2014).

While significant progress has been made in producing various climate datasets, continuous effort is needed to further improve the quality of existing historical, reanalysis and reconstructed datasets, either by advancing our existing tools or by seeking new statistical techniques to overcome the known limitations. For documenting and understanding decadal scale variability, our existing satellite and in situ data records are too short and there remains an imperative for sustained in situ and satellite observations into the future. Since sea level represents the vertically integrated effects of ocean heat and freshwater uptakes, changes in surface and subsurface ocean circulations as well as their associated redistributions of heat, freshwater and mass, continued observations at the surface and subsurface are needed in order to understand the causes for decadal sea level variability, including those induced by climate modes. Recent advances in global observing system development—such as the Argo network that provides vertical temperature and salinity profiles in the upper 2000 m and satellite missions that provide accurate measurements of sea level (e.g., Jason series), ocean vector winds (e.g., international scatterometer missions) and sea ice extent/concentration—have substantially enhanced our capability to conduct sea level and climate research. These satellite and in situ measurements systems combine to provide indispensible observational resources for studying regional decadal variability. Sustaining and enhancing these observing systems, including enhancing deep ocean observations to measure hydrographic profiles below 2000 m, to obtain reliable and continuous data records are essential to ensure our future progress and success in understanding and predicting decadal sea level variability on regional scales.

Acknowledgments Part of this work was completed when Weiqing Han was visiting NCAR during her sabbatical leave in spring and the University of Hamburg in summer 2016. She thanks CCR/NCAR for providing partial summer salary support and CliSAP of the University of Hamburg for providing travel support. W. Han is supported by National Science Foundation (NSF) AGS 1446480 and NSF OCE 1558736. G.A. Meehl and A. Hu are partly supported by the Regional and Global Climate Modelling Program (RGCM) of the U.S. Department of Energy's Office of Science (BER), Cooperative Agreement No. DE-FC02-97ER62402, and the National Science Foundation. D. Stammer is supported in part through the German RACE project funded by the BMBF and the European NaCLIM project. This work contributes to the DFG-funded Excellence Cluster CliSAP of the University of Hamburg. B. Hamlington and P. Thompson are supported by NSF OCE 1558736. Jessica Kenigson is supported by NASA Harriett G. Jenkins Graduate Fellowship Program Grant NNX13AR74H. H. Palanisamy is funded by the CNES-CLS doctoral fellowship. The National Center for Atmospheric Research is sponsored by the National Science Foundation. The paper is an outcome of the ISSI Workshop on "Integrative Study of Sea Level" during February 2–6, 2015, in Bern, Switzerland.

References

Abram NJ, Gagan MK, Cole J, Hantoro WS, Mudelsee M (2008) Recent intensification of tropical climate variability in the Indian Ocean. Nat Geosci 1:849–853. doi:10.1038/ngeo357

Alexander MA (2010) Extratropical air-sea interaction, SST variability and the pacific decadal oscillation (PDO). In: Sun D, Bryan F (eds) Climate dynamics: why does climate vary. AGU Monograph #189, Washington, pp 123–148

Alexander MA, Blade I, Newman M, Lanzante JR, Lau NC, Scott JD (2002) The atmospheric bridge: the influence of ENSO teleconnections on air-sea interaction over the global oceans. J Clim 15:2205–2231. doi:10.1175/1520-0442(2002)015<2205:TABTIO>2.0.CO;2

Andres M, Gawarkiewicz GG, Toole JM (2013) Interannual sea level variability in the western North Atlantic: regional forcing and remote response. Geophys Res Lett 40:5915–5919. doi:10.1002/2013GL058013

Ashok K, Chan W-L, Motoi T, Yamagata T (2004) Decadal variability of the Indian Ocean dipole. Geophys Res Lett 31:L24207. doi:10.1029/2004GL021345

Balmaseda MA, Vidard A, Anderson DL (2008) The ECMWF ocean analysis system: ORA-S3. Mon Weather Rev 136:3018–3034. doi:10.1175/2008MWR2433.1

Barnston AG, Livezey RE (1987) Classification, seasonality and persistence of low-frequency atmospheric circulation patterns. Mon Weather Rev 115:1083–1126. doi:10.1175/1520-0493(1987)115<1083:CSAPOL>2.0.CO;2

Behera SK, Yamagata T (2001) Subtropical SST dipole events in the southern Indian Ocean. Geophys Res Lett 28:327–330. doi:10.1029/2000GL011451

Behera S, Yamagata T (2010) Imprint of the El NiñoModoki on decadal sea level changes. Geophys Res Lett 37:L23702. doi:10.1029/2010GL045936

Bindoff NL, Willebrand J, Artale V, Cazenave A, Gregory J et al (2007) Observations: oceanic climate change and sea level, ch5. In: Solomon S, Qin D, Manning M et al (eds) Climate change 2007: the physical science basis, contribution of working group I to the fourth assessment report of the intergovernmental panel on climate change. Cambridge University Press, Cambridge, pp 385–432

Bingham RJ, Hughes CW (2009) Signature of the Atlantic meridional overturning circulation in sea level along the east coast of North America. Geophys Res Lett 36:L02603. doi:10.1029/2008GL036215

Boon JD (2012) Evidence of sea level acceleration at U.S. and Canadian tide stations, Atlantic Coast, North America. J Coast Res 285:1437–1445

Booth BB, Dunstone NJ, Halloran PR, Andrews T, Bellouin N (2012) Aerosols implicated as a prime driver of twentieth-century North Atlantic climate variability. Nature 484(7393):228–232. doi:10.1038/nature10946

Bromirski PD, Miller AJ, Flick RE, Auad G (2011) Dynamical suppression of sea level rise along the Pacific coast of North America: indications for imminent acceleration. J Geophys Res 116:C07005. doi:10.1029/2010JC006759

Cai W, Cowan T, Sullivan A (2009) Recent unprecedented skewness towards positive Indian Ocean Dipole occurrences and its impact on Australian rainfall. Geophys Res Lett 36:L11705. doi:10.1029/2009GL037604

Cai W, Zheng X-T, Weller E, Collins M, Cowan T, Lengaigne M, Yu W, Yamagata T (2013) Projected response of the Indian Ocean Dipole to greenhouse warming. Nat Geosci 6:999–1007. doi:10.1038/NGEO2009

Calafat FM, Chambers D, Tsimplis MN (2012) Mechanisms of decadal sea level variability in the Eastern North Atlantic and the Mediterranean Sea. J Geophys Res 117:C09022. doi:10.1029/2012JC008285

Calafat FM, Chambers DP, Tsimplis MN (2013) Inter-annual to decadal sea-level variability in the coastal zones of the Norwegian and Siberian Seas: the role of atmospheric forcing. J Geophys Res Oceans 118:1287–1301. doi:10.1002/jgrc.20106

Calafat FM, Chambers DP, Tsimplis MN (2014) On the ability of global sea level reconstructions to determine trends and variability. J Geophys Res Oceans 119:1572–1592

Carton JA, Giese BS, Grodsky SA (2005) Sea level rise and the warming of the oceans in the Simple Ocean Data Assimilation (SODA) ocean reanalysis. J Geophys Res 110:C09006. doi:10.1029/2004JC002817

Chambers DP, Merrifield MA, Nerem RS (2012) Is there a 60-year oscillation in global mean sea level? Geophys Res Lett 39:L18607. doi:10.1029/2012GL052885

Chaudhuri Ayan H, Gangopadhyay Avijit, Bisagni James J (2011) Response of the Gulf Stream transport to characteristic high and low phases of the North Atlantic Oscillation. Ocean Model 39:220–232

Christiansen B, Schmith T, Thejll P (2010) A surrogate ensemble study of sea level reconstructions. J Clim 23:4306–4326. doi:10.1175/2010JCLI3014.1

Church JA, Gregory JM, White NJ, Platten SM, Mitrovica JX (2011) Understanding and projecting sea level change. Oceanography 24(2):130–143

Church JA, Clark PU, Cazenave A, Gregory JM, Jevrejeva S, Levermann A, Merrifield MA, Milne GA, Nerem RS, Nunn PD, Payne AJ, Pfeffer WT, Stammer D, Unnikrishnan AS (2013) Sea level change. In: Stocker TF, Qin D, Plattner G-K, Tignor M, Allen SK, Boschung J, Nauels A, Xia Y, Bex V, Midgley PM (eds) Climate change 2013: the physical science basis. Contribution of working group I to the fifth assessment report of the intergovernmental panel on climate change. Cambridge University Press, Cambridge

Chylek P, Foland CK, Dijkstra HA, Lesins G, Dubey MK (2011) Ice-core data evidence for a prominent near 20 year time-scale of the Atlantic Multidecadal Oscillation. Geophys Res Lett 38:L13704. doi:10.1029/2011GL047501

Clarke AJ, Lebedev A (1997) Interannual and decadal changes in equatorial wind stress in the Atlantic, Indian, and Pacific Oceans and the eastern ocean coastal response. J Clim 10(7):1722–1729

Curry R, McCartney M (2001) Ocean gyre circulation changes associated with the North Atlantic Oscillation. J Phys Oceanogr 31:3374–3400

Danabasoglu G (2008) On multidecadal variability of the Atlantic meridional overturning circulation in the community climate system model version 3. J Clim 21:5524–5544

Delworth TL, Mann ME (2000) Observed and simulated multidecadal variability in the Northern Hemisphere. Clim Dyn 16:661–676. doi:10.1007/s003820000075

Deser C, Phillips AS, Hurrell JW (2004) Pacific interdecadal climate variability: linkages between the tropics and North Pacific during boreal winter since 1900. J Clim 17:3109–3124

Di Lorenzo E et al (2008) North Pacific Gyre Oscillation links ocean climate and ecosystem change. Geophys Res Lett 35:L08607. doi:10.1029/2007GL032838

Di Nezio PN, Gramer LJ, Johns WE, Meinen CS, Baringer MO (2009) Observed interannual variability of the Florida Current: wind forcing and the North Atlantic Oscillation. J Phys Oceanogr 39(3):721–736

Ding Q, Steig EJ, Battisti DS, Wallace JM (2012) Influence of the tropics on the Southern Annular Mode. J Clim 25(18):6330–6348

Doblas-Reyes FJ, Balmaseda MA, Weisheimer A, Palmer TN (2011) Decadal climate prediction with the European Centre for Medium-Range Weather Forecasts coupled forecast system: impact of ocean observation. J Geophys Res 116:D19111. doi:10.1029/2010JD015394

Dong L, Zhou T, Chen X (2014) Changes of Pacific decadal variability in the twentieth century driven by internal variability, greenhouse gases, and aerosols. Geophys Res Lett 41:8570–8577. doi:10.1002/2014GL062269

Du Y, Xie SP (2008) Role of atmospheric adjustments in the TIO warming during the 20th century in climate models. Geophys Res Lett 35:L08712. doi:10.1029/2008GL033631

Dunne RP, Barbosa SM, Woodworth PL (2012) Contemporary sea level in the Chagos Archipelago, central Indian Ocean. Glob Planet Change 82–83:25–37

Enfield DB, Mestas-Nunez AM, Trimble PJ (2001) The Atlantic Multidecadal Oscillation and its relationship to rainfall and river flows in the continental U.S. Geophys Res Lett 28:2077–2080

England MH et al (2014) Recent intensification of wind-driven circulation in the Pacific and the ongoing warming hiatus. Nat Clim Change 4:222–227. doi:10.1038/nclimate2106

Ezer T (2013) Sea level rise, spatially uneven and temporally unsteady: why the U.S. East Coast, the global tide gauge record, and the global altimeter data show different trends. Geophys Res Lett 40:5439–5444. doi:10.1002/2013GL057952

Ezer T, Corlett WB (2012) Is sea level rise accelerating in the Chesapeake Bay? A demonstration of a novel new approach for analyzing sea level data. Geophys Res Lett 39:L19605. doi:10.1029/2012GL053435

Ezer T, Atkinson LP, Corlett WB, Blanco JL (2013) Gulf Stream's induced sea level rise and variability along the U.S. mid-Atlantic coast. J Geophys Res Oceans 118:685–697. doi:10.1002/jgrc.20091

Folland C, Renwick JA, Salinger MJ, Mullan AB (2002) Relative influences of the Interdecadal Pacific Oscillation and ENSO on the South Pacific Convergence Zone. Geophys Res Lett 29:1643. doi:10.1029/2001GL014201

Forget G, Ponte RM (2015) The partition of regional sea level variability. Prog Oceanogr 137:173–195

Frankcombe LM, Dijkstra HA (2009) Coherentmultidecadal variability in NorthAtlantic sea level. Geophys Res Lett 36:L15604. doi:10.1029/2009GL039455

Frankcombe LM, Von der Heydt AS, Dijkstra HA (2010) North Atlantic multidecadal climate variability: an investigation of dominant time scales and processes. J Clim 23:3626–3638

Frankcombe LM, McGregor S, England MH (2015) Robustness of the modes of Indo-Pacific sea level variability. Clim Dyn 45:1281–1298. doi:10.1007/s00382-014-2377-0

Fukumori I, Wang O (2013) Origins of heat and freshwater anomalies underlying regional decadal sea level trends. Geophys Res Lett 40:563–567. doi:10.1002/grl.50164

Garreaud R, Battisti D (1999) Interannual (ENSO) and interdecadal (ENSO-like) variability in the Southern Hemisphere tropospheric circulation. J Clim 12:2113–2123

Goddard L, Baethgen W, Kirtman B, Meehl GA (2009) The urgent need for improved models and predictions. EOS 90:343

Goddard PB, Yin J, Griffies SM, Zhang S (2015) An extreme event of sea-level rise along the Northeast coast of North America in 2009–2010. Nat Commun 6:6346. doi:10.1038/ncomms7346

Gong D, Wang S (1999) Definition of Antarctic oscillation index. Geophys Res Lett 26:459–462

Hakkinen S (2000) Decadal air-sea interaction in the North Atlantic based on observations and modeling results. J Clim 13:1195–1219

Hakkinen S (2001) Variability in sea surface height: a qualitative measure for the meridional overturning in the North Atlantic. J Geophys Res 106:13837–13848. doi:10.1029/1999JC000155

Hakkinen S, Rhines PB (2004) Decline of subpolar North Atlantic gyre circulation during the 1990s. Science 304:555–559. doi:10.1126/science.1094917

Hamlington BD, Leben R, Strassburg MW, Nerem RS, Kim KY (2013) Contribution of the Pacific decadal oscillation to global mean sea level trends. Geophys Res Lett 40:5171–5175

Hamlington BD, Leben R, Nerem RS, Han W, Kim KY (2011) Reconstructing sea level using cyclostationary empirical orthogonal functions. J Geophys Res. doi:10.1029/2011JC007529

Hamlington BD, Strassburg MW, Leben R, Han W, Nerem RS, Kim KY (2014) Uncovering the anthropogenic warming-induced sea level rise signal in the Pacific Ocean. Nat Clim Change 4:782–785

Han W, Webster PJ (2002) Forcing mechanisms of sea-level interannual variability in the bay of Bengal. J Phys Oceanogr 32:216–239

Han W, Meehl GA, Rajagopalan B, Fasullo J, Hu A, Lin J, Large W, Wang J, Quan X, Trenary L, Wallcraft A, Shinoda T, Yeager S (2010) Patterns of Indian Ocean sea level change in a warming climate. Nat Geosci 3:546–550. doi:10.1038/NGEO901

Han W, Meehl GA, Hu A, Alexander M, Yamagata T, Yuan D, Ishii M, Pegion P, Zheng J, Hamlington B, Quan X-W, Leben R (2014a) Intensification of decadal and multi-decadal sea level variability in the western tropical Pacific during recent decades. Clim Dyn 43:1357–1379

Han W, Vialard J, McPhaden MJ, Lee T, Masumoto Y, Feng M, de Ruijter WPM (2014b) Indian Ocean decadal variability: a review. Bull Am Meteorol Soc 95:1679–1703. doi:10.1175/BAMS-D-13-00028.1

Henry O, Prandi P, Llovel W, Cazenave A, Jevrejeva S, Stammer D, Meyssignac B, Koldunov N (2012) Tide gauge-based sea level variations since 1950 along the Norwegian and Russian coasts of the Arctic Ocean: contribution of the steric and mass components. J Geophys Res 117:C06023. doi:10.1029/2011JC007706

Hong BG, Sturges W, Clarke AJ (2000) Sea level on the U.S. east coast: decadal variability caused by open ocean wind-curl forcing. J Phys Oceanogr 30:2088–2098. doi:10.1175/1520-0485(2000)030<2088:SLOTUS>2.0.CO;2

Hu A, Meehl G, Han W, Yin J (2009) Transient response of the MOC and climate to potential melting of the Greenland Ice Sheet in the 21st century. Geophys Res Lett 36:L10707. doi:10.1029/2009GL037998

Huck T, Vallis GK (2001) Linear stability analysis of the threedimensional thermally-driven ocean circulation: application to interdecadal oscillations. Tellus A 53:526–545

Hurrell JW (1995) Decadal trends in the North Atlantic Oscillation: regional temperatures and precipitation. Science 269:676–679

Hurrell JW, Meehl GA, Bader D, Delworth T, Kirtman B, Wielicki B (2009) A unified approach to climate system prediction. Bull Am Meteorol Soc 90:1819–1832

Ihara C, Kushnir Y, Cane MA (2008) Warming trend of the Indian Ocean SST and Indian Ocean dipole from 1880 to 2004. J Clim 21:2035–2046

Jevrejeva S, Moore JC, Grinsted A, Woodworth PL (2008) Recent global sea level acceleration started over 200 years ago? Geophys Res Lett 35:L08715. doi:10.1029/2008GL033611

Kay JE, Deser C, Phillips A, Mai A, Hannay C, Strand G, Arblaster JM, Bates SC, Danabasoglu G, Edwards J, Holland M, Kushner P, Lamarque JF, Lawrence D, Lindsay K, Middleton A, Munoz E, Neale R, Oleson K, Polvani L, Verten-stein M (2015) The Community Earth System Model (CESM) large ensemble project: a community resource for studying climate change in the presence of internal climate variability. Bull Am Meteorol Soc. doi:10.1175/BAMS-D-13-00255.1

Kenigson J, Han W (2014) Detecting and understanding the accelerated sea level rise along the east coast of the United States during recent decades. J Geophys Res 119:8749–8766

Kerr RA (2000) A North Atlantic climate pacemaker for the centuries. Science 288(5473):1984–1986

Klein SA, Soden BJ, Lau NC (1999) Remote sea surface temperature variations during ENSO: evidence for a tropical atmospheric bridge. J Clim 12:917–932

Knudsen MF, Jacobsen BH, Seidenkrantz MS, Olsen J (2014) Evidence for external forcing of the Atlantic Multidecadal Oscillation since termination of the Little Ice Age. Nat Commun 5:3323. doi:10.1038/ncomms4323

Köhl A, Stammer D (2008) Variability of the meridional overturning in the North Atlantic from the 50-year GECCO state estimation. J Phys Oceanogr 38:1914–1930

Köhl A, Stammer D, Cornuelle B (2007) Interannual to decadal changes in the ECCO global snthesis. J Phys Oceanogr 37:313–337

Kolker AS, Hameed S (2007) Meteorologically driven trends in sea level rise. Geophys Res Lett 34:L23616. doi:10.1029/2007GL031814

Kopp RE (2013) Does the mid-Atlantic United States sea level acceleration hot spot reflect ocean dynamic variability? Geophys Res Lett 40:3981–3985. doi:10.1002/grl.50781

Kopp RE, Hay CC, Little CM, Mitrovica JX (2015) Geographic variability of sea-level change. Curr Clim Change Rep 1:192–204. doi:10.1007/s40641-015-0015-5

Kripalani RH, Kumar P (2004) Northeast monsoon rainfall variability over south peninsular India vis-a-vis Indian Ocean dipole mode. Int J Climatol 24:1267–1282

Lee T, McPhaden MJ (2008) Decadal phase change in large-scale sea level and winds in the Indo-Pacific region at the end of the 20th century. Geophys Res Lett 35:L01605. doi:10.1029/2007GL032419

Levitus S, Antonov JI, Boyer TP, Baranova OK, Garcia HE, Locarnini RA, Mishonov AV, Reagan JR, Seidov D, Yarosh ES, Zweng MM (2012) World ocean heat content and thermosteric sea level change (0–2000 m), 1955–2010. Geophys Res Lett 39:L10603. doi:10.1029/2012GL051106

Li Y, Han W (2015) Decadal sea level variations in the Indian Ocean investigated with HYCOM: roles of climate modes, ocean internal variability and stochastic wind forcing. J Clim 28:9143–9165. doi:10.1175/JCLI-D-15-0252.1

Limpasuvan V, Hartmann DL (1999) Eddies and the annular modes of climate variability. Geophys Res Lett 26:3133–3136

Liu Z (2012) Dynamics of interdecadal climate variability: an historical perspective. J Clim 25:1963–1995. doi:10.1175/2011JCLI3980.1

Llovel W, Lee T (2015) Importance and origin of halosteric contribution to sea level change in the southeast Indian Ocean during 2005–2013. Geophys Res Lett 42:1148–1157. doi:10.1002/2014GL062611

Lombard A, Cazenave A, Le Traon P-Y, Ishii M (2005) Contribution of thermal expansion to present-day sea-level change revisited. Glob Planet Change 47:1–16
Lozier MS, Roussenov V, Mark S, Reed C, Williams RG (2010) Opposing decadal changes for the North Atlantic meridional overturning circulation. Nat Geosci 3:728–734
Luo JJ, Sasaki W, Masumoto Y (2012) Indian Ocean warming modulates Pacific climate change. Proc Natl Acad Sci 109:18701–18706
Makowski JK, Chambers DP, Bonin JA (2015) Using ocean bottom pressure from the gravity recovery and climate experiment (GRACE) to estimate transport variability in the southern Indian Ocean. J Geophys Res Ocean 120(6):4245–4259. doi:10.1002/2014JC010575
Mann ME, Steinman BA, Miller SK (2014) On forced temperature changes, internal variability, and the AMO. Geophys Res Lett 41:3211–3219. doi:10.1002/2014GL059233
Mantua NJ, Hare SR, Zhang Y, Wallace JM, Francis RC (1997) A Pacific interdecadal climate oscillation with impacts on salmon production. Bull Am Meteorol Soc 78:1069–1079
Marcos M, Tsimplis MN, Calafat FM (2012) Inter-annual and decadal sea level variations in the north-western Pacific marginal seas. Prog Oceanogr 105:4–21
Marshall GJ (2003) Trends in the southern annular mode from observations and reanalyses. J Clim 16:4134–4143
McCarthy GD, Haigh ID, Hirschi JJ-M, Grist JP, Smeed DA (2015) Ocean impact on decadal Atlantic climate variability revealed by sea-level observations. Nature 521(7553):508–510. doi:10.1038/nature14491
McGregor S, Sen Gupta A, England MH (2012) Constraining wind stress products with sea surface height observations and implications for Pacific Ocean sea-level trend. J Clim 25:8164–8176
McGregor S, Timmermann A, Stuecker MF, England MH, Merrifield M, Jin F-F, Chikamoto Y (2014) Recent Walker circulation strengthening and Pacific cooling amplified by Atlantic warming. Nat Clim Change 4:888–892. doi:10.1038/nclimate2330
Meehl GA, Arblaster JM (2011) Decadal variability of Asian-Australian Monsoon-ENSO-TBO relationships. J Clim 24:4925–4940
Meehl GA, Arblaster JM (2012) Relating the strength of the tropospheric biennial oscillation (TBO) to the phase of the Interdecadal Pacific Oscillation (IPO). Res Lett, Geophys. doi:10.1029/2012GL053386
Meehl GA, Hu A (2006) Megadroughts in the Indian monsoon region and southwest North America and a mechanism for associated multidecadal Pacific sea surface temperature anomalies. J Clim 19:1605–1623
Meehl GA, Goddard L et al (2009) Decadal prediction: can it be skillful? Bull Am Meteorol Soc 90:1467–1484
Meehl GA et al (2014) Decadal climate prediction: an update from the trenches. Bull Am Meteorol Soc 95:243–267. doi:10.1175/BAMS-D-12-00241.1
Meinen CS, Baringer MO, Garcia RF (2010) Florida current transport variability: an analysis of annual and longer-period signals. Deep Sea Res I 57:835–846
Merrifield MA (2011) A shift in western tropical Pacific sea level trends during the 1990s. J Clim 24:4126–4138
Merrifield MA, Maltrud ME (2011) Regional sea level trends due to a Pacific trade wind intensification. Geophys Res Lett 38:L21605. doi:10.1029/2011GL049576
Merrifield MA, Thompson PR, Lander M (2012) Multidecadal sea level anomalies and trends in the western tropical Pacific. Geophys Res Lett 39(13):2–6. doi:10.1029/2012GL052032
Meyssignac B, Salas y Melia D, Becker M, Llovel W, Cazenave A (2012) Tropical Pacific spatial trend patterns in observed sea level: internal variability and/or anthropogenic signature? Clim Past Discuss 8:349–389
Miller L, Douglas BC (2007) Gyre-scale atmospheric pressure variations and their relation to 19th and 20th century sea level rise. Geophys Res Lett 34:L16602. doi:10.1029/2007GL030862
Milne GA, Gehrels WR, Hughes CW, Tamisiea ME (2009) Identifying the causes of sea-level change. Nat Geosci 2:471–478
Minobe S (1997) A 50–70 year climatic oscillation over the North Pacific and North America. Geophys Res Lett 24:683–686
Moon J-H, Song YT, Bromirski PD, Miller AJ (2013) Multidecadal regional sea level shifts in the Pacific over 1958–2008. J Geophys Res Oceans 118:7024–7035. doi:10.1002/2013JC009297
Moon J-H, Song YT, Lee H (2015) PDO and ENSO modulations intensified decadal sea level variability in the tropical Pacific. J Geophys Res 120:8229–8237. doi:10.1002/2015JC011139
National Research Council (2012) Sea-level rise for the coasts of California, Oregon, and Washington: past, present, and future. The National Academies Press, Washington. http://www.nap.edu/catalog.php?record_id=13389

Newman M, Compo GP, Alexander MA (2003) ENSO-forced variability of the Pacific decadal oscillation. J Clim 16:3853–3857

Newman M, Alexander MA et al (2016) The Pacific decadal oscillation, revisited. J Clim 29:4399–4427. doi:10.1175/JCLI-D-15-0508.1

Nidheesh AG, Lengaigne M, Vialard J, Unnikrishnan AS, Dayan H (2013) Decadal and long-term sea level variability in the tropical Indo-Pacific Ocean. Clim Dyn 41:381–402. doi:10.1007/s00382-012-1463-4

Palanisamy H, Meyssignac B, Cazenave A, Delcroix T (2015) Is anthropogenic sea level fingerprint already detectable in the Pacific Ocean? Environ Res Lett 10:084024. doi:10.1088/1748-9326/10/8/084024

Penduff T, Juza M, Barnier B, Zika J, Dewar WK, Treguier A-M, Molines J-M, Audiffren N (2011) Sea level expression of intrinsic and forced ocean variabilities at interannual time scales. J Clim 24:5652–5670. doi:10.1175/JCLI-D-11-00077.1

Piecuch CG, Ponte RM (2012) Buoyancy-driven interannual sea level changes in the southeast tropical Pacific. Geophys Res Lett 39:L05607. doi:10.1029/2012GL051130

Piecuch CG, Ponte RM (2013) Buoyancy-Driven Interannual Sea Level Changes in the Tropical South Atlantic. J Phys Oceanogr 43:533–547. doi:10.1175/JPO-D-12-093.1

Piecuch CG, Ponte RM (2015) Inverted barometer contributions to recent sea level changes along the northeast coast of North America. Geophys Res Lett 42(14):5918–5925. doi:10.1002/2015GL064580

Pohlmann H, Jungclaus JH, Köhl A, Stammer D, Marotzke J (2009) Initializing decadal climate predictions with the GECCO oceanic synthesis: effects on the North Atlantic. J Clim 22:3926–3938

Polkova I, Köhl A, Stammer D (2015) Predictive skill for regional sea surface height and mechanisms for predictability. J Clim 28:7407–7419. doi:10.1175/JCLI-D-14-00811.1

Power S, Casey T, Folland C, Colman A, Mehta V (1999) Interdecadal modulation of the impact of ENSO on Australia. Clim Dyn 15:319–324

Proshutinsky A, Ashik I, Häkkinen S, Hunke E, Krishfield R, Maltrud M, Maslowski W, Zhang J (2007) Sea level variability in the Arctic Ocean from AOMIP models. J Geophys Res 112:C04S08. doi:10.1029/2006JC003916

Proshutinsky A, Pavlov V, Bourke RH (2001) Sea level rise in the Arctic Ocean. Geophys Res Lett 28:2237–2240

Proshutinsky A, Ashik IM, Dvorkin EN, Häkkinen S, Krishfield RA, Peltier WR (2004) Secular sea level change in the Russian sector of the Arctic Ocean. J Geophys Res 109:C03042. doi:10.1029/2003JC002007

Qiu B (2002) Large-scale variability in the midlatitude subtropical and subpolar North Pacific Ocean: observations and causes. J Phys Oceanogr 32:353–375

Qiu B, Chen S (2012) Multidecadal sea level and gyre circulation variability in the northwestern tropical Pacific Ocean current off the Philippines. J Phys Oceanogr 42:193–206

Rao SA, Behera SK, Masumoto Y, Yamagata T (2002) Interannual variability in the subsurface tropical Indian Ocean. Deep-Sea Res II 49:1549–1572

Richter K, Nilsen JE, Drange H (2012a) Contributions to sea level variability along the Norwegian coast for 1960–2010. J Geophys Res 117:C05038. doi:10.1029/2011JC007826

Richter K, Segtnan OH, Furevik T (2012b) Variability of the Atlantic inflow to the Nordic Seas and its causes inferred from observations of sea surface height. J Geophys Res 117:C04004. doi:10.1029/2011JC007719

Roemmich D, Gilson J, Davis R, Sutton P, Wijffels S, Riser S (2007) Decadal spinup of the South Pacific subtropical gyre. J Phys Oceanogr 37:162–173

Saji NH, Goswami BN, Vinayachandran PN, Yamagata T (1999) A dipole mode in the tropical Indian Ocean. Nature 401:360–363

Sallenger AH, Doran KS, Howd PA (2012) Hotspot of accelerated sea-level rise on the Atlantic coast of North America. Nat Clim Change 2(12):884–888. doi:10.1038/nclimate1597

Scafetta N (2014) Multi-scale dynamical analysis (MSDA) of sea level records versus PDO, AMO, and NAO indexes. Clim Dyn 43:175–192. doi:10.1007/s00382-013-1771-3

Schlesinger ME, Ramankutty N (1994) An oscillation in the global climate system of period 65–70 years. Nature 367(6465):723–726. doi:10.1038/367723a0

Schloesser F, Furue R, McCreary JP, Timmermann A (2012) Dynamics of the Atlantic meridional overturning circulation. Part 1: buoyancy-forced response. Prog Oceanogr 101(1):33–62. doi:10.1016/j.pocean.2012.01.002

Schloesser F, Furue R, McCreary JP, Timmermann A (2014) Dynamics of the Atlantic meridional overturning circulation. Part 2: forcing by winds and buoyancy. Prog Oceanogr 120:154–176. doi:10.1016/j.pocean.2013.08.007

Schneider N, Cornuelle B (2005) The forcing of the Pacific decadal oscillation. J Clim 18:4355–4373

Schwarzkopf F, Böning C (2011) Contribution of Pacific wind stress to multi-decadal variations in upper-ocean heat content and sea level in the tropical south Indian Ocean. Geophys Res Lett 38:L12602. doi:10.1029/2011GL047651

Sérazin G, Penduff T, Grégorio S, Barnier B, Molines J-M, Terray L (2015) Intrinsic variability of sea level from Global Ocean simulations: spatiotemporal scales. J Clim 28(10):4279–4292. doi:10.1175/JCLI-D-14-00554.1

Sevellec F, Fedorov AV (2013) The leading, interdecadal eigenmode of the Atlantic Meridional Overturning Circulation in a realistic ocean model. J Clim 26:2160–2183

Shankar D, Shetye SR (1999) Are interdecadal sea level changes along the Indian coast influenced by variability of monsoon rainfall? J Geophys Res 104:26031–26042

Shankar D, Aparna SG, McCreary JP, Suresh I, Neetu S, Durand F, Shenoi SSC, Al Saafani MA (2010) Minima of interannual sea-level variability in the Indian Ocean. Prog Oceanogr 84:225–241

Si Z, Xu Y (2014) Influence of the Pacific Decadal Oscillation on regional sea level rise in the Pacific Ocean from 1993 to 2012. Chin J Oceanol Limnol 32:1414–1420. doi:10.1007/s00343-014-3363-4

Song Q, Vecchi G, Rosati A (2007) Indian Ocean variability in the GFDL coupled climate model. J Clim 20:2895–2916

Stammer D, Cazenave A, Ponte RM, Tamisiea ME (2013) Causes for contemporary regional sea level changes. Annu Rev Mar Sci 5:21–46. doi:10.1146/annurev-marine-121211-172406

Sturges W, Douglas BC (2011) Wind effects on estimates of sea level rise. J Geophys Res 116:C06008. doi:10.1029/2010JC006492

Sturges W, Hong BG (1995) Wind forcing of the Atlantic thermocline along 32 N at low frequencies. J Phys Oceanogr 25:1706–1715

Suzuki R, Behera SK, Iizuka S, Yamagata T (2004) Indian Ocean subtropical dipole simulated using a coupled general circulation model. J Geophys Res 109:C09001. doi:10.1029/2003JC001974

Thompson DWJ, Wallace JM (1998) The Arctic Oscillation signature in the wintertime geopotential height and temperature fields. Geophys Res Lett 25(9):1297–1300

Thompson DWJ, Wallace JM (2000) Annular modes in the extratropical circulation, Part I: month-to-month variability. J Clim 13:1000–1016

Thompson PR, Merrifield MA (2014) A unique asymmetry in the pattern of recent sea level change. Geophys Res Lett 41(21):1–9. doi:10.1002/2014GL061263

Thompson PR, Mitchum GT (2014) Coherent sea level variability on the North Atlantic western boundary. J Geophys Res Oceans 119:5676–5689. doi:10.1002/2014JC009999

Thompson PR, Merrifield MA, Wells JR, Chang CM (2014) Wind-driven coastal sea level variability in the northeast Pacific. J Clim 27:4733–4751. doi:10.1175/JCLI-D-13-00225.1

Thompson PR, Piecuch CG, Merrifield MA, McCreary JP, Firing E (2016) Forcing of recent decadal variability in the Equatorial and North Indian Ocean. J Geophys Res Oceans. doi:10.1002/2016JC012132

Timmermann A, McGregor S, Jin FF (2010) Wind effects on past and future regional sea level trends in the southern Indo-Pacific. J Clim 23:4429–4437

Ting M, Kushnir Y, Seager R, Li C (2009) Forced and internal twentieth-century SST trends in the North Atlantic. J Clim 22:1469–1481. doi:10.1175/2008JCLI2561.1

Tozuka T, Luo J, Masson S, Yamagata T (2007) Decadal modulations of the Indian Ocean dipole in the SINTEX-F1 coupled GCM. J Clim 20:2881–2894

Trenary L, Han W (2012) Intraseasonal-to-interannual variability of South Indian Ocean sea level and thermocline: remote versus local forcing. J Phys Oceanogr 42:602–627

Trenary L, Han W (2013) Local and remote forcing of decadal sea level and thermocline depth variability in the South Indian Ocean. J Geophys Res 118:381–398. doi:10.1029/2012JC008317

Trenberth KE, Shea DJ (2006) Atlantic hurricanes and natural variability in 2005. Geophys Res Lett 33:L12704. doi:10.1029/2006GL026894

Trenberth KE et al (2007) Surface and atmospheric climate change. In: Solomon S, Qin D, Manning M et al (eds) Climate change 2007: the physical science basis, contribution of working group I to the fourth assessment report of the intergovernmental panel on climate change. Cambridge University Press, Cambridge, pp 235–336

Tsimplis MN, Shaw AGP (2008) The forcing of mean sea level variability around Europe. Glob Planet Change 63:196–202. doi:10.1016/j.gloplacha.2007.08.018

Tsimplis MN, Shaw AGP, Flather RA, Woolf DK (2006) The influence of the North Atlantic Oscillation on the sea level around the northern European coasts reconsidered: the thermosteric effects. Philos Trans R Soc A 364(1841):845–856. doi:10.1098/rsta.2006.1740

Venegas SA, Mysak LA, Straub DN (1998) An interdecadal climate cycle in the South Atlantic and its links to other ocean basins. J Geophys Res 103(C11):24723–24736

Vianna ML, Menezes VV (2011) Double-celled Subtropical Gyre in the South Atlantic Ocean: means, trends and interannual changes. J Geophys Res 116:C03024. doi:10.1029/2010JC006574
Vianna ML, Menezes VV (2013) Bidecadal sea level modes in the North and South Atlantic Oceans. Geophys Res Lett 40:5926–5931. doi:10.1002/2013GL058162
Vimont D (2005) The contribution of the interannual ENSO cycle to the spatial pattern of decadal ENSO-like variability. J Clim 18:2080–2092
Volkov DL, Pujol M-I (2012) Quality assessment of a satellite altimetry data product in the Nordic, Barents, and Kara seas. J Geophys Res 117:C03025. doi:10.1029/2011JC007557
Von der Heydt AS, Dijkstra HA (2007) Localization of multidecadal variability. Part I: cross-equatorial transport and interbasin exchange. J Phys Oceanogr 37:2401–2414
Wakelin SL, Woodworth PL, Flather RA, Williams JA (2003) Sea-level dependence on the NAO over the NW European continental shelf. Geophys Res Lett 30(7):1403. doi:10.1029/2003GL017041
Webster PJ, Moore AM, Loschnigg JP, Leben RR (1999) Coupled ocean-atmosphere dynamics in the Indian Ocean during 1997–1998. Nature 401:356–360
Williams RG, Roussenov V, Smith D, Lozier MS (2014) Decadal evolution of ocean thermal anomalies in the North Atlantic: the effects of Ekman, overturning, and horizontal transport. J Clim 27:698–719. doi:10.1175/JCLI-D-12-00234.1
Wittenberg AT (2004) Extended wind stress analyses for ENSO. J Clim 17(13):2526–2540
Wittenberg AT (2009) Are historical records sufficient to constrain ENSO simulations? Geophys Res Lett 36:L12702. doi:10.1029/2009GL038710
Woodworth PL, Pouvreau N, Wöppelmann G (2010) The gyre-scale circulation of the North Atlantic and sea level at brest. Ocean Sci 6:185–190. doi:10.5194/os-6-185-2010
Woodworth PL, Maqueda MAM, Roussenov VM, Williams RG, Hughes CW (2014) Mean sea-level variability along the northeast American Atlantic coast and the roles of the wind and the overturning circulation. J Geophys Res Oceans 119:8916–8935. doi:10.1002/2014JC010520
Woolf D, Shaw AGP, Tsimplis MN (2003) The influence of the North Atlantic Oscillation on sea level variability in the North Atlantic region. Glob Atmos Ocean Syst 9(4):145–167. doi:10.1080/10236730310001633803
Xie SP, Hu K, Hafner J, Tokinaga H, Du Y, Huang G, Sampe T (2009) Indian Ocean capacitor effect on Indo-western Pacific climate during the summer following El Nino. J Clim 22:730–747
Yan Z, Tsimplis MN, Woolf D (2004) An analysis of relationship between the North Atlantic Oscillation and sea level changes in NW Europe. Int J Climatol 24:743–758. doi:10.1002/joc.1035
Yan L, Du Y, Zhang L (2013) Southern ocean SST variability and its relationship with ENSO on inter-decadal time scales. J Ocean Univ China 12:287–294
Yeager S, Danabasoglu G (2014) The origins of late 20th century variations in the large-scale North Atlantic circulation. J Clim 27:3222–3247. doi:10.1175/JCLI-D-13-00125.1
Yin J, Goddard PB (2013) Oceanic control of sea level rise patterns along the East Coast of the United States. Geophys Res Lett 40:5514–5520. doi:10.1002/2013GL057992
Yin J, Schlesinger ME, Stouffer RJ (2009) Model projections of rapid sea-level rise on the northeast coast of the United States. Nat Geosci 2(4):262–266
Yuan Y, Chan CLJ, Zhou W, Li CY (2008) Decadal and interannual variability of the indian ocean dipole. Adv Atmos Sci 25:856–866
Zhang R (2008) Coherent surface-subsurface fingerprint of the Atlantic meridional overturning circulation. Geophys Res Lett 35:L20705. doi:10.1029/2008GL035463
Zhang X, Church JA (2012) Sea level trends, interannual and decadal variability in the Pacific Ocean. Geophys Res Lett 39:L21701. doi:10.1029/2012GL053240
Zhang Y, Wallace JM, Battisti DS (1997) ENSO-like interdecadal variability: 1900–93. J Clim 10:1004–1020
Zhang R, Delworth TL, Sutton R et al (2013) Have aerosols caused the observed Atlantic multidecadal variability? J Atmos Sci 70(4):1135–1144. doi:10.1175/JAS-D-12-0331.1
Zhuang W, Feng M, Du Y, Schiller A, Wang D (2013) Low-frequency sea level variability in the southern Indian Ocean and its impacts on the oceanic meridional transports. J Geophys Res 118:1302–1315. doi:10.1002/jgrc.20129

Surv Geophys (2017) 38:251–275
DOI 10.1007/s10712-016-9390-2

CrossMark

Arctic Sea Level During the Satellite Altimetry Era

A. Carret[1] · J. A. Johannessen[2] · O. B. Andersen[3] · M. Ablain[4] · P. Prandi[4] · A. Blazquez[1] · A. Cazenave[1]

Received: 2 May 2016 / Accepted: 17 October 2016 / Published online: 14 November 2016

Abstract Results of the sea-level budget in the high latitudes (up to 80°N) and the Arctic Ocean during the satellite altimetry era. We investigate the closure of the sea-level budget since 2002 using two altimetry sea-level datasets based on the Envisat waveform retracking: temperature and salinity data from the ORAP5 reanalysis, and Gravity Recovery And Climate Experiment (GRACE) space gravimetry data to estimate the steric and mass components. Regional sea-level trends seen in the altimetry map, in particular over the Beaufort Gyre and along the eastern coast of Greenland, are of halosteric origin. However, in terms of regional average over the region ranging from 66°N to 80°N, the steric component contributes little to the observed sea-level trend, suggesting a dominant mass contribution in the Arctic region. This is confirmed by GRACE-based ocean mass time series that agree well with the altimetry-based sea-level time series. Direct estimate of the mass component is not possible prior to GRACE. Thus, we estimated the mass contribution from the difference between the altimetry-based sea level and the steric component. We also investigate the coastal sea level with tide gauge records. Twenty coupled climate models from the CMIP5 project are also used. The models lead us to the same conclusions concerning the halosteric origin of the trend patterns.

Keywords Arctic Ocean · Sea-level change · Steric sea level · Satellite altimetry · Ocean mass · CMIP5 models

✉ A. Carret
alice.carret@legos.obs-mip.fr

✉ A. Cazenave
anny.cazenave@legos.obs-mip.fr

1 LEGOS, CNES, CNRS, IRD, UPS, 14 avenue Édouard Belin, 31400, Toulouse, France

2 NERSC, Thormøhlens gate 47, 5006 Bergen, Norway

3 DTU, Elektrovej building 327 +328 +371 and Ørsteds Plads building 348, 2800 Kongens Lyngby, Denmark

4 CLS, 11 rue Hermès, Parc Technologique du Canal, 31520 Ramonville Saint-Agne, France

1 Introduction

During recent decades, the Arctic region (see Fig. 1) has warmed at a rate about twice the rest of the globe (Rhein et al. 2013). This primarily results from human-induced climate change, with strong amplification of anthropogenic warming in this region. Feedback mechanisms mostly result from decreasing albedo due to sea ice melting (Vaughan et al. 2013). But the Arctic region is also affected by a number of other global warming-related phenomena, notably Greenland ice sheet mass loss (e.g., Shepherd et al. 2012; Velicogna et al. 2014); Alaskan glacier melting (Gardner et al. 2013); permafrost melting with potential release of carbon dioxide and methane and decrease in snow cover extent (Vaughan et al. 2013); decrease in Arctic lakes ice cover (Kouraev et al. 2004); and the drying of Siberian lakes (Smith et al. 2005).

Studies of high-latitude and Arctic Ocean sea-level changes have mostly focused on the use of tide gauge data along the Russian and Norwegian coastlines (Proshutinsky et al. 2001, 2004, 2007, 2011; Henry et al. 2012), as the presence of sea ice has limited the use of altimetry. However, Cheng et al. (2015) recently developed a dedicated reprocessed along-track algorithm to produce a new dataset as 3-day gridded sea-level anomaly maps from September 1992 to October 2014 using ERS-1/2, Envisat, and Cryosat-2 altimetry data with coverage up to 82°N (hereinafter called DTU—Technical University of Denmark—product). A major improvement in data coverage was thus produced, providing 4–10 times the amount of data in regions such as the Beaufort Gyre compared with standard altimetry products. In result, estimation of sea-level changes from satellite altimetry in the Arctic

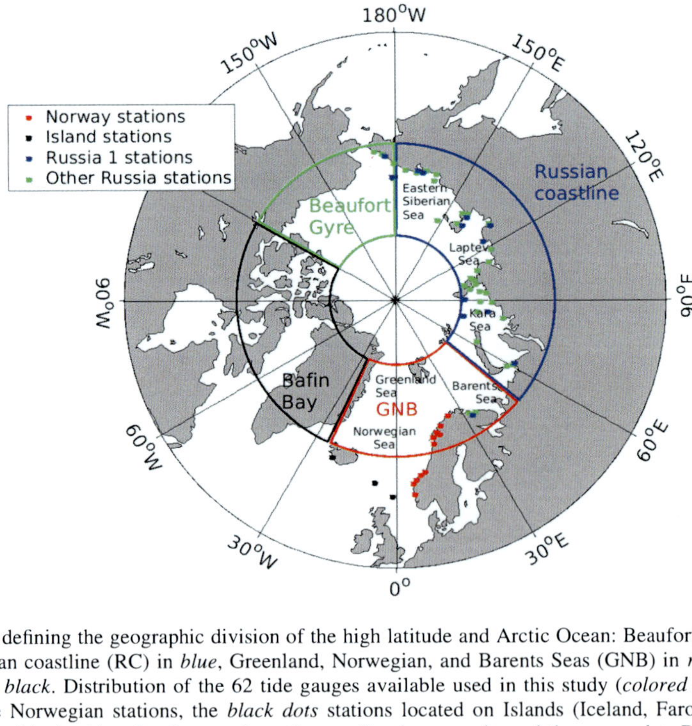

Fig. 1 Map defining the geographic division of the high latitude and Arctic Ocean: Beaufort Gyre (BG) in *green*, Russian coastline (RC) in *blue*, Greenland, Norwegian, and Barents Seas (GNB) in *red*, and Baffin Bay (BB) in *black*. Distribution of the 62 tide gauges available used in this study (*colored dots*). The *red dots* indicate Norwegian stations, the *black dots* stations located on Islands (Iceland, Faroe Islands, and Shetland Islands), the *blue dots* long and continuous Russian records, and the *green dots* Russian records ending in the 1990s

Ocean has become more reliable. In the meantime, a new sea-level product over the Arctic Ocean has been produced from a joint effort between CLS (Collecte Localisation Satellites, France) and PML (Plymouth Laboratory, UK). Basically, altimetry waveforms have been classified by surface types (Valladeau et al. 2015). Over the frozen ocean surfaces, areas with open water such as leads and polynyas have been identified. In these areas, a specific waveform retracking (Poisson et al. in preparation) and post-processing have been developed to optimize and improve the sea-level estimation. The new product is available over the Envisat period from 2003 to 2010 and is called CLS/PML.

The goal of this study is to estimate the steric and mass contributions to the observed mean sea-level change at different timescales and in different parts of the Arctic Ocean. For that purpose, we compare newly available altimetry-based sea-level datasets over the Arctic region. Using ocean reanalysis from ECMWF (European Centre for Medium-Range Weather Forecasts) as well as gravimetry data from GRACE (Gravity Recovery And Climate Experiment) for estimating the steric and ocean mass components, we also study the high-latitude and Arctic Ocean sea-level budget over the last decade (using the DTU and CLS/PML previously presented products). Next, we analyze tide gauge data to estimate coastal mean sea-level variations in the Arctic. Finally, we compare the satellite observations with CMIP5 (Coupled Model Intercomparison Project Phase 5, Taylor et al. 2012) coupled climate models. Section 2 describes the data used in this study. Section 3 focuses on spatial patterns of sea-level trends of the different datasets and also provides regional averages. In Sect. 4, the steric and mass contributions in terms of time series are analyzed. Coastal sea-level variations are analyzed in Sect. 5. Finally, Sect. 6 investigates the capability of CMIP5 models to reproduce observed variations, followed by a discussion and conclusions in Sect. 7.

2 Data Description

2.1 Satellite Altimetry-Based Mean Sea-Level Data

We used two new altimetry datasets produced by CLS/PML and DTU based on dedicated retracking and/or reprocessing of the altimetry waveforms in the high-latitude seas and Arctic Ocean. Version 3 of the DTU dataset was built using ERS 1/2, Envisat, and Cryosat-2 data and spanning the period from 1992 to 2014 (Andersen and Piccioni 2016). In contrast, the CLS/PML dataset includes only Envisat data and spans from 2002 to 2010. Both datasets are referenced to the mean sea surface (MSS) DTU2013 (Andersen et al. 2015) and consist of gridded datasets every week. The grid spacing is of 0.5° for DTU and 2° for CLS/PML with a common spatial coverage from 66°N to 80°N (see the black dashed lines in Fig. 2a). The datasets are extracted for the time period from 2003 to 2010 in order to avoid a drift reported in the global Envisat mean sea level (Ollivier and Guibbaud 2012).

The DTU dataset is based on the reprocessing of the multi-mission altimetry for the 1992–2014 period. Rather than performing a retracking of the data, Cheng et al. (2015) demonstrated that in large parts of the interior of the Arctic Ocean, the number of altimetric observations can increase by up to several times if the standard editing criteria were substituted with editing criteria tailored to high-latitude Arctic conditions without degrading the accuracy of the derived sea-level anomalies. Even though an increase in data coverage of four- to ten fold could be achieved in the interior of the Arctic Ocean, a far from complete coverage could be reached. Besides upgrading the editing criteria, residual

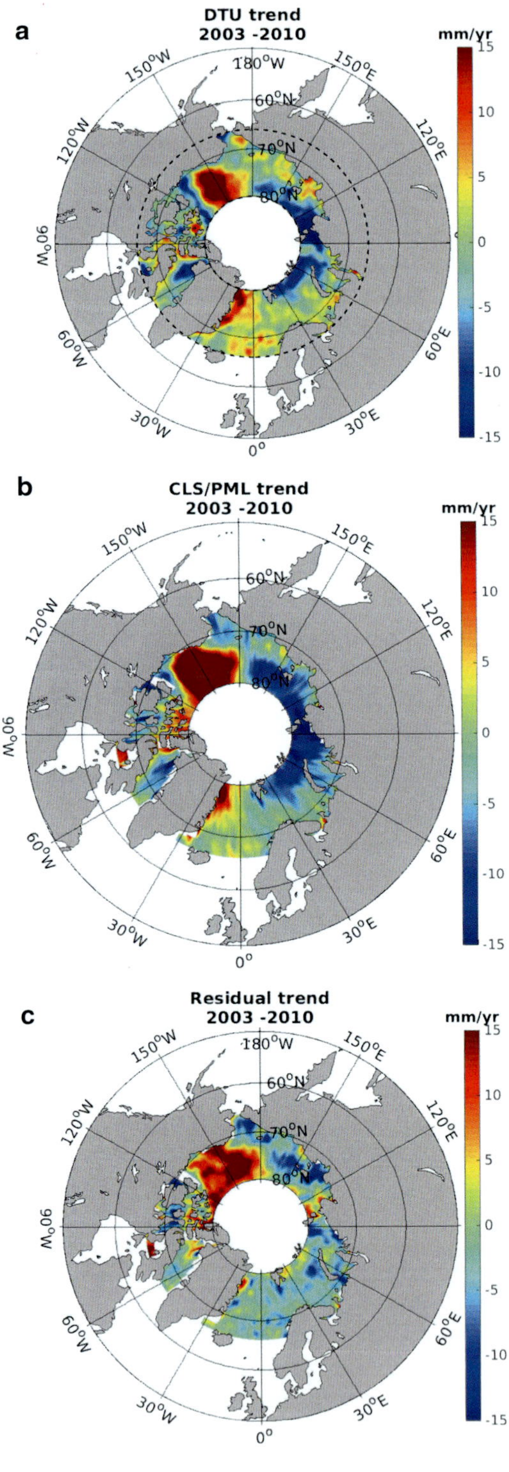

Fig. 2 **a** Spatial trend patterns in the sea level in mm/year for the period 2003–2010 for the DTU dataset. The *black dashed lines* indicate the boundaries (66°N and 80°N) of the studied region. **b** Spatial trend patterns in the sea level in mm/year for the period 2003–2010 for the CLS/PML dataset. **c** Spatial trend patterns in the sea level in mm/year for the period 2003–2010 for the residual dataset (CLS/PML-DTU). Note that the differences over the Beaufort Gyre between CLS and DTU are indeed around 10 mm/year even if both datasets seem to present in Fig. 1a, b the same trends, reaching more than 15 mm/year. Indeed, the maximum of the *colorbar* is reached, but the DTU trends are around 15/20 mm/year, whereas the CLS trends are around 25/30 mm/year

orbit errors were determined by processing global tracks and minimizing the crossovers with simultaneous TOPEX/Jason data at low latitudes.

The CLS/PML processing is based on the reprocessing of 20-Hz altimetry measurements. The first step consists in classifying the altimeter waveforms by ocean surface types in order to separate frozen ocean areas from open water corresponding to leads and polynyas. Usual retracking algorithms do not work well over these surfaces due to the very peaky shapes of the echoes. Thus, a new retracking algorithm has been developed (Poisson et al. in preparation) allowing for a better retrieval of the altimetric parameters. Sea level is then calculated using the best geophysical corrections dedicated to high latitudes (DTU2013 for the MSS, FES-Finite Element Solution-2014 for ocean tide, atmospheric correction derived from ERA-interim reanalyzes, http://www.ecmwf.int/en/research/climate-reanalysis/era-interim). After editing spurious sea-level data, weekly 2° resolution sea-level grids have been computed using a weighted 4-week moving average from the 20-Hz along-track measurements.

2.2 Reanalysis-Based Steric Data

To estimate the steric contribution to the Arctic Ocean sea level, we used temperature and salinity data coming from the Ocean ReAnalysis Pilot 5 (ORAP5) reanalysis (Zuo et al. 2015). Assimilated data are coming from the EN3 database. The EN3 database gathers temperature and salinity data from WOD05 (World Ocean Database 2005), GTSPP (Global Temperature and Salinity Profile Project), Argo and the ASBO (Arctic Synoptic Basin-wide Observations) projects. The WOD05 datasets come from ocean station data, CTD (conductivity–temperature–depth), XBT (expendable bathythermographs), moored buoys, drifting buoys, profiling floats, and gliders (Boyer et al. 2006). The EN3 database is developed by the Met Office/Hadley Centre UK (http://www.metoffice.gov.uk/hadobs/en3/en3_data_sources.html). The surface forcing fields come from the ERA-Interim atmospheric reanalysis with data assimilation (Dee et al. 2011). ORAP5 data can be downloaded at https://reanalyses.org/ocean/overview-current-reanalyses. The steric sea-level change is computed since 1979, by integrating density variations from the sea surface to the bottom. The thermosteric and halosteric sea-level contributions are also estimated using a monthly climatology (obtained by averaging monthly data from 1979 to 2014) at a spatial resolution of 0.25°.

2.3 GRACE-Based Ocean Mass Data

Direct estimation of the Arctic Ocean mass component is possible since 2002 thanks to the GRACE space gravimetry mission. Different versions of this product corresponding to distinct data processing are available. We used three of these monthly products, notably: (1) the GRACE CSR (Center for Space Research) Tellus solution (Chambers 2012; Chambers and Bonin 2012; Chambers and Willis 2010), processed by Don Chambers, supported by the National Aeronautics and Space Administration (NASA) MeaSUREs Program, and available at http://grace.jpl.nasa.gov/data/get-data/monthly-mass-grids-ocean/; (2) the GRACE MASCONS (Mass Concentration blocks) solution (Watkins et al. 2015; Wiese 2015) supported by the NASA MeaSUREs Program and available at http://grace.jpl.nasa.gov/data/get-data/jpl_global_mascons/; (3) and a new product developed at LEGOS (Laboratoire d'Études en Géophysique et Océanographie Spatiales) (Blazquez et al. in preparation). Processing of the GRACE CSR data is based on the Release-05 (RL05) spherical harmonics (Chambers and Bonin 2012), while the processing steps of the GRACE MASCONS product rely on the Level-1 GRACE observations, processed at JPL (Jet Propulsion Laboratory). Both products include the C20 (the C20 is a degree 2 term of the gravity field spherical harmonic expansion, it measures the oblateness of the Earth) correction from Cheng et al. (2011) and spherical harmonic coefficients of degree 1 from Swenson et al. (2008). They use the Glacial Isostatic Adjustment (GIA) correction from Geruo et al. (2013). The CSR solution applies a destriping filter, a 500-km-wide Gaussian filter, and a land hydrology leakage correction. For the MASCONS solution, a Coastline Resolution Improvement (CRI) filter is applied, separating the land and ocean mass signal. Finally, the LEGOS solution is based on a forward modeling approach (e.g., Chen et al. 2015). This product is an ensemble of 30 solutions from 5 processing centers (CSR, GFZ—GeoforschungsZentrum-, JPL, GRGS—Groupe de Recherche de Géodésie Spatiale-, and TUG—Graz University of Technology). 2 GIA models are used from Peltier et al. (2015) and Geruo et al. (2013). In addition, 3 ocean models are used to correct the first degree of the gravity field. All the solutions use the C20 correction from Cheng et al. (2011) and a destriping filter from Kusche et al. (2009). A leakage correction over a 300-km-wide zone off the coastlines is also applied, based on comparison with observation-based ocean mass estimates (details are given in Blazquez et al. in preparation). Here, we used the average of the 30 solutions.

2.4 CMIP5 Model Dataset

The 20 available CMIP5 simulations are selected for comparison with the altimetry-based sea level for the high latitude and the Arctic Ocean. A brief overview of these 20 different models is provided in Table 1. The following three variables are compared: the sea surface height (SSH) above the geoid, the steric sea level, and the thermosteric sea level in the whole water column. The CMIP5 historical runs from 1850 to 2005 were extended until 2015 using the RCP8.5 experiment. All gridded fields are projected onto a $1^\circ \times 1^\circ$ map. The data are also corrected for model drift using pre-industrial control runs. The drift removed for each grid cell is a quadratic function for the SSH and is computed for a 150-year time span beginning when the control run is branched to the historical run. In contrast, the steric and thermosteric variables are corrected for a linear drift computed for the whole pre-industrial experiment time span.

3 Sea Level During the Satellite Altimetry Era

As we are interested in interannual to decadal timescales, seasonal cycles (annual and semiannual) are removed from all time series. In addition, a 6-month running mean filter further is applied at each grid mesh of all datasets to smooth the time series.

3.1 Altimetry-Based Sea Level Over the 2002–2010 Time Span: Spatial Trend Patterns

To obtain an overall view of the high-latitude and Arctic Ocean sea-level variations, we first compare the spatial sea-level trend patterns of the two gridded altimetry datasets (Fig. 2), choosing a common space and time coverage: from 66°N to 80°N over the Envisat era from July 2003 to August 2010. As expected, small-scale features are more visible in the DTU data (Fig. 2a) compared to the CLS/PML data (Fig. 2b). Positive trends are

Table 1 Model names, their institution, and the differences between the versions of each institution

Institution	Model name	Components changing according to the version
CSIRO-BOM	ACCESS1-0	Land surface; atmosphere
	ACCESS1-3	
Commonwealth Scientific and Industrial Research Organisation in collaboration with the Queensland Climate Change Centre of Excellence (CSIRO-QCCCE)	CSIRO-Mk3-6-0	
Canadian Center for Climate Modelling	CanESM2	
NCAR	CCSM4	
Centre National de Recherches Météorologiques and Centre Européen de Recherche et Formation Avancées en Calcul Scientifique	CNRM-CM5	
NOAA	GFDL-CM3	Atmosphere; aerosol; atmospheric chemistry; ocean biogeochemistry
	GFDL-ESM2M	
University of Tokyo, National Institute for Environmental Studies, Japan Agency for Marine Earth Science and Technology	MIROC5	Atmosphere; ocean; ocean biogeochemistry; atmospheric chemistry
	MIROC-ESM	
	MIROC-ESM-CHEM	
Max Planck Institute for Meteorology	MPI-ESM-LR	Ocean and atmosphere resolution
	MPI-ESM-MR	
Meteorological Research Institute	MRI-CGCM3	
Norwegian Climate Center	Nor-ESM1-ME	Ocean biogeochemistry
	NorESM1-M	
UK Met Office	HadGEM2-CC	Carbon cycle
	HadGEM2-ES	Earth system
IPSL	IPSL-CM5A-LR	Resolution
	IPSL-CM5A-MR	

noticed along the Russian coastline and in the Greenland, Norwegian, and Barents Seas for the DTU dataset (Fig. 2a) while in these regions the CLS/PML dataset shows smaller values (Fig. 2b). Distinct patterns of positive trends are found for both datasets along the Greenland coasts and in the Beaufort Gyre area while negative trends are depicted north of the Kara and Laptev Seas. However, important differences concern the extent of the positive trend pattern in the Beaufort Gyre. For the CLS/PML dataset, a large and distinct positive trend pattern is depicted in the Beaufort Sea. In Sea. By way of contrast, the trend pattern of the DTU dataset is much more narrow. On the other hand, it reveals an extended positive trend signal up to the coastline in the Eastern Siberian Sea. The residual trend map

(Fig. 2c) is mainly displaying differences located in the Beaufort Gyre and in the Eastern Siberian Sea, reaching more than 10 mm/year. These differences are probably encountered due to the fact that the two datasets have a distinctly different spatial and temporal resolution. The seasonal sea ice changes imply that only a fraction of the data will be available in the interior of the Arctic Ocean compared to the ice-free regions in the North Atlantic Ocean (see Cheng et al. 2012) where the agreement is also better. Seasonal data coverage introduces a bias in the computation of trends and illustrates the difficulties of processing data in the presence of sea ice. The interannual variability is large in the Arctic, and the computation of trends on such a short period is subject to large uncertainties which generally explains the differences in the ice-free regions.

Ablain et al. (2015) estimated an error budget for the global mean sea-level change from satellite altimetry at climatic scales (more than 10 years) and concluded that the uncertainty on the global mean trend is 0.5 mm/year but can reach 2–3 mm/year when considering regional averages. In the Arctic Ocean, Prandi et al. (2012) estimated the regional trend uncertainty to 1.3 mm/year from 17-year-long altimetry records, and Cheng et al. (2015) found a similar value using 20 years of data. More recently, Armitage et al. (2016) used a processing similar to CLS/PML, using a different retracking algorithm, and estimated an uncertainty of 1.1 mm on monthly regional SSH averages.

3.2 Steric and Mass Trends During the 2002–2010 Period

To investigate the origin of the spatial trend patterns observed by altimetry, we analyzed the steric data covering the same time span and region (from 66°N to 80°N). The spatial patterns derived from the ORAP5 reanalysis are shown in Fig. 3a together with the halosteric (Fig. 3b) and thermosteric (Fig. 3c) contributions. The steric trends in the Beaufort Gyre region and along the Greenland coastlines patterns are very similar with those derived from the altimetry data, suggesting that the sea-level change seems to have a dominant steric origin in these regions. On the other hand, the trends along Siberia and the northern coast of Norway are not consistent with the altimetry-based sea-level trends, suggesting the importance of the non-steric sea-level contribution. This is in agreement with Volkov et al. (2013) who found especially that in the Barents Sea, the ocean mass signal is important. In the Arctic Ocean, the spatial trends in sea level are mainly due to the halosteric component (Fig. 3b), in agreement with the spatial pattern of the model-based sea level derived by Koldunov et al. (2014). The halosteric origin of these distinct sea-level trends is particularly strong along the East Greenland coast and in the Beaufort Gyre, implying a decrease in salinity. Morison et al. (2012) explain the salinity decrease in the Beaufort Gyre by an increase in Eurasian runoff which is redirected toward the Canada basin, whereas Koldunov et al. (2014) suggest that it results from anomalous Ekman pumping. Along the East Greenland coast, the salinity decrease can be linked to melting ice from the Greenland ice sheet. Concerning the differences observed with altimetry, especially over the Beaufort Gyre, the halosteric trends are in closer agreement with the CLS/PML trends. The thermosteric trends, on the other hand, are quite small almost everywhere except in the Norwegian and Greenland Seas and in the eastern Barents Sea. The positive trends observed in these regions might be related to the pathway of the transport of the warm Atlantic water. Indeed, shallow and warm currents are entering the Arctic Ocean through the Norwegian Sea followed by a separation between the inflow via the West Spitsbergen Current through the Fram Strait and the inflow passing through the Barents Sea where the water is cooled (Rudels 2012). These small thermosteric trends are in agreement with Yin et al. (2010) and result from a smaller thermal expansion coefficient

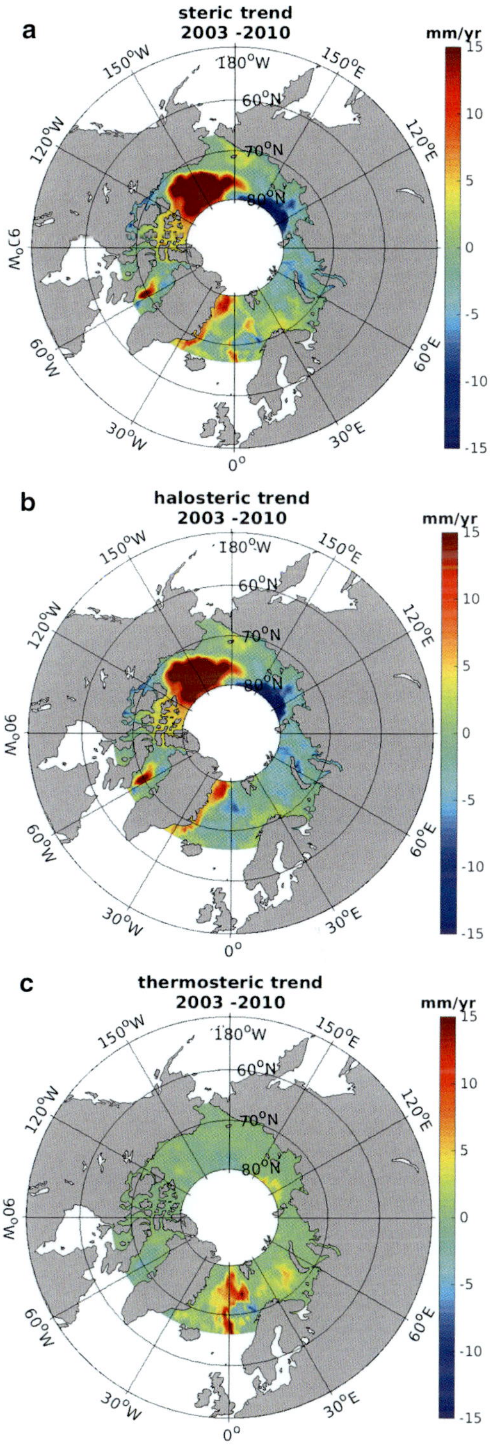

Fig. 3 **a** Spatial trend patterns of steric sea level in mm/year over 2003–2010. **b** Spatial trend patterns of halosteric sea level in mm/year over 2003–2010. **c** Spatial trend patterns of thermosteric sea level in mm/year over 2003–2010

expected in low temperature environments. As stated in Volkov et al. (2013), the Norwegian Coastal Current (NCC) is influenced by the continental runoff which explains the positive halosteric trend pattern visible along the Norwegian coastline in Fig. 3b. However, the positive thermosteric component observed in the Norwegian and Greenland Seas and in the eastern Barents Sea is balanced by the negative halosteric trends and thus does not clearly appear on the steric trend map. Along (and offshore) the Russian coastline, halosteric trends are mostly negative, especially at 150°E, suggesting ice formation in these regions.

To illustrate how non-steric effects contribute to the trend patterns, the trend maps of altimetry-based sea level minus the steric component are shown for the DTU (Fig. 4a) and for the CLS/PML (Fig. 4b) datasets. In the Baffin Bay and in the Greenland, Norwegian, and Barents Seas, the DTU and CLS/PML residual trends are reasonably similar. The

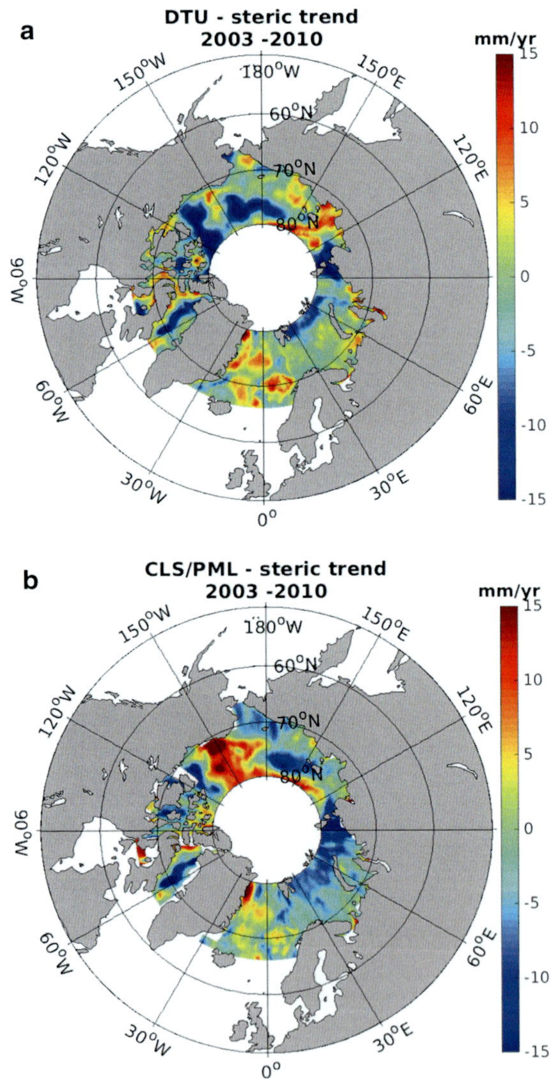

Fig. 4 **a** Observed residual trend patterns in mm/year from DTU altimetry minus steric sea level over 2003–2010. **b** Observed residual trend patterns in mm/year from CLS/PML altimetry minus steric sea level over 2003–2010

positive trends noticed in these seas suggest increased runoff and ice melting. However, for the remaining regions, differences are huge and the trend patterns have sometimes a reversed sign. Regarding absolute values, the non-steric signal is strong along the East Greenland coast and over the Beaufort Gyre corresponding to regions where the halosteric trend signals are also important. One should bear in mind that these maps are reflecting differences between trends and not the trend of the differences. As such, they may be flawed by uncertainties on the respective trends.

To further analyze these non-steric residuals, the GRACE-based sea-level trends are examined (Fig. 5). The ocean mass trend patterns of the three GRACE products display distinct differences from one solution to another. Nonetheless, the three maps display positive trends over the Beaufort Gyre and the Norwegian Sea, but they mostly disagree in other regions. The comparison with the residual maps (i.e., altimetry minus steric trends, see Fig. 4a, b) does not provide conclusive evidence toward the best solution as the residual maps resolve smaller scale phenomena. In the Arctic Ocean, all the GRACE solutions appear to be very different as compared to the rest of the oceanic domain. This clearly needs further investigation.

3.3 Regional Averages and Interannual Variability

The seasonal to interannual variability of the altimetry-based sea level and steric sea level is shown in Fig. 6. For each time step, we computed an average of the altimetric and steric data over 66°N–80°N, applying a cosine latitude weighting. The two altimetry-based sea-level curves agree well. The correlation coefficient is 0.72. Note that Ollivier and Guibbaud (2012) reported a drift in the global Envisat mean sea level. This was corrected in the DTU dataset.

In terms of regional average, the steric component is small compared to altimetry-based sea level. The distinct trend patterns noticed in Fig. 3a offset each other, leading to a small weighted averaged steric contribution. This suggests that, in terms of regional averages in the Arctic region (66°N–80°N), most of the sea-level contribution comes from mass variations which supports the results of Volkov (2014) who found that the signal in the Nordic and Barents Seas is mainly due to ocean mass signal and suggested to extend these results to the whole Arctic Ocean.

4 Sea-Level Budget During the Satellite Altimetry Era

4.1 Comparison Between Altimetry-Based Sea Level with the Steric Component Removed and GRACE-Based Sea Level Over the 2002–2010 Time Span

As mentioned, the GRACE data are available from 2002 to 2015. This allows us to compare time series from the three GRACE datasets to the altimetry-based sea-level datasets, for which the steric component has been deducted. The steric component in the high latitude and Arctic Ocean is small, as already noted, suggesting a dominant mass contribution to the sea-level changes. The time series of the regionally averaged (over the region from 66°N to 80°N) GRACE-based ocean mass component and the regionally averaged altimetry-based mean sea level with the steric effects removed are shown in Fig. 7 for the three GRACE datasets. The CSR solution (Fig. 7a) agrees well with the CLS/

Fig. 5 **a** GRACE trend patterns in mm/year for the CSR solution during 2003–2010. **b** GRACE trend patterns in mm/year for the MASCONS solution during 2003–2010. **c** GRACE trend patterns in mm/year for the LEGOS solution during 2003–2010

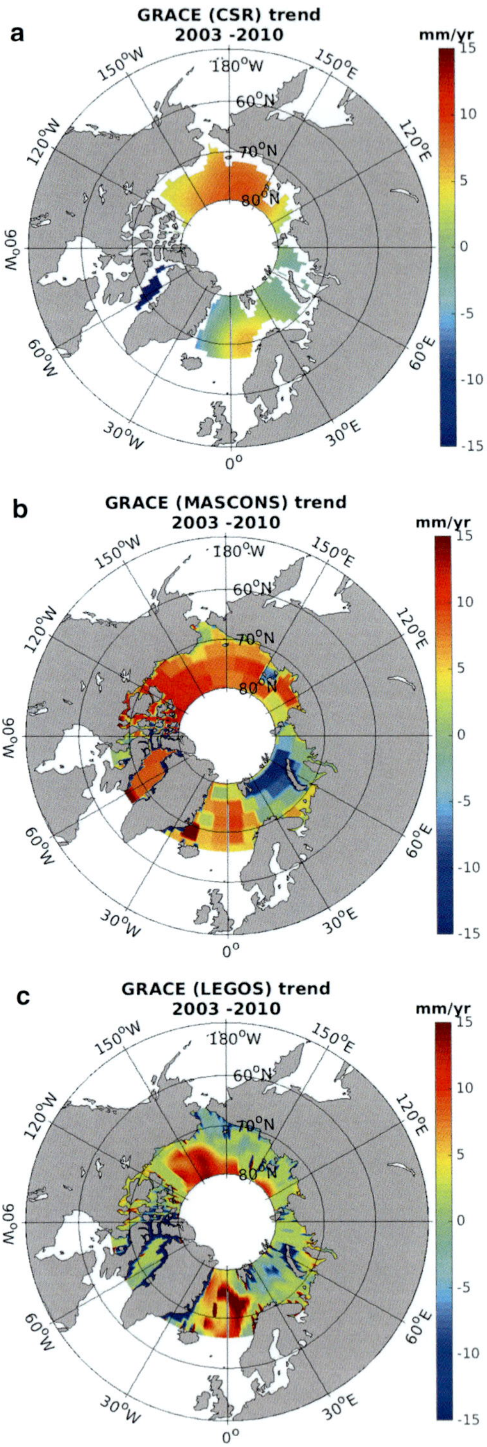

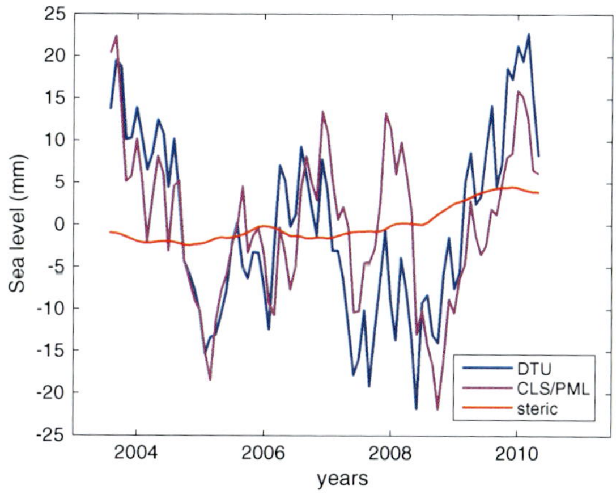

Fig. 6 Time series of altimeter-based sea level in mm from DTU (*blue*) and CLS/PML (*pink*). Data are averaged for each time step and weighted by the latitude cosine. The steric sea level from ORAP5 reanalysis is superimposed (*red*) for comparison

PML sea level with a correlation of 0.73 while the correlation with the DTU data is lower (0.54). The LEGOS solution (Fig. 7b) is also agreeing quite well with a correlation of 0.65 to both CLS/PML and DTU. In contrast, the MASCONS solution (Fig. 7c) is in somewhat less agreement with the two altimetry-based sea-level time series with a correlation of 0.36 for CLS/PML and 0.26 for DTU. The different correlations seem to mainly result from the early part of the period (2003–2005) and to a lesser extent from the end of the time span. Note that the regional average for the CSR solution is different from the others as there are no data in the 300-km zone near the coasts. Moreover, a slight attenuation of the GRACE-based sea level can be noted, coming from the GRACE reprocessing.

4.2 Analysis by Sectors Over 2003–2010

In order to refine the regional average analysis, we divided the Arctic region into four sectors, corresponding to the different oceanographic conditions. We choose to divide the region (see Fig. 1) into an ice-free sector covering the Greenland, Norwegian, and Barents Seas (GNB sector) where the sea-level budget is expected to be closed, the Russian Coastline (RC sector), the Beaufort Gyre (BG) where most of the discrepancies are found and where the altimetry products disagree, and finally, the Baffin Bay (BB sector) where there is particularly a lack of data due to the CSR solution (no data in the 300 km near the coasts). For each of these 4 sectors, the time series of the averaged DTU and CLS/PML altimetry-based sea level (after removing the steric component) are compared to the CSR GRACE product which gave the best correlation coefficient in the Arctic region. The results are presented in Fig. 8. Note that the results for the BB sector are not shown, although the results will be discussed.

In the GNB sector (Fig. 8a), the GRACE-based ocean mass component agrees well with the altimetry-based sea level minus the steric component with a correlation coefficient of 0.86 with CLS/PML and 0.72 with DTU. A strong correlation of 0.71 is also found between the two altimetry products in this ice-free region. In contrast, the agreement with the CSR solution is not so good in the RC sector (Fig. 8b), and the correlation coefficient between the altimetry products drops to 0.45. As observed in the trend pattern maps in Fig. 2a, b, the altimetry products largely disagree over the Beaufort Gyre with a correlation coefficient −0.14. However, the comparison with the GRACE-based ocean mass in this region as displayed in Fig. 8c gives a better correlation for the

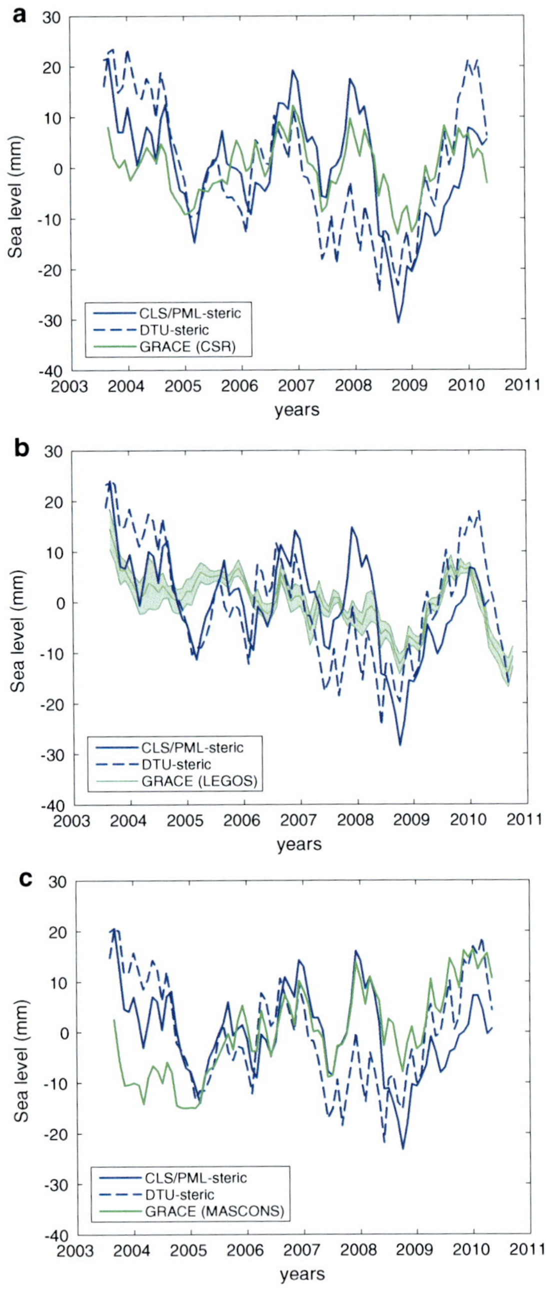

Fig. 7 **a** Averaged altimetry-based sea level in mm corrected for the steric contribution for CLS/PML (*blue*) and DTU (*dashed blue*). GRACE-based ocean mass component (*green*) for the CSR data. **b** Averaged altimetry-based sea level in mm corrected for the steric contribution for CLS/PML (*blue*) and DTU (*dashed blue*). GRACE-based ocean mass component (*green*) for the LEGOS data. **c** Averaged altimetry-based sea level in mm corrected for the steric contribution for CLS/PML (*blue*) and DTU (*dashed blue*). GRACE-based ocean mass component (*green*) for the MASCONS data

CLS/PML product (correlation coefficient of 0.48 compared to −0.33 for the DTU dataset). In the Baffin Bay, the agreement between the altimetry datasets is high (0.73), and the comparison to the CSR solution remains quite reasonable (correlation coefficients of 0.55 for the CLS/PML dataset and 0.47 for the DTU dataset). We therefore

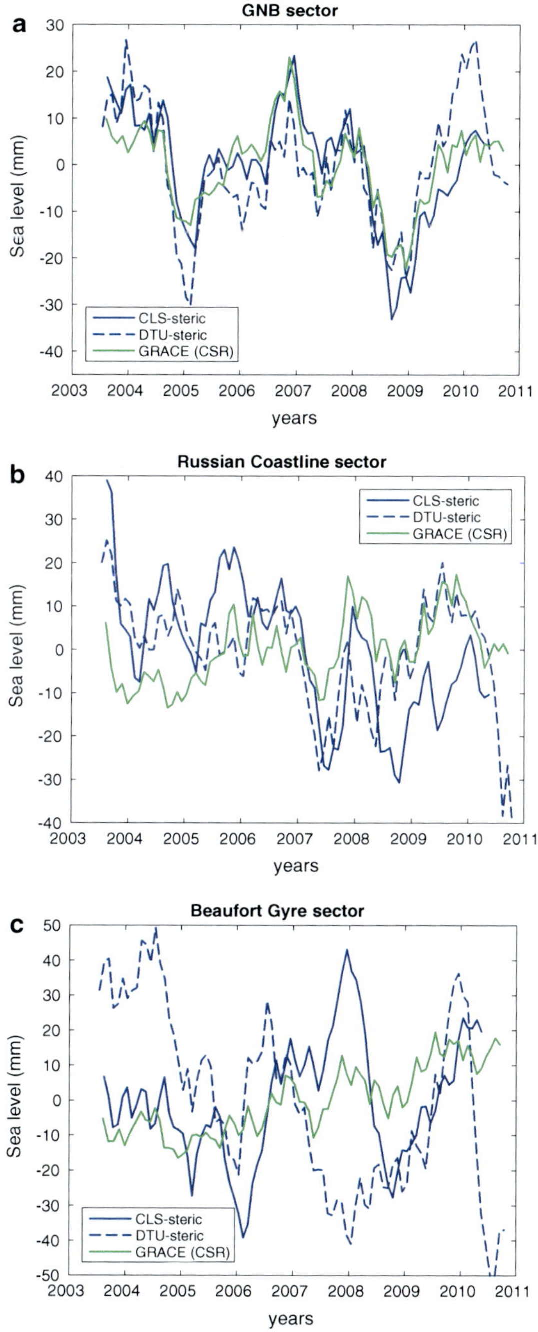

Fig. 8 **a** Altimetry-based sea level minus the steric component in mm for CLS/PML (*blue*) and DTU (*blue dashed*) averaged on the GNB sector. GRACE-based mass component in mm is represented in *green* for the CSR solution. **b** Altimetry-based sea level minus the steric component in mm for CLS/PML (*blue*) and DTU (*blue dashed*) averaged on the RC sector. GRACE-based mass component in mm is represented in *green* for the CSR solution. **c** Altimetry-based sea level minus the steric component in mm for CLS/PML (*blue*) and DTU (*blue dashed*) averaged on the BG sector. GRACE-based mass component in mm is represented in *green* for the CSR solution

conclude that the main differences observed between the total averaged altimetry-based sea level with the steric component removed and the GRACE-based ocean mass (Fig. 7a) result from the Russian coast region.

4.3 Arctic Sea Level Over 1992–2014

The high-latitude and Arctic Ocean (from 66°N to 80°N) sea-level change was also estimated for the time period 1992–2014 using the DTU dataset (Fig. 9a) and ORAP5 reanalysis for the steric component (Fig. 9b). In contrast to the spatial trend patterns in DTU altimetry-based sea level for 2002–2010 shown in Fig. 2a, the trend patterns displayed in Fig. 9a show positive rates almost everywhere, although the spatial variations are distinct. Furthermore, the positive trend patterns along the East Greenland coast seen over 2003–2010 are accompanied by positive trends along the West Greenland coast (see Fig. 9a). This may indicate recent changes in this region as both positive trend patterns are not visible for the 1992–2010 time span (not shown). The positive trend pattern in the Beaufort Gyre is still present. As for the shorter time span, the Beaufort Gyre trends are essentially steric in origin (Fig. 9b). Except for this region, the steric trend map shows weak trends in the Barents Sea sector along the coast of northern Norway that are poorly correlated with the altimetry trend patterns in the DTU trend map.

The steric sea level also displays a dominant halosteric component, e.g., in the Beaufort Gyre region. Since a direct estimate of the mass component is not possible before 2002, we estimated it by computing the residual trend patterns (Fig. 9c). In fact, the residual trend map represents, in addition to ocean mass redistribution, other non-steric components (e.g., freshwater input). The residual map shows positive trends almost everywhere. If attributed to mass redistribution alone, this implies positive trends in the whole Arctic for the mass component. The strongest signals are located in the Greenland and Norwegian Seas.

On a 20-year time span, the strong interannual variability is also accompanied by an increasing average Arctic sea level as revealed in Fig. 10. The DTU time series clearly shows an upward trend with some peaks starting in 1996 and 2003. Over the entire period from 1992 to 2014, the altimetric sea-level trend amounts to 2.10 ± 0.63 mm/year. In contrast, the averaged steric contribution is minor, although a slight increase can be noticed since 2008.

5 Coastal Sea Level from Tide Gauges Data (Norwegian and Siberian Sectors)

The tide gauge records from the Revised Local Reference dataset from Permanent Service for Mean Sea Level (PSMSL) used by Henry et al. (2012) and consisting of 11 stations along the Norwegian coast, 48 along Siberia, and 3 island (Iceland, Faroe Islands, and Shetland Islands) tide gauges (Holgate et al. 2013; PSMSL 2016) are revisited (see the colored dots in Fig. 1). As in Henry et al. (2012), the Russian tide gauges are divided into 2 sets according to the continuity of the records, notably: (1) records which end after 2000 (called Russian 1 dataset, the blue dots in Fig. 1) and (2) a combination of Russian 1 dataset with records ending in the 1990s (called Russian 2 datasets, the blue and green dots in Fig. 1). The end time of the tide gauge series is 2013 for most cases. Following Henry et al. (2012), we linearly interpolated time series if data presented gaps of less than 3 years, removed the annual and semiannual cycles, and smoothed the data with a running mean filter.

The time series are corrected for the GIA using the Peltier correction (with deglaciation history ICE-5G and mantle viscosity structure VM2—Peltier 2004). The GIA rates are between −3.07 and 1.17 mm/year with higher values along the East Greenland coastlines

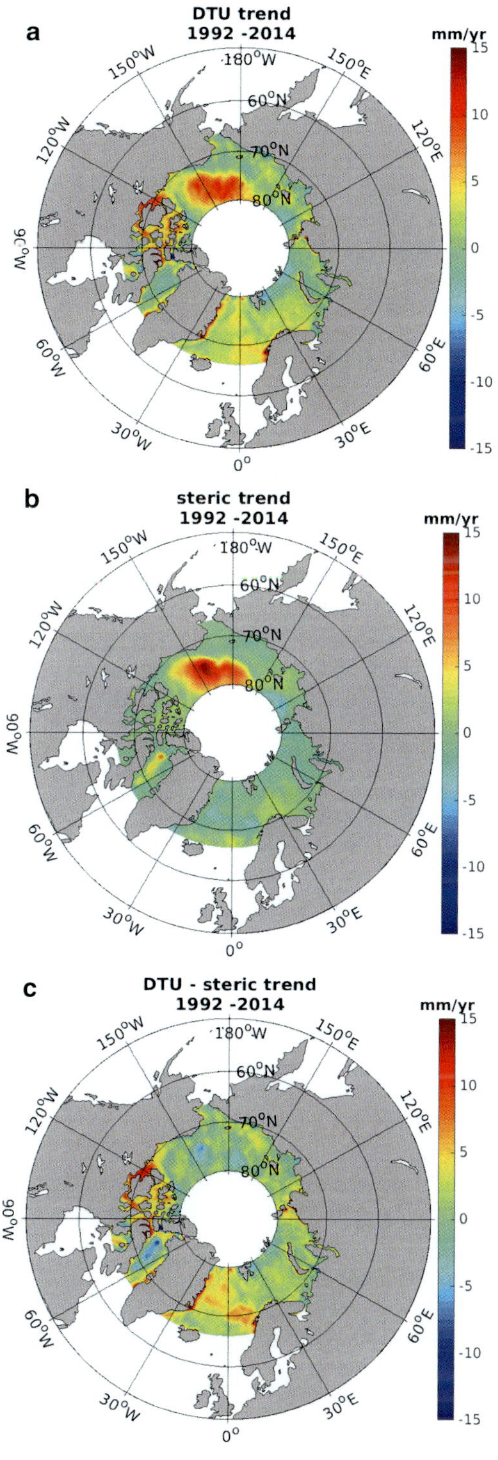

Fig. 9 **a** Spatial trend patterns in DTU-based sea level in mm/year over 1992–2014. **b** Spatial trend patterns in steric sea level in mm/year over 1992–2014. **c** Residual map of DTU-steric in mm/year over 1992–2014

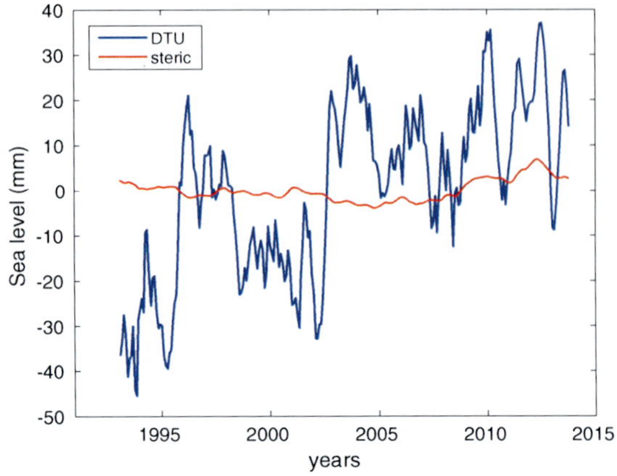

Fig. 10 DTU-based sea level (*blue*) in mm for the 1992–2014 time span. Steric sea level in mm from ORAP5 reanalysis is superimposed in *red*

and lower values mainly along the Norwegian coastlines. Moreover, in order to correct for the inverse barometer (IB) effect, the NCEP (National Centers for Environmental Prediction) reanalysis surface pressure data provided by the Physical Sciences Division (PSD) of the Earth System Research Laboratory (ESRL) of the National Oceanic and Atmospheric Administration (NOAA) are downloaded from their Web site http://www.esrl.noaa.gov/psd/. The data are available as monthly grids, and the nearest grid point for each tide gauge is selected to compute the IB correction. On average, IB trends are in the order of 0.15 mm/year. The corresponding average of tide gauge-based sea-level time series corrected for the IB effect and the GIA is shown in Fig. 11 for the Norwegian coastal stations together with the Russian 1 and Russian 2 datasets. In general, all the datasets show an increase in sea level since 1980, consistent with other studies (Henry et al. 2012; Proshutinsky et al. 2001). Since the beginning of the tide gauge records, trends of 1.58 ± 0.23 mm/year for the entire Arctic, 0.60 ± 0.14 mm/year for the Norwegian sector, 2.24 ± 0.24 mm/year for the Russian 1, and 2.10 ± 0.20 mm/year for the Russian 2 sector are found. The Russian trends are higher than those found by Proshutinsky et al. (2004) for the 1954–1989 time span and the 1954–1999 period, indicating an acceleration in sea-level rise. During the altimetry era (1992–2014), the rate of sea-level rise for the tide gauges of the three sectors amounts to 1.79 mm/year, a value smaller than the altimetric sea-level trend of 2.10 mm/year, reported in Sect. 4.3.

The coastal mean sea level (CMSL) obtained by averaging all the tide gauge records of the 3 sectors and superimposed on the NOAA Arctic Oscillation (AO) index available at http://www.cpc.ncep.noaa.gov/products/precip/Cwlink/daily_ao_index/ao.shtml is shown in Fig. 12 for the period 1950–2014. Except for a few short periods (1960–1963, 1977–1980, 1986–1988, 2009–2010) and a longer one (between 2002 and 2008), the AO is highly correlated with the tide gauge-based sea-level record with a correlation coefficient of 0.61. A positive Arctic Oscillation Index is associated with a lower than average sea-level pressure and stronger wind stress (Volkov and Landerer 2013a) resulting in more ocean mass along the coast. This is confirmed by Fig. 13, where the Arctic CMSL from tide gauges is represented along with the steric mean sea level from ORAP5 reanalysis interpolated at each tide gauge station. Indeed, the steric mean sea level is small, thus indicating mass-related effects at the coast.

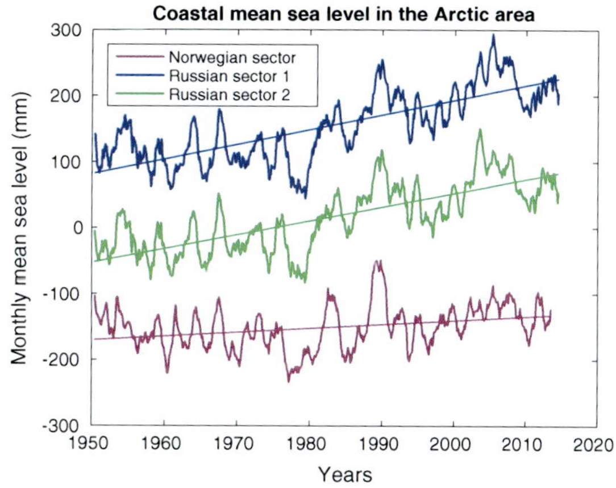

Fig. 11 Time series of coastal sea level from tide gauge data for the Norwegian, Russian 1 and Russian 2 sectors for the period 1950–2014. The *trend lines* are superimposed

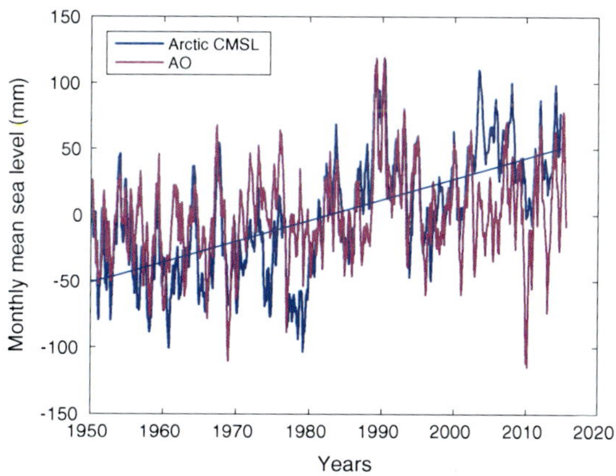

Fig. 12 Time series of the mean coastal sea level in mm (*blue*) in mm between 1950 and 2014. The AO index (*pink*) is superimposed

6 Sea-Level Comparison Between CMIP5 Models and Altimetry Data

In addition to in situ and remote observations, CMIP5 model data are used to investigate the veracity of the climate models in the Arctic region. For the 1992–2014 time span, the spatial trends for 20 CMIP5 models (not shown) revealed highly different patterns, even from one version of a given model to another. Moreover, the positive trend pattern over the Beaufort Gyre as found in the altimetry maps is not reproduced by all CMIP5 simulations. Hence, only those models showing a pattern similar to the altimetry in this particular region were selected. In addition, models that displayed the highest RMSE between their steric trend map and the ORAP5 steric trend map were also disregarded. This approach led to a down selection of 6 models: CCSM4, CSIRO-Mk3-6-0, GFDL-CM3, MIROC-ESM, MPI-ESM-MR, and NorESM1-ME. The spatial trends for these six models are represented in Fig. 14. For most of the models, the spatial trends are smaller than in the observations especially over the Beaufort Gyre. Indeed, the internal variability is poorly reproduced by the models. The ensemble mean trend map for the six selected models is presented in Fig. 15a together with the steric (Fig. 15b) and thermosteric (Fig. 15c) trend maps, which are also provided by the model simulations. The steric trend map is clearly displaying similar patterns as the ensemble mean map, although it shows higher amplitudes and

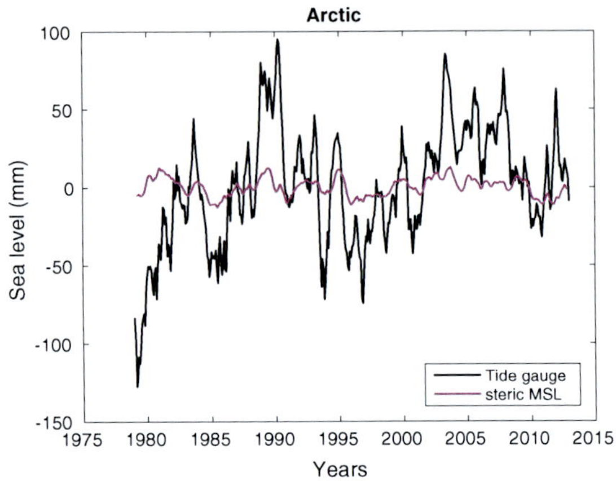

Fig. 13 Coastal mean sea level in mm for all Arctic tide gauges (*black*) and steric component interpolated at the stations (*pink*) between 1979 and 2014

displays different patterns in the Greenland and Norwegian Seas (note that the color scale is not the same as in Fig. 14). The latter regional amplification and pattern difference are due to the thermosteric contribution as Fig. 15c displays positive trends in these regions and also maybe to the ocean mass contribution. Moreover, the halosteric contribution that is dominating the trend pattern in the Beaufort Gyre (depicted from comparing Fig. 15b, c) clearly supports the conclusion drawn above from the inspection of the altimetry and ORAP5 maps that the sea-level trends mainly come from the halosteric contribution.

7 Discussion and Conclusions

The Arctic region is strongly affected by global warming and thus requires careful monitoring. Among the essential Arctic climate variables, sea level remains one of the least known because of the presence of sea ice cover that limits the full capacity and use of altimetry. Specific algorithms are required to separate SSH (sea surface height) values from echoes contaminated by sea ice. Here, we present new results of the high-latitude and Arctic Ocean sea-level variations over the satellite altimetry era based on higher-quality altimetry data and improved coverage of the Arctic Ocean. Two new altimetry-based sea-level products were compared and, in general, found to be in good agreement both in terms of spatial trend patterns and regional averages. However, some local differences in trend patterns are noticed over the Beaufort Gyre. These may result from the presence of blocks of ice trapped in the circulation of the Gyre for several years (Rigor et al. 2002) leading to discrepancies between two sea-level products due to different strategies in the data processing. Concerning the GRACE (Gravity Recovery and Climate Experiment) ocean mass signals in the Arctic, we found better agreement with altimetry-based sea level corrected for steric effects when considering the LEGOS (Laboratoire d'Études en Géophysique et Océanographie Spatiales) and CSR (Center for Space Research) solutions rather than the MASCONS (Mass Concentration blocks) solution.

Results for the thermosteric and halosteric components indicate significant regional differences in terms of trends. In terms of regional averages for the 2003–2010 time span, both components show positive trends of 0.84 mm/year (halosteric) and 0.59 mm/year (thermosteric), indicating a decrease in salinity and an increase in temperature. The corresponding increase in freshwater input is described in Morison et al. (2012). The positive

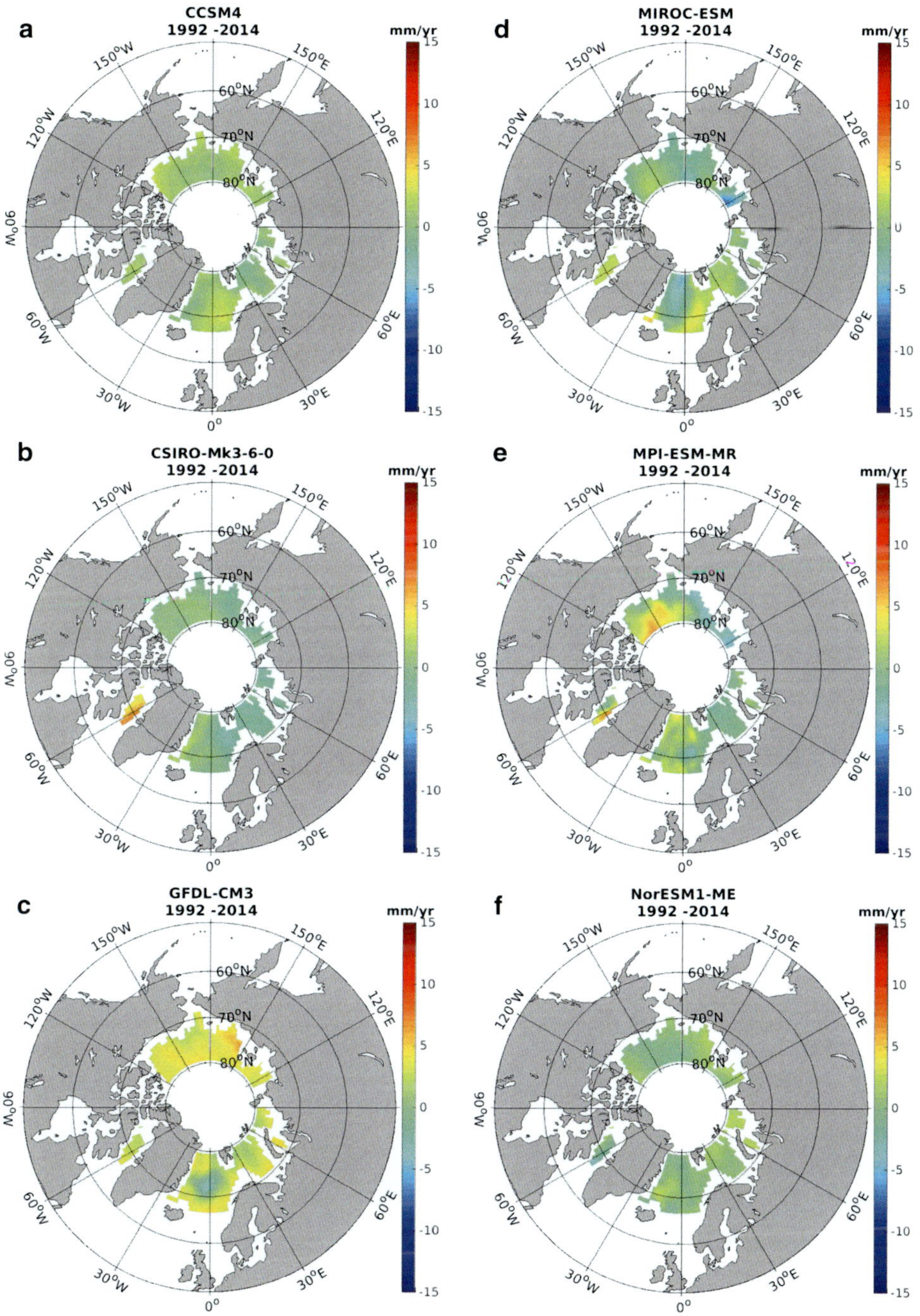

Fig. 14 Spatial trend patterns of sea surface height in mm/year for the 6 selected CMIP5 model simulations during the 1992–2014 period

Fig. 15 **a** Ensemble mean sea-level trend in mm/year from the selected six CMIP5 simulations. **b** Ensemble mean steric sea-level trend in mm/year from the selected six CMIP5 simulations. **c** Ensemble mean thermosteric sea-level trend in mm/year from the selected six CMIP5 simulations

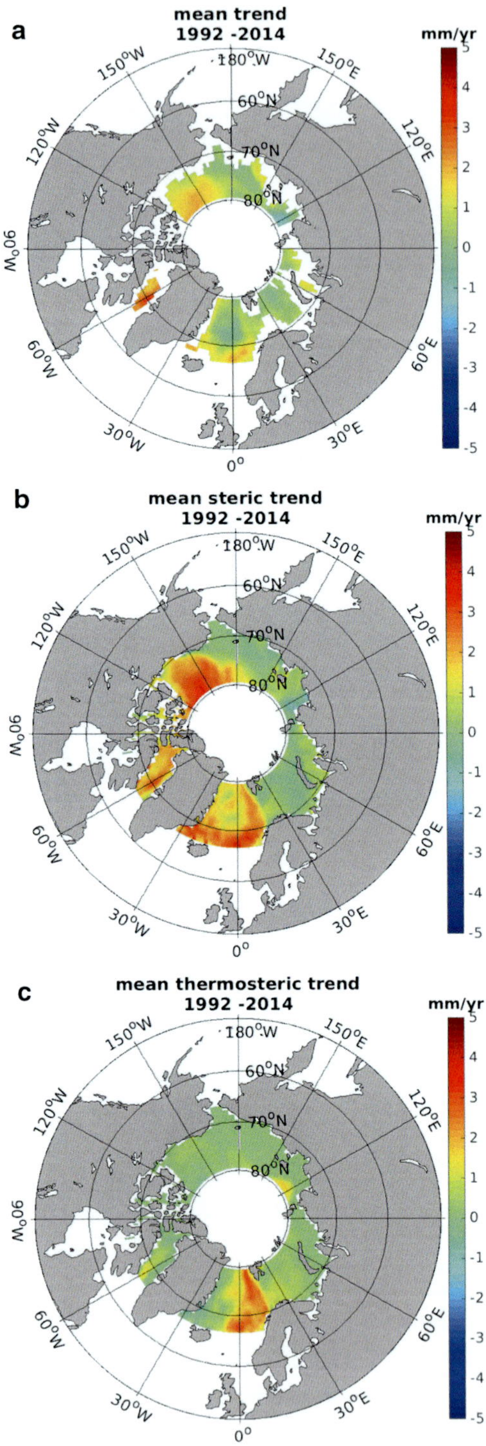

trend patterns in the thermosteric trend map are found in the Norwegian Sea and may correspond to input of warm water from the Atlantic Ocean.

In terms of regional sea-level variations, we observed positive trend patterns over the Beaufort Gyre and along the east coast of Greenland and negative trends along the Siberia shelf. For the 2003–2010 period, this almost results in a trend cancellation, as the regional averages are dominated by the interannual variability. For longer periods, since 1992 or even 1950, upward trends are nevertheless displayed by altimetry and tide gauges. For the tide gauge period (1950–2014), the total coastal sea-level trend amounts to 1.58 ± 0.23 mm/year, a value slightly lower than the global mean average for the same time span but with strong variability according to the location of tide gauges since the Russian trends are higher. Besides, these trends hide a period with no significant changes in the Arctic sea level followed by a strong increase during the whole altimetry period (1992–2014).

The sea-level budget indicates a dominant mass contribution in the GNB (Greenland, Norwegian, and Barents Seas) sector where the correlation between datasets is significant and the RMSEs (root mean square errors) are low. However, this is not clear over the Beaufort Gyre where the steric contribution is much more important and displays values of the same order as the mass component. This important contribution is due to the strong decrease in salinity. The low correlation obtained in that region, on the other hand, can be explained by errors in altimetry data due to the presence of sea ice.

The use of CMIP5 (Coupled Model Intercomparison Project Phase 5) models is subject to high variations from one model to another. As CMIP models do not reproduce interannual variability phasing well, the 20 years of altimetry records may be too short to allow for a significant comparison between the model-based and observation-based sea levels. However, a careful selection of some models and the use of ensemble means enable us to obtain trend patterns rather similar to those derived from satellite altimetry.

Acknowledgments We thank three anonymous reviewers for their comments that helped us to improve the manuscript. Alice Carret is supported by the MONARCH project of the 7th Framework Programme of the European Union. This paper is an overview of the oral presentation delivered at the International Space Science Institute (ISSI) workshop on "Integrative study of the mean sea level and components" in February 2015.

References

Ablain M, Cazenave A, Larnicol G, Balmaseda M, Cipollini P, Faugère Y, Fernandes MJ, Henry O, Johannessen JA, Knudsen P, Andersen O, Legeais J, Meyssignac B, Picot N, Roca M, Rudenko S, Scharffenberg MG, Stammer D, Timms G, Benveniste J (2015) Improved sea level record over the satellite altimetry era (1993–2010) from the Climate Change Initiative project. Ocean Sci 11:67–82. doi:10.5194/os-11-67-2015

Andersen OB, Piccioni G (2016) Recent Arctic sea level variations from satellites. Front Mar Sci. doi:10.3389/fmars.2016.00076

Andersen O, Knudsen P, Stenseng L (2015) The DTU13 MSS (mean sea surface) and MDT (mean dynamic topography) from 20 years of satellite altimetry. IAG Symp. doi:10.1007/1345_2015_182

Armitage TWK, Bacon S, Ridout AL, Thomas SF, Aksenov Y, Wingham DJ (2016) Arctic sea surface height variability and change from satellite radar altimetry and GRACE, 2003–2014. J Geophys Res Oceans 121:4303–4322. doi:10.1002/2015JC011579

Boyer TP, Antonov JI, Garcia HE, Johnson DR, Locarnini RA, Mishonov AV, Pitcher MT, Baranova OK, Smolyar IV (2006) World Ocean Database 2005. In: Levitus S (ed) NODC Atlas NESDIS 60. U.S. Government Printing Office, Washington

Blazquez A, Meyssignac B, Lemoine JM, Cazenave A, Berthier E (in preparation) Uncertainty in GRACE estimates of the mass redistributions at the Earth surface and impact on the sea level budget

Chambers DP (2012) GRACE monthly ocean mass grids Netcdf release 5.0. Ver. 5.0. PO.DAAC, CA, USA. doi:10.5067/TEOCN-0N005. Accessed 15 Dec 2015

Chambers DP, Bonin JA (2012) Evaluation of release 05 time-variable gravity coefficients over the ocean. Ocean Sci 8:859–868. doi:10.5194/os-8-859-2012

Chambers DP, Willis JK (2010) A global evaluation of ocean bottom pressure from GRACE, OMCT, and steric-corrected altimetry. J Ocean Atmos Technol 27:1395–1402. doi:10.1175/2010jtecho738.1

Chen JL, Wilson CR, Li J, Zhang Z (2015) Reducing leakage error in GRACE-observed long term ice mass change: a case study in West Antarctica. J Geod 89:925–940. doi:10.1007/s00190-015-0824-2

Cheng M, Ries JC, Tapley BD (2011) Variations of the Earth's figure axis from satellite laser ranging and GRACE. J Geophys Res. doi:10.1029/2010JB000850

Cheng Y, Andersen OB, Knudsen P (2012) First evaluation of MyOcean altimetric data in the Arctic Ocean. Ocean Sci Discuss 9:291–314. doi:10.5194/osd-9-291-2012

Cheng Y, Andersen OB, Knudsen P (2015) An improved 20-year Arctic Ocean altimetric sea level data record. Mar Geod 38(2):146–162. doi:10.1080/01490419.2014.954087

Dee DP, Uppala SM, Simmons AJ, Berrisford P, Poli P, Kobayashi S, Andrae U, Balmaseda MA, Balsamo G, Bauer P, Bechtold P, Beljaars M, Van de Berg L, Bidlot J, Bormann N, Delsol C, Dragani R, Fuentes M, Geer AJ, Haimberger L, Healy SB, Hersbach H, Holm EV, Isaksen L, Kallberg P, Köhler M, Matricardi M, McNally AP, Monge-Sanz BM, Morcrette JJ, Park BK, Peubey C, de Rosnay P, Tavolato C, Thépaut JN, Vitart F (2011) The ERA-Interim reanalysis: configuration and performance of the data assimilation system. Q J R Meteorol Soc 137:553–597. doi:10.1002/qj.828

Gardner AS, Moholdt G, Graham Cogley J, Wouters B, Arendt AA, Wahr J, Berthier E, Hock R, Pfeffer WT, Kaser G, Ligtenberg SRM, Bolch T, Sharp MJ, Hagen JO, van den Broeke MR, Paul F (2013) A reconciled estimate of glacier contributions to sea level rise, 2003–2009. Science 340:852–857. doi:10.1126/science.1234532

Geruo A, Wahr J, Zhong S (2013) Computations of the viscoelastic response of a 3-D compressible Earth to surface loading: an application to Glacial Isostatic Adjustment in Antarctica and Canada. Geophys J Int 192:557–572. doi:10.1093/gji/ggs030

Henry O, Prandi P, Llovel W, Cazenave A, Jevrejeva S, Stammer D, Meyssignac B, Koldunov N (2012) Tide gauge-based sea level variations since 1950 along the Norwegian and Russian coasts of the Arctic Ocean: contribution of the steric and mass components. J Geophys Res. doi:10.1029/2011JC007706

Holgate SJ, Matthews A, Woodworth PL, Rickards LJ, Tamisiea ME, Bradshaw E, Foden PR, Gordon KM, Jevrejeva S, Pugh J (2013) New data systems and products at the permanent service for mean sea level. J Coast Res 29(3):493–504. doi:10.2112/JCOASTRES-D-12-00175.1

Koldunov NV, Serra N, Köhl A, Stammer D, Henry O, Cazenave A, Prandi P, Knudsen P, Andersen OB, Gao Y, Johannessen J (2014) Multimodel simulations of Arctic Ocean sea surface height variability in the period 1970–2009. J Geophys Res Oceans 119:8936–8954. doi:10.1002/2014JC010170

Kouraev AV, Papa F, Mognard NM, Buharizin PI, Cazenave A, Crétaux JF, Dozortseva J, Remy F (2004) Sea ice cover in the Caspian and Areal Seas from historical and satellite data. J Mar Syst 47:89–100. doi:10.1016/j.jmarsys.2003.12.011

Kusche J, Schmidt R, Petrovic S, Rietbroek R (2009) Decorrelated GRACE time-variable gravity solutions by GFZ, and their validation using a hydrological model. J Geod 83(10):903–913. doi:10.1007/s00190-009-0308-3

Morison J, Kwok R, Peralta-Ferriz C, Alkire M, Rigor I, Andersen R, Steele M (2012) Changing Arctic Ocean freshwater pathways. Nature 481:66–70. doi:10.1038/nature10705

Ollivier A, Guibbaud M (2012) Envisat RA-2/MWR ocean data validation and cross-calibration activities. Yearly report. Technical note CLS.DOS/NT/12.021, contract SALP-RP-MA-EA-22062-CLS

Peltier WR (2004) Global glacial isostasy and the surface of the ice-age earth: the ICE-5G (VM2) model and GRACE. Ann Rev Earth Planet Sci 32:111–149. doi:10.1146/annurev.earth.32.082503.144359

Peltier WR, Argus DF, Drummond R (2015) Space geodesy constrains ice-age terminal deglaciation: the global ICE-6G_C (VM5a) model. J Geophys Res Solid Earth 120:450–487. doi:10.1002/2014JB011176

Permanent Service for Mean Sea Level (PSMSL) (2016) Tide gauge data. http://www.psmsl.org/data/obtaining/. Retrieved Nov 2016

Poisson JC et al (in preparation) Extending sea level estimation into the Arctic Ocean Using ENVISAT Altimetry

Prandi P, Ablain M, Cazenave A, Picot N (2012) A new estimation of mean sea level in the Arctic Ocean from satellite altimetry. Mar Geod 35(sup1):61–81. doi:10.1080/01490419.2012.718222

Proshutinsky A, Pavlov V, Bourke RH (2001) Sea level rise in the Arctic Ocean. Geophys Res Lett 28(11):2237–2240. doi:10.1029/2000GL012760

Proshutinsky A, AshiI M, Dvorkin EN, Häkkinen S, Krishfield RA, Peltier WR (2004) Secular sea level change in the Russian sector of the Arctic Ocean. J Geophys Res. doi:10.1029/2003JC002007

Proshutinsky A, Ashik I, Häkkinen S, Hunke E, Krishfield R, Maltrud M, Maslowski W, Zhang J (2007) Sea level variability in the Arctic Ocean from AOMIP models. J Geophys Res. doi:10.1029/2006JC003916

Proshutinsky A, Timmermans ML, Ashik I, Beszcynska-Moeller A, Carmack E, Eert J, Frolov I, Itoh M, Kikuchi T, Krishfield R, McLaughlin F, Rabe B, Schauer U, Shimada K, Sokolov V, Steele M, Toole J, Williams W, Woodgate R, Zimmermann S (2011) The Arctic Ocean. Bull Am Meteorol Soc 92(-supplement 6):S145–S148

Rhein M, Rintoul SR, Aoki S, Campos E, Chambers D, Feely RA, Gulev S, Johnson GC, Josey SA, Kostianoy A, Mauritzen C, Roemmich D, Talley LD, Wang F (2013) Observations: Ocean. In: Stocker TF, Qin D, Plattner G-K, Tignor M, Allen SK, Boschung J, Nauels A, Xia Y, Bex V, Midgley PM (eds) Climate Change 2013: the physical science basis. Contribution of working group I to the fifth assessment report of the Intergovernmental Panel on Climate Change. Cambridge University Press, Cambridge

Rigor IG, Wallace JM, Colony RG (2002) Response of sea ice to the Arctic Oscillation. J Clim 15:2648–2663. doi:10.1175/1520-0442(2002)015<2648:rositt>2.0.co;2

Rudels B (2012) Arctic Ocean circulation and variability—advection and external forcing encounter constraints and local processes. Ocean Sci 8:261–286. doi:10.5194/os-8-261-2012

Shepherd A, Ivins ER, Geruo A, Barletta VR, Bentley MJ, Bettadpur S, Briggs KH, Bromwich DH, Forsberg R, Galin N, Horwarth M, Jacobs S, Joughin I, King MA, Lenaerts JTM, Li J, Ligtenberg SRM, Luckman A, Luthcke SB, McMillan M, Meister R, Milne G, Mouginot J, Muir A, Nicolas JP, Paden J, Payne AJ, Pritchard H, Rignot E, Rott H, Sørensen LS, Scambos TA, Scheuchl B, Schrama EJO, Smith B, Sundal AV, van Angelen JH, van de Berg WJ, van den Broeke MR, Vaughan DG, Velicogna I, Wahr J, Whitehouse PL, Wingham DJ, Yi D, Young D, Zwally HJ (2012) A reconciled estimate of ice-sheet mass balance. Science 338(6111):1183–1189. doi:10.1126/science.1228102

Smith LC, Sheng Y, MacDonald GM, Hinzman LD (2005) Disappearing Arctic lakes. Science 308:1429. doi:10.1126/science.1108142

Swenson SC, Chambers DP, Wahr J (2008) Estimating geocenter variations from a combination of GRACE and ocean model output. J Geophys Res Solid Earth. doi:10.1029/2007JB005338

Taylor KE, Stouffer RJ, Meehl GA (2012) An overview of CMIP5 and the experiment design. Bull Am Meteorol Soc 93:485–498. doi:10.1175/BAMS-D-11-00094.1

Valladeau G, Thibaut P, Picard B, Poisson JC, Tran N, Picot N, Guillot A (2015) Using SARAL/AltiKa to improve Ka-band altimeter measurements for coastal zones, hydrology and ice: the PEACHI prototype. Mar Geod 38(S1):124–142. doi:10.1080/01490419.2015.1020176

Vaughan DG, Comiso JC, Allison I, Carrasco J, Kaser G, Kwok R, Mote P, Murray T, Paul F, Ren J, Rignot E, Solomina O, Steffen K, Zhang T (2013) Observations: cryosphere. In: Stocker TF, Qin D, Plattner G-K, Tignor M, Allen SK, Boschung J, Nauels A, Xia Y, Bex V, Midgley PM (eds) Climate Change 2013: the physical science basis. Contribution of working group I to the fifth assessment report of the Intergovernmental Panel on Climate Change. Cambridge University Press, Cambridge

Velicogna I, Sutterley TC, van den Broeke MR (2014) Regional acceleration in icemass loss from Greenland and Antarctica using GRACE time-variable gravity data. Geophys Res Lett. doi:10.1002/2014GL061052

Volkov DL (2014) Do the North Atlantic winds drive the nonseasonal variability of the Arctic Ocean sea level? Geophys Res Lett. doi:10.1002/2013GL059065

Volkov DL, Landerer FW (2013) Nonseasonal fluctuations of the Arctic Ocean mass observed by the GRACE satellites. J Geophys Res Oceans 118:6451–6460. doi:10.1002/2013JC009341

Volkov DL, Landerer FW, Kirillov SA (2013) The genesis of sea level variability in the Barents Sea. Cont Shelf Res 66:92–104. doi:10.1016/j.csr.2013.07.007

Watkins MW, Wiese DN, Yuan D-N, Boening C, Landerer FW (2015) Improved methods for observing Earth's time variable mass distribution with GRACE using spherical cap mascons. J Geophys Res Solid Earth. doi:10.1002/2014JB011547

Wiese DN (2015) GRACE monthly global water mass grids netcdf release 5.0. Ver. 5.0. PO.DAAC, CA, USA. doi:10.5067/TEMSC-OCL05. Accessed 15 Dec 2015

Yin J, Griffies SM, Stouffer RJ (2010) Spatial variability of sea level rise in twenty-first century projections. J Clim. doi:10.1175/2010JCLI3533,1

Zuo H, Balmaseda MA, Mogensen K (2015) The new eddy-permitting ORAP5 ocean reanalysis: description, evaluation and uncertainties in climate signals. Clim Dyn. doi:10.1007/s00382-015-2675-1

Surv Geophys (2017) 38:277–293
DOI 10.1007/s10712-016-9391-1

CrossMark

Phenological Responses to ENSO in the Global Oceans

M.-F. Racault[1,2,3,4] · **S. Sathyendranath**[1,2,4] · **N. Menon**[4,5] · **T. Platt**[1,4]

Received: 29 April 2016 / Accepted: 17 October 2016 / Published online: 9 November 2016

Abstract Phenology relates to the study of timing of periodic events in the life cycle of plants or animals as influenced by environmental conditions and climatic forcing. Phenological metrics provide information essential to quantify variations in the life cycle of these organisms. The metrics also allow us to estimate the speed at which living organisms respond to environmental changes. At the surface of the oceans, microscopic plant cells, so-called phytoplankton, grow and sometimes form blooms, with concentrations reaching up to 100 million cells per litre and extending over many square kilometres. These blooms can have a huge collective impact on ocean colour, because they contain chlorophyll and other auxiliary pigments, making them visible from space. Phytoplankton populations have a high turnover rate and can respond within hours to days to environmental perturbations. This makes them ideal indicators to study the first-level biological response to environmental changes. In the Earth's climate system, the El Niño–Southern Oscillation (ENSO) dominates large-scale inter-annual variations in environmental conditions. It serves as a natural experiment to study and understand how phytoplankton in the ocean (and hence the organisms at higher trophic levels) respond to climate variability. Here, the ENSO influence on phytoplankton is estimated through variations in chlorophyll concentration, primary production and timings of initiation, peak, termination and duration of the growing period. The phenological variabilities are used to characterise phytoplankton responses to changes in some physical variables: sea surface temperature, sea surface height and wind. It is reported that in oceanic regions experiencing high annual

✉ M.-F. Racault
mfrt@pml.ac.uk

1 Plymouth Marine Laboratory (PML), Prospect Place, The Hoe, Plymouth, UK

2 National Centre for Earth Observation (NCEO), PML, Plymouth PL1 3DH, UK

3 ESA Living Planet Fellowship, PML, Plymouth PL1 3DH, UK

4 INDO-European Research Facilities, Studies on MARine Ecosystem and CLIMate in India (INDO-MARECLIM), NERCI, Kochi, Kerala 682016, India

5 Nansen Environmental Research Centre India (NERCI), 6A, Oxford Business Centre (6th Floor), Sreekandath Road, Ravipuram, Kochi, Kerala 682016, India

variations in the solar cycle, such as in high latitudes, the influence of ENSO may be readily measured using annual mean anomalies of physical variables. In contrast, in oceanic regions where ENSO modulates a climate system characterised by a seasonal reversal of the wind forcing, such as the monsoon system in the Indian Ocean, phenology-based mean anomalies of physical variables help refine evaluation of the mechanisms driving the biological responses and provide a more comprehensive understanding of the integrated processes.

Keywords ESA climate change initiative · Ocean colour · Phytoplankton · Phenology · El Niño–southern oscillation · Climate impact

1 Introduction

Phytoplankton are microscopic unicellular algae living in the upper layer of oceans. Through chlorophyll and associated pigments that they contain, phytoplankton carry out photosynthesis, which contribute to the oceanic uptake of the CO_2 emitted to the atmosphere every year. This CO_2 sink is part of a very active, natural carbon cycle, through which phytoplankton in the surface layer of the ocean fix CO_2 into organic matter, some of which subsequently sinks below the mixed layer. Through this process, phytoplankton help to modulate the increase in atmospheric CO_2 that results from the burning of fossil fuels. Moreover, phytoplankton are at the base of the food chain and transfer energy to higher trophic levels. This transfer of energy has a knock-on effect on fisheries and dependent human societies especially in highly productive and coastal upwelling regions. Thus, phytoplankton are key players in the planetary carbon cycle and provide important services to the society.

Several metrics have been developed to quantify variations in phytoplankton populations. These metrics form ecological indicators, which provide systematic and objective information about the state of the marine ecosystem (Platt and Sathyendranath 2008). Analysis of a suite of indicators belonging to different ecosystem attributes (i.e. composition, structure, functioning) can help ensure that different modes of variability within the ecological system are represented and thus enable comprehensive assessment of the ecosystem state (Racault et al. 2014). In situ or remote-sensed measurements of chlorophyll concentration provide key information about the structure of phytoplankton populations. Using chlorophyll concentration and irradiance observations, algorithms can be implemented to estimate primary production (PP), providing a measure of the rate of conversion of inorganic carbon in CO_2 to organic carbon by photosynthesis, which is key to assess ecosystem functioning. Also based on surface chlorophyll concentration, phenological algorithms can be applied to estimate the specific timings of important events in the phytoplankton growing period and provide further information on the ecosystem functioning.

Ocean-colour sensors on satellites can provide estimates of chlorophyll concentration at high spatial and temporal resolutions and at global scale. Because they provide data consistently and frequently and over long periods of time, they are suitable for computations of several ecological indicators (including PP and phenology) and for studying inter-annual variations and long-term trends in the state of the marine ecosystem. However, ocean-colour sensors do have a finite lifespan, and differences in instrument design and algorithms make it difficult to compare data from multiple sensors. When overlapping data are available from two or more sensors, such data can be used to establish inter-sensor bias and correct for it. The ESA Ocean Colour CCI (OC-CCI) has merged ocean-colour data using the Sea-viewing Wide Field-of-View Sensor (SeaWiFS 1997–2010), the Moderate-Resolution Imaging Spectroradiometer (MODIS 2002-present) and the MEdium-Resolution Imaging Spectrometer (MERIS 2002–2012) to provide the first 17-year (1997 to present) global scale, high-quality, bias-corrected and error-characterised data record of ocean colour (Sathyendranath et al. 2016). Furthermore, implementation of the coupled ocean–atmosphere POLYMER atmospheric correction algorithm (MERIS period) has increased significantly the coverage of chlorophyll observations (Steinmetz et al. 2011; Racault et al. 2015; Sathyendranath et al. 2016). The improvements realised in the OC-CCI products will help us to enhance evidence for, and improve confidence in, our understanding of the impacts of climate variability and change on the marine ecosystem.

Recent research has shown that variations in the abundance and phenology of phytoplankton populations can profoundly alter: (1) the efficiency of the biological pump, with inevitable impact of the global carbon cycle and (2) the interactions across trophic levels (Edwards and Richardson 2004), which can engender trophic mismatch with deleterious impact on the survival of commercially important fish and crustacean larvae (Platt et al. 2003; Koeller et al. 2009; Lo-Yat et al. 2011). The high turnover rate of phytoplankton, which is tightly coupled to environmental perturbations, makes them ideal indicators to study the first-level biological response to environmental changes. The main drivers of variations in phytoplankton populations include light and nutrient availability, which may be modulated by stratification, mixing, upwelling and riverine inputs. Perturbations in these physical processes may be characterised using observations of sea surface temperature (SST), net heat flux, winds, rainfall and sea surface height (SSH) and are broadly related to large-scale patterns of climate variability.

In the Earth's climate system, the El Niño–Southern Oscillation (ENSO) is a dominant driving force of climate variability, involving warm (El Niño) and cold (La Niña) episodes with a typical periodicity of 2–7 years (McPhaden et al. 2006). The occurrence of ENSO episodes is characterised by anomalous changes in trade wind intensity and SST in the tropical Pacific. The planetary influence of ENSO is induced through a complex suite of ocean–atmosphere feedbacks, tropical–extratropical interactions and atmospheric teleconnections. The specific influence of each of these remote forcing mechanisms is not fully determined yet, but they can severely disrupt temperature and rainfall patterns, storm tracks and cyclone trajectories (Cai et al. 2015), with knock-on effects on crops, vector diseases (Martinez-Urtaza et al. 2016), and on marine ecosystem composition, structure and functioning (e.g., Behrenfeld et al. 2001; Jackson et al. 2011; Masotti et al. 2011; Racault et al. 2012).

This paper provides an overview of phytoplankton responses to the ENSO mode of climate variability based on a suite of ecological indicators (i.e. chlorophyll concentration, primary production, phenological metrics) estimated using ESA CCI ocean-colour observations in the global oceans (Sathyendranath et al. 2016). It explores and discusses the use of phenological metrics to identify occurrences of changes in environmental conditions and specifically of ENSO-related changes in environmental conditions. The environmental variables are based on ESA CCI-SST (Merchant et al. 2014) and CCI-SSH (Ablain et al. 2015) observations, and ECMWF ERA Interim reanalysis products of winds. Finally, the phenological responses to ENSO are used as a framework to help us understand further the mechanisms driving variability in phytoplankton populations.

2 Phenology Based on Ocean-Colour Observations

Phenological metrics of timings of initiation, peak and termination, and duration of the phytoplankton growing period (Platt and Sathyendranath 1996; Platt and Sathyendranath 2008; Racault et al. 2012) can be calculated based on relative changes in the concentration of chlorophyll. In the present paper, the analysis uses Level-3 ESA OC-CCI chlorophyll dataset at 5-day temporal resolution and 1 × 1 degree spatial resolution over the period 1998–2009. The resolution has been chosen to minimise gaps in the data while retaining maximum resolution in time. A schematic representation of the phenological method is presented in Fig. 1. The phenological algorithm permits us to estimate up to two

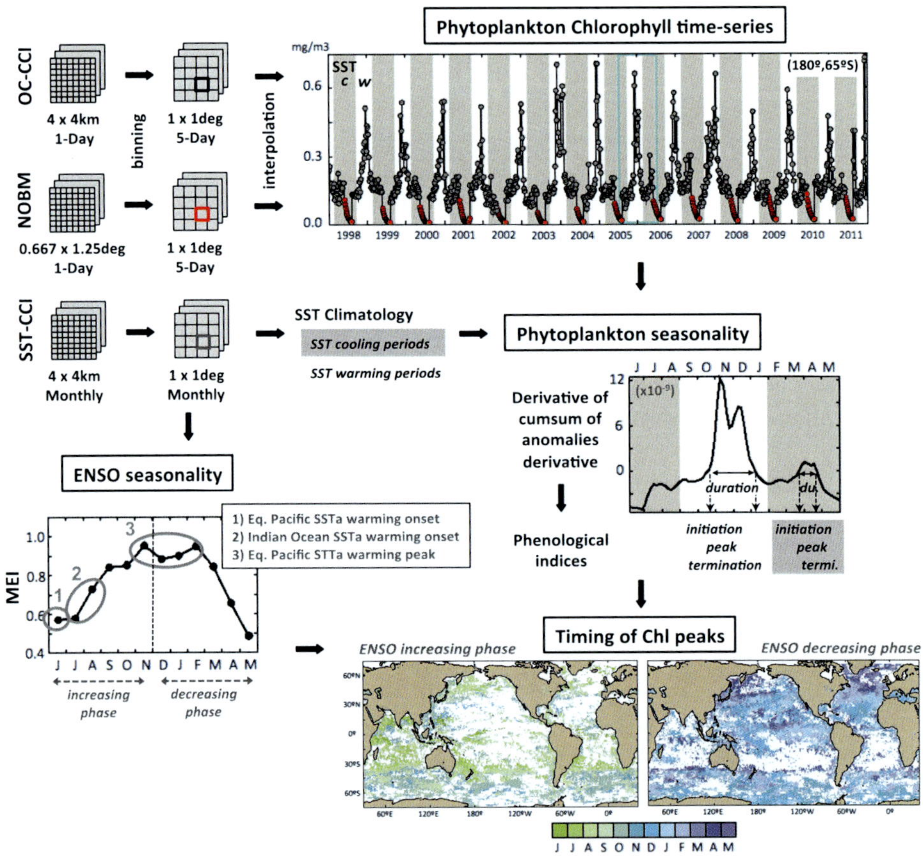

Fig. 1 Schematic diagram of the phenology algorithm. The methodology to estimate two chlorophyll peaks per annual cycle has been based on Racault et al. (2015) and developed further. Ocean Colour Climate Change Initiative data were used over the period 1998–2009 at 5-day resolution. NOBM model data of chlorophyll concentration were estimated at 5-day resolution and used to fill persistent missing data due to high solar zenith angle at higher latitudes (*red dots* in the Chlorophyll time-series). Climatology of Sea Surface Temperature SST-CCI data were used to provide time boundaries to identify chlorophyll peaks in the annual cycle. Multivariate ENSO Index MEI was used to separate the annual cycle in two phases (increasing and decreasing ENSO phases). The plot of the derivative of the cumulative sum of chlorophyll anomalies and maps of timing of chlorophyll peak during ENSO increasing and decreasing phases are displayed here as an example, for the 12-month period between June 2005 to May 2006

phytoplankton growing periods per year. It is based on a threshold criterion (i.e. chlorophyll long-term median plus 5 %) and calculation of the derivative of the cumulative sum of chlorophyll anomalies. The latter method had been initially developed at regional scale (i.e. the Red Sea) based on OC-CCI climatology at 8-day resolution (Racault et al. 2015). Here, the method has been further developed to be compatible with multi-annual chlorophyll time-series in the global oceans, and at an improved 5-day temporal resolution. Furthermore, construction of complete chlorophyll time-series (no data gaps) was achieved: (1) by applying linear interpolation to fill missing data due to cloud cover and (2) by inserting NASA Ocean Biogeochemical Model (NOBM, Gregg and Casey 2007; Gregg and Rousseaux 2014) chlorophyll data to fill persistent missing data due high solar zenith angle in winter at high latitudes. Such gap-free chlorophyll time-series are required to compute cumulative sums of anomalies and then estimate phenological indices. The integration of the NOBM data did not introduce bias in the time-series used to estimate the phenological analysis. This was apparent in the filled chlorophyll time-series and in the calculation of the cumulative sum (see panel "Phytoplankton time-series" in Fig. 1). In addition, to account for the large variability in timing of occurrence of the phytoplankton growing periods in the global oceans (Racault et al. 2012), SST seasonal cycle was used to define specific time intervals during which phenological indices were estimated: (1) during SST warming phase and (2) during SST cooling phases (white and grey shaded areas, respectively, in Fig. 1). This is a further improvement compared to the initial algorithm of Racault et al. (2015), which was based on fixed delineation of SST periods.

The phenological metrics and threshold criterion were estimated on a pixel-by-pixel basis. The timings of initiation and termination of phytoplankton growth are defined when the cumulative sum of chlorophyll changes sign (i.e. going from negative to positive and vice versa, respectively, Fig. 1). The sign changes correspond to the time when the chlorophyll concentrations rise above and fall below the relative threshold criterion (Siegel et al. 2002; Racault et al. 2012). Finally, the duration of the growing period is estimated as the time elapsed between initiation and termination. Given that phytoplankton response to ENSO has been demonstrated in the tropics, subtropics and during austral and boreal summer periods at high latitudes in the South and North Hemispheres, respectively (Behrenfeld et al. 2001, Yoder and Kennelly 2003, Messié and Chavez 2012), information about ENSO seasonal cycle was also taken into account in the development of the phenological algorithm. In particular, the time boundaries of the annual cycles were delineated over the period from June (of year t) to May (of year $t + 1$) (i.e. spanning two calendar years). This 12-month delineation period was chosen to follow the seasonality of ENSO activity, generally peaking in the month of November to January (higher SST anomalies in the Equatorial Pacific, Fig. 1). The timings of chlorophyll peaks and phytoplankton growing periods were classified based on phases of increasing and decreasing ENSO anomaly.

In subpolar regions, nutrients are generally replenished during the winter season through enhanced mixing of the water column, and phytoplankton growth is primarily limited by light availability, which may be enhanced in spring when net heat flux becomes positive and the water-column stratifies. In these conditions, the timing of chlorophyll peak follows the latitudinal increase in light availability (Fig. 1) from the months of ~July to November in the Southern Hemisphere (i.e. ENSO increasing phase) and from the months of ~January to May in the Northern Hemisphere (i.e. ENSO decreasing phase). In the tropics and subtropics, light is plentiful all-year round, and phytoplankton growth is primarily limited by nutrient availability, which is enhanced by water-column mixing following environmental perturbations. In these regions, the timing of phytoplankton growth is not seasonal and may occur throughout the course of an annual cycle (Fig. 1).

The phenology indices presented in this paper were compared with, and shown to be consistent with, those in the literature, for instance results based on in situ or satellite observations at regional scale for North Atlantic (González Taboada and Anadón 2014; Cole et al. 2015), Southern Ocean (Thomalla et al. 2011; Carranza and Gille 2015) and North Pacific (Sasaoka et al. 2011), as well as previously published results for the global oceans (Racault et al. 2012). The increased resolution of the chlorophyll composites from 8 to 5 days was made possible by the significant improvement in data coverage in the OC-CCI product. Initially, the indices were estimated both at 5- and 8-day resolutions using the improved phenological algorithm, and the results from these two resolutions were seen to be consistent with each other. Hence, only the analysis at 5-day resolution is shown here.

The present phenological algorithm further allows us to provide, in the global oceans, estimates of the probability that: (1) the main chlorophyll peak occurs during increasing or decreasing ENSO phase (Fig. 2a, b) and (2) two chlorophyll peaks occur each year (Fig. 2c). North and South Hemispheres seasonalities are apparent in this analysis: the main growing period (defined by the peak with the highest amplitude) is shown to occur during the months of June to November (i.e. JJASON in Fig. 2a) in the South Hemisphere, whereas it occurs during the months of December to May (i.e. DJFMAM in Fig. 2b) in the North Hemisphere. Interestingly, regions showing high probability to have a main chlorophyll peak in DJFMAM (i.e. predominantly found in the North Hemisphere) can also be found in the South Hemisphere, in the Pacific Ocean tropics and subtropics, off the east coast of Madagascar and in the Mozambique Channel, and along the west and northwest coast of Australia. The probability to have two chlorophyll peaks per year is almost zero in the tropics and subtropics, while it increases almost symmetrically towards higher latitudes in the North and South Hemispheres (Fig. 2c). Moreover, it is noteworthy that in high-latitude regions, the probability to have two chlorophyll peaks per year reaches values of ~0.5, indicating that two peaks in chlorophyll are only observed in ~half of the years during the period 1998–2009 (i.e. approximately half of the years present two peaks and the other half present one peak per year). The latter probability estimates, which are based on satellite observations, were compared and showed consistency with the latitudinal variations in the occurrence of phytoplankton blooms obtained in the North Atlantic using a model based on simple theoretical assumptions (Platt et al. 2009). This model demonstrates that the main driver explaining the variations in the probability of occurrence of two

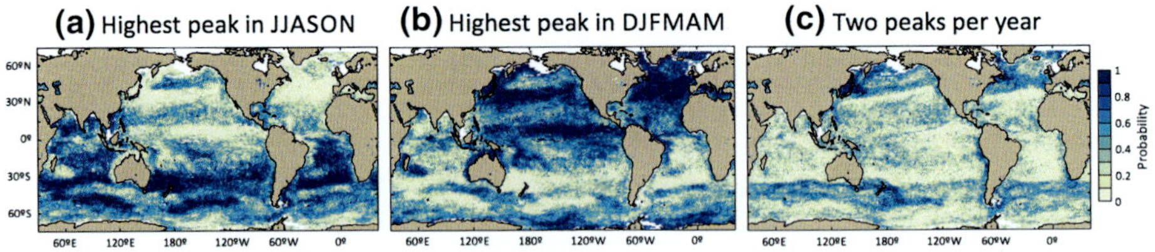

Fig. 2 Probability of occurrence of chlorophyll peaks based on the phenological algorithm implemented using OC-CCI chlorophyll data at 5-day resolution. **a** Probability that the main chlorophyll peak occurs during the months of June to November JJASON (i.e. ENSO increasing phase, see Fig. 1); **b** Probability that the main chlorophyll peak occurs during the months of December to May DJFMAM (i.e. ENSO decreasing phase, see Fig. 1); and **c** Probability to have two chlorophyll peaks per year. If there are two chlorophyll peaks in 1 year, the main chlorophyll peak is defined by the peak with the higher amplitude (i.e. higher maximum chlorophyll value)

chlorophyll peaks per year is the latitudinal variations in the strength and periodicity of the initial forcing (i.e. variations in the magnitude in the total daily irradiance). Another approach based on 1000 a posteriori simulations from a model fitted to remote-sensed observations of chlorophyll concentration (15 consecutive seasonal cycles from 1998/1999 to 2012/2013) has permitted assessment of the probability of detecting different peaks in chlorophyll concentration and their timing in the Atlantic Ocean (between 15°S and 80°N; González Taboada and Anadón 2014). The authors showed higher probability of occurrence of two chlorophyll peaks per year in the North Atlantic subtropical region, which is consistent with the analysis presented in this paper.

The processes driving the inter-annual variability in phytoplankton phenology have been investigated at regional and global scales (e.g., Henson et al. 2009; Thomalla et al. 2011; Racault et al. 2012; Brody et al. 2013). Relationships have been demonstrated between the timing of phytoplankton growth and the timing of light availability and water-column stratification at high latitudes, and between the timing of phytoplankton growth and the timing of deepening of the mixed layer depth (as a proxy for water-column mixing and nutrient availability) in the tropics and subtropics. In the following sections, we focus on the influence of ENSO on phytoplankton phenological variability and review some of the possible driving processes available from satellite observations of SSH and SST.

3 ENSO Impact on Chlorophyll, Primary Production and Phenology

ENSO activity, consisting of irregular El Niño and La Niña episodes, can profoundly impact marine ecosystem indicators. During the 1997–1999 El Niño/La Niña transition period, phytoplankton biomass increased by 10 % globally (Behrenfeld et al. 2001), and new production (dependent on new nitrogen) varied by more than a factor of two in the Equatorial Pacific (Turk et al. 2001). Short-term variability (less than one decade) in chlorophyll concentration, primary production and phenology of phytoplankton populations have been shown to correlate with ENSO variability in the Equatorial regions and in the global oceans albeit with marked regional differences (Yoder and Kennelly 2003; Behrenfeld et al. 2006; Vantrepotte and Mélin 2011; Chavez et al. 2011; Messié and Chavez 2012; Racault et al. 2012; Raitsos et al. 2015). Regional variability may weaken or enhance long-term trends, which may be further modulated by decadal oscillations in physical oceanographic conditions. In particular, the influence of ENSO and regional climate oscillations on SST and phytoplankton has been investigated in the North Atlantic (e.g., ENSO and North Atlantic Oscillation, Lee et al. 2008), in the North Pacific (e.g., ENSO and Pacific Decadal Oscillation, Di Lorenzo et al. 2008; Martinez et al. 2009), in the Indian Ocean (e.g., ENSO and IOD, Saji et al. 1999; Brewin et al. 2012; Currie et al. 2013), and in the Southern Ocean (e.g., ENSO and Southern Annular Mode, Soppa et al. 2016).

One of the most widely used environmental indices for assessing specifically the impact of El Niño on ocean biology is the Multivariate ENSO Index (MEI) (Wolter and Timlin 1993) (Table 1). The MEI encapsulates short-term variations of coupled ocean–atmosphere processes rooted in the tropical Pacific. Specifically, the MEI is defined as the first seasonally varying principal component of six atmosphere–ocean variable fields in the tropical Pacific basin (i.e. sea level pressure, zonal and meridional wind components, sea surface and air temperatures, and total cloudiness).

Linear regression analyses between MEI and annual mean chlorophyll concentration anomalies, between MEI and annual mean primary production anomalies, and between

Table 1 ENSO characteristics during the period of study 1998–2009

	Years										
	98/99	99/00	00/01	01/02	02/03	03/04	04/05	05/06	06/07	07/08	08/09
MEI sign	–	–	–	+	+	+	+	–	+	–	–
Event type	LN	LN	LN		CP EN	CP EN		LN	CP EN	LN	LN

The El Niño and La Niña events are defined based on variations of the Multivariate ENSO Index by ±0.5 Standard Deviation. The classification of Central Pacific El Niño events is based on Yu et al. (2012) and Radenac et al. (2012)

EN El Niño, *CP* Central Pacific, *LN* La Niña

MEI and phenological metric anomalies can be used to characterise some of the variations in phytoplankton populations associated with ENSO. In this paper, linear regression analyses are performed based on annual mean of MEI values based on the Wolter and Timlin (1993) dataset, the ESA OC-CCI project chlorophyll dataset (Level 3, Mapped, 1 × 1 degree and monthly resolutions) (Sathyendranath et al. 2016), the Transboundary Waters Assessment Programme (GEF-TWAP) primary production dataset (Level 4, Mapped, 1 × 1 degree and monthly resolutions) and the phenological datasets presented in Sect. 2. The phenological indices can be estimated only when a complete seasonal cycle of chlorophyll is available. As the OC-CCI data record begins in September 1997 (with the SeaWiFS mission), the indices could not be estimated for the global oceans, during the 1997–1998 extreme El Niño event. However, the indices could be estimated during the following ENSO events over the period June 1998 to May 2009. Furthermore, the period of study (1998–2009) is marked by a major La Niña event in 1998/1999, a moderate La Niña event in 2007/2008 and a period of quasi-continuous positive MEI values between 2002 and 2007, which are characterised by anomalous SST warming in the central equatorial Pacific (Table 1). The meridional position (i.e. central or eastern Pacific) and amplitude of the SST anomalies may be classified as different extreme types of El Niño (Capotondi et al. 2015): the Eastern Pacific (EP El Niño), also referred to as the "typical" or canonical El Niño, characterised by anomalous SST warming in the eastern tropical Pacific; and the Central Pacific (CP El Niño), variously referred to as El Niño Modoki (Pseudo El Niño; Ashok et al. 2007; Kao and Yu 2009), warm-pool El Niño (Kug et al. 2009), or dateline El Niño (Larkin and Harrison 2005), and characterised by ocean warming anomalies occurring in the central tropical Pacific. The influence of these two extreme types of El Niño can lead to significantly different perturbations of environmental conditions and biological responses (e.g., Ashok and Yamagata 2009, Yu et al. 2012, Gierach et al. 2012, Radenac et al. 2012). Finally, the period of study from 1998 to 2009 was also chosen as it spans the availability of all of the different data records: OC-CCI chlorophyll (1997-present), TWAP primary production (1998–2010), phenology (1998–2009) and MEI (1950-present).

3.1 ENSO Impact on Phytoplankton Phenology

During positive phase of the MEI (Fig. 3a–c), the timings of initiation, peak and termination show delays of from ~25 to 40 days (positive anomalies) in tropical and extratropical regions of the central and eastern Pacific Ocean, in the subtropical regions of the Indian Ocean, in tropical and subtropical regions of the Atlantic Ocean, and also towards higher latitudes, between 40 and 50°N in the North Atlantic, and between 20 and 50°S in the western side of the South Atlantic. Conversely, the timings of the phytoplankton

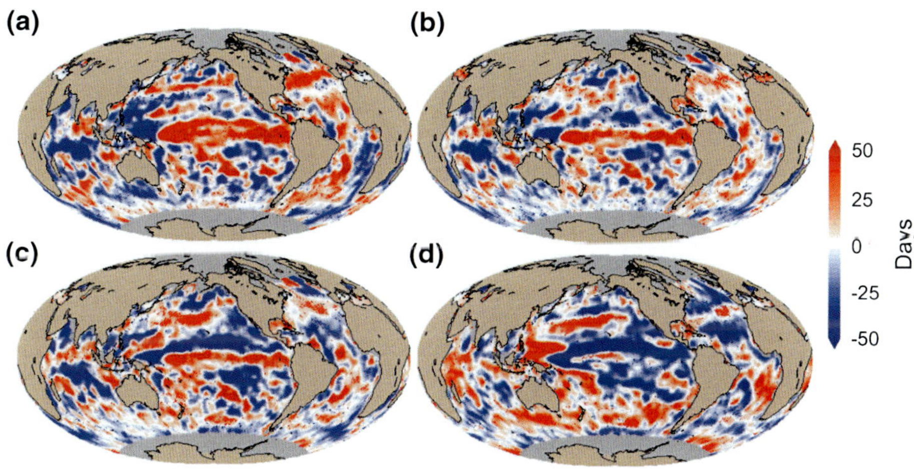

Fig. 3 ENSO impact on phytoplankton phenology estimated using linear regression analysis between **a** MEI and anomalies of timing of initiation of phytoplankton growing period, **b** MEI and anomalies of timing of peak, **c** MEI and anomalies of timing of termination, and **d** MEI and anomalies of duration. Increase and decrease are indicated by positive (*red*) and negative (*blue*) anomalies, respectively. The phenological metrics were estimated over the period 1998–2009 based on OC-CCI chlorophyll data. In all panels, *red and blue stippling* indicates where the linear regression coefficients are significant at the 90 % confidence level

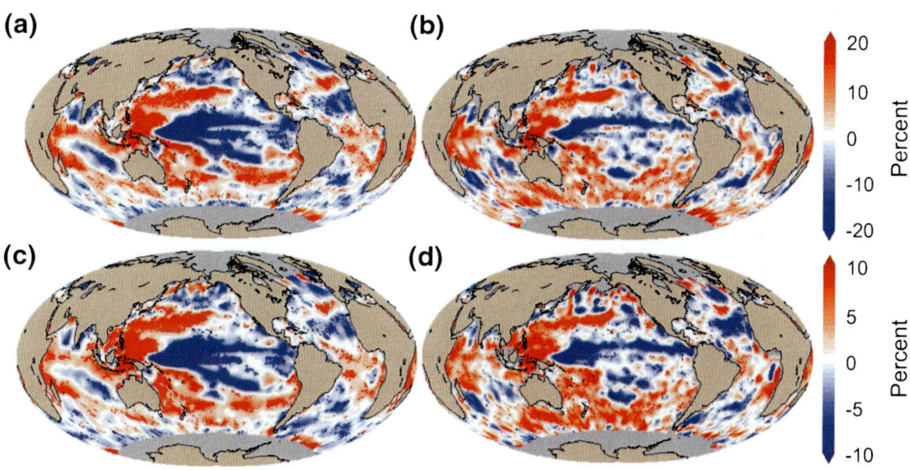

Fig. 4 ENSO impact on annual and phenology-based chlorophyll concentration and primary production estimated using linear regression analysis between **a** MEI and annual mean anomalies of chlorophyll concentration, **b** MEI and mean anomalies of chlorophyll concentration over the duration of the phytoplankton growing period, **c** MEI and annual mean anomalies of primary production and **d** MEI and mean anomalies of primary production over the duration of the phytoplankton growing period. Increase and decrease are indicated by positive (*red*) and negative (*blue*) anomalies, respectively. Chlorophyll data are on OC-CCI data product, and primary production data are from TWAP (based on the algorithm of Platt and Sathyendranath (1988) were analysed during the period 1998–2009. In all panels, *red and blue stippling* indicates where the linear regression coefficients are significant at the 90 % confidence level

growing period are observed to occur earlier (between −15 and −30 days, negative anomalies) in the eastern equatorial region of the Pacific Ocean, in the equatorial region of the Indian Ocean and in large regions of the Southern Ocean and the eastern side of the South Atlantic Ocean. It is noteworthy that higher anomalies are observed in the timing of initiation compared with the timings of peak and termination. Furthermore, the regional changes observed in response to ENSO are coherent between the ecosystem indices: when initiation of the phytoplankton growing period is delayed, the timing of peak is also delayed and the timing of termination is advanced, leading to shorter duration and lower mean annual chlorophyll concentration and primary production (Figs. 3a, d, 4a, c).

3.2 ENSO Impact on Chlorophyll and Primary Production

During positive phase of the MEI (Fig. 4a, c), annual mean chlorophyll concentration and primary production anomalies show marked decreases in tropical and extratropical regions of the central and eastern Pacific Ocean, in the subtropical and subpolar regions of the North Atlantic Ocean (between 10 and 20°N and polewards of 40°N) and in the subtropical region of the Indian Ocean. Conversely, increases in chlorophyll concentration and primary production are observed in the western Pacific Ocean, as well as over large regions of the Southern Ocean, and the equatorial region of the Indian Ocean. ENSO-related changes in chlorophyll concentration are larger, varying by ±20 % compared with the changes observed in primary production, varying by ±10 %. The changes are expressed in per cent rather than in absolute values because chlorophyll concentration can span three orders of magnitude and, hence, the values can be more readily interpreted and compared when expressed in relative terms.

To refine estimations of ENSO-related changes in phytoplankton, mean anomalies of chlorophyll concentration and primary production are also calculated specifically during the time interval between the initiation and termination of the main phytoplankton growing period each year. Globally, phytoplankton responses show similar pattern between the annual mean and phenology-based mean, indicating that the ENSO mode of variability dominates the observed annual variations in chlorophyll and primary production. Regionally, phenology-based responses present larger increases in chlorophyll and primary production in the Southern Ocean, and the Indian Ocean, and spatially more defined delineation of increases in the Atlantic Ocean. The decreases in phytoplankton concentration and production observed in the central and eastern tropical and subtropical Pacific Ocean are also more limited in extent (constrained to the equatorial Pacific region) when using the phenology-based estimates. The latter estimates further highlight larger decreases in the subtropical North Atlantic Ocean, and a marked decrease in phytoplankton of the southwest coast of Madagascar, which is quite prominent compared with the increases in chlorophyll and primary production observed in the surrounding waters. Regional and local differences between annual and phenology-based mean anomalies of chlorophyll and primary production are also observed in the Gulf of Guinea large marine ecosystem.

3.3 Emergent Properties in Ocean-Colour Indices

Relationships between ENSO-related responses shown in ecological indicators can be explored using linear regression analysis between relative changes in duration of phytoplankton growth and chlorophyll concentration, and between duration and primary production. Each ecological indicator is estimated using climatologies of positive and negative MEI years over the period 1998–2009. In the latter period, positive MEI years include

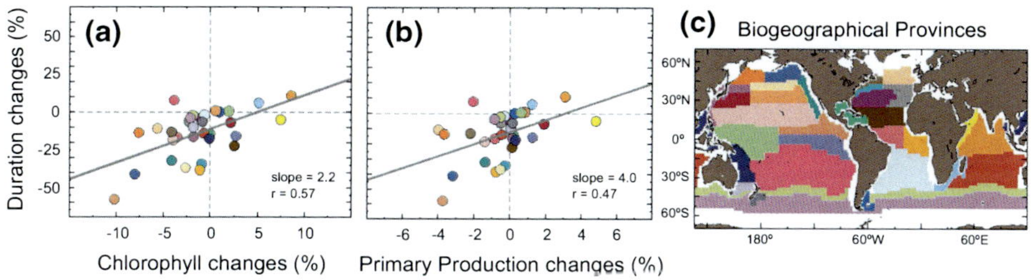

Fig. 5 Emergent properties in ecological indicator responses to ENSO activity. **a** Relationship between relative changes in chlorophyll concentration and duration of phytoplankton growing period between years of positive ENSO and years of negative ENSO indices; **b** Relationship between relative changes in primary production and duration of phytoplankton growing period between years of positive ENSO and years of negative ENSO indices. The *colour of each dot* indicates the changes for each biogeochemical province (Longhurst 1998) delineated in panel **c**)

2001/2002 to 2004/2005, 2006/2007 and negative MEI years include 1998/1999 to 2000/2001, 2005/2006, 2007/2008, 2008/2009 (Table 1). The relative difference between responses to positive and negative MEI is computed for the annual mean chlorophyll concentration, annual mean primary production and the duration of phytoplankton growing period. The regression analysis is performed first on a pixel-by-pixel basis, and then the results are averaged in each biogeographical province (Fig. 5). This averaging procedure allows us to weight evenly the influence of ENSO in the tropics, subtropics and subpolar provinces. The partitioning of the provinces is based on physical, chemical and biological oceanographic knowledge and provides comprehensive geographical units supporting scientific findings interpretation and extrapolation (Longhurst 1998).

Based on the linear regression analyses, the relative changes observed in annual mean chlorophyll can explain 57 % of the relative changes in duration, and the relative changes observed in annual mean primary production can explain 47 % of the relative changes in duration (Fig. 5a, b, respectively, $p < 0.01$). The sign and magnitude of the slopes are positive and greater than one, such that increases in MEI-associated changes in duration are accompanied by a twofold increase in the response of chlorophyll to MEI, and increases observed in MEI-associated changes in duration are accompanied by a fourfold increase in the response of primary production to MEI. Largest MEI-related increases in chlorophyll, primary production and duration are observed in the Indian Ocean Monsoon Gyre province and in the Agulhas and Somali Current Large Marine Ecosystems province, whereas the largest decreases in the indicator values are observed in the eastern Equatorial Pacific, subtropical and subpolar North Pacific regions (Fig. 5a, b). The emergence of linear relationships amongst ENSO-responses of indicators, which are initially measured in different units (i.e. mgChl m^{-3} for chlorophyll concentration, mgC m^{-2} $year^{-1}$ for primary production, and days for duration), can be particularly useful to analyse and compare indicators estimated from non-continuous data records, and when inter-sensor bias correction cannot be quantified (for instance, to compare changes in phytoplankton population between non-overlapping ocean-colour sensors CZCS (1978–1986) and contemporary sensors, starting in 1997 with SeaWiFS, and follow-on sensors in 2002 with MODIS or MERIS).

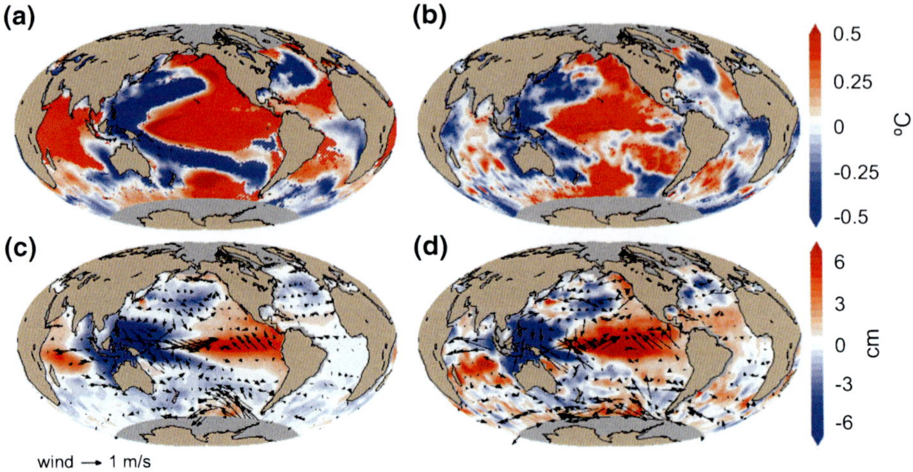

Fig. 6 ENSO influence on annual and phenological mean physical conditions estimated using linear regression analysis between **a** MEI and annual mean anomalies of SST, **b** MEI and mean anomalies of SST over the duration of the phytoplankton growing period, **c** MEI and annual mean anomalies of SSH and mean wind vectors and **d** MEI and mean anomalies of SSH and mean wind vectors over the duration of the phytoplankton growing period. SST and SSH datasets are from ESA CCI program, and wind data are from ECMWF ERA Interim reanalysis. Increase and decrease are indicated by positive (*red*) and negative (*blue*) anomalies, respectively. In all panels, *red and blue stippling* indicates where the linear regression coefficients are significant at the 90 % confidence level over the entire period 1998–2009

4 Phenology-Based Biophysical Responses to ENSO

Mechanistic understanding of the biological responses to ENSO activity is difficult to assess because they result from complex biophysical interactions. In this section, similarly to the analysis carried out with the ocean-colour data products, the influence of ENSO on oceanic physical variables are investigated using linear regression analyses between MEI and annual mean SST anomalies, between MEI and annual mean wind anomalies and between MEI and annual mean SSH anomalies (Fig. 6). Since sea level reflects on the integral effect of surface and subsurface processes (e.g., ocean warming or cooling), SSH anomaly can be used as a good indicator for upwelling anomaly for very large upwelling regions that extend a long way offshore, and large-scale open-ocean upwelling such as in the Equatorial Pacific (Fu and Cazenave 2001). In addition, as variations in phytoplankton populations are tightly coupled to changes in environmental conditions, phytoplankton responses can be used as sentinels, or indicators of other, less obvious, changes occurring in the environment. In this context, phenological metrics are used to estimate ENSO-related changes in physical conditions that occur specifically during the phytoplankton growing period. For this approach, anomalies in the physical variables are averaged over the duration of the main growing period (characterised by the highest chlorophyll peak in the year, Fig. 2), so as to provide refined (more targeted) detection of the period and magnitude of physical changes occurring in the environment.

The influence of ENSO estimated using annual mean and phenology-related mean physical variables present consistent patterns in the Pacific and Atlantic Oceans and show marked differences in the Indian Ocean (Fig. 6). During positive MEI phases, large decrease is observed in the west Pacific Ocean, forming a V-shape rooted in the Equatorial region, and extending to the subtropics and towards higher latitudes in subpolar regions. In contrast, the central and eastern equatorial Pacific Ocean present large increases in SST and SSH. The latter patterns are coherent with the observed influence of Central Pacific El Niño events (Ashok and Yamagata 2009, Gierach et al. 2012; Table 1), showing enhanced easterly trade winds in the east and westerlies in the west, which push warmer, nutrient-poor waters to the central-western Equatorial Pacific (Fig. 6c, d), and cause a deepening of the thermocline, enhanced stratification, a subsequent decrease in chlorophyll and primary production (Fig. 4), and a delay and shortening of the timing of initiation and duration of the phytoplankton growing period, respectively (Fig. 3). It is further noteworthy that the phytoplankton response is not limited to this "wind convergence" region, but rather extends across the equatorial Pacific from the central-western basin to the eastern basin. This is because the westerly wind anomalies in the western-central Pacific cause equatorial Ekman convergence, increasing sea level (Fig. 6c, d) and deepening the thermocline (Palanisamy et al. 2015). These signals propagate eastward as equatorial Kelvin waves, overcoming the effects of upwelling favourable easterly wind anomalies in the eastern basin. The increased sea level and deepened thermocline signals weaken the mean upwelling in these regions, which limit the transport of nutrients to the surface mixed layer, and thus reduce the chlorophyll concentration and primary production.

In the tropical regions of the Atlantic Ocean, during positive phases of the MEI, patterns of increasing SST and SSH and enhanced Equatorial easterlies are observed in the east (Fig. 6a, c), bringing nutrient-poor waters to around 15°N (Lübbecke and McPhaden 2012), and causing a decrease in chlorophyll and primary production (Fig. 4), as well as a delay and shortening of the timing of initiation and duration of phytoplankton growth (Fig. 3). Although similar biological responses are observed towards higher latitudes in the North Atlantic (Figs. 3, 4), the underlying mechanism driving these responses is different: in higher latitudes, phytoplankton growth may be reduced and delayed, when water-column mixing is too high and light availability too low. During positive phases of the MEI over the period of study, enhanced cyclonic wind is observed in the North Atlantic (Fig. 6d), enhancing water divergence, which brings cooler-deeper water to the surface and decreasing SST and SSH (Fig. 6). In these conditions, water-column stratification is reduced, which subsequently delays phytoplankton growth, and decreases chlorophyll, primary production and duration of the growing period (Figs. 3, 4).

In the Indian Ocean, the ENSO-related changes estimated using annual means show increases in SST and SSH across the western and eastern basin (Fig. 6a, c), whereas based on the phenological mean, a weakly significant dipole pattern of decreasing SST in the west and increasing SST in the east is observed, together with a band of decreasing SSH along the equator (Fig. 6b, d). In the wind forcing responses, based on annual mean anomalies, the south tropical Indian Ocean is dominated by easterlies (Fig. 6c), whereas based on phenological mean anomalies, the forcing is characterised by strengthened easterly trade winds blowing both above and below the Equator (Fig. 6d). Over the south tropical Indian Ocean, increased SSH and SST appear in the western and central basin during positive ENSO phase (Fig. 6). This region is the mean upwelling zone of the Indian Ocean that is characterised by the thermocline ridge (i.e. TRIO region; Trenary and Han 2012). The high sea level (deepened thermocline) is primarily forced by easterly wind (Fig. 6c, d), which reduces the mean upwelling and thus reduce chlorophyll and primary production (Fig. 4a, c). In this basin, the phenology-based analysis appears to reveal coherent changes in wind forcing and responses of phytoplankton populations, which would have been more difficult to interpret from the analysis based on annual mean observations. Thus, phenological analysis can be suggested as a useful approach to help identify physical processes driving variations in phytoplankton populations in oceanic regions characterised by complex climate forcing (such as the monsoon in the Indian Ocean).

5 Conclusions

Phytoplankton responses to ENSO, noticeable in the global oceans but with marked regional differences, are characterised by variations in timing of growing period between ±30 days, chlorophyll concentration between ±20 % and primary production between ±10 %. Such variations may have profound impact on the carbon cycle (i.e. changes in export production) and the functioning of the marine ecosystem (i.e. trophic match/mismatch). The patterns of variations in duration of phytoplankton growing period are shown to co-vary with variations in chlorophyll concentration and with variations in primary production. These emergent properties between ocean-colour metrics based on relative patterns (i.e. duration) and absolute changes in concentration and production rates (i.e. chlorophyll and primary production) may provide a relevant and alternative approach to support comparisons of ocean-colour products estimated from different sensors for which inter-calibration and bias correction processes cannot be carried out.

The analysis of phenology-based estimates of SST, SSH and wind data is shown to help refine the evaluation and understanding of the mechanisms of impact of ENSO on phytoplankton inter-annual variability at global and regional scales. The present overview highlights that in oceanic regions where ENSO may influence a climate system tied to annual variations in the solar cycle, such as in high latitudes, annual means of physical variables may be useful metrics to understand the mechanisms driving the regional biological variability. However, in oceanic regions where ENSO influences a climate system characterised by a seasonal reversal of the wind forcing, such as the monsoon in the Indian Ocean, which can drive phytoplankton responses that are equal in strength but opposite in direction, then the estimation of phenology-based mean of physical variables may be necessary to evaluate the mechanisms driving the biological responses and provide a more comprehensive understanding of the integrated processes. Thus, phenological studies on ENSO impact have broader implications for climate research: above and beyond understanding biological responses to shorter-term climate variability, they may be used to help us improve predictions on the impact of climate change on the marine ecosystem.

Acknowledgements This paper is an outcome of the ISSI Workshop "Integrative Study of Sea Level Budget," held in Bern in February 2015. This work is a contribution to the European Space Agency Ocean Colour Climate Change Initiative and Living Planet Fellowship Programme, to the FP7 EC INDO-MARECLIM project and to the NERC's UK National Centre for Earth Observation. The authors thank the ESA CCI teams for providing OC-CCI chlorophyll data, SL-CCI sea level data and NCEO-ESA-SST-CCI sea surface temperature data. The authors further acknowledge TWAP for providing primary production data. The authors would like to acknowledge James Dingle, Thomas Jackson and the NERC Earth Observation Data Acquisition and Analysis Service (NEODAAS) for support with OC datasets. We acknowledge the two reviewers for providing constructive and insightful comments on our manuscript.

References

Ablain M, Cazenave A, Larnicol G, Balmaseda M, Cipollini P, Faugère Y, Fernandes MJ, Henry O, Johannessen JA, Knudsen P, Andersen O, Legeais J, Meyssignac B, Picot N, Roca M, Rudenko S, Scharffenberg MG, Stammer D, Timms G, Benveniste J (2015) Improved sea level record over the satellite altimetry era (1993–2010) from the climate change initiative project. Ocean Sci 11:67–82

Ashok K, Yamagata T (2009) The El Niño with a difference. Nature 461:481–484

Ashok K, Behera SK, Rao SA, Weng H, Yamagata T (2007) El Niño Modoki and its possible teleconnection. J Geophys Res 112:C11007

Behrenfeld MJ, Randerson JT, McClain CR, Feldman GC, Los SO, Tucker CJ, Falkowski PG, Field CB, Frouin R, Esaias WE, Kolber DD, Pollack NH (2001) Biospheric primary production during an ENSO transition. Science 291:2594–2595

Behrenfeld MJ, O'Malley RT, Siegel DA, McClain CR, Sarmiento JL, Feldman GC, Milligan AJ, Falkowski PG, Letelier RM, Boss MS (2006) Climate-driven trends in contemporary ocean productivity. Nature 444:752–755

Brewin RJW, Hirata T, Hardman-Mountford NJ, Lavender SJ, Sathyendranatha S, Barlow R (2012) The influence of the Indian ocean dipole on inter annual variations in phytoplankton size structure as revealed by earth observation. Deep Sea Res Part II 77–80:117–127

Brody SR, Lozier MS, Dunne JP (2013) A comparison of methods to determine phytoplankton bloom initiation. J Geophys Res 118:1–13

Cai W, Santoso A, Wang G, Yeh S, An S, Cobb KM, Collins M, Guilyardi E, Jin F, Kug J, Lengaigne M, McPhaden MJ, Takahashi K, Timmermann A, Vecchi G, Watanabe M, Wu L (2015) ENSO and greenhouse warming. Nat Clim Change 5:849–859

Capotondi et al (2015) Understanding ENSO diversity. Am Meteorol Soc. doi:10.1175/BAMS-D-13-00117.1

Carranza MM, Gille ST (2015) Southern Ocean wind-driven entrainment enhances satellite chlorophyll-a through the summer. J Geophys Res: Oceans 120:304–323

Chavez FP, Messié M, Pennington JT (2011) Marine primary production in relation to climate variability and change. Ann Rev Mar Sci 3:227–260

Cole HS, Henson S, Martin AP, Yool A (2015) Basin-wide mechanisms for spring bloom initiation: how typical is the North Atlantic? ICES J Mar Sci 72:2029–2040

Currie JC, Lengaigne M, Vialard J, Kaplan DM, Aumont O, Naqvi SWA, Maury O (2013) Indian Ocean dipole and El Niño/southern oscillation impacts on regional chlorophyll anomalies in the Indian Ocean. Biogeosciences 10:6677–6698

Di Lorenzo E, Schneider N, Cobb KM, Franks PJS, Chhak K, Miller AJ, McWilliams JC, Bograd SJ, Arango H, Curchitser E, Powell TM, Rivière P (2008) North Pacific Gyre Oscillation links ocean climate and ecosystem change. Geophys Res Lett 35:L08607

Edwards M, Richardson AJ (2004) Impact of climate change on marine pelagic phenology and trophic mismatch. Nature 430:881–884

Fu L-L, Cazenave A (eds) (2001) Satellite altimetry and earth sciences: a handbook of techniques and applications. Academic Press, San Diego, p 463

Gierach MM, Lee T, Turk D, McPhaden MJ (2012) Biological response to the 1997–98 and 2009–10 El Niño events in the equatorial Pacific Ocean. Geophys Res Lett 39:L10602

González Taboada F, Anadón R (2014) Seasonality of North Atlantic phytoplankton from space: impact of environmental forcing on a changing phenology (1998–2012). Glob Change Biol. doi:10.1111/gcb.12352

Gregg WW, Casey NW (2007) Sampling biases in MODIS and SeaWiFS ocean chlorophyll data. Remote Sens Environ 111:25–35

Gregg WW, Rousseaux CS (2014) Decadal trends in global pelagic ocean chlorophyll: a new assessment integrating multiple satellites, in situ data, and models. J Geophys Res. doi:10.1002/2014JC010158

Henson SA, Dunne JP, Sarmiento JL (2009) Decadal variability in North Atlantic phytoplankton blooms. J Geophys Res 114:C04013

Jackson T, Bouman HA, Sathyendranath S, Devred E (2011) Regional-scale changes in diatom distribution in the Humboldt upwelling system as revealed by remote sensing: implications for fisheries. ICES J Mar Sci 68:729–736

Kao H-Y, Yu J-Y (2009) Contrasting Eastern-Pacific and Central-Pacific types of ENSO. J Clim 22:615–632

Koeller P, Fuentes-Yaco C, Platt T, Sathyendranath S, Richards A, Ouellet P, Orr D, Skúladóttir U, Wieland K, Savard L, Aschan M (2009) Basin-scale coherence in phenology of shrimps and phytoplankton in the North Atlantic Ocean. Science 324:791–793

Kug J-S, Jin F-F, An S-I (2009) Two types of El Niño events: cold tongue El Niño and warm pool El Niño. Am Meteorol Soc. doi:10.1175/2008JCLI2624.1

Larkin NK, Harrison DE (2005) On the definition of El Niño and associated seasonal average US weather anomalies. Geophys Res Lett. doi:10.1029/2005GL022738

Lee SK, Enfield DB, Wang C (2008) Why do some El Niños have no impact on tropical North Atlantic SST? Geophys Res Lett 35:L16705

Longhurst A (1998) Ecological geography of the sea. Academic Press, California

Lo-Yat A, Simpson SD, Meekan M, Lecchini D, Martinez E, Glazin R (2011) Extreme climatic events reduce ocean productivity and larval supply in a tropical reef ecosystem. Glob Change Biol 17:1695–1702

Lübbecke JF, McPhaden MJ (2012) On the inconsistent relationship between Pacific and Atlantic Niños. J Clim 25:4294–4303

Martinez E, Antoine D, D'Ortenzio F, Gentili B (2009) Climate-driven basin-scale decadal oscillations of oceanic phytoplankton. Science 326:1253–1256

Martinez-Urtaza J, Trinanes J, Gonzalez-Escalona N, Baker-Austin C (2016) Is El Niño a long-distance corridor for waterborne disease? Nat Microbiol 1:1–3

Masotti I, Moulin C, Alvain S, Bopp L, Antoine D (2011) Large scale shifts in phytoplankton groups in the Equatorial Pacific during ENSO cycles. Biogeosciences 8:539–550

McPhaden MJ, Zebiak SE, Glantz MH (2006) ENSO as an integrating concept in earth science. Science 314:1740–1745

Merchant CJ, Embury O, Roberts-Jones J, Fiedler E, Bulgin CE, Corlett GK, Good S, McLaren A, Rayner N, Morak-Bozzo S, Donlon C (2014) Sea surface temperature datasets for climate applications from Phase 1 of the European Space Agency Climate Change Initiative (SST CCI). Geosci Data J 1:179–191

Messié M, Chavez FP (2012) A global analysis of ENSO synchrony: the oceans' biological response to physical forcing. J Geophys Res 117:C09001

Palanisamy H, Cazenave A, Delcroix T, Meyssignac B (2015) Spatial trend patterns in the Pacific Ocean sea level during the altimetry era: the contribution of thermocline depth change and internal climate variability. Ocean Dyn. doi:10.1007/s10236-014-0805-7

Platt T, Sathyendranath S (1988) Oceanic primary production: estimation by remote sensing at local and regional scales. Science 241:1613–1620

Platt T, Sathyendranath S (1996) Modelling primary production? Aquabiology 18:378–380

Platt T, Sathyendranath S (2008) Ecological indicators for the pelagic zone of the ocean from remote sensing. Remote Sens Environ 112:3426–3436

Platt T, Fuentes-Yaco C, Frank K (2003) Spring algal bloom and larval fish survival. Nature 423:398–399

Platt T, White III, Zhai L, Sathyendranath S, Roy S (2009) The phenology of phytoplankton blooms: ecosystem indicators from remote sensing. Ecol Model 220:3057–3069

Racault M-F, Le Quéré C, Buitenhuis E, Sathyendranath S, Platt T (2012) Phytoplankton phenology in the global ocean. Ecol Indic 14:152–163

Racault MF, Platt T, Sathyendranath S, Ağirba E, Martinez Vicente V, Brewin R (2014) Plankton indicators and ocean observing systems: support to the marine ecosystem state assessment. J Plankton Res. doi:10.1093/plankt/fbu016

Racault MF, Raitsos DE, Berumen ML, Brewin RJW, Platt T, Sathyendranath S, Hoteit I (2015) Phytoplankton phenology indices in coral reef ecosystems: application to ocean-color observations in the Red Sea. Remote Sens Environ 160:222–234

Radenac MH, Léger F, Singh A, Delcroix T (2012) Sea surface chlorophyll signature in the tropical Pacific during eastern and central Pacific ENSO events. J Geophys Res 117:C04007

Raitsos DE, Yi X, Platt T, Racault MF, Brewin RJW, Pradhan Y, Papadopoulos PV, Sathyendranath S, Hoteit I (2015) Monsoon oscillations regulate fertility of the Red Sea. Geophys Res Lett 42:855–862

Saji NH, Goswami BN, Vinayachandran PN, Yamagata T (1999) A dipole mode in the tropical Indian Ocean. Nature 401:360–363

Sasaoka K, Chiba S, Saino T (2011) Climatic forcing and phytoplankton phenology over the subarctic North Pacific from 1998 to 2006, as observed from ocean color data. Geophys Res Lett. doi:10.1029/2011GL048299

Sathyendranath S, Brewin RJW, Brockmann C, Brotas V, Ciavatta S, Chuprin A, Couto AB, Doerffer R, Dowell M, Grant M, Groom S, Horseman A, Jackson T, Krasemann H, Lavender S, Martinez Vicente V, Mélin Moore TS, Müller D, Regner P, Roy S, Steinmetz F, Swinton J, Taberner M, Thompson A, Valente A, Zühlke M, Brando VE, Feldman G, Franz B, Frouin R, Gould Jr RW, Hooker S, Kahru M, Mitchell MG, Muller-Karger F, Sosik HM, Voss KJ, Werdell J, Platt T (2016) Creating an ocean-colour time series for use in climate studies: the experience of the ocean-colour climate change initiative. (Manuscript)

Siegel DA, Doney SC, Yoder JA (2002) The North Atlantic spring phytoplankton bloom and Sverdrup's critical depth hypothesis. Science 296:730–733

Soppa MA, Volker C, Astrid B (2016) Diatom phenology in the Southern Ocean: mean patterns, trends and the role of climate oscillations. Remote Sens. doi:10.3390/rs8050420

Steinmetz F, Deschamps PY, Ramon D (2011) Atmospheric correction in presence of sun glint: application to MERIS. Opt Expr 19:9783–9800

Thomalla SJ, Fauchereau N, Swart S, Monteiro PMS (2011) Regional scale characteristics of the seasonal cycle of chlorophyll in the Southern Ocean. Biogeosciences 8:2849–2866

Trenary LL, Han W (2012) Intraseasonal-to-interannual variability of south indian ocean sea level and thermocline: remote versus local forcing. J Phys Oceanogr 42:602–627

Turk D, McPhaden MJ, Busalacchi AJ, Lewis M (2001) Remotely sensed biological production in the equatorial Pacific. Science 293:471–474

Vantrepotte V, Mélin F (2011) Inter-annual variations in the SeaWiFS global chlorophyll a concentration (1997–2007). Deep Sea Res Part I Oceanogr Res Pap 58:429–441

Wolter K, Timlin MS (1993) Monitoring ENSO in COADS with a seasonally adjusted principal component index. In: Proceedings of 17th climate diagnostics workshop (Norman, Oklahoma) 52–57 (NOAA/NMC/CAC, NSSL, Oklahoma Climate Survey, CIMMS and the School of Meteorology, Univ. Oklahoma, 1993)

Yoder J, Kennelly M (2003) Seasonal and ENSO variability in global ocean phytoplankton chlorophyll derived from 4 years of SeaWiFS measurements. Glob Biogeochem Cycles. doi:10.1029/2002GB001942

Yu JY, Zou Y, Kim ST, Lee T (2012) The changing impact of El Niño on US winter temperatures. Geophys Res Lett 39:L15702

Part III
Sea Level Closure Budget at Global and Regional Scale

Surv Geophys (2017) 38:295–307
DOI 10.1007/s10712-016-9405-z

The Twentieth-Century Sea Level Budget: Recent Progress and Challenges

S. Jevrejeva[1] · A. Matthews[1] · A. Slangen[2]

Received: 5 January 2016 / Accepted: 25 November 2016 / Published online: 8 December 2016

Abstract For coastal areas, given the large and growing concentration of population and economic activity, as well as the importance of coastal ecosystems, sea level rise is one of the most damaging aspects of the warming climate. Huge progress in quantifying the cause of sea level rise and closure of sea level budget for the period since the 1990s has been made mainly due to the development of the global observing system for sea level components and total sea levels. We suggest that a large spread (1.2 ± 0.2–1.9 ± 0.3 mm year^{-1}) in estimates of sea level rise during the twentieth century from several reconstructions demonstrates the need for and importance of the rescue of historical observations from tide gauges, with a focus on the beginning of the twentieth century. Understanding the physical mechanisms contributing to sea level rise and controlling the variability of sea level over the past few 100 years are a challenging task. In this study, we provide an overview of the progress in understanding the cause of sea level rise during the twentieth century and highlight the main challenges facing the interdisciplinary sea level community in understanding the complex nature of sea level changes.

Keywords Sea level rise · Sea level budget · Observing system · Data archeology

1 Introduction

For delicate coastal ecosystems, small islands and fast-growing coastal cities (Hallegate et al. 2013; Jevrejeva et al. 2014), sea level rise is one of the most dangerous aspects of climate change (IPCC 2013). Global sea level rise is an integral measure of warming

✉ S. Jevrejeva
sveta@noc.ac.uk

[1] National Oceanography Centre, Joseph Proudman building, 6 Brownlow Street, Liverpool L3 5DA, UK

[2] Institute for Marine and Atmospheric Research Utrecht (IMAU), Utrecht University, Princetonplein 5, 3584 CC Utrecht, The Netherlands

climate (Munk 2002; Church et al. 2013; Jevrejeva et al. 2010), reflecting alterations in the dynamics and thermodynamics of the atmosphere, ocean and cryosphere as a response to changes in radiative forcing. Understanding the physical mechanisms contributing to sea level rise and controlling the variability of sea level over the past few 100 years are a challenging task (Munk 2002; Church et al. 2013). The primary climate-related contributors to twentieth-century sea level rise are ice loss of land-based glaciers and ice sheets in Greenland and Antarctica, and thermal expansion of the oceans (Church et al. 2013). In addition, there is a non-climatic contributor—changes in water storage on land due to groundwater mining and the construction of reservoirs (Church et al. 2013). However, the relative contributions from these components to twentieth-century sea level rise are still not well understood, and the closure of sea level budget is a subject of debate (Bindoff et al. 2007; Church et al. 2013; Gregory et al. 2013; Moore et al. 2011; Jevrejeva et al. 2008, 2012).

There are two main methods of estimating global-mean sea level rise. Firstly, an estimate can be obtained by adding together the cumulative effect of the main contributors to sea level rise: melting of ice in glaciers, ice loss from the Greenland and the Antarctic ice sheets, thermal expansion and changes in land water storage. Secondly, global sea level rise can be estimated using observations from tide gauges, complemented since 1993 with satellite altimeter measurements. If these two estimates agree (within an uncertainty range), then we call the sea level budget closed.

The main motivation to improve our understanding of the twentieth-century sea level budget is described by Munk (2002), suggesting that "…Sea level is important as a metric for climate change as well as in its own right. We are in the uncomfortable position of extrapolating into the next century without understanding the past." Quantifying the cause of past sea level rise is important for future sea level rise projections as the conventional approach to project sea level rise is based on simulation of individual sea level components, such as ocean thermal expansion and ice mass loss from glaciers and the ice sheets, and then sum them up (Meehl et al. 2007; Church et al. 2013).

Over the past 10–20 years, the sea level community has made huge progress in understanding present-day sea level rise, mainly due to unique information about changes in global and regional sea levels from space missions (Cazenave and Nerem 2004; Cazenave et al. 2009; Leuliette and Scharroo 2010; Cazenave and Llovel 2010; Cazenave et al. 2014). Since 1992, satellite altimetry measurements have provided a continuous and near-global record of modern-day sea level change, suggesting the rate of 3.2 ± 0.4 mm year^{-1} global sea level rise for the period 1993–2012 (Cazenave et al. 2012; Boening et al. 2012), which notably exceeds the estimate of 1.7 [1.5 1.9] mm year^{-1} sea level rise for the twentieth century (Church et al. 2013). In 2002, a pair of satellites, called the Gravity Recovery and Climate Experiment (GRACE), were launched to make monthly observations of changes in Earth's gravity field, providing estimates of mass loss from ice sheets in Greenland, Antarctica, and glaciers (Shepherd et al. 2012; Jacob et al. 2012) and tracking water mass movements at unique spatial scales (Leuliette and Willis 2011; Church et al. 2013). Data from ICESat (Ice, Cloud, and land Elevation Satellite), a satellite mission for measuring glaciers and ice sheet mass balance during the period 2003–2009, contributed to our understanding of ice mass changes from the cryosphere and its contribution to sea level rise (Schutz et al. 2005; Neckel et al. 2014). In addition, the development in the ARGO network (a series of autonomous floats that sink and ascend, monitoring temperature and salinity in the top 1000–2000 m of the ocean, with more than 3000 floats since 2000) has contributed to improved understanding of the ocean role in sea level rise (von Schuckmann and Le Traon 2011). These simultaneous measurements from satellite altimetry, GRACE and ARGO provide observational constraints on closure of the sea level budget for the period since 2003 (Dieng et al. 2015; Cazenave et al. 2014; Boening et al. 2012). In terms of global averages, the sum of global ocean mass from GRACE and global thermosteric sea level change from ARGO is roughly equal, within uncertainties, to the total sea level change observed by satellite altimetry (Dieng et al. 2015; Cazenave et al. 2014; Boening et al. 2012).

However, sea level budget cannot be assessed the same way for the twentieth century, mainly due to the lack of observational data sets for individual sea level components. There is no shortage of excellent publications (e.g., Cazenave et al. 2014) and review papers (Dieng et al. 2015; Leuliette and Willis 2011) on the topic of sea level budget during the period since 2003; however, there are only a limited number of studies (e.g., Gregory et al. 2013; Jevrejeva et al. 2012; Moore et al. 2011) about sea level budget during the twentieth century.

In this paper, we present a summary of the progress in understanding of the cause of sea level rise during the twentieth century and highlight the main challenges facing the interdisciplinary sea level community in understanding the complex nature of sea level changes.

2 Progress in Understanding the Twentieth-Century Sea Level Budget

Since the 1990s each of the Intergovernmental Panel on Climate Change (IPCC) reports has produced an assessment of the twentieth-century sea level rise. The First Assessment Report (FAR) (Houghton et al. 1990) provided the foundation for our current understanding of sea level change. FAR concluded that there was observational evidence that sea level had risen at an average rate of 1.0–2.0 mm year^{-1} during the twentieth century, and that the rate had increased compared to the eighteenth and nineteenth centuries. The causal factors that could explain the twentieth-century sea level rise were ocean thermal expansion and ice mass loss from glaciers and the margins of the Greenland ice sheet (Warrick and Oerlemans 1990). The sea level budget for the twentieth century from FAR is presented in Table 1.

The Second Assessment Report (SAR) introduced additional contributions from surface water and groundwater storage, which was labeled as "very uncertain and speculative" (Warrick et al. 1996). In addition, SAR discussed a contribution from ice sheets suggesting that there was "simply insufficient evidence, either from models or from data, to say whether the average mass balances have been positive or negative" (Warrick et al. 1996). SAR concluded that the difficulty in reconciling the past change in sea level components emphasizes the uncertainties in projections of future sea level rise.

Table 1 Estimated contributors to sea level rise over the twentieth century (in cm) from FAR (Warrick and Oerlemans 1990; based on Table 9.8 in the FAR)

Contributor	Low	Best estimate	High
Thermal expansion	2	4	6
Glaciers	1.5	4	7
Greenland ice sheet	1	2.5	4
Antarctic ice sheet	−5	0	5
Total	−0.5	10.5	22
Observed	10	15	20

The global sea level budget for 1910–1990 was analyzed in the Third Assessment Report (TAR) (Church et al. 2001) and for the period 1961–2003 in the Fourth Assessment Report (AR4) (Bindoff et al. 2007), where the individual contributions summed to less than the observed rate of sea level rise. For example, the AR4 assessed the mean observational rate for 1961–2003 as 1.8 ± 0.5 mm year^{-1}, and the sum of the budget terms as 1.1 ± 0.5 mm year^{-1} (Bindoff et al. 2007; Hegerl et al. 2007). However, the large uncertainties in estimates of the contributions of individual components and total sea level demonstrate the difficulties in closing the sea level budget.

The Fifth Assessment Report (AR5) concluded that the observational sea level budget cannot be rigorously assessed for 1901–1990 or 1971–2010 (Table 2), due to insufficient observational information to estimate ice sheet contributions with high confidence before the 1990s, and in addition ocean data sampling is too sparse to permit an estimate of global-mean thermal expansion before the 1970s (Church et al. 2013).

Several publications about the sea level budget for the historical time period are focused on the second part of the twentieth century. In a study by Jevrejeva et al. (2008) the sea level budget for the period 1955–2003 was analyzed, and the observed sea level rise rate of 1.6 mm year^{-1} was partially explained by the 0.41 mm year^{-1} contribution from thermal expansion and 0.75 mm year^{-1} due to ice loss from glaciers and ice sheets in Greenland and Antarctica, suggesting that 25% of the sea level rise (0.44 mm year^{-1}) was associated with the so-called unexplained contribution. That unexplained component was described as a combination of a long-term trend and variability that was likely caused by underestimating the contribution from ice masses (the linear trend component) and decadal variability associated with the hydrological cycle and changes in continental water storage contribution (Jevrejeva et al. 2008). Domingues et al. (2008) presented a sea level budget for 1963–2003 with an improved estimate of the contribution from upper-ocean thermal expansion and suggested a possible contribution from a deep-ocean component. The sum of contributors 1.5 ± 0.4 mm year^{-1} was in good agreement with the 1.6 ± 0.2 mm year^{-1} estimate of global sea level rise from Church and White (2006). For the second part of the twentieth century several studies demonstrated that the sea level budget could be closed by climate-related contributors, assuming some contribution from ice sheets, in a study by Moore et al. (2011), or with a small −0.1 ± 0.2 mm year^{-1} contribution from change in land water storage (Church et al. 2011). Nevertheless, large uncertainties in sea level components and in observed total sea level still remain (Church et al. 2013; Gregory et al. 2013) largely due to the lack of observational data to estimate contributions from the Greenland and Antarctica ice sheets.

Gregory et al. (2013) published an overview of estimates of individual sea level contributions over the twentieth century, partly data-based and partly model-based, and sea level rise estimates from several global sea level reconstructions. The range of possible sea levels obtained by combining all individual estimates in various combinations (total 144 combinations) suggested that the observed sea levels lie at the very edge of the range and a residual trend is needed to make up for the discrepancy, selecting the largest or smallest estimates for individual contributors. Gregory et al. (2013) concluded that if the residual trend can be interpreted as a long-term Antarctic contribution, an ongoing response to climate change over previous millennia, the budget can be satisfactorily closed. Arguably, results from Gregory et al. (2013) demonstrate that the only possibility to close the sea level budget is to select the most sensitive models and the largest individual estimates.

Recently published results by Slangen et al. (2016) suggest that for the period 1900–2005 the sum of modeled contributors of sea level rise (125.22 ± 21.97 mm) agreed with observed ensemble (174 ± 71 mm) within 2σ uncertainties (Table 2), implying

Table 2 Global-mean sea level budget (mm year^{-1}) over the twentieth century (two time intervals) from observations and model-based contributions, based on Table 13.1 in Church et al. (2013) with updated estimates from the recent publications by Slangen et al. (2016), Marzeion et al. (2015) and Hay et al. (2015)

Source	1901–1990 (Church et al. 2013)	1971–2010 (Church et al. 2013)	1900–2005 (Slangen et al. 2016)	1902–2005 (Marzeion et al. 2015)
Observed contributions				
Thermal expansion	–	0.8 [0.5 1.1]		
Glaciers except in Greenland and Antarctica	0.54 [0.47 0.61]	0.62 [0.25 0.99]		80.4 ± 21.1 mm[b] 63.2 ± 7.9 mm[c]
Glaciers in Greenland	0.15 [0.10 0.19]	0.06 [0.03 0.09]		
Greenland ice sheet	–	–		
Antarctic ice sheet	–	–		
Land water storage	−0.11 [−0.16 −0.06]	0.12 [0.03 0.22]		
Total contributors	–	–		
Observed sea level rise	1.5 [1.3 1.7] 1.2 [1.0 1.4][d]	2.0 [1.7 2.3]	174 ± 71 mm	
Modeled contributions				
Thermal expansion	0.37 [0.06 0.67]	0.96 [0.51 1.41]	36.7 ± 18.8 mm	
Glaciers except in Greenland and Antarctica	0.63 [0.37 0.89]	0.62 [0.41 0.84]	69.6 ± 7.1 mm	
Glaciers in Greenland	0.07 [−0.02 0.16]	0.10 [0.05 0.15]		
Greenland ice sheet			14.1 ± 2.9 mm	
Antarctic ice sheet			7.8 ± 8.8 mm	
Total including land water storage	1.0 [0.5 1.4]	1.8 [1.3 2.4]	125.2 ± 22.0 mm[e]	
Residual[a]	0.5 [0.1 1.0]	0.2 [−0.4 0.8]		

Uncertainties in brackets are 5–95%

[a] Observed GMSL rise- modeled thermal expansion- modeled glaciers- observed land water storage (see Church et al. 2013, Table 13.1)

[b] Estimate for global integrated glacier mass change reconstructions (excluding Antarctic periphery) updated from Leclercq et al. (2011)

[c] Estimate for global integrated glacier mass change reconstructions (excluding Antarctic periphery) updated from Marzeion et al. (2012)

[d] Estimate of 1.2 ± 0.2 mm year^{-1} for total sea level rise from 1901 to 1990 (Hay et al. 2015)

[e] Total including ice sheet/deep-ocean contributions of 13.8 ± 23.7 mm

progress in closing the twentieth-century budget. However, the sum of best estimates is still smaller than the observed rise. Of the four time series used to construct the observation ensemble (Church and White 2011; Ray and Douglas 2011; Jevrejeva et al. 2014; Hay et al. 2015), three are within the modeled range and only the largest observed sea level rise (Jevrejeva et al. 2014) is outside the modeled range (Slangen et al. 2016).

Studies of historical sea level budget by Moore et al. (2011) and Jevrejeva et al. (2012) and the sea level budget over the twentieth century by Gregory et al. (2013) demonstrated that progress has been made toward accounting for the long-term sea level changes. A study by Mitrovica et al. (2015) has demonstrated that use of the lowest estimates of sea level components of the twentieth century obtained from the AR5 IPCC report (Church et al. 2013), improved modeling of the GIA process and the correction of the eclipse record for a signal due to angular momentum exchange between the fluid outer core and the mantle reconciles all three Earth rotation observations discussed in Munk (2002) as an enigma. Nevertheless, there is a substantial gap in our knowledge about the contribution from the main components to the twentieth-century sea level rise, in particular for the first half of the century. All these studies acknowledged that there are still large uncertainties in estimates of global sea level rise, its components and how these components relate to climate forcing.

3 Challenges to Improve the Historical Records

It might never be possible to determine contributions from sea level components to the twentieth-century sea level rise to the same accuracy as has been archived for the past 10–20 years. However, it remains important to understand better the magnitude and uncertainties of the physical processes that contributed to sea level rise and variability during the twentieth century.

One of the challenges is to explain the observed temporal and spatial variability in sea level records from tide gauges, which provide instrumental data prior to the satellite altimetry and are widely used to estimate global sea level rise during the twentieth century. Individual tide gauge observations (Douglas 1997), global sea level reconstructions using tide gauge data (Gornitz et al. 1982; Jevrejeva et al. 2006; Grinsted et al. 2007; Jevrejeva et al. 2008; Merrifield et al. 2009; Wenzel and Schroter 2010; Ray and Douglas 2011; Jevrejeva et al. 2014), reconstructions that jointly use satellite altimetry and tide gauge records (Church and White 2006, 2011) and a reconstruction which combines tide gauge records with physics-based and model-derived geometries of the contributing processes (Hay et al. 2015) provide a wide range (from 1.2 ± 0.2 to 1.9 ± 0.3 mm year$^{-1)}$ of estimates of global sea level rise during the twentieth century. There is a good agreement between estimates of global sea level rise from several sea level reconstructions (Church and White 2011; Jevrejeva et al. 2008; Hay et al. 2015) for the past 60 years shown in Hay et al. (2015); however, considerable differences are demonstrated for the first part of the twentieth century. Large differences in global sea level rise estimates could be explained by the use of different methods, selection of different tide gauges and choice of vertical land movement corrections (Jevrejeva et al. 2014; Hay et al. 2015; Hamlington and Thompson 2015; Thompson et al. 2016). It seems that in time with sufficient enough coverage of tide gauge data, for example during the last 20 years, all reconstructions are in good agreement with estimates from satellite altimetry, e.g., the rate of 3.1 ± 0.6 mm year^{-1} from tide gauge-based reconstruction is almost the same as 3.2 ± 0.4 mm year^{-1} calculated from satellite altimetry (Jevrejeva et al. 2014). In a study by Hamlington and Thompson (2015) the impact of tide gauge selection is explored by calculating global-mean trends using selected tide gauge data sets in recent sea level reconstructions by Church and White (2011), Ray and Douglas (2011) and Hay et al. (2015). The calculated trends over 1900–2013 from the original reconstructions were: 1.95 ± 0.24 mm year^{-1} for Church and White (2011), 1.82 ± 0.13 mm year^{-1} for Ray and Douglas (2011) and 1.34 ± 0.25 mm year^{-1} for Hay et al. (2015). However, the Hay et al. (2015) reconstruction included a large number of high-latitude stations. Hamlington and Thompson (2015) recalculated the reconstructions without tide gauges from Scandinavia, Alaska and the western coast of Canada for all reconstructions and, in addition, excluding the high-latitude gauges in Hay et al. (2015), resulting in estimates of 2.01 ± 0.12 mm year^{-1} (Ray and Douglas 2011), 2.12 ± 0.18 mm year^{-1} (Church and White 2011) and 2.13 ± 0.19 mm year^{-1} (Hay et al. 2015). This suggests that the differences between estimates of the twentieth-century sea level rise in these three studies are not entirely due to distinct methods, but at least partially due to the selection of tide gauge records. In addition, a recently published study using long tide gauge records concludes that it is highly unlikely that the rate of global average sea level rise was <1.4 mm year^{-1} during the twentieth century, while the most likely value was closer to 1.7 mm year^{-1} (Thompson et al. 2016).

Going back in time, estimates of sea level rise are based only on a limited number of tide gauges (Fig. 1), most of them located in Europe and North America, with only few tide gauges in Southern Hemisphere available from the 1900s (Holgate et al. 2013).

Figure 2 shows the last year of the data available from stations that existed during the period 1895–1905. Most of the long-term records overlapping with satellite altimetry observations are in Europe and USA, with only two records from the Southern Hemisphere. Our understanding of sea level rise for the first part of the twentieth century is based largely on the North American and European records with some information available from Australia, Argentina and New Zealand (www.psmsl.org).

Data archeology could improve spatial and temporal coverage of historical sea level observations, as many historical tide gauge data exist in non-digital form (Pouvreau 2008; Holgate et al. 2013; Caldwell 2012; Talke and Jay 2013; Bradshaw et al. 2015), mostly paper-based data sets. These data could contribute greatly to the extension of existing sea level records as far back as possible in order to permit a better understanding of the timescales of sea level rise and variability.

The location of the data uncovered by an extensive search of US and Canadian archives for North American and Pacific Tidal Data (Talke and Jay 2013; Caldwell 2012) and data held in French archives (Pouvreau 2008) are presented in Fig. 3. The color of the data point indicates the length of the record, and the shape indicates the earliest year of data found. It would take some time to digitize these data and make them available for the scientific community. Several publications with extended records for Marseille (Wöppelmann et al. 2014), Brest (Wöppelmann et al. 2008), Cadiz (Marcos et al. 2011) and particularly valuable records from Southern Hemisphere on Saint Paul Island in the Indian Ocean (Testut et al. 2010) have already contributed to our understanding of past sea level changes.

The sea level data archeology community is actively looking for improvements in technology, such as faster automated digitization of tide gauge charts and automatic

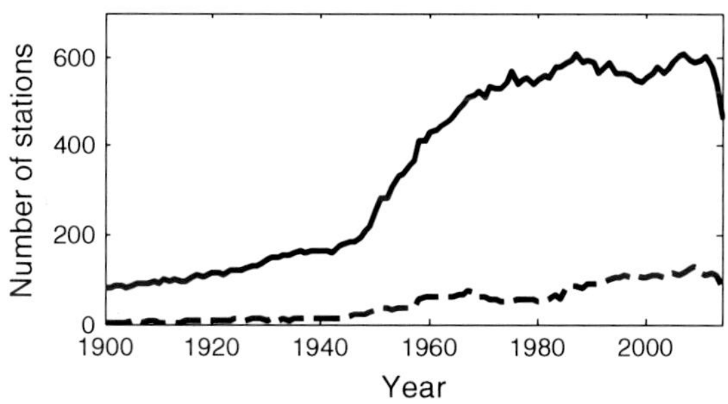

Fig. 1 Data availability as a function of time, number of tide gauges in the Northern Hemisphere represented by *solid line* and by *dashed line* in the Southern Hemisphere

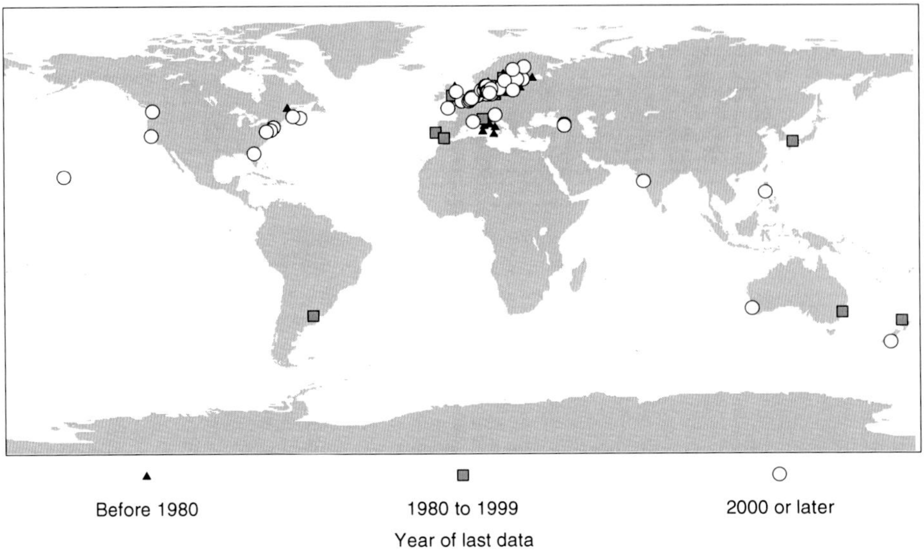

Fig. 2 Distribution of long-term research quality (RLR) records with sea level measurements started in 1900 (www.psmsl.org)

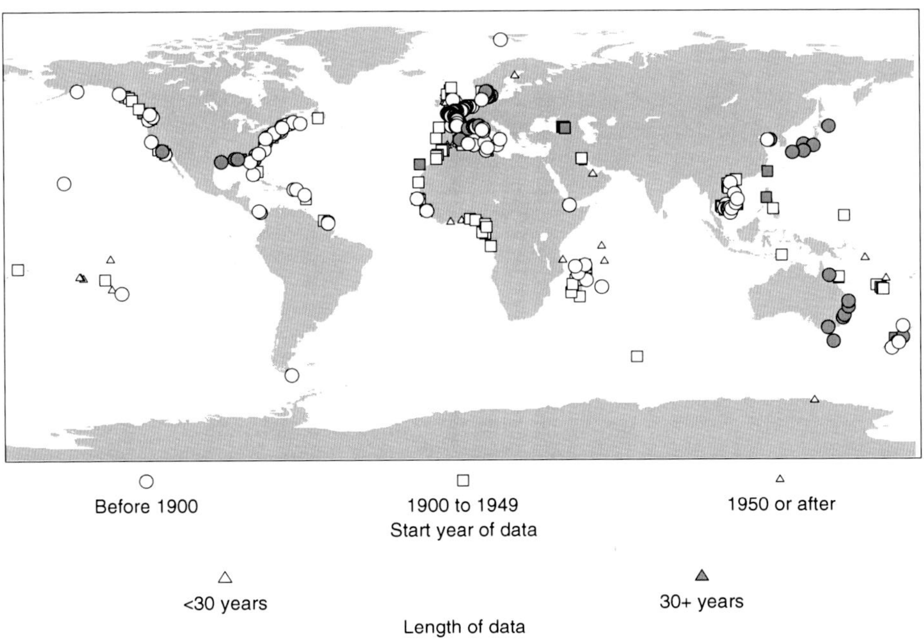

Fig. 3 Data uncovered by Caldwell (2012), Talke and Jay (2013) and Pouvreau (2008). The *color* of the *data point* indicates the length of the record, and the *shape* indicates the earliest year of data found

transcribing of handwritten ledgers. The Global Sea Level Observing System (GLOSS) Group of Experts (GE) is taking the first steps in coordinating the efforts and sharing the knowledge in sea level data archeology (Bradshaw et al. 2015).

The lack of long-term tide gauge records is not the only limitation of utilizing the historical data sets. Tide gauges are attached to the land, which can move vertically and introduce highly localized signals in tide gauge measurements (Douglas 1997; Holgate et al. 2013; Church et al. 2013; Wöppelmann and Marcos 2016; Thompson et al. 2016). One way to remove the impact of vertical land movement on estimates of global-mean sea level rise in the twentieth century is to measure the land motion component using the global positioning system (GPS) and remove it from tide gauge records. Estimates of vertical land movement from GPS have been used for individual tide gauge locations (Wöppelmann et al. 2009; Becker et al. 2012; King et al. 2012); however, the number of available GPS sites close to the tide gauge locations is limited to 100–300 globally (see Figure 1, in King et al. 2012), and most of the GPS sites are in Europe, North America and Japan. The polar regions, long coastal lines in South America, Africa, Southeast Asia, coastal areas of Indian Ocean and a large part of Australian coast are not covered by the GPS observations. GPS-derived vertical land movement corrections are available only for 10% of more than 1300 tide gauge records available for sea level studies from PSMSL (King et al. 2012; Jevrejeva et al. 2014). In addition, the rates of land motion from the global positioning system are obtained from a relatively short times series (<10 years), and these corrections might be less applicable in regions where the recent land motion might not represent that for the past 10–100 years. The lack of information about corrections for geophysical and anthropogenic signals over a range of spatial scales presented in tide gauge data (e.g., vertical land movement due to earthquakes, groundwater extraction and sedimentation) lead to the dismissal of some tide gauge records for estimates of long-term changes.

The long-term tide gauge records in Europe and North America in Fig. 2 are contaminated by the vertical land movement due to glacial isostatic adjustment (GIA). The selection of GIA corrections is important for these historical sea level records. Modeled GIA corrections are available for each tide gauge location and have been used in all sea level reconstructions (Church and White 2011; Ray and Douglas 2011; Jevrejeva et al. 2006, 2008, 2014; Hay et al. 2015; Wöppelmann and Marcos 2016; Thompson et al. 2016). Figure 4 shows the difference ranging from -4 up to 5 mm year^{-1} between the GIA corrections from ICE 6G and ICE 5G, and from ICE 6G and ICE 4G in more than 1000 tide gauge locations. Large uncertainties introduced by the choice of GIA corrections in the long-term trend for individual tide gauge records, regions and global reconstructions have been explored in studies by Jevrejeva et al. (2014) and Hay et al. (2015), and these uncertainties have been discussed as one of the challenges in the assessment of the twentieth-century sea level rise (Wöppelmann and Marcos 2016).

Unlike the time series available for steric sea level from CMIP5 model simulations and reconstructions of the contributions from mountain glaciers (Leclercq et al. 2011; Marzeion et al. 2012, 2015), the large ice sheets have no continuous extensive records or model simulations of ice mass loss during the twentieth century. One of the challenges to close the budget of the twentieth century is to estimate how much ice sheets in Greenland and Antarctica have contributed to the twentieth-century sea level rise.

The contribution to sea level rise from the Greenland ice sheet and its response to the climate forcing during the twentieth century remain contentious (Gregory et al. 2013; Church et al. 2013). Mitrovica et al. (2001) estimated a contribution of 0.6 mm year^{-1} from Greenland during the twentieth century by analyzing the regional pattern of global sea level rise from tide gauges in comparison with regional patterns expected from modeled ice mass loss from Antarctica and mountain glaciers. Using historical aerial images over the past 80 years Bjork et al. (2012) concluded that many land-terminated glaciers underwent a more rapid retreat in the 1930s than in the 2000s, with additional

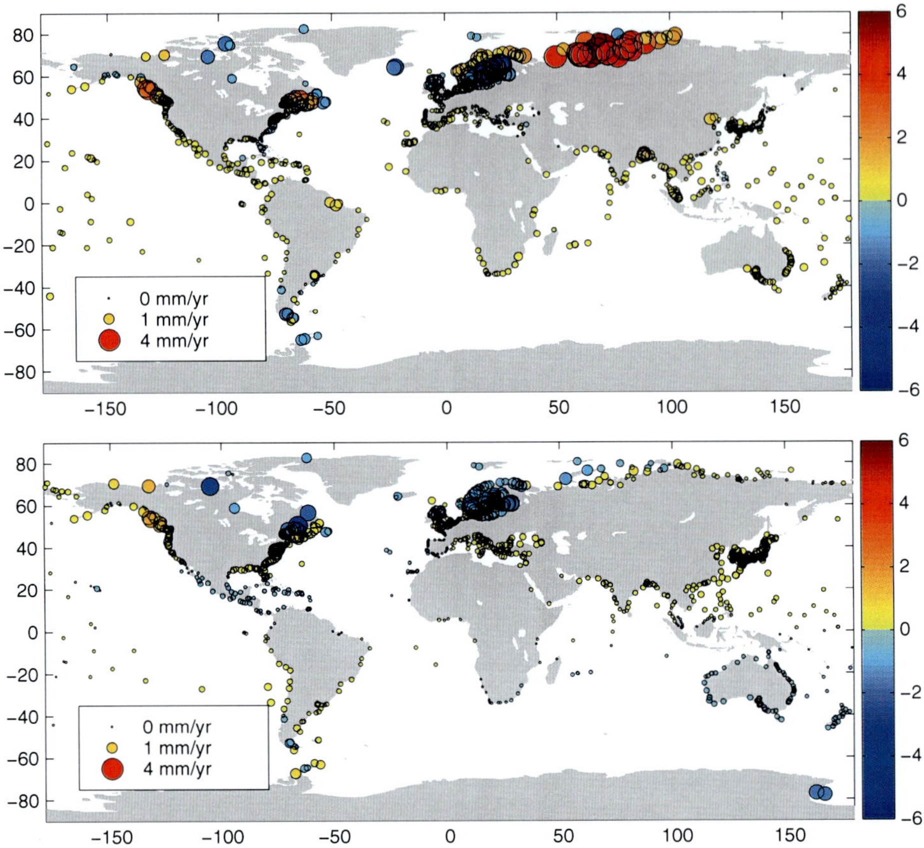

Fig. 4 Maps of differences between GIA corrections from ICE 6G and ICE 4G (*top*) and ICE 6G and ICE 5G (*bottom*) at individual locations of tide gauges. *Color bar* in mm year^{-1} and *circle size* in mm

contributions from marine-terminating glaciers and the ice sheet. Recently published observation-based findings (Kjeldsen et al. 2015) show the Greenland ice sheet contributed at least 25.0 ± 9.4 millimeters of global-mean sea level rise during the twentieth century, providing observation-based evidence of considerable mass loss from the Greenland ice sheet and minimizing the unexplained residuals in global sea level rise during the twentieth century. Continuous time series of mass loss from the Greenland ice sheet (Kjeldsen et al. 2015) contribute enormously to our gap in knowledge regarding the ice sheet response to the climate forcing.

4 Outlook

Much progress has been made over the past few decades in identifying the physical processes contributing to twentieth-century sea level rise, although large uncertainties remain. Tremendous steps forward in understanding the cause of sea level rise and closure of sea level budget for the period since the 1990s are mainly due to the development of a global observing system for sea level components and total sea levels. The observational data sets available through satellite monitoring have played a dominant role in the rate of progress.

It is crucial to maintain the current level of sea level observations (both satellite and in situ systems); longer time series and improvement of models will contribute to interpretation of sea level components and their changes as the planet continues to adjust to the warming climate.

For the twentieth-century budget and an understanding of the main contributors to global sea level rise, the remaining challenges require a coordinated and sustained multidisciplinary effort by oceanographers, geodesists, glaciologists, climate, ice sheet and solid Earth modelers to provide reliable estimates and realistic error bars for sea level and its main contributors. Understanding of the key processes contributing to the twentieth-century sea level rise and variability, such as the response of the ice sheets and glaciers to changing climate forcing, the role of the ocean (e.g., heat uptake by the ocean, role of deep ocean, ocean dynamics), the interaction between the ocean and ice sheets and the redistribution of ocean mass due to gravitational forcing are the main challenges for the decade to come.

Acknowledgements This paper is a result of the ISSI Workshop on Integrative Study of Sea Level, held in Bern, Switzerland, February 2–6, 2015. We would like to thank anonymous reviewers for helpful comments that improved our manuscript. This publication has received funding from the European Union's Seventh Programme for Research, Technological Development and Demonstration under Grant Agreement No: FP7-ENV-2013-Two-Stage-603396-RISES-AM. A. M. and S. J. partially supported by the Natural Environment Research Council National Capability funding. A. S. was supported by the NWO-Netherlands Polar Programme.

References

Becker M et al (2012) Sea level variations at tropical Pacific islands since 1950. Glob Planet Change 80–81:85–98

Bindoff NL et al (2007) Observations: oceanic climate change and sea level. In: Solomon S et al (eds) Climate change 2007: the physical science basis. Cambridge University Press, Cambridge, pp 385–432

Bjork et al (2012) An aerial view of 80 years of climate-related glacier fluctuations in southeast Greenland. Nat Geosci 5:427–432

Boening C et al (2012) The 2011 La Niña: so strong, the oceans fell. Geophys Res Lett 39:L19602. doi:10.1029/2012GL053055

Bradshaw E et al (2015) Sea level data archaeology and the Global Sea Level Observing System (GLOSS). GeoResJ 6:9–16

Caldwell P (2012) Tide gauge data rescue. In: Duranti L, Shaffe E (eds) Proceedings of the memory of the world in the digital age: digitization and preservation. Vancouver 2012, pp 134–149

Cazenave A, Nerem RS (2004) Present-day sea level change: observations and causes. Rev Geophys 42:RG3001. doi:10.1029/2003RG000139

Cazenave A, Llovel W (2010) Contemporary sea level rise. Annu Rev Mar Sci 2:145–173

Cazenave A et al (2009) Sea level budget over 2003–2008: a reevaluation from GRACE space gravimetry, satellite altimetry and Argo. Glob Planet Change 65:83–88

Cazenave A et al (2012) Estimating ENSO influence on the global mean sea level, 1993–2010. Mar Geodesy 35:82–97. doi:10.1080/01490419.2012.718209

Cazenave A et al (2014) The rate of sea-level rise. Nat Clim Change 4:358–361

Church JA, White NJ (2006) A 20th century acceleration in global sea-level rise. Geophys Res Lett 33:L01602. doi:10.1029/2005GL024826

Church JA, White NJ (2011) Sea-level rise from the late 19th to the early 21st century. Surv Geophys 32:585–602

Church JA et al (2001) Changes in sea level. In: Houghton JT, Ding Y, Griggs DJ, Noquer M, van der Linden PJ, Dai X, Maskell K, Johnson CA (eds) Climate change 2001: the scientific basis. Contribution of Working Group I to the Third Assessment Report of the Intergovernmental Panel on Climate Change. Cambridge University Press, Cambridge, pp 639–693

Church JA et al (2011) Revisiting the Earth's sea-level and energy budgets from 1961 to 2008. Geophys Res Lett 38:L18601. doi:10.1029/2011GL048794

Church JA et al (2013) Sea level change. In: Stocker TF, Qin D, Plattner G-K, Tignor M, Allen SK, Boschung J, Nauels A, Xia Y, Bex V, Midgley PM (eds) Climate change 2013, the physical science basis. Contribution of Working Group I to the Fifth Assessment Report of the Intergovernmental Panel on Climate Change. Cambridge University Press, Cambridge

Dieng HB et al (2015) The sea level budget Since 2003: inference on the deep ocean heat content. Surv Geophys 36:209–229

Domingues CM et al (2008) Improved estimates of upper-ocean warming and multi-decadal sea level rise. Nature 453:1090–1093

Douglas BC (1997) Global sea rise: a redetermination. Surv Geophys 18:270–292

Grinsted A et al (2007) Observational evidence for volcanic impact on sea level and the global water cycle. *PNAS* 104:19730–19734. doi:10.1073/pnas.0705825104

Gornitz V et al (1982) Global sea level trend in the past century. Science 215:1611–1614. doi:10.1126/science.215.4540.1611

Gregory JM et al (2013) Twentieth-century global-mean sea level rise: is the whole greater than the sum of the parts? J Clim. doi:10.1175/JCLI-D-12-00319.1

Hallegatte S et al (2013) Future flood losses in major coastal cities. Nat Clim Change 3:802–806. doi:10.1038/nclimate1979

Hamlington B, Thompson P (2015) Considerations for estimating the 20th century trend in global mean sea level. Geophys Res Lett 42:4102–4109. doi:10.1002/2015GL064177

Hay C et al (2015) Probabilistic reanalysis of twentieth-century sea-level rise. Nature 517:481–484. doi:10.1038/nature14093

Hegerl GC et al (2007) Understanding and attributing climate change. In: Solomon S, Qin D, Manning M, Chen Z, Marquis M, Averyt KB, Tignor M, Miller HL (eds) Climate change 2007: the physical science basis. Contribution of Working Group I to the Fourth Assessment Report of the Intergovernmental Panel on Climate Change. Cambridge University Press, Cambridge, pp 663–745

Holgate et al (2013) New data systems and products at the permanent service for mean sea level. J Coast Res 29:493–504

Houghton et al (1990) Climate change 1990: the science of climate change. Cambridge University Press, Cambridge

Intergovernmental Panel on Climate Change (IPCC) (2013) Summary for policymakers. In: Stocker TF, Qin D, Plattner G-K, Tignor M, Allen SK, Boschung J, Nauels A, Xia Y, Bex V, Midgley PM (eds) Climate change 2013: the physical science basis. Contribution of Working Group I to the Fifth Assessment Report of the Intergovernmental Panel on Climate Change. Cambridge University Press, Cambridge

Jacob T et al (2012) Recent contributions of glaciers and ice caps to sea level rise. Nature. doi:10.1038/nature10847

Jevrejeva S et al (2006) Nonlinear trends and multi-year cycle in sea level records. J Geophys Res 111 **(2005JC003229)**. doi:10.1029/2005JC003229

Jevrejeva S et al (2008) Relative importance of mass and volume changes to global sea level rise. J Geophys Res 113:D08105. doi:10.1029/2007JD009208

Jevrejeva S et al (2010) How will sea level respond to changes in natural and anthropogenic forcings by 2100? Geophys Res Lett 37:L07703 **(2010GL042947)**

Jevrejeva S et al (2012) Potential for bias in 21st century semiempirical sea level projections. J Geophys Res 117:D20116. doi:10.1029/2012JD017704

Jevrejeva S et al (2014) Upper limit for sea level projections by 2100. Environ Res Lett 9:104008

King MA et al (2012) Regional biases in absolute sea-level estimates from tide gauge data due to residual unmodeled vertical land movement. Geophys Res Lett 39:L14604

Kjeldsen KK et al (2015) Spatial and temporal distribution of mass loss from the Greenland Ice Sheet since AD 1900. Nature 528:396–400

Leclercq PW, Oerlemans J, Cogley JG (2011) Estimating the glacier contribution to sea-level rise over the period 1800–2005. Surv Geophys 32:519–535. doi:10.1007/s10712-011-9121-7

Leuliette EW, Scharroo R (2010) Integrating Jason-2 into a multiple-altimeter climate data record. Mar Geodesy 33:504

Leuliette EW, Willis JK (2011) Balancing the sea level budget. Oceanography 24:122–129

Marcos M et al (2011) The long sea level record at Cadiz (southern Spain) from 1880 to 2009. J Geophys Res 116(C12):1978–2012

Marzeion B et al (2012) Past and future sea-level changes from the surface mass balance of glaciers. Cryosphere 6:1295–1322

Marzeion B, Leclercq PW, Cogley JG, Jarosch AH (2015) Brief communication: global reconstructions of glacier mass change during the 20th century are consistent. Cryosphere 9:2399–2404. doi:10.5194/tc-9-2399-2015
Meehl GA et al (2007) Global climate projections. In: Solomon S, Qin D, Manning M, Chen Z, Marquis M, Averyt KB, Tignor M, Miller HL (eds) Climate change 2007: contribution of Working Group I to the Fourth Assessment Report of the IPCC. Cambridge University Press, Cambridge
Merrifield MA et al (2009) An anomalous recent acceleration of global sea level rise. J Clim 22:5772–5781. doi:10.1175/2009JCLI2985.1
Mitrovica JX et al (2001) Recent mass balance of polar ice sheets inferred from patterns of global sea-level change. Nature 409:1026–1029
Mitrovica JX et al (2015) Reconciling past changes in Earth rotation with 20th century global sea-level rise: resolving Munk's enigma. Sci Adv 1(11), Article e1500679
Moore JC et al (2011) The historical sea level budget. Ann Glac 52:59
Moore JC et al (2013) Semi-empirical and process-based global sea level projections. Rev Geophys. doi:10.1002/rog.20015
Munk W (2002) Twentieth century sea level: an enigma. Proc Natl Acad Sci USA 99:6550–6555
Neckel et al (2014) Glacier mass changes on the Tibetan Plateau 2003–2009 derived from ICESat laser altimetry measurements. Environ Res Lett 9:014009. doi:10.1088/1748-9326/9/1/014009
Peltier WR (2001) Global glacial isostatic adjustment and modern instrumental records of relative sea level history. In: Douglas BC, Kearney MS, Leatherman SP (eds) Sea level rise. Elsevier, New York, pp 65–93
Peltier WR (2004) Global glacial isostasy and the surface of the ice-age earth: the ICE-5G (VM2) model and GRACE. Annu Rev Earth Planet Sci 32:111–149
Peltier WR et al (2015) Space geodesy constrains ice age terminal deglaciation: the global ICE-6G_C (VM5a) model. J Geophys Res Solid Earth. doi:10.1002/2014JB011176
Pouvreau N (2008) Trois cents ans de mesures marégraphiques en France: outils, méthodes et tendances des composantes du niveau de la mer au port de Brest. Université de La Rochelle. Ph.D. thesis
Ray RD, Douglas BC (2011) Experiments in reconstructing twentieth-century sea levels. Prog Oceanogr 91:496–515. doi:10.1016/j.pocean.2011.07.021
Schutz BE et al (2005) Overview of the ICESat Mission. Geophys Res Lett 32:L21S01. doi:10.1029/2005GL024009
Shepherd A et al (2012) A reconciled estimate of ice-sheet mass balance. Science 338:1183–1189
Slangen A et al (2016) Anthropogenic forcing dominates global mean sea-level rise since 1970. Nat Clim Change 6:701–705. doi:10.1038/NCLIMATE2991
Talke SA, Jay DA (2013) Nineteenth century North American and Pacific tidal data: lost or just forgotten? J Coast Res 29(6a):118–127
Testut L, Miguez BM, Wöppelmann G, Tiphaneau P, Pouvreau N, Karpytchev M (2010) Sea level at Saint Paul Island, southern Indian Ocean, from 1874 to the present. J Geophys Res (1978–2012) 115(C12028). doi:10.1029/2010JC006404
Thompson et al (2016) Are long tide gauge records in the wrong place to measure global mean sea level rise? Geophys Res Lett. doi:10.1002/2016GL070552
von Schuckmann K, Le Traon PY (2011) How well can we derive Global Ocean Indicators from Argo data? Ocean Sci 7:783–791
Warrick RA, Oerlemans J (1990) Sea level rise. In: Climate change, The IPCC Scientific Assessment, pp 260–281
Warrick RA et al (1996) Changes in sea level. In: Houghton JT, Meira LG, Callander A, Harris N, Kattenberg A, Maskell K (eds) Climate change 1995: the science of climate change. Contribution of WGI to the Second Assessment Report of the Intergovernmental Panel on Climate Change. Cambridge University Press, Cambridge, pp 359–405
Wenzel M, Schroter J (2010) Reconstruction of regional mean sea level anomalies from tide gauges using neural networks. J Geophys Res. doi:10.1029/2009JC005630
Wöppelmann G, Marcos M (2016) Vertical land motion as a key to understanding sea level change and variability. Rev Geophys 54:64–92. doi:10.1002/2015RG000502
Wöppelmann G et al (2008) Tide gauge datum continuity at Brest since 1711: France's longest sea-level record. Geophys Res Lett 35:L22605. doi:10.1029/2008GLO35783
Wöppelmann G et al (2009) Rates of sea-level change over the past century in a geocentric reference frame. Geophys Res Lett 36:L12607. doi:10.1029/2009GL038720
Wöppelmann G et al (2014) Rescue of the historical sea level record of Marseille (France) from 1885 to 1988, and its extension back to 1849–1851. J Geodesy 88:869–885

Surv Geophys (2017) 38:309–327
DOI 10.1007/s10712-016-9381-3

Evaluation of the Global Mean Sea Level Budget between 1993 and 2014

Don P. Chambers[1] · Anny Cazenave[2,3] · Nicolas Champollion[3] · Habib Dieng[2] · William Llovel[4] · Rene Forsberg[5] · Karina von Schuckmann[6] · Yoshihide Wada[7]

Received: 21 December 2015 / Accepted: 22 July 2016 / Published online: 17 August 2016

Abstract Evaluating global mean sea level (GMSL) in terms of its components—mass and steric—is useful for both quantifying the accuracy of the measurements and understanding the processes that contribute to GMSL rise. In this paper, we review the GMSL budget over two periods—1993 to 2014 and 2005 to 2014—using multiple data sets of both total GMSL and the components (mass and steric). In addition to comparing linear trends, we also compare the level of agreement of the time series. For the longer period (1993–2014), we find closure in terms of the long-term trend but not for year-to-year variations, consistent with other studies. This is due to the lack of sufficient estimates of the amount of natural water mass cycling between the oceans and hydrosphere. For the more recent period (2005–2014), we find closure in both the long-term trend and for month-to-month variations. This is also consistent with previous studies.

Keywords Sea level · Ocean mass · Steric sea level · Climate change

✉ Don P. Chambers
donc@usf.edu

1 College of Marine Science, University of South Florida, 140 7th Ave. S, MSL119, St. Petersburg, FL 33701, USA

2 Laboratoire d'Etudes en Géophysique et Océanographie Spatiales (LEGOS), 18 avenue Edouard BELIN, 31401 Toulouse Cedex 9, France

3 International Space Science Institute (ISSI), Hallerstrasse 6, 3012 Bern, Switzerland

4 Centre Européen de Recherche et de Formation Avancée en Calcul Scientifique (CERFACS), URA1875, 42 Avenue Gaspard Coriolis, 31057 Toulouse Cedex 1, France

5 Division of Geodynamics, Technical University of Denmark, Elektrovej Building 327, room 022, 2800 Kgs., Lyngby, Denmark

6 MERCATOR-Ocean, 10 Rue Hermès, 31520 Ramonville-Saint-Agne, France

7 Department of Physical Geography, Utrecht University, P.O. Box 80115, 3508, TC, Utrecht, The Netherlands

1 Introduction

Sea level varies on a variety of spatial and temporal scales. Regionally, fluctuations in wind and currents can cause large deviations in sea level away from the global mean for periods of years to decades (e.g., Miller and Douglas 2007; Sturges and Douglas 2011; Chambers et al. 2012; Calafat and Chambers 2013; Palanisamy et al. 2015). One example of regional sea level change, from the tide gauge at Key West, Florida, is shown in Fig. 1, compared with an estimate of global mean sea level (GMSL) reconstructed from tide gauge records.

These large regional fluctuations, however, reflect mainly dynamical redistributions of heat and mass in the ocean and thus should average to zero when integrated over the global ocean. While the tide gauge network before the 1960s might not be sufficient to completely average these effects (see Fig. 8 in Calafat et al. 2014), satellite observations of sea surface height allow us to almost completely average out these internal variations and detect the smaller GMSL signal. These records show that GMSL has been rising at a rate between 2.8 and 3.6 mm year^{-1} between January 1993 and December 2014 (90 % confidence bands), with significant low-frequency variability superimposed (e.g., Nerem et al. 2010; Church et al. 2013; Ablain et al. 2015).

However, in order to predict future sea level rise, it is not sufficient to just determine the rate of rise of GMSL. One also needs to understand the mechanisms driving GMSL variability, and how they are changing in time. The primary mechanisms leading to current GMSL rise are: (1) water mass lost from ice sheets, glaciers, and ice caps that is gained by the oceans (e.g., Shepherd et al. 2012; Gardner et al. 2013), (2) volume (density) change due to thermal expansion as the oceans warm (e.g., Domingues et al. 2008; Levitus et al. 2012), and (3) changes in land water storage (e.g., Wada et al. 2012). Salinity changes due to land ice melt, river runoff, and changes in evaporation/precipitation increase have only second-order effects on the GMSL (e.g., Gregory and Lowe 2000). But in practice, as observations are not exactly global, salinity changes to density should also be accounted for when data are available. The combined effect of ocean temperature and salinity is

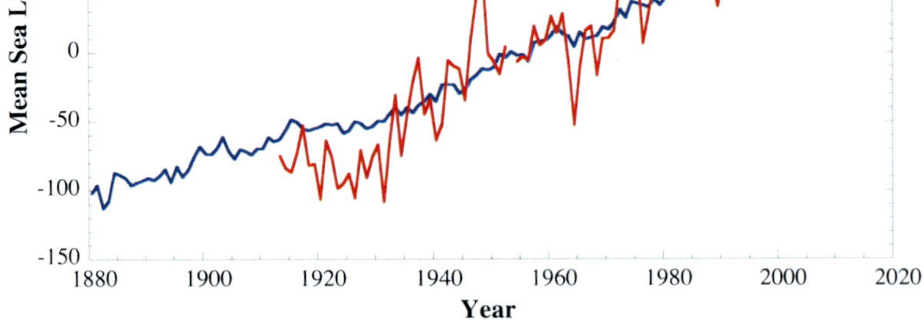

Fig. 1 Yearly averaged sea level change recorded by tide gauges at Key West, Florida compared to GMSL estimate from Church and White (2011). Tide gauge is from the Permanent Service for Mean Sea Level in Liverpool, UK

called the *steric* component, with the thermal contribution denoted *thermosteric*, and salinity contribution *halosteric*.

Because GMSL and estimates of the steric and mass components have different uncertainties and the potential for systematic errors, one often investigates the sea level budget to see how well it closes:

$$\mathrm{GMSL}(t) = \mathrm{GMSL}_{\mathrm{mass}}(t) + \mathrm{GMSL}_{\mathrm{steric}}(t). \tag{1}$$

At any particular time, t, the residual ($\mathrm{GMSL}(t) - \mathrm{GMSL}_{\mathrm{mass}}(t) - \mathrm{GMSL}_{\mathrm{steric}}(t)$) is unlikely to be exactly zero due to random and short-period errors. However, over the long-term, the residual differences should be small. When they are not, it indicates a problem in one or more of the terms in Eq. (1).

In addition, the mass and steric components are often subdivided into the various contributors. For mass, this includes separate estimates for contributions from the Greenland and Antarctica ice sheets, as well as from glaciers and ice caps. Exchanges of water mass between the oceans and the continents related to natural variability contribute significantly to seasonal and interannual contributions (e.g., Chambers et al. 2004; Llovel et al. 2011; Fasullo et al. 2013; Cazenave et al. 2014, Dieng et al. 2015a; Reager et al. 2016; Rietbroek et al. 2016), while storing water behind dams and extracting water from aquifers lead to non-negligible trends in GMSL (Chao et al. 2008; Konikow 2011; Pokhrel et al. 2012; Wada et al. 2012; Wada 2015; Dieng et al. 2015a).

Thermosteric changes are often separated into the upper ocean above 700 m, the layer between 700 and 1000 m, and the deeper ocean (e.g., Domingues et al. 2008; Purkey and Johnson 2010; Levitus et al. 2012). This is mainly an artifact of the older observing system, with substantially more observations in the upper ocean so that yearly averages could be obtained, whereas for the deeper layers, more temporal averaging is needed to extract the trend in the thermosteric component of GMSL. The halosteric component due to salinity changes is poorly known before the advent of the Argo observing system in the early 2000s (e.g., Durack et al. 2013), so typically, the halosteric contribution is neglected in sea level budget studies that include data from the 1990s (e.g., Domingues et al. 2008; Church et al. 2011). For this study, we also only consider the thermosteric sea level for the longer time period (1993–2015) and the steric sea level for the shorter time period (2005–2014).

One way to quantify the closure of the sea level budget is computing trends in GMSL and the various components over various periods of time, and summing these up to see if they match within the uncertainty (e.g., Church et al. 2011, 2013). However, with the advent of global measurements of ocean mass from the GRACE satellite mission (e.g., Johnson and Chambers 2013; Llovel et al. 2014) and ocean temperatures and salinity above 2000 m depth from Argo autonomous profiling floats (Llovel et al. 2014; von Schuckmann et al. 2014), one can now look at the closure on monthly time scales since 2005 (e.g., Dieng et al. 2015b, c for a recent review).

In this paper, we will review the closure of the sea level budget not only in terms of trends, but also in terms of the temporal variability of GMSL and its components. We utilize numerous estimates of GMSL from altimetry, as well as several estimates of the mass components and thermosteric change. These data sets have slightly different temporal sampling and filtering applied. To be consistent, we will utilize a common sampling and filtering scheme to all data. Trends are computed over the same time periods, and uncertainty is computed accounting for correlated signals in the residuals. We assess the closure of the sea level budget on two different time periods: January 1993 to December

2014 (a mixture of all measurements), and January 2005 to December 2014 (GRACE/Argo/altimetry only).

Section 2 will discuss the specific datasets used, filtering applied, and methods used to compute trends and uncertainty. Section 3 will summarize the level of closure of the sea level budget for the three time periods, and Sect. 4 will discuss the results.

2 Data and Methods

2.1 Satellite Altimetry

Products from six processing groups are available for the altimetry-based sea level time series:

1. Validation and Interpretation of Satellite Oceanographic (AVISO; http://www.aviso.altimetry.fr/en/data/products/ocean-indicators-products/actualitesindicateurs-des-oceansniveau-moyen-des-mersindexhtml.html);
2. University of Colorado (CU Release 5; http://sealevel.colorado.edu/).
3. National Oceanographic and Atmospheric Administration (NOAA; http://www.star.nesdis.noaa.gov/sod/lsa/SeaLevelRise/LSA_SLR_timeseries_global.php).
4. Goddard Space Flight Center (GSFC version 2; http://podaac-ftp.jpl.nasa.gov/dataset/MERGED_TP_J1_OSTM_OST_GMSL_ASCII_V2)
5. Commonwealth Scientific and Industrial Research Organization (CSIRO; www.cmar.csiro.au/sealevel/sl_data_cmar.html).
6. The CCI sea level data (ftp.esa-sealevel-cci.org/Products/SeaLevel-ECV/V1_11092012/), Ablain et al. (2015).

The first five sea level data sets are based on TOPEX/Poseidon, Jason-1 and Jason-2 data averaged over the 66°S–66°N domain, except for the CSIRO data averaged over 65°S–65°N. The CCI dataset is based primarily on TOPEX/Poseidon, Jason-1 and Jason-2, but also includes data from the Envisat, ERS-1, and ERS-2 altimeter missions after they have been adjusted to remove orbit error and biases relative to TOPEX/Poseidon, Jason-1 and Jason-2. For each product, a set of instrumental and geophysical corrections is applied (details are given on the websites of each data set). In addition, the effect of glacial isostatic adjustment (GIA) using the estimate proposed by Peltier (2004) is accounted for in each sea level time series except for the NOAA data set. Thus, we corrected the latter for the GIA effect, by adding 0.3 mm year^{-1} to the GMSL time series (Peltier 2004).

The sea level time series are obtained either by directly averaging the along-track sea surface height data (e.g., CU) or by first gridding the unevenly distributed along-track data and then performing grid averaging (e.g., AVISO and NOAA). In all cases, an area weighting is applied. In addition to the geographical averaging method, other differences exist between the GMSL data sets because of the applied geophysical and instrumental corrections and the number of satellites considered. Discussion on these differences can be found in Masters et al. (2012), Henry et al. (2014), and Ablain et al. (2015). Details on the exact corrections applied to the altimetry data are detailed on the webpages of each group.

Five of the time series used in this study cover the period January 1993–December 2014, but one (the CCI product) ends in December 2013. Figure 2 shows the GMSL time series, after removing an annual and semiannual sinusoids and smoothing with a 3-month running mean to reduce a 60-day erroneous signal (e.g., Masters et al. 2012). At shorter time scales (from sub-seasonal to multi-annual) significant discrepancies of several mm are

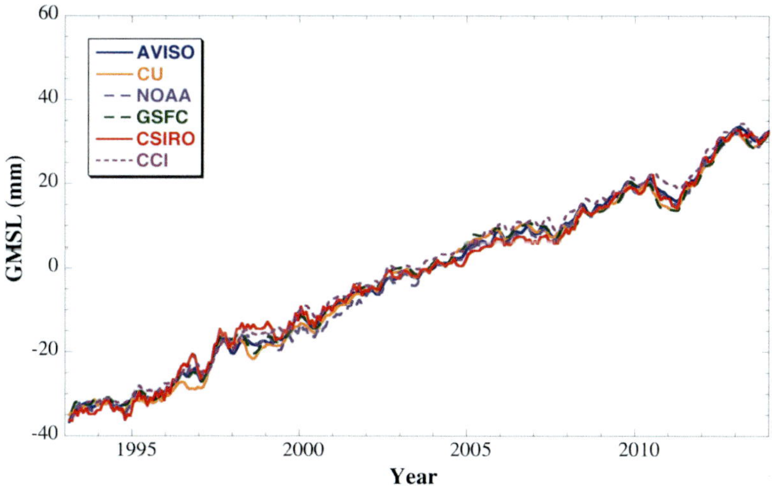

Fig. 2 GMSL from altimetry calculated by six different centers. Annual and semi-annual sinusoids have been estimated and removed, and a 3-month running mean filter has been applied

observed between the different GMSLs, especially between 2005 and 2008, and between mid-2010 and mid-2011, when there was a significant drop in GMSL related to changes in water storage over Australia and South America (Fasullo et al. 2013).

2.2 Steric Sea Level

Before the early 2000s, information on the steric sea level component comes from temperature data, due to the lack of global salinity data, coming primarily from expendable bathythermographs (XBTs), some mechanical bathythermographs (MBTs) and a much smaller number of conductivity-temperature-depth casts (CTDs) (e.g., Levitus et al. 2012; Abraham et al. 2013). The depth ranges of these instruments are very different, with XBTs mostly going only as deep as 700 m (although some go as deep as 1000–1500 m), while CTDs often make measurements to the sea floor. In the beginning of the 2000s, with the advent of the Argo program of autonomous floats (Roemmich et al. 2009), more measurements are available at more regular time intervals and also more globally. The current average density is approximately 1 float for every 3° × 3° grid over the ocean. However, the depth to which Argo floats reach has also changed over time, starting at 1000 m, but now extending to 2000 m.

Computing steric sea level anomalies from these disparate data is not a trivial matter, but many groups have done so, using different interpolation and mapping methods, as well as different corrections for the XBT fall rate biases that have only recently been discovered (e.g., Gouretski and Koltermann 2007; Wijffels et al. 2008). Some groups use all available temperature data, while others restrict the estimate to only Argo data.

To quantify the steric component of GMSL change, we will consider different datasets. For the longer period (1993–2015), we will use two that merge XBTs, MBTs, CTDs, and Argo data, and provide the thermosteric sea level anomalies as a time series. They are:

1. NOAA data set, at https://www.nodc.noaa.gov/OC5/3M_HEAT_CONTENT (Levitus et al. 2012). The data are available as yearly averages of global mean thermosteric sea

level anomalies from the surface to 700 m depth, and 5-year running averages from the surface to 2000 m depth.

2. Domingues et al. (2008) version 3.1 dataset available from http://www.cmar.csiro.au/sealevel/thermal_expansion_ocean_heat_timeseries.html. The data are available as global mean thermosteric sea level anomalies from the surface to 700 m depth at yearly time steps but with a 3-year running mean filter applied.

To compare the two time series, we have applied a 3-year running mean to the NOAA time series (Fig. 3). Although the two estimates have similar long-term trends, there are substantial differences outside the authors' estimated standard errors at interannual periods. The biggest differences occur between 2000 and 2006, when the observing system is transitioning from mainly XBTs to mainly Argo floats. It has been shown that the different mapping techniques are highly sensitive to the mixture of the XBT/Argo data during this transition, partly due to small, unknown biases between different instrument types (Lyman and Johnson 2008; Lyman et al. 2010).

We use the Levitus et al. (2012) estimate of thermosteric sea level to 2000 m, available as 5-year running means, to estimate the thermosteric component between 700 and 2000 m depth. We reconstruct the signal between 700 and 2000 m by subtracting 5-year averages of the Levitus et al. (2012) 0–700 m time series. Uncertainty is that reported by the authors (after subtracting the uncertainty from 0 to 700 m assuming no correlation).

The only observations of deep thermosteric contributions are trends computed from deep hydrographic sections (Purkey and Johnson 2010; Kouketsu et al. 2011), estimated to be 0.11 ± 0.10 mm year^{-1} [uncertainty 95 % as reported by Purkey and Johnson (2010)] between approximately 1995–2005. We assume this value represents the trend for the entire period between 1993.0 and 2015.0.

After 2005, sufficient Argo floats are available to compute steric sea level anomalies from only these data (von Schuckmann et al. 2014; Roemmich et al. 2015). For this study, we utilize four different gridded datasets, providing temperature and salinity down to 2000 m depth at monthly intervals. They are:

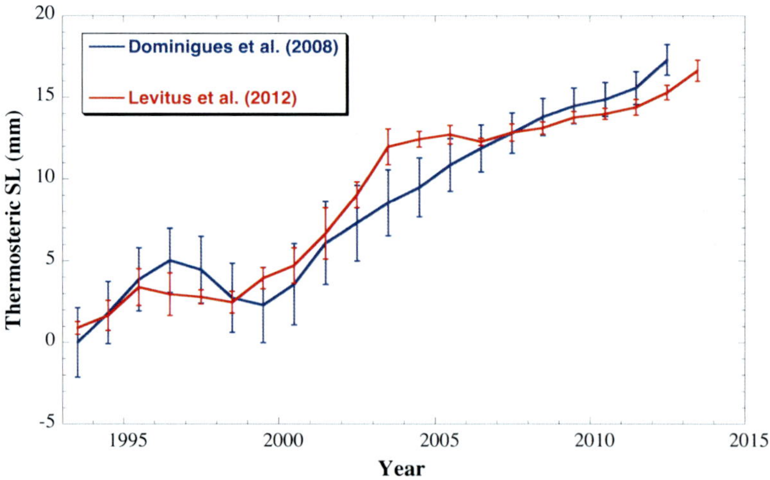

Fig. 3 Global mean thermosteric sea level contribution for upper ocean (0–700 m) from two analyses along with one standard error bars as computed by authors (Domingues et al. 2008; Levitus et al. 2012)

1. The International Pacific Research Center (IPRC; http://apdrc.soest.hawaii.edu/projects/Argo/data/Documentation/gridded-var.pdfhttp://apdrc.soest.hawaii.edu/projects/Argo/data/gridded/On_standard_levels/index-1.html)
2. The Japan Agency for Marine-Earth Science and Technology (JAMSTEC; ftp://ftp2.jamstec.go.jp/pub/argo/MOAA_GPV/Glb_PRS/OI/). Updated from Hosoda et al. (2008).
3. The SCRIPPS Institution of Oceanography (SCRIPPS; http://sio-argo.ucsd.edu/RG_Climatology.html). Updated from Roemmich et al. (2009).
4. The estimate from Karina von Schuckmann (KvS). Updated from Von Schuckmann and Le Traon (2011).

For the IPRC, JAMSTEC, and SCRIPPS data, we computed steric sea level time series from the surface down to 2000 m at monthly interval on a $1° \times 1°$ grid for the period of January 2005 to December 2014 by integrating the density anomalies (defined as differences between the density estimate and a reference density at 0 °C and 35.16504 absolute salinity using the equation of state of seawater TEOS10 (http://www.teos-10.org/index.htm) at each standard depth. The KvS time series was computed using a similar methodology more fully described in von Schuckmann et al. (2009).

The Argo-based estimates of steric sea level since 2005 show similar decadal trends, but significantly different monthly and interannual variations (Fig. 4). Differences are of the order of 5 mm over some periods (2005 and 2010–2011, for example), but closer to 3 mm over other periods. The standard deviation of the differences ranges from a low of 1.5 mm between IPRC and JAMSTEC, to 2.9 mm between Scripps and JAMSTEC. IPRC and JAMSTEC also have the highest correlation of 0.92, while the correlation between Scripps and JAMSTEC, while still significant, is only 0.63. See Dieng et al. (2015b, c) for a detailed discussion on these differences.

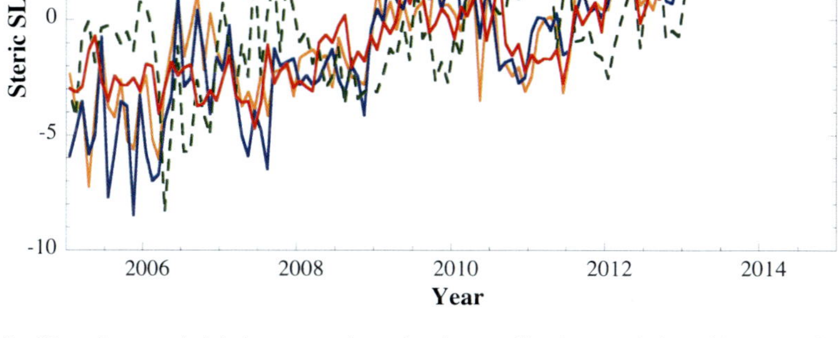

Fig. 4 Monthly estimates of global mean steric sea level anomalies (seasonal sinusoids removed) computed from Argo data from four processing centers

2.3 Satellite Gravity

The GRACE mission measures the Earth gravity field every month, generally released in terms of normalized spherical harmonic coefficients (Tapley et al. 2004), or more recently as gridded mass concentrations, or mascons (Watkins et al. 2015). One can convert the spherical harmonics into ocean mass variations in terms of an equivalent water thickness using either an averaging kernel approach (e.g., Chambers et al. 2004; Johnson and Chambers 2013), or more recently, by simply averaging the gridded mascons over the ocean domain (Watkins et al. 2015).

We utilize both approaches in this study and also use data from the three main processing centers—the Center for Space Research (CSR), the Helmholtz Centre Potsdam, German Research Centre for Geosciences (GFZ), and the Jet Propulsion Laboratory (JPL). JPL produces both spherical harmonics (SH) and mascons. Each center uses slightly different processing and analysis strategies, but many models and methods are similar. Thus, while comparison of the ocean mass from the various centers is instructive for quantifying a level of uncertainty, any systematic errors will not be obvious.

The data utilized are:

1. Averaging kernels from CSR, GFZ and JPL SH coefficients using the method described in Johnson and Chambers (2013), available from https://dl.dropboxusercontent.com/u/31563267/ocean_mass_orig.txt.
2. Averaging the JPL mascons (Watkins et al. 2015) over the global oceans, using an ocean mask that extends to the coastlines. The data are available from: http://grace.jpl.nasa.gov. Note that these data include a small signal from the global mean atmospheric pressure over the ocean, as explained in Johnson and Chambers (2013). This has been estimated and removed using the atmosphere model used to dealias the GRACE gravity data.

The differences between global mean ocean mass from the three centers and between spherical harmonics and mascons are small (Fig. 5). The biggest differences of order 5 mm

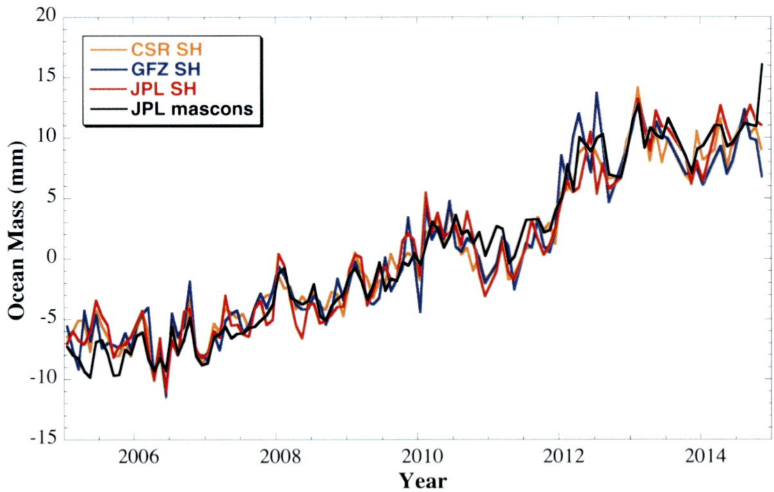

Fig. 5 Monthly estimate of global mean ocean mass anomalies (seasonal sinusoids removed) from GRACE computed by three different processing centers based on spherical harmonics (SH) and mascons

occur at the end and beginning of the records. The overall standard deviation of monthly differences is 1.6 mm.

2.4 Contributions from Ice Sheets and Glaciers

We consider contributions from the largest ice sheets (Greenland and Antarctica) separately from the glaciers and ice caps (GICs). The time series for Greenland and Antarctic mass balance comes from, for the longer period back to 1993, the synthesis of Shepherd et al (2012), based on surface mass balance models, synthetic aperture radar data, altimetry, and gravimetry (available from http://imbie.org/data-downloads/). For the shorter period since 2003, we use the estimates based only on GRACE computed by the group at the Technical University of Denmark (Sørensen and Forsberg 2010; Barletta et al. 2013).

GIC data come from an analysis of global glacier models driven by gridded climate observations, which has been shown to be consistent with extrapolations of in situ mass balance measurements (Marzeion et al. 2015).

2.5 Contributions from Land Hydrology

Ignoring the contribution from glaciers on continents, which is treated separately (Sect. 2.3), the land water contribution to sea level variation includes groundwater depletion, water impoundments behind dams, storage loss of endorheic lakes and wetlands, deforestation, and changes in soil moisture, permafrost and snow (i.e., natural water stores) (Church et al. 2011). Natural water storage change mostly varies with decadal climate variation and with insignificant trend on time periods greater than several decades (Ngo-Duc et al. 2005; Llovel et al. 2011), but can contribute to the trend on shorter periods (Cazenave et al. 2014). However, decomposing this signal is still a matter of significant research and fraught with uncertainties before GRACE observations are available starting in 2002, so we do not consider it for this study for the longer period 1993–2015.

Instead, we consider only estimates of groundwater depletion, water impoundment, deforestation, and the loss from large endorheic lakes. The contribution of groundwater depletion to GMSL is estimated using a flux-based method, i.e., calculating the difference between grid-based groundwater recharge (natural recharge and return flow from irrigation as additional recharge) and groundwater pumping (Wada et al. 2012; Wada 2015). This method, however, overestimates groundwater depletion for humid regions of the world. In order to correct the estimate, a global multiplicative correction factor is applied to the original estimate. The correction factor is based on a comparison between regionally reported groundwater depletion rates and simulated groundwater depletion rates (over 30 regions; Wada et al. 2012). An uncertainty analysis is performed with a Monte Carlo simulation, generating 100 equiprobable realizations of groundwater recharge and 100 equiprobable realizations of groundwater pumping, thus resulting in 10,000 possible realizations of groundwater depletion (assuming errors in groundwater recharge and groundwater abstraction to be independent) (Wada et al. 2012).

Water impoundment behind dams including additional storage in surrounding groundwater (through seepage) is based on the dataset of Chao et al. (2008). As this dataset only covers the period 1900–2007, it has been updated to include recently built dams including the Three Gorges dam and 250 other large dams up to the year 2011 (Wada et al. 2012). After the year 2011, the data are extrapolated. Deforestation rates are estimated from three different sources, averaged, and converted into a contribution to GMSL (Wada et al. 2012). Wetland loss rate is estimated for USA where reported data are available, and then extrapolated the rate to rest of the world (Wada et al. 2012). Storage loss from endorheic basins is estimated only for the Aral and Caspian Seas (Wada et al. 2012).

2.6 Temporal Filtering and Combining Similar Data

The datasets previously discussed are provided with a range of temporal sampling, from monthly to yearly. Moreover, some have been filtered over longer times than they are sampled. For instance, the upper ocean thermosteric sea level estimate from Domingues et al. (2008) is provided at yearly time steps, but has had a 3-year running mean applied. The upper ocean thermosteric estimates from Levitus et al. (2012), on the other hand, are yearly averages. Thus, direct averaging of the two will lead to spurious differences related to the different smoothing applied.

Since one cannot unfilter a dataset, we are forced to utilize the longest filtering period among the datasets in order to make the time series as uniform as possible, and reduce the effect of unfiltered higher-frequency variability in some data. This means that yearly sampling and a 3-year running mean filter is applied to the time series that extend back to 1993, including the altimetry (Sect. 2.1), thermosteric (Sect. 2.2), ice contributions (Sect. 2.3), and hydrology components (Sect. 2.4). For the period 2005–2015, a monthly average is used, but the seasonal variation is estimated and removed by fitting a sinusoid term with annual (1 cycle per year, cpy) and semi-annual (2 cpy) frequencies using ordinary least squares in order to focus on only the interannual and longer variations. A 3-month running mean is also applied to be consistent with the smoothing used with the altimetry time series (Sect. 2.1).

When multiple datasets are available (e.g., altimetry GMSL, Argo thermosteric variations, GRACE ocean mass) the time series is averaged to compute an ensemble mean. Uncertainty is computed from the standard deviation of the residuals of the individual time series with the ensemble mean. This is assumed to be the standard error at each time step. For time series without multiple estimates, the uncertainty from the authors of the data is used. The total thermosteric signal is reconstructed from the ensemble average of the upper ocean time series (Fig. 3), the estimate from 700 to 2000 m based on Levitus et al. (2012), and the deep warming trend from Purkey and Johnson (2010). Standard errors in each component are added assuming they are uncorrelated by using a root-sum-square (RSS).

The satellite altimetry and GRACE observations also have a likelihood of unknown systematic errors that will affect the trend estimate. For altimetry, this arises from drifts and biases in the different instruments and the difficulty of detecting it through calibration with tide gauges, which also have vertical land motion that is often poorly measured (Mitchum 2000; Ablain et al. 2015; Watson et al. 2015). The full range of possible drift errors has previously been estimated to be between ± 0.4 mm year^{-1} (Mitchum 2000) and ± 0.5 mm year^{-1} (Ablain et al. 2015). More recently, Watson et al. (2015) found a higher possible change of 0.6 mm year^{-1} for the combined record. Here we use the value of ± 0.6 mm year^{-1} to be most conservative. This uncertainty is used for all time periods, even though Watson et al. (2015) argue it is considerably less for the Jason-1 and Jason-2 altimeters (post 2002). For GRACE, the uncertainty arises from uncertainty in the glacial isostatic adjustment (GIA) correction and has been estimated to be ± 0.3 mm year^{-1} (Chambers et al. 2010). These uncertainty values are added to those determined from the internal statistics (Sect. 2.6) using an RSS.

2.7 Fitting Trends and Computing Uncertainty

While trends do not give a complete picture of the sea level budget closure, they are a useful tool to detect imbalance and have been frequently used as a measure of the sea level budget closure (e.g., Church et al. 2011, 2013). We will fit a bias plus a trend ($a_0 + a_1 t$) model to each time series using ordinary least squares (OLS). Uncertainty estimates from ordinary least squares, however, assumes: (1) The uncertainty is proportional to the standard deviation of the residuals about the fit, (2) the uncertainty is proportional to 1/sqrt(N), where N is the number of points, and (3) the N points are statistically uncorrelated. In practice, these assumptions are rarely all true. For example, by temporally smoothing data, the points are not uncorrelated, and assumption (3) is violated. Assumption (1) is based on uncertainty arising from internal, unmodeled variability. But if this is smaller than the standard error of the estimate, the uncertainty in the trend will be underestimated.

There are several ways to deal with this issue and account for uncertainty properly. For small numbers of points (<20 or so), using Monte Carlo estimates with a colored noise model is not preferred, as it is difficult to compute the autocovariance of the data with such limited samples. This is the case for the longer time span when we have applied a 3-year mean filter. Instead, it is better to estimate the effective degrees of freedom (eDOF), which is the number of statistically independent observations minus the number of model parameters estimated (in the case of the 3-year smoothed data, it is only a bias + trend, so 2). In the case of the 3-year smoothed data between 1993 and 2015, we assume the points are uncorrelated after three years. The effective degrees of freedom for each data set are given in Table 1, noting they vary because the time lengths are slightly different.

Once the eDOF is known, it is straightforward to estimate the corrected uncertainty by

$$\sigma_{\text{corr}} = \sigma_{\text{OLS}} \sqrt{\frac{N}{e\text{DOF}}}, \tag{2}$$

Table 1 Estimated trends in GMSL and components between approximately January 1993 and December 2015

Quantity	Period	Trend (mm year^{-1})	Temporal averaging	Effective DOF
GMSL	1993–2015	3.19 ± 0.63[a]	3-year running means	5
Thermosteric				
0–700 m	1992.0–2014.0	0.85 ± 0.2	3-years	5
700–2000 m	1992.0–2014.0	0.24 ± 0.07	5-years	3
Below 2000 m	~1995–2005	0.11 ± 0.1	Trend only	N/A
Total thermosteric	~1992–~2014	1.20 ± 0.23	Sum of component trends	N/A
Antarctica	1992.0–2011.0	0.22 ± 0.14	3-years	4
Greenland	1992.0–2011.0	0.37 ± 0.28	3-years	4
Glaciers/ice caps	1992.0–2013.0	0.76 ± 0.30	3-years	5
Hydrology	1992.0–2013.0	0.45 ± 0.16	3-years	5
Total mass	~1992–~2013	1.8 ± 0.46		
Sum of components		3.00 ± 0.52		

Exact time period for each representative time series is given. Uncertainty is 90 % confidence except for the thermosteric below 2000 m, which is 95 % as estimated by Purkey and Johnson (2010)

[a] Includes uncertainty in knowing systematic drifts of ±0.6 mm year^{-1} (added as RSS)

where σ_{OLS} is the standard error from OLS based on the residuals and assuming N uncorrelated observations, and σ_{corr} is the corrected uncertainty by accounting for the smaller number of independent observations. However, the corrected uncertainty may still be too small, if the standard deviation of the residuals about the fit (used to scale the covariance matrix in OLS) is smaller than the prescribed standard errors for each observation (σ_{obs}). In this case, scaling the covariance matrix using the observation errors such as in weighted least squares (WLS) is better. Thus, to fully compute the most conservative uncertainty, we calculate both σ_{OLS} (based on the residuals to the fit) and σ_{WLS} (based on the observation errors) and derive σ_{corr} based on the greater of the two:

$$\sigma_{corr} = \begin{bmatrix} \sigma_{OLS}\sqrt{\dfrac{N}{eDOF}}, & \text{if } \sigma_{OLS} > \sigma_{WLS} \\ \sigma_{WLS}\sqrt{\dfrac{N}{eDOF}}, & \text{if } \sigma_{WLS} > \sigma_{OLS} \end{bmatrix}. \tag{3}$$

For the monthly sampled datasets between 2005 and 2015, we will use instead a Monte Carlo simulation based on a set of 10,000 simulated time series residuals that have an autocovariance similar to the true residuals. To do this, we use an auto-regression (AR) model to impose correlations to an initial random time series. An AR(p) model estimates values (y) at some time, t, based on p earlier times scaled by coefficients that had correlation:

$$y(t) = a_1 y(t-1) + a_2 y(t-2) + a_3 y(t-3) + \cdots + a_p y(t-p) + \varepsilon(t) \tag{4}$$

where $\varepsilon(t)$ is random noise with a prescribed variance.

The coefficients (a) are determined using the Yule-Walker algorithm, based on the one-sided autocovariance (i.e., where negative lags are treated the same as positive lags in the computation, assuming symmetry):

$$\begin{bmatrix} a_1 \\ a_2 \\ a_3 \\ \vdots \\ a_p \end{bmatrix} = \begin{bmatrix} R_0 & R_1 & \cdots & R_{p-1} \\ R_1 & R_0 & \cdots & R_{p-2} \\ \vdots & \vdots & \ddots & \vdots \\ R_{p-1} & R_{p-2} & \cdots & R_0 \end{bmatrix}^{-1} \begin{bmatrix} R_1 \\ R_2 \\ R_3 \\ \vdots \\ R_p \end{bmatrix} \tag{5}$$

$$\sigma_\varepsilon^2 = R_0 - \sum_{k=1}^{p} a_k R_k$$

R_0 is the autocovariance at lag=0, R_1 is the autocovariance at lag $= 1$, etc., and σ_ε is the standard deviation of the random noise needed to match the covariance at lag $= 0$.

Starting from 10,000 random time series with a standard deviation equal to that of the residuals, we derive and use the coefficients of an AR(3) model and create a 10,000 different colored noise models so that the covariance to lag-3 matches that of the original residuals. For the monthly sampled time series from 2005 to 2015, we do not consider the uncertainty of the observations, as the standard deviation of the residuals is higher than the observation error for all time series (altimetry, GRACE, Argo). We then fit trends to these simulated residuals, and the standard deviation of the 10,000 sample trends is used as one standard error for the trend uncertainty. This will properly inflate the uncertainty to account for correlation in the time series. The degrees of freedom are computed from the autocorrelation of the time series, based on dividing the full time length in months by the decorrelation time, and rounding down. The decorrelation time is computed as twice the lag at which the autocorrelation drops below 0.5. The effective degrees of freedom are reduced by 6 (bias + trend, plus the previously estimated annual/semi-annual sinusoid).

All uncertainties are scaled to 90 % confidence, assuming a two-tailed t-distribution and accounting for the eDOF.

3 Results and Analysis

3.1 1993 to 2013

Figure 6 and Table 1 summarize the results of our assessment of the sea level budget from January 1993 to approximately December 2013. The end date is approximate, because several of the time series end earlier (notably the estimated from the Greenland and Antarctica ice sheets). The majority of the data have end dates in 2013.

Based on assessment of the trends over the 20-year interval (Table 1), the sea level budget closes within the uncertainty. The trend in GMSL is 3.19 ± 0.63 mm year^{-1}, while the trend in the sum of the components is 3.00 ± 0.52 mm year^{-1}. Thus, we have confidence that we understand the various contributors to GMSL rise over the last 20 years, at least within our current ability to measure them. The largest single contributor has been thermal expansion, explaining about 40 % of the trend (comparing relative to the sum of all components). The upper ocean alone explains about 28 % of the trend, with about 8 % coming from the middle layers, and 4 % from the deep ocean below 2000 m.

The contributors that combine to increase ocean mass, however, explain 60 % of the trend. Thus, the mass component of sea level rise between 1993 and 2014 was roughly 50 % greater than thermal expansion. Of the contributors, the glaciers and ice caps outside of Greenland and Antarctica had the largest effect (~25 % of total GMSL), hydrology the

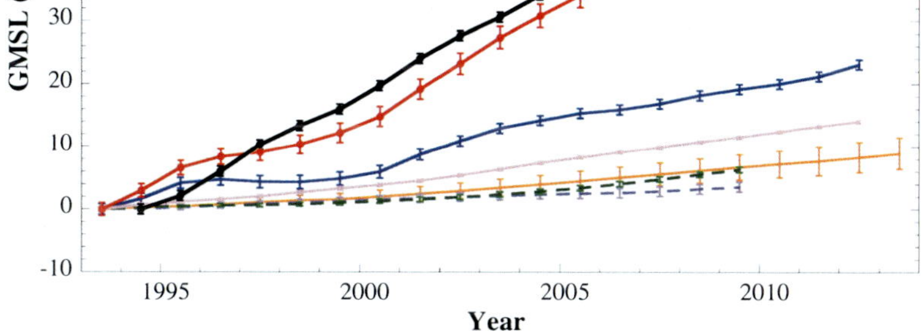

Fig. 6 Three-year running means of GMSL from altimetry, its components, and the sum of the components from 1993.0 to 2014.0. Uncertainty bars are one standard error as described in text

next (~15 %), then Greenland (~12 %), and finally Antarctica (~7 %). However, the contribution from Greenland has accelerated in recent years, as shown by several recent studies (e.g., Shepherd et al. 2012; Schrama et al. 2014; Velicogna et al. 2014; Yi et al. 2015). This is evidenced in Fig. 6 by the increasing separation between Greenland and Antarctica contributions. By 2010, the contribution to GMSL from Greenland has equaled the amount estimated from hydrological sources.

Although trends agree well, the time series of GMSL from altimetry and the sum of the components do not agree that well (Fig. 6). They disagree significantly at low frequencies. This is likely because our hydrological estimate does not include natural climate fluctuations in water cycling between the oceans and continents (Sect. 2.4). It is known that these signals on interannual (3- to 5-year time scales) can be of order 10 mm or so (e.g., Fasullo et al. 2013; Cazenave et al. 2014). These variations will be reflected in the GMSL estimate from altimetry, but not in the sum of the components.

3.2 2005 to 2014

Our ability to balance the sea level budget improves significantly after 2005 (Fig. 7; Table 2). Now, by measuring ocean mass directly from the satellite gravity measurements, we can observe similar low-frequency variability, for example, the significant drop of approximately 5 mm in 2011, followed by a subsequent rise of 16 mm between 2011 and 2013, followed by another 5 mm drop in late 2013 (Fig. 7). These are all related to exchanges of water mass between the oceans and continents and have only small steric sea level signatures. The perturbation in 2011 has been linked to anomalous rainfall over mainly Australia, with a lesser contribution from South America (Fasullo et al. 2013).

As with the 20-year period, the sea level budget over the last 10 years closes to within the uncertainty (Table 2). The rate in GMSL from altimetry is nearly unchanged from the 20-year estimate (3.17 ± 0.67 vs. 3.19 ± 0.63 mm year^{-1}). Although a few recent studies have found lower rates of GMSL over the last 10 years or so, these are affected by the

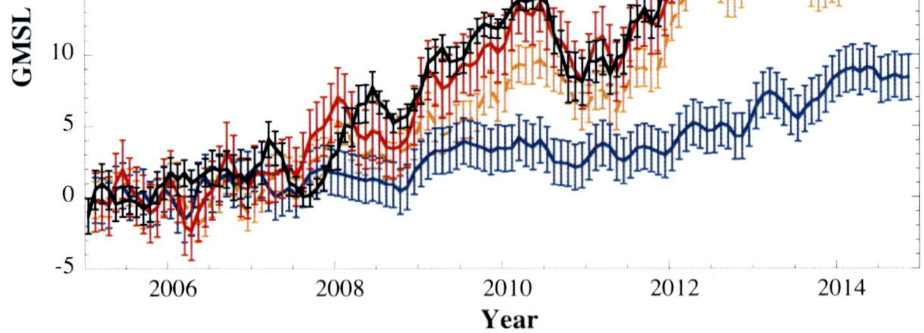

Fig. 7 Three-month running means of GMSL from altimetry, ocean mass from GRACE, and the thermosteric component from Argo for 2005.0–2015.0 (seasonal sinusoids removed). Time series are ensemble means and uncertainty *bars* are one standard error as described in text

Table 2 Estimated trends in GMSL and components between approximately January 2005 and December 2014 and from the representative time series

Quantity	Period	Trend (mm year^{-1})	Temporal averaging	Effective DOF
GMSL	2005.0–2015.0	3.17 ± 0.67[a]	3-month running mean	10
Thermosteric				
0–2000 m	2005.0–2015.0	0.86 ± 0.11	3-month running mean	10
Below 2000 m	~1995–2005	0.11 ± 0.1	Trend only	N/A
Total thermosteric	~2005–~2015	0.97 ± 0.15	Sum of component trends	N/A
Mass	2005.0–2015.0	2.11 ± 0.36[b]	3-month running mean	10
Sum of components		3.08 ± 0.39		

Exact time period for each representative time series is given. Uncertainty is 90 % confidence except for the thermosteric below 2000 m, which is 95 % as estimated by Purkey and Johnson (2010)

[a] Includes uncertainty in knowing systematic drifts of ±0.6 mm year^{-1} (added as RSS)

[b] Includes uncertainty in GIA of ±0.3 mm year^{-1} (added as RSS)

large interannual variability since 2011 (Cazenave et al. 2014). Earlier studies based on a scaling of El Niño indices suggested at least 15 years of data are necessary to distinguish longer term GMSL rise from that related to internal, natural variability (Nerem et al. 1999). Thus, one needs to be cautious of over-interpreting small changes in trends with short records.

The total steric contribution to GMSL is slightly smaller in the last decade than over the 20 years, while the mass component is slightly higher, although the means agree within uncertainty. Although we do not partition the ocean mass component into individual components from 2005 to 2014 due to limited degrees of freedom in the glacier and hydrology data, we do compare the relative contribution of Greenland and Antarctica mass loss (as measured by GRACE) to mean ocean mass (Fig. 8).

Over the last decade, the trend in Antarctica mass loss accounts for 3–27 % (central value 16 %) of the trend in global ocean mass, while Greenland accounts for 21–40 % (central value 32 %), when uncertainty is included in the possible spread. Greenland and Antarctica now account for about for 18–43 % (central value 28 %) of total GMSL rise, approximately the same amount as thermal expansion (25–56 %, central value 37 %). The contribution from glaciers and hydrology make up the remaining 35 %. Compare that to the period that includes the 1990s before the ice sheets began losing mass at an accelerated rate, when Greenland and Antarctica accounted for only ~20 % of the GMSL rate, while glaciers and hydrology accounted for ~40 %, based on the central values.

4 Conclusions

The results of this study are in agreement with previous estimates of the sea level budget (e.g., Church et al. 2011, 2013; Llovel et al. 2014; von Schuckmann et al. 2014; Dieng et al. 2015a, b, c). The main differences are in the slightly different time periods we study, along with what we consider more robust uncertainty estimates. The overall conclusion is that the sea level budget closes on both the longer (1993–2014) and shorter period (2005–2014) within the uncertainty. This gives us high confidence in the disparate measurements that go into the budget calculation. It also gives us increasing confidence that we

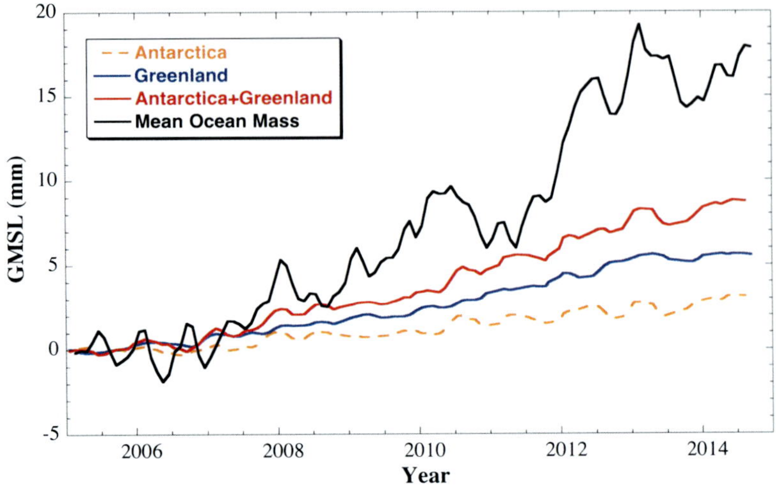

Fig. 8 Three-month running means of global mean ocean mass from GRACE, and the contributions from Greenland and Antarctica, also measured by GRACE

can partition the sources driving the observed rise in GMSL of ~3.2 mm year^{-1}. From the sea level budget exercise, we know that thermal expansion drives about 40 % of the signal, with the majority of the expansion in the upper 700 m of the water column. Temperature changes in the deeper ocean, while smaller, still contribute significantly. Of the remaining 60 %, glaciers and ice caps outside of Greenland and Antarctica contribute the most (~25 % of total GMSL), Greenland and Antarctica ice sheets the next most (19 %), and hydrology the next (~15 %).

Over the last decade, the contributions from Greenland and Antarctica have accelerated considerably. Since 2005, Greenland has contributed 21 % of GMSL rise, while Antarctica has contributed 11 %. Combined, this is nearly a third of GMSL rise, and increase of nearly 68 % compared to the 20-year trend. The trend in GMSL over the two time intervals is roughly the same value (~3.2 mm year^{-1}). Thus, only by measuring the components of GMSL separately are we able to deduce changes in the mechanisms responsible for GMSL change, which allows us to better understand the processes.

These estimates, however, are based on the center of the possible spread of the trends. Unfortunately, we are still limited in reducing the spread due to potential systematic error in the observing system. The two largest uncertainties that have been documented are the inability to constrain the drift in altimeters to better than ±0.6 mm year^{-1} and the GIA uncertainty on gravimetry estimates of ocean mass (±0.3 mm year^{-1}). As these are systematic, they will not be reduced by longer time series, unlike the error arising from unmodeled internal variability, which is reduced by a factor of 1/sqrt(n). Thus, concerted and continued efforts are needed to understand, measure, and correct systematic uncertainty in all portions of the observing system. Although the sea level budget closes to less than 0.2 mm year^{-1} (Tables 1, 2), this is just as likely to be a fortuitous cancellation of systematic errors as a real indication of accuracy.

Finally, it is vital to continue the observations of not only GMSL, but also the steric and mass components. Greenland and Antarctica contributions can be most directly observed via space-based gravimetry measurements, which provide an important constraint on other types of measurements (altimetry measurements of topography or input–output methods).

The independent Argo measurements confirm a slightly lower steric contribution, leaving no significant change in GMSL. In addition, expansion of the Argo measurements below 2000-m depth is important, as the only information we have from the deeper ocean are from more limited CTD casts. Although these are highly precise instruments, and deep ocean temperature/salinity are more correlated over larger regions, the low data availability results in uncertainty of order 50 %. Although the deep ocean is only a small contributor to sea level rise, as more heat is sequestered into the deeper ocean due to deep water formation, it may play an increasing role, considering that the volume of the deep ocean (below 2000 m) is nearly half that of the total ocean.

Acknowledgments This paper is the outcome of an International Space Science Institute Workshop (ISSI) "Integrative Study of Sea Level", and ISSI provided travel funding for the authors to attend the workshop. The Argo data were collected and made freely available by the International Argo Program and the national programs that contribute to it (http://www.argo.ucsd.edu and http://argo.jcommops.org). The Argo Program is part of the Global Ocean Observing System. DPC was supported by NASA Grant NNX12AL28G and internal funding from the University of South Florida. WL was supported by a CNES post-doctoral fellowship carried out at CERFACS. HD is supported by a doctoral fellowship of the European Space Agency within the Climate Change Initiative (CCI) Programme.

References

Ablain M, Cazenave A et al (2015) Improved sea level record over the satellite altimetry era (1993–2010) from the Climate Change Initiative Project. Ocean Sci 11:67–82. doi:10.5194/os-11-67-2015

Abraham JP et al (2013) A review of global ocean temperature observations: implications for ocean heat content estimates and climate change. Rev Geophys 51:450–483. doi:10.1002/rog.20022

Barletta VR, Sørensen LS, Forsberg R (2013) Scatter of mass changes estimates at basin scale for Greenland and Antarctica. Cryosphere 7:1411–1432. doi:10.5194/tc-7-1411-2013

Calafat FM, Chambers DP (2013) Quantifying recent acceleration in sea level unrelated to internal climate variability. Geophys Res Lett. doi:10.1002/grl.50731

Calafat FM, Chambers DP, Tsimplis MN (2014) On the ability of global sea level reconstructions to determine trends and variability. J Geophys Res Oceans 119:1572–1592. doi:10.1002/2013JC009298

Cazenave A, Dieng H, Meyssignac B, von Schuckmann K, Decharme B, Berthier E (2014) The rate of sea level rise. Nat Clim Chang 4:358–361. doi:10.1038/NCLIMATE2159

Chambers DP, Merrifield MA, Nerem RS (2012) Is there a 60-year oscillation in global mean sea level? Geophys Res Lett 39:L18607. doi:10.1029/2012GL052885

Chambers DP, Wahr J, Nerem RS (2004) Preliminary observations of global ocean mass variations with GRACE. Geophys Res Ltrs 31:L13310. doi:10.1029/2004GL020461

Chao BF, Wu YH, Li YS (2008) Impact of artificial reservoir water impoundment on global sea level. Science 320:212–214. doi:10.1126/science.1154580

Church JA, White NJ (2011) Sea-level rise from the late 19th to the early 21st century. Surv Geophys 32(4–5):585–602. doi:10.1007/s10712-011-9119-1

Church JA, White NJ, Konikow LF, Domingues CM, Cogley JG, Rignot E, Gregory JM, van den Broeke MR, Monaghan AJ, Velicogna I (2011) Revisiting the Earth's sea-level and energy budgets from 1961 to 2008. Geophys Res Lett 38:L18601. doi:10.1029/2011GL048794

Church JA, Clark PU, Cazenave A, Gregory JM, Jevrejeva S, Levermann A, Merrifield MA, Milne GA, Nerem RS, Nunn PD, Payne AJ, Pfeffer WT, Stammer D, Unnikrishnan AS (2013) Sea level change. In: Stocker TF, Qin D, Plattner G-K, Tignor M, Allen SK, Boschung J, Nauels A, Xia Y, Bex V, Midgley PM (eds) Climate Change 2013: The Physical Science Basis. Contribution of Working Group I to the Fifth Assessment Report of the Intergovernmental Panel on Climate Change. Cambridge University Press, Cambridge, UK and New York, NY, USA

Dieng H, Champollion N, Cazenave A, Wada Y, Schrama E, Meyssignac B (2015a) Total land water storage change over 2003–2013 estimated from a global mass budget approach. Environ Res Lett **(in press)**

Dieng H, Palanisamy H, Cazenave A, Meyssignac B, von Schuckmann K (2015b) The sea level budget since 2003: inference on the deep ocean heat content. Surv Geophys 36:1. doi:10.1007/s10712-015-9314-6

Dieng H, Cazenave A, von Shuckmann K, Ablain M, Meyssignac B (2015c) Sea level budget over 2005–2013: missing contributions and data errors. Ocean Sci 11:789–802. doi:10.5194/os-11-789-2015

Domingues CM, Church JA, White NJ, Gleckler PJ, Wijffels SE, Barker PM, Dunn JR (2008) Improved estimates of upper-ocean warming and multi-decadal sea-level rise. Nature 453:1090–1093
Durack PJ, Wijffels SE, Boyer TP (2013) Long-term salinity changes and implications for the global water cycle. In: Siedler G, Griffies SM, Gould J, Church JA (eds) Ocean circulation and climate, a 21st century perspective, 2nd edn. International Geophysics, Academic, Elsevier, Oxford, pp 727–757
Fasullo JT, Boening C, Landerer FW, Nerem RS (2013) Australia's unique influence on global mean sea level in 2010–2011. Geophys Res Lett 40(16):4368–4373. doi:10.1002/grl.50834
Gardner AS, Moholdt G, Cogley JG, Wouters B, Arendt AA, Wahr J, Berthier E, Hock R, Pfeffer WT, Kaser G, Ligtenberg SRM, Bolch T, Sharp MJ, Hagen JO, van den Broeke MR, Paul F (2013) A reconciled estimate of glacier contributions to sea level rise: 2003 to 2009. Science 340:852–857
Gouretski VV, Koltermann KP (2007) How much is the ocean really warming? Geophys Res Lett. doi:10.1029/2006GL027834
Gregory JM, Lowe JA (2000) Predictions of global and regional sea-level rise using AOGCMs with and without flux adjustment. Geophys Res Lett 27:3069–3072
Henry O, Ablain M, Meyssignac B, Cazenave A, Masters D, Nerem S, Leuliette E, Garric G (2014) Investigating and reducing differences between the satellite altimetry-based global mean sea level time series provided by different processing groups. J Geod 88:351–361. doi:10.1007/s00190-013-0687-3
Hosoda S et al (2008) A monthly mean dataset of global oceanic temperature and salinity derived 344 from Argo float observations. JAMSTEC Rep Res Dev 8:47–59
Johnson GF, Chambers DP (2013) Ocean bottom pressure seasonal cycles and decadal trends from GRACE release-05: ocean circulation implications. J Geophys Res Oceans. doi:10.1002/jgrc.20307
Konikow LF (2011) Contribution of global groundwater depletion since 1900 to sea-level rise. Geophys Res Lett 38:L17401. doi:10.1029/2011GL048604
Kouketsu S et al (2011) Deep ocean heat content changes estimated from observation and reanalysis product and their influence on sea level change. J Geophys Res Oceans 116:C03012
Levitus S, Antonov JI, Boyer TP, Baranova OK, Garcia HE, Locarnini RA, Mishonov AV, Reagan JR, Seidov D, Yarosh ES, Zweng MM (2012) World ocean heat content and thermosteric sea level change (0–2000 m), 1955–2010. Geophys Res Lett 39:L10603. doi:10.1019/2012GL051106
Llovel W, Becker M, Cazenave A, Jevrejeva S, Alkama R, Decharme B, Douville H, Ablain M, Beckley B (2011) Terrestrial waters and sea level variations on interannual time scale. Glob Planet Chang 75:76–82. doi:10.1016/j.gloplacha.2010.10.008
Llovel W, Willis JK, Landerer FW, Fukumori I (2014) Deep-ocean contribution to sea level and energy budget not detectable over the past decade. Nat Clim Chang. doi:10.1038/NCLIMATE2387
Lyman JM, Johnson GC (2008) Estimating annual global upper-ocean heat content anomalies despite irregular in situ ocean sampling. J Clim 21:5629–5641
Lyman JM, Good SA, Gouretski VV, Ishii M, Johnson GC, Palmer MD, Smith DA, Willis JK (2010) Robust warming of the global upper ocean. Nature 465:334–337. doi:10.1038/nature09043
Marzeion B, Leciercq PW, Cogley JG, Jarosch AH (2015) Brief communication: global glacier mass loss reconstructions during the 20th century are consistent. Cryosphere Discuss 9:3807–3820. doi:10.5194/tcd-9-3807-2015. www.the-cryosphere-discuss.net/9/3807/2015/
Masters D, Nerem RS, Choe C, Leuliette E, Beckley B, White N, Ablain M (2012) Comparison of global mean sea level time series from TOPEX/Poseidon, Jason-1, and Jason-2. Mar Geod 35:20–41
Miller L, Douglas BC (2007) Gyre-scale atmospheric pressure variations and their relation to 19th and 20th century sea level rise. Geophys Res Lett 34:L16602. doi:10.1029/2007GL030862
Mitchum GT (2000) An improved calibration of satellite altimetric heights using tide gauge sea levels with adjustment for land motion. Marine Geodesy 23:145–166
Nerem RS, Chambers DP, Choe C, Mitchum GT (2010) Estimating mean sea level change from the TOPEX and Jason altimeter missions. Mar Geod 33(Supplement 1):435–446. doi:10.1080/01490419.2010.491031
Nerem RS, Chambers DP, Leuliette E, Mitchum GT, Giese BS (1999) Variations in global mean sea level during the 1997–98 ENSO event. Geophys Res Ltrs 26:3005–3008
Ngo-Duc T, Laval K, Polcher J, Lombard A, Cazenave A (2005) Effects of land water storage on global mean sea level over the past 50 years. Geophys Res Lett 32:L09704. doi:10.1029/2005GL022719
Palanisamy H, Cazenave A, Delcroix T, Meyssignac B (2015) Spatial trend patterns in Pacific Ocean sea level during the altimetry era : the contribution of thermocline depth change and internal climate variability. Ocean Dyn. doi:10.1007/s10236-014-0805-7
Peltier WR (2004) Global glacial isostasy and the surface of the ice-age Earth: the ICE-5G (VM2) model and GRACE. Annu Rev Earth Planet Sci 32:111–149

Pokhrel YN, Hanasaki N, Yeh PJ-F, Yamada T, Kanae S, Oki T (2012) Model estimates of sea level change due to anthropogenic impacts on terrestrial water storage. Nat Geosci 5:389–392. doi:10.1038/ngeo1476

Purkey SG, Johnson GC (2010) Warming of global abyssal and deep Southern Ocean waters between the 1990 s and 2000 s: contributions to global heat and sea level rise budgets. J Clim 23:6336–6351

Reager JT, Gardner AS, Famiglietti JS, Wiese DN, Eicker A, Lo M-H (2016) A decade of sea level rise slowed by climate driven hydrology. Science 351:699–703. doi:10.1126/science.aad8386

Rietbroek R, Brunnabend SE, Kushche J, Schröter J, Dahle C (2016) Revisiting the contemporary sea-level budget on global and regional scales. Proc Natl Acad Sci 113:1504–1509. doi:10.1073/pnas.1519132113

Roemmich D, Johnson GC, Riser S, Davis R, Gilson J, Owens WB, Garzoli SL, Schmid C, Ignaszewski M (2009) The Argo program: observing the global ocean with profiling floats. Oceanography 22(2):34–43. doi:10.5670/oceanog.2009.36

Roemmich D, Church J, Gilson J, Monselesan D, Sutton P, Wijffels S (2015) Unabated planetary warming and its ocean structure since 2006. Nat Clim Chang 5:240–245

Schrama EJO, Wouters B, Rietbroek R (2014) A mascon approach to assess ice shett and glacier mass balance and their uncertainties from GRACE data. J Geophys Res Solid Earth 119:6048–6066. doi:10.1002/2013JB010923

Shepherd A et al (2012) A reconciled estimate of ice-sheet mass balance. Sci 338(6111):1183–1189

Sørensen LS, Forsberg R (2010) Greenland ice sheet mass loss from GRACE monthly models. In: Gravity, Geoid and Earth Observation. Springer. (International Association of Geodesy Symposia; No. 135), pp 527–532. doi:10.1007/978-3-642-10634-7_70

Sturges W, Douglas BC (2011) Wind effects on estimates of sea level rise. J Geophys Res 116:C06008. doi:10.1029/2010JC006492

Tapley BD, Bettadpur S, Watkins M, Reigber C (2004) The gravity recovery and climate experiment: mission overview and early results. Geophys Res Lett 31:L09607. doi:10.1029/2004GL019920

Velicogna I, Sutterley TC, van den Broeke MR (2014) Regional acceleration in ice mass loss from Greenland and Antarctica using Grace time variable gravity data. Res Lett Geophys. doi:10.1002/2014GL061052

Von Schuckmann K, Le Traon PY (2011) How well can we derive Global Ocean indicators from Argo data? Ocean Sci 7(6):783–791. doi:10.5194/os-7-783-2011

Von Schuckmann K, Gaillard F, Le Traon P-Y (2009) Global hydrographic variability patterns during 2003–2008. J Geophys Res 114:C09007. doi:10.1029/2008JC005237

Von Schuckmann K, Sallée JB, Chambers D, Le Traon PY, Cabanes C, Gaillard C, Speich S, Hamon M (2014) Consistency of the current global ocean observing systems from an Argo perspective. Ocean Sci 10:547–557. doi:10.5194/os-10-547-2014

Wada Y (2015) Modelling groundwater depletion at regional and global scales: Present state and future prospects. Surv Geophys. doi:10.1007/s10712-015-9347-x, Special Issue: ISSI Workshop on Remote Sensing and Water Resources

Wada Y, van Beek LPH, Sperna-Weiland FC, Chao BF, Wu Y-H, Bierkens MFP (2012) Past and future contribution of global groundwater depletion to sea-level rise. Geophys Res Lett 39:L09402. doi:10.1029/2012GL051230

Watkins MM, Wiese DN, Yuan D-N, Boening C, Landerer FW (2015) Improved methods for observing Earth's time variable mass distribution with GRACE. J Geophys Res Solid Earth. doi:10.1002/2014JB011547

Watson CS, White NJ, Church JA, King MA, Burgette RJ, Legresy B (2015) Unabated global mean sea-level rise over the satellite altimeter era. Nat Clim Chang 5(6):565. doi:10.1038/nclimate2635

Wijffels SE, Willis J, Domingues CM, Barker P, White NJ, Gronell A, Ridgway K, Church JA (2008) Changing expendable bathythermograph fall rates and their impact on estimates of thermosteric sea level rise. J Clim 21:56575672. doi:10.1175/2008JCLI2290.1

Yi S, Sun W, Heki K, Qian A (2015) An increase in the rate of global mean sea level rise since 2010. Geophys Res Lett. doi:10.1002/2015GL063902

Part IV
Detection & Attribution

Surv Geophys (2017) 38:329–348
DOI 10.1007/s10712-016-9373-3

Internal Variability Versus Anthropogenic Forcing on Sea Level and Its Components

Marta Marcos[1] · Ben Marzeion[2] · Sönke Dangendorf[3] · Aimée B. A. Slangen[4,5] · Hindumathi Palanisamy[6] · Luciana Fenoglio-Marc[7]

Received: 30 November 2015 / Accepted: 18 May 2016 / Published online: 30 May 2016

Abstract In this paper we review and update detection and attribution studies in sea level and its major contributors during the past decades. Tide gauge records reveal that the observed twentieth-century global and regional sea level rise is out of the bounds of its natural variability, evidencing thus a human fingerprint in the reported trends. The signal varies regionally, and it partly depends on the magnitude of the background variability. The human fingerprint is also manifested in the contributors of sea level for which observations are available, namely ocean thermal expansion and glaciers' mass loss, which dominated the global sea level rise over the twentieth century. Attribution studies provide evidence that the trends in both components are clearly dominated by anthropogenic forcing over the second half of the twentieth century. In the earlier decades, there is a lack of observations hampering an improved attribution of causes to the observed sea level rise. At certain locations along the coast, the human influence is exacerbated by local coastal activities that induce land subsidence and increase the risk of sea level-related hazards.

Keywords Mean sea level rise · Thermal expansion · Glaciers melting · Detection and attribution · Land subsidence

✉ Marta Marcos
marta.marcos@uib.es

1 Mediterranean Institute for Advanced Studies (UIB-CSIC), Mallorca, Spain

2 University of Bremen, Bremen, Germany

3 University of Siegen, Siegen, Germany

4 Commonwealth Scientific and Industrial Research Organisation, Hobart, Australia

5 Institute for Marine and Atmospheric Research, Utrecht University, Utrecht, The Netherlands

6 Laboratoire d'Etudes en Géophysique et Océanographie Spatiales, Toulouse, France

7 University of Bonn, Bonn, Germany

1 Introduction

Long-term sea level observations from tide gauges distributed worldwide provide evidence of global sea level rise during the twentieth century. Depending on how these tide gauge records are combined, global sea level rise rates range between 1.3 and 2.0 mm/year for the twentieth century (Jevrejeva et al. 2008; Church and White 2011; Hay et al. 2015), with uncertainties that have been estimated at about an order of magnitude smaller. These rates for the instrumental era contrast with negligible changes reported for the past 2000 years from sea level proxies based on salt-marsh sediments (Kemp et al. 2011) and have been identified as the largest during at least the last twenty-seven centuries (Kopp et al. 2016). Superimposed on the centennial trend, global sea level exhibits shorter-term variability at inter-annual up to multi-decadal timescales, which is generally assumed to be the response to natural variations of the climate system. For example, some of the most prominent features in global sea level curves correspond to strong El Niño episodes (e.g., Han et al. this issue and references therein). The presence of these oscillations opens the debate of to what extent the reported trends are impacted by internal climate variability or, conversely, they can be attributed to external drivers related to anthropogenic forcing.

To address the role of internal variability in sea level, the persistence of sea level changes at long timescales has been discussed and characterized in recent works (Bos et al. 2013; Dangendorf et al. 2014a, b; Becker et al. 2014). Sea level variability has a strong long-term memory that implies an enhancement of the uncertainty in the observed trends and, consequently, a reduction of their statistical significance. The detection of an external forcing is therefore constrained by the spectral characteristics of the time series. In addition to this, Dangendorf et al. (2015) demonstrated that the different contributors to sea level present diverse spectral properties and that these should be characterized separately for an accurate assessment of the statistical significance of the observed trends. On the basis of this procedure, Dangendorf et al. (2015) concluded that at least 45 % of the observed global sea level rise during the twentieth century is of anthropogenic origin.

Detection and attribution (D&A) of anthropogenic signals in sea level change and associated properties is still a rapidly developing field. In the Intergovernmental Panel for Climate Change (IPCC) terminology, detection is meant to demonstrate a statistical change in the climate system, while attribution consists of the evaluation of the contribution of multiple causes to such a change (Hegerl et al. 2010). In the IPCC AR5 chapter on D&A (Bindoff et al. 2013), there was no formal assessment of total sea level change or its contributors. However, there have been some works addressing this issue. Jevrejeva et al. (2009) used a statistical model that showed that anthropogenic forcing was responsible for up to 70 % of sea level rise since 1900. More recently, Kopp et al. (2016) estimated this fraction as about 50 %. Along the same line, Slangen et al. (2016) found an increasing anthropogenic contribution to global sea level rise, due to increased concentrations of greenhouse gases (GHGs) becoming dominant after 1970, according to climate models.

At the global scale the two major contributors to rising sea levels are ocean thermal expansion and water mass addition from glacier and ice sheet mass loss. Although the physical relationship between the increase in ocean heat content and thermal expansion was appointed to an anthropogenic contribution to sea level rise (e.g., Barnett et al. 2005; Gleckler et al. 2012), and similarly the observed glacier mass loss and thus its contribution to sea level rise were attributed to human influence, the anthropogenic contributions could not yet be quantified at the time the last report of the IPCC was published. More recent attempts have shed some light on the anthropogenic origin of these two main contributors to global sea level rise and have quantified the human fingerprint for the periods in which there are independent observations (Marcos and Amores 2014; Slangen et al. 2014, 2015; Marzeion et al. 2014).

At the regional scale, D&A studies on sea level are highly challenging. One main reason is that the internal climate variability introduces strong changes in regional sea level on timescales from years to decades which makes the signal-to-noise ratio very unfavourable to detect an external signal above the unforced internal variability (Richter and Marzeion 2014; Monselesan et al. 2015; Palanisamy et al. 2015a). This effect is more important in regions with large sea level variance. For example, the internal sea level variability related to El Niño Southern Oscillation (ENSO) and to the Pacific Decadal Oscillation (PDO) climate modes in the Pacific Ocean is of the order of ±10–20 cm and can, therefore, mask sea level changes due to any externally forced signal (Palanisamy et al. 2015a).

In the present paper, we review the most recent works that deal with D&A studies on sea level and its contributors at both global and regional scales. We first discuss and update the estimates of anthropogenic contribution to the twentieth-century sea level rise as observed by tide gauges (Sect. 2). We then separately assess the two most important contributors to global sea level change during the past century, namely ocean thermal expansion and glacier mass loss. Anthropogenic impacts are quantified on the basis of earlier works and, in the case of glaciers, we also include updated results (Sect. 3). In Sect. 4, the ability to separate the human impact from internal sea level variations at the regional scale is discussed, paying special attention to the Pacific Ocean, where the strongest sea level trends have been observed during the last two decades. For completeness, we also discuss here the local effects of vertical land motion that are a consequence of human activities, since these are relevant in terms of coastal impacts of sea level rise (Sect. 5). Finally, the different assessments are discussed and set in a common framework in the concluding section.

2 Detection of Anthropogenic Twentieth-Century Sea Level Rise

Sea level records show temporally correlated noise whose spectrum can be properly described by a power-law model (Agnew 1992; Barbosa et al. 2008). The particular shape of the spectrum of sea level variability may alter the magnitude of a trend estimated via ordinary least squares regression even on centennial timescales and therefore significantly enhance the uncertainty of linear trends. Indeed, Bos et al. (2014) found that accounting for temporal correlations increased the uncertainties in linear trends of monthly tide gauge data by a factor up to 2 with respect to classical models using white noise or AR(1) processes. An accurate assessment of these uncertainties is a key factor to determine the statistical significance of the observed linear sea level rise rates and, consequently, to detect an external forcing.

Different approaches can be used to characterize the long-term correlations in sea level time series. Barbosa et al. (2008) applied three different methods, namely parametric statistical tests for stationarity, wavelet analysis for assessing scaling features, and generalized least squares to long tide gauge records in the North Atlantic. They concluded that the low-frequency structure of tide gauge records shows diverse features, indicating the local or regional character of their long-term persistence. More recently, Dangendorf et al. (2014a) and Becker et al. (2014) applied a second-order detrended fluctuation analysis

(DFA2; Kantelhardt et al. 2001) to evaluate the persistence in long tide gauge records. DFA2 computes the fluctuation function whose slope α (on a log–log scale) is equivalent to the Hurst exponent of a stochastic power-law process and represents the temporal correlation features of the time series. Hence, long-term correlated processes are identified when $\alpha > 0.5$. The advantage of DFA2 lies in its robustness even in the presence of non-stationarities (Kantelhardt et al. 2001), which are from a palaeo-perspective without any doubt visible in sea level over the past century (e.g., Kemp et al. 2011; Gehrels and Woodworth 2013).

Becker et al. (2014) estimated the fraction of the observed linear trends at tide gauges and for global sea level (Church and White 2011; Jevrejeva et al. 2006) that cannot be explained by the long-term memory of sea level variations and that is thus linked to an external forcing. This is done through the probability of observing a relative increase larger than a given value that can be derived from Monte Carlo simulations and a fitted Student's t distribution (Dangendorf et al. 2015). They concluded that, at the global scale, the anthropogenic contribution to sea level rise during the twentieth century is at least between 1 and 1.3 mm/year, depending on the global sea level curve considered in the analysis. In other words, more than 50 % of the observed global sea level rise rate, with a 99 % confidence level. Becker et al. (2014) showed that for individual tide gauges the strength of the detectable external contribution (including anthropogenic forcing but also other sources such as vertical land motion) in local sea level varies significantly, in agreement with the finding by Barbosa et al. (2008).

Dangendorf et al. (2014a) investigated the sources of the long-term memory with respect to the individual contributors to modern sea level rise. The authors demonstrated that long-term memory is clearly visible in the steric component, which underlines the large inertia of the ocean (Levermann et al. 2013; Wunsch and Heimbach 2014), but also in land-based ice. This confirms findings from independent analyses, for instance, from glaciers, as shown in the next section. However, while there is no doubt about the influence of steric and mass variations on global scales and at many open ocean sites, along the coasts, where tide gauges are situated, local wind and pressure-driven redistribution processes or ocean circulation changes are known to be the dominant contributor to the observed sea level variability (e.g., Wakelin et al. 2003; Marcos and Tsimplis 2007; Calafat and Chambers 2013; Thompson et al. 2014; Piecuch and Ponte 2015). This is, for example, the case in the North Sea, where local atmospheric forcing is a major driver of sea level variability (Dangendorf et al. 2014a, b). Dangendorf et al. (2015) argued that, since this atmospheric component is a short-term correlated process that can be accurately represented by an AR(1) model, it should be addressed separately from the slowly varying steric and mass components when assessing the persistence in local sea level. Otherwise, natural linear trends can be underestimated by up to 1 mm/year locally on centennial timescales. However, on global scales such redistribution effects cancel out. Based on a model assessment, Dangendorf et al. (2015) further concluded that the anthropogenic contribution to global sea level rise ranges between 0.7 and 1.3 mm/year, confirming the human fingerprint in twentieth-century global sea level rise identified by Becker et al. (2014). These numbers, representing 45 % of the observed global sea level rise, are in good agreement with Slangen et al. (2016) who obtained an anthropogenic contribution of 37 ± 38 %, despite the different methods applied. They are also consistent with Kopp et al. (2016), whose estimate was about 50 %.

One remaining issue from the discussed detection studies by Becker et al. (2014) and Dangendorf et al. (2015) is the unresolved effect of vertical land motion of the sites in which the tide gauges are grounded. To update the two analyses and to tackle this issue, we evaluate here the observed sea level changes on a regional scale by using the tide gauge data set recently published by Wöppelmann et al. (2014). Wöppelmann et al. (2014) updated available GPS-based estimates of vertical land motion and subtracted them from individual stations around the world. They further merged the individual records, whenever possible, into regional averages to get a more or less homogeneous picture over sea level changes around the world. These regional series are shown in Fig. 1 (left side), while information about their temporal availability and their regional composition is provided in Table 1. In agreement with Bos et al. (2014), Becker et al. (2014) and Dangendorf et al. (2014a) we find evidence for long-term correlations in all investigated series. The fluctuation functions of the individual series all have a slope well above 0.5 (Fig. 1, right side) leading to Hurst exponents ranging from 0.62 in the Baltic Sea up to 0.94 in the Pacific (Table 1). The regional variability in Hurst exponents mainly reflects the varying influence of barotropic adjustments of the ocean to local or regional winds (Dangendorf et al. 2015). While in some regions, such as the North Sea, the regional averaging can reduce this bias compared to earlier studies (Becker et al. 2014; Dangendorf et al. 2014a, b), in the Baltic Sea the Hurst exponents still remain on a comparably low level (Table 1). This is probably related to the fact that, in the mostly enclosed Baltic Sea, sea level adjusts on a basin scale

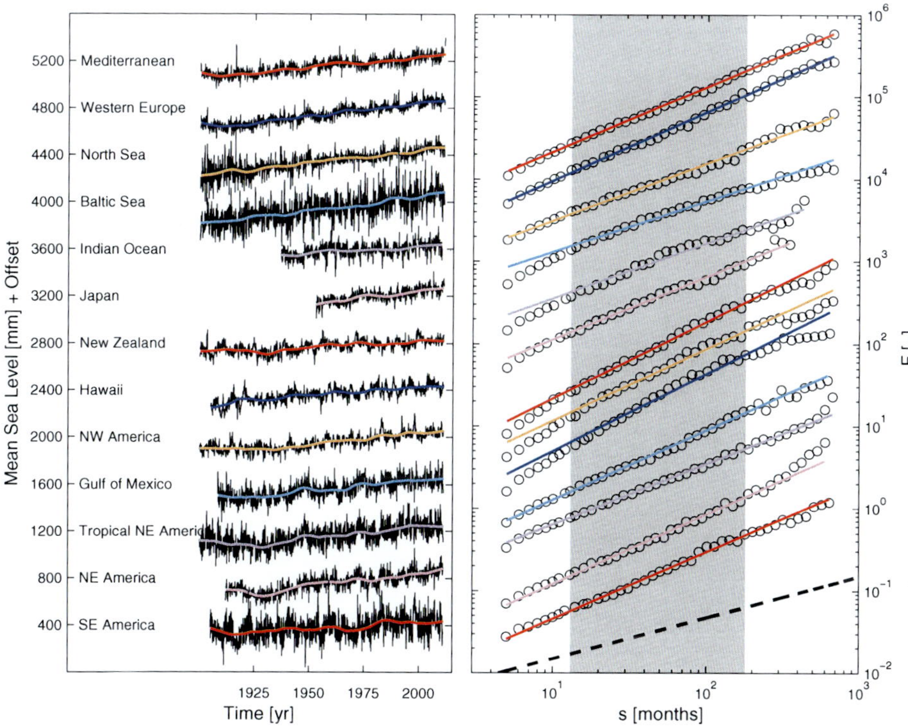

Fig. 1 **a** Regional sea level records from Wöppelmann et al. (2014), built with GPS-corrected deseasoned tide gauge time series (*black curves*). *Coloured lines* represent a nonlinear trend derived from a LOWESS filter with a cut-off period of 15 years. **b** Fluctuation functions calculated with DFA2. The *black dashed line* marks the exponent $\alpha = 0.5$, i.e. uncorrelated noise. The *grey-shaded area* displays the window for which the α value has been estimated ($13 \leq s \leq 339$ months)

Table 1 Anthropogenic contribution to observed regional sea level rise

Region	No. of stations	Period (year)	Alpha (−)	Observed trend (mm/year)	Significance (%)	Minimum anthropogenic contribution (mm/year)	Minimum anthropogenic contribution (%)
Mediterranean	3	1900–2010	0.79	1.72	1.00	1.21	70.39
Western Europe	5	1900–2010	0.83	2.24	1.00	1.66	74.15
North Sea	12	1900–2010	0.69	2.16	1.00	1.74	80.53
Baltic Sea	12	1900–2010	0.62	2.39	1.00	1.92	80.62
Indian Ocean	2	1937–2009	0.67	1.26	1.00	0.59	46.73
Japan	2	1953–2010	0.76	2.32	1.00	1.27	54.84
New Zealand	3	1900–2010	0.93	1.03	0.99	–	–
Hawaii	2	1905–2010	0.94	1.61	1.00	0.33	20.47
NW America	8	1900–2010	0.87	1.64	1.00	0.82	49.88
Gulf of Mexico	5	1908–2010	0.84	1.85	1.00	0.76	40.90
Tropical NE America	4	1900–2010	0.74	1.72	1.00	0.97	56.58
NE America	12	1912–2010	0.85	2.12	1.00	1.12	52.64
SE America	2	1905–2010	0.81	1.03	0.97	–	–

For each region, shown here are the number of stations used to build the curve, the period covered by the resulting time series, the Hurst exponent, the observed linear trend, its statistical significance, and the minimum anthropogenic contribution (in absolute value and percentage of the total trend) to wind stress forcing. Hence, the regional averaging only barely reduces the strong short-term variability visible in local records (Hünicke et al. 2015).

Based on ordinary least squares in combination with the updated Hurst exponents we follow Becker et al. (2014) and Dangendorf et al. (2015) and estimate the linear trends as well as their related uncertainties for each individual region. For each series the estimated trend, its significance S, the maximum natural contribution ($P = 0.99$) and the resulting minimum anthropogenic contribution ($P = 0.99$) are shown in Table 1 and Fig. 2. In 9 out of 11 regions we find statistically significant trends ($S > 0.99$) over the twentieth century meaning that the observed slope clearly exceeds the value, which can be maximally expected by natural variations. The two regions where no significant trend ($P = 0.99$) can be detected are New Zealand and southeast America (Table 1 and Fig. 2). However, it is important to notice that in both regions the trends would become significant using a less stringent confidence level of 95 %. The largest minimum anthropogenic contribution is found in the North Atlantic (North Sea, Baltic Sea, Mediterranean, Western Europe, and NE America) where the signals vary between 1.12 and 1.92 mm/year (Fig. 2; Table 1). This is equal to a minimum anthropogenic contribution of roughly 52–81 % of the

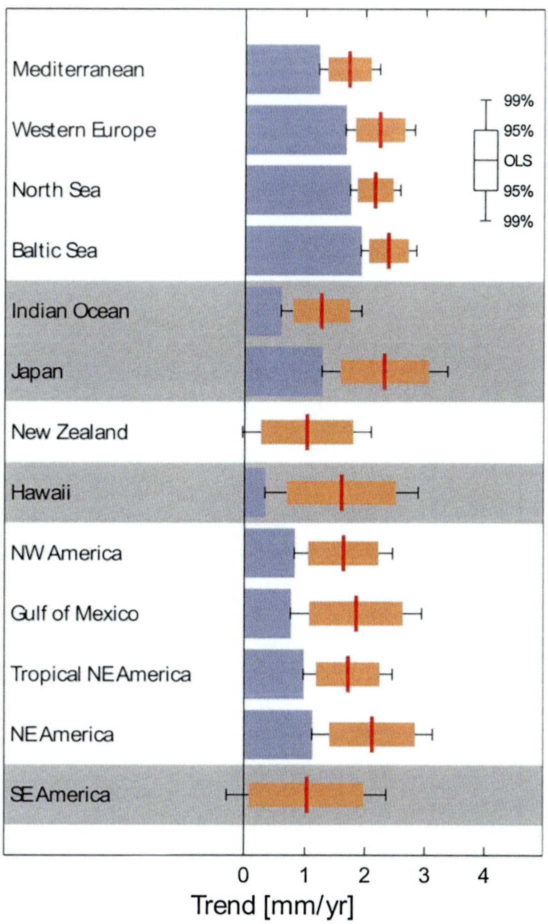

Fig. 2 Observed regional sea level linear trends and their maximum natural contributions (at 95 and 99 % confidence levels) are represented in *box-plots*. The minimum anthropogenic regional sea level trends (at the 99 % confidence level) are also *plotted as blue bars*

observed linear trends over the twentieth century in these regions (Table 1). Slightly smaller numbers are found in the Pacific, the Indian Ocean, and the Gulf of Mexico with values scattering between 0.33 mm/year at Hawaii and 0.82 mm/year along coast of NW America. The corresponding percentage of the minimum anthropogenic contribution ranges between roughly 20 and 50 %. These numbers are in general accordance with the results obtained by Becker et al. (2014) and underpin the finding that in the Indo-Pacific region a considerable portion of the observed sea level changes are related to climate internal modes (Meyssignac et al. 2012; Hamlington et al. 2014; Palanisamy et al. 2015a; see also Sect. 4 in this paper). Summarizing, we find evidence for an anthropogenic signature in the majority of the global ocean. However, our results suggest hemispheric differences in the minimum anthropogenic contribution with larger values obtained for the Northern Hemisphere. These differences may be related to both the hemispheric differences obtained in the amplitudes of the trends (Wöppelmann et al. 2014) and also differences in the role of internal variability.

3 Detection and Attribution of Human Impact on Global Sea Level Components

3.1 Ocean Thermal Expansion

Long-term sea level studies rely on a number of historical tide gauge records that, although sparse, cover most of the twentieth-century period. Conversely, when the ocean thermal expansion is studied, any investigation is constrained by the limited availability of ocean temperature observations. Only since about 1950 was the upper ocean layer extensively monitored, although it was not until late 1960s that widely used expendable bathythermographs (XBTs) replaced earlier instrumentation (Lyman and Johnson 2014). This is the reason why D&A studies in ocean warming have been limited to the most recent decades (Barnett et al. 2001, 2005; Gleckler et al. 2012; Marcos and Amores 2014; Slangen et al. 2014). To make up for this short observational period and the small number and uneven distribution of hydrographic data, numerical ocean models have successfully been applied as a key tool for attributing the observed changes in ocean rising temperatures to a human origin.

Ocean thermal expansion, or thermosteric sea level, integrates changes in ocean temperatures through the entire water column. Marcos and Amores (2014) investigated the human fingerprint in thermosteric sea level in the top 700 m of the ocean since 1970. They adopted this shorter period as a trade-off between the spatial coverage and the length, in order to improve the observational coverage while keeping the period long enough so as to detect an external forcing in the warming signal. They used ocean temperatures from numerical simulations of the Coupled Model Intercomparison Project Phase 5 (CMIP5) together with hydrographic databases (Ishii and Kimoto 2009; Levitus et al. 2012). They applied a signal-to-noise (S/N) maximizing empirical orthogonal function (EOF) analysis, formulated by Venzke et al. (1999), to characterize the common response of an ensemble of numerical realizations to a predefined external forcing. Very briefly, this technique separates the time-varying forced signal from the internal variability of the system in an ensemble of numerical simulations, thus eliminating the climate noise (see Marcos and Amores 2014, for a full description of the algorithm). Based on the comparison between the forced signals in historical simulations and in natural forcing-only simulations, Marcos and Amores (2014) concluded that, since 1970, 87 % (with 95 % confidence interval of

72–100 %) of the observed warming-related sea level rise in the 0–700 m of the global ocean is of anthropogenic origin.

Compared to Marcos and Amores (2014), Slangen et al. (2014) extended the analysis of global mean thermosteric sea level to a slightly longer period (1957–2005), to the full ocean depth, and to 5 different CMIP5 historical experiments: historical (all-forcing), natural-only, anthropogenic-only, GHG-only, and aerosol-only. Based on the comparison with different hydrographic data sets, they found that the sea level response to anthropogenic GHG forcing explains most of the magnitude of the observed differences, while the natural forcing is required to explain temporal variability. Optimal regression (Allen and Tett 1999) was used to find the best fit of the modelled responses to external forcing with observed changes, in the presence of internal climate variability. For the 1957–2005 period, they found a scaling factor of 1.08 ± 0.13 for the anthropogenic-only experiment, leading to an anthropogenic contribution of 93 % (1/1.08). Considering the same period as in Marcos and Amores (2014), that is 1970–2005, they identify the anthropogenic fraction as 89 %. This means that, despite the differences in analysis method, time period, depth, and the use of a dedicated anthropogenic experiment (instead of a difference between historical and natural-only in Marcos and Amores 2014), the results of both studies agree very well and attribute the vast majority of observed thermosteric change in the late twentieth century to anthropogenic forcing.

3.2 Glaciers

As with hydrographic ocean observations, there are also limitations in glacier measurements. Direct, field-based measurements of glacier mass change are costly and often unfeasible in remote high mountain terrain, where many of the world's glaciers are located. The sampling of glacier mass change measurements is therefore very sparse (in the global mean, a density of 1.2 measurements per 1000 km^2 per year), very uneven (in the Alps, where less than 0.3 % of the world's glacierized area is located, the density is 84.6 measurements per 1000 km^2 per year), and probably not representative (see the supplement of Gardner et al. 2013, for more details). To obtain continuous time series long enough to allow D&A statistics, the uneven and temporally varying sampling has to be taken into account in the upscaling from local measurements to global estimates (Cogley 2009).

Remote sensing allows a better spatial coverage of glacier change measurements, but has other limitations: mass change estimates of glaciers based on volumetric measurements (e.g., from differencing of digital elevation models) need to make assumptions about the density of ice, snow, and firn, and about potential changes in density during the covered period. Gravimetric measurements need to separate the glacier mass change signal from other mass changes, e.g., in the hydrologic system, of tectonic origin, or from ocean mass redistribution (Gardner et al. 2013). Wide spatial coverage of glaciers using these methods is also limited to the recent past, and is thus not able to create continuous time series long enough to allow statistical separation of internal variability and systematic change.

The detection of systematic trends in glacier mass is further complicated by the long timescales of glacier response to changes in the climatic forcing. Since the adaptation of a glacier's geometry to a change in forcing may take between decades and centuries, low-frequency variability in glacier mass change time series can be expected even if there is no corresponding low-frequency variability in the forcing (Burke and Roe 2014). Glacier length change records are available for much longer periods than glacier mass change measurements and, with a few exceptions, show a consistent retreat of glaciers at least during the twentieth century, in some regions also earlier (Leclercq et al. 2014). While these data may harbour the potential to detect systematic changes in glacier behaviour during the observation period (without necessarily attributing them to individual causes), a joint analysis of the collected length change records has not been done. For individual glaciers, such a detection is impeded through the large variability of glacier length that can be explained by random climate forcing alone (Roe and Baker 2014).

Detection of systematic changes to the glaciers mass budget, and in particular attribution of these changes to specific climate forcings, has therefore mostly relied on a combination of observational data with climate and glacier models. Based on a highly idealized glacier model, Oerlemans (1988) concluded that roughly half of glacier retreat since the end of the Little Ice Age (LIA) is attributable to anthropogenic GHG. Reichert et al. (2002) modelled the behaviour of two glaciers in Norway and Switzerland, using atmospheric forcing obtained from a coupled general circulation model (GCM). They found that the observed retreats of those glaciers exceeded what could be expected from internal variability alone.

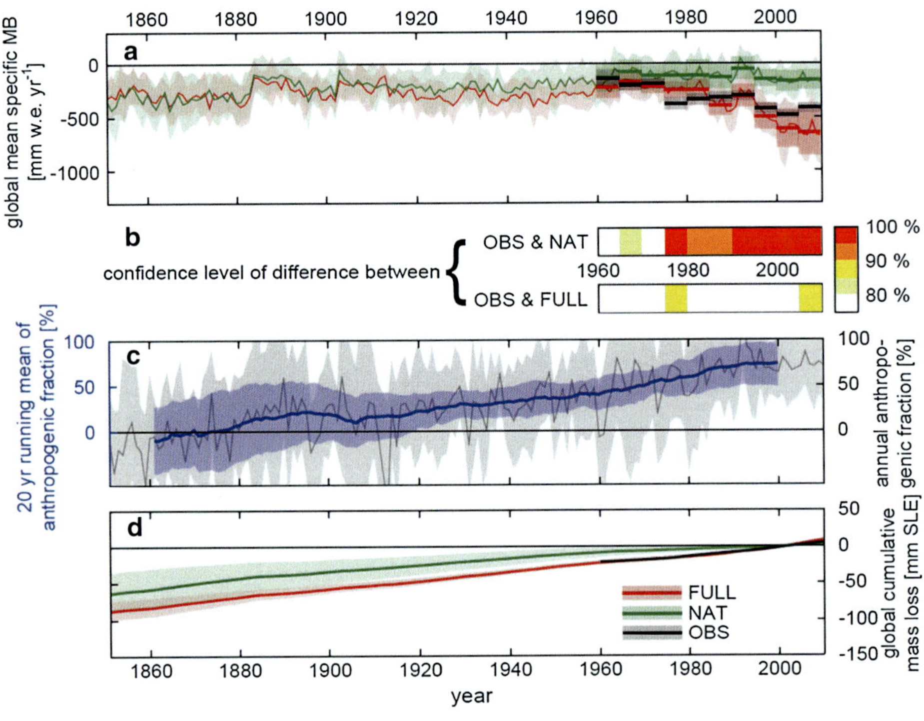

Fig. 3 **a** Global mean specific mass balance time series (*thin lines* are the ensemble means, and *shading* indicates one ensemble standard deviation) and pentadal means (*thick lines* are the ensemble means, and shading indicates one standard error). *Green* NAT results, *red* FULL results, *black* observations. **b** Confidence level of the difference between interpolated observations (OBS) updated from Cogley (2009) and model results for the NAT and FULL models for each pentad. **c** Anthropogenic fraction of total glacier mass loss, annual values (*grey*), and running mean over 20-year periods (*blue*). The *solid line* is the ensemble mean, and *shading* indicates one ensemble standard deviation. **d** Glacier contribution to global mean sea level rise, relative to the mean of 1991–2010. Modelled results include modelled glacier area change, and observations assume constant glacier area. The *solid lines* are the ensemble means, and *shading* indicates one ensemble standard deviation. Updated from Marzeion et al. (2014)

Marzeion et al. (2014) modelled mass changes of glaciers on the global scale for the period 1850–2010, based on two different climate reconstruction scenarios: first, an ensemble of GCMs using natural forcing (i.e. of solar and volcanic origin) as the only boundary condition was used to force the glacier model. Then they used the same ensemble of GCMs, additionally including anthropogenic forcing (i.e. aerosols, land use change, and GHG) in the boundary conditions. Figure 3 shows an update of their results (based on a more current version of the glacier inventory used to initialize the glacier model). On the global scale, for the past three decades, modelled glacier mass change agrees with observations if—and only if—anthropogenic forcing is included in the boundary conditions of the GCMs used to force the glacier model. Before this period, the difference between the two reconstruction scenarios is too small compared to the uncertainties of both the observations and the model results to robustly detect the anthropogenic impact (Fig. 3a, b). Nevertheless, the results indicate that the anthropogenic fraction of glacier mass loss has been increasing more or less steadily since the early twentieth century and has reached 71 ± 40 % for the period 1991 to 2010 (Fig. 3c). While the majority of glacier mass loss since the end of the LIA is thus attributable to natural causes (Fig. 3d), anthropogenic forcing is now dominating the glaciers' contribution to sea level rise.

4 Anthropogenic Signal in Regional Sea Level Trend Patterns

Regional sea level variability is significantly larger than globally averaged sea level changes due to the internal climate variations (e.g., Church et al. 2013; Stammer et al. 2013). For example, the regional sea level changes in the tropical Pacific Ocean are governed by natural climate modes such as ENSO and PDO/IPO at inter-annual and decadal timescales, respectively (Zhang and Church 2012; Han et al. 2013; Hamlington et al. 2014a; Han et al. this issue). This implies that, at the regional scale, there may be a low signal-to-noise ratio that hampers the detection of an external forcing on sea level patterns.

Several studies have demonstrated the spatially heterogeneous response of sea level to an external forcing, either under future emission scenarios (Bordbar et al. 2015; Yin et al. 2010; Bouttes et al. 2014; Bouttes and Gregory 2014) or in the past. In particular, Slangen et al. (2015) carried out a detection study of human imprints on sea level patterns. Using the same set of numerical experiments as in Slangen et al. (2014), they evaluated the modelled response of sea level to external forcings over the periods 1861–2005. They found that, similar to the global mean thermosteric change, the regional sea level pattern is dominated by the response to anthropogenic forcing. In addition, the modelled historical sea level response shows patterns that are distinctly different from the response to only natural forcing or internal variability. However, a formal D&A analysis could not be carried out, since observations of sea level are not available before 1993 (the satellite era).

Recently, in line with D&A research, a number of "Time of Emergence" (ToE) studies have been performed on regional sea level variability (Lyu et al. 2014; Jordà 2014; Richter and Marzeion 2014; Bilbao et al. 2015). ToE is defined as the time when the ratio of the climate change signal to the noise of natural variability exceeds a particular threshold and emerges from the natural climate variability at regional scale. Lyu et al. (2014) found that, relative to the 1986–2005 reference period, the externally forced trend would be detectable in both steric and dynamic sea level by early to mid-2040s in 50 % of all the oceans, under representative concentration pathways (RCP) 4.5 and RCP8.5 scenarios.

Likewise, Richter and Marzeion (2014) concluded that the externally forced signal is detectable in the early 2030s relative to 1990 in 50 % of the world oceans. These regions include the South Atlantic Ocean, Arctic Ocean, eastern Pacific Ocean, and most parts of the Indian Ocean. Furthermore, Jordà (2014) has shown that, on average, it would require a minimum period of 40 years to identify the externally forced signal at the regional scale. However, in regions with strong decadal and inter-annual sea level variability, the emergence time increases up to 60–80 years, in agreement with Richter and Marzeion (2014) and Lyu et al. (2014). More recently Bilbao et al. (2015) have used the pattern scaling method to study the ToE of regional sea level change. The assumption underlying this method is that the local response of a climate variable (sea level, in this case) is linearly related to a global variable known as the predictor (e.g., global mean surface air temperature, global mean sea surface temperature, and ocean volume mean temperature), with the geographical pattern of the change being independent of the forcing. Choosing mean ocean temperature as predictor, Bilbao et al. (2015) showed that the local sea level change will emerge first in the northern latitudes of the Southern Ocean and in the Tropical Atlantic, where the unforced internal variability is smaller. In contrast, at southern latitudes of the Southern Ocean, it may not emerge until after 2100. Interestingly, they noted that the local time of emergence is independent of the future emission scenarios, since the local sea level signal emerges in the early twenty-first century when the scenarios have not yet significantly varied. Overall, all these works point out that in regions of high internal variability, the trend due to externally forced signal is masked during longer time spans than in regions of low internal variability. The former is the case, for example, of the tropical Pacific Ocean that is discussed in detail in the following.

4.1 The Case of the Pacific Ocean

The Pacific Ocean is the region where the largest sea level trends over the altimetry period are observed and it has, therefore, been a subject of keen interest among the climate and sea level community. Sea level trends in the Pacific Ocean since 1993 (Fig. 4) display two distinct patterns: (1) a V-shaped broad-scale positive trend pattern extending from about 30°–50°N in the central basin to the western equatorial Pacific and then 30°–50°S in the central–eastern basin and (2) a well-pronounced strong dipole-like pattern with positive trends in the western tropical Pacific (with two relative maxima near 10°N and 10°S) and negative trends in the central–eastern tropical Pacific (with relative minima trapped in the equatorial band) limited to about 30°N and 20°S. Several studies (Carton et al. 2005; Köhl et al. 2007; Bindoff et al. 2007; Levitus et al. 2009; Timmermann et al. 2010; Merrifield 2011; Merrifield and Maltrud 2011; Merrifield et al. 2012; McGregor et al. 2012; Becker et al. 2012; Nidheesh et al. 2013; Stammer et al. 2013; Palanisamy et al. 2015b) have shown that the strong dipole-like pattern with a positive (negative) trend in the western (eastern) tropical Pacific during the altimetry era is mainly of thermosteric (0–700 m) origin due to surface wind stress-driven vertical thermocline movement. This spatial pattern has also been related to ENSO events. While the strong dipole-like sea level trend pattern in the tropical Pacific has been related to the ENSO internal climate variability, the V-shaped broad-scale positive trend pattern extending towards the entire Pacific Ocean has been linked with the decadal IPO/PDO climate variability (Zhang and Church 2012; Han et al. 2013; England et al. 2014; Hamlington et al. 2014). The reader is referred to Han et al. (this issue) for a detailed review on sea level spatial trend patterns and internal climate modes.

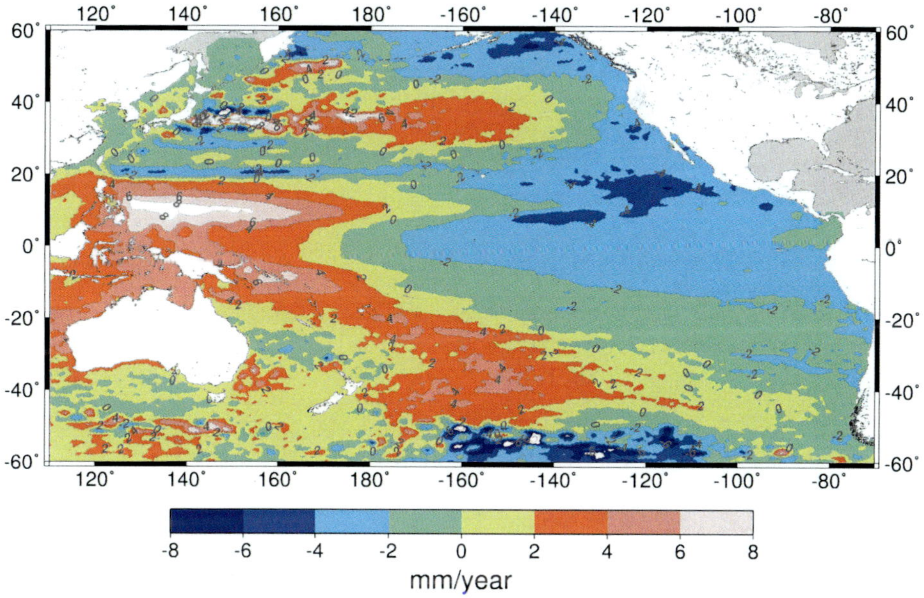

Fig. 4 Observed altimetry-based Pacific Ocean sea level spatial trend pattern over 1993–2013 (uniform global mean has been removed)

The observed response of the tropical Pacific is the highest climate-related regional sea level trend variability not only during the altimetry era, but also at longer timescales (for the past 6 decades, at least) (Becker et al. 2012, Meyssignac et al. 2012). Understanding the origin and mechanisms involved and separating the roles of internal climate variability and anthropogenic signal in the Pacific Ocean sea level spatial trend patterns is a very challenging task that has been addressed in a number of earlier works. As demonstrated by Zhang and Church (2012) the contribution of ENSO and IPO internal climate modes (inter-annual and decadal modes, respectively) to regional sea level variability patterns in the Pacific Ocean during the altimetry era explains about 60 % of the total sea level variance. This raises the question whether the observed regional sea level trend pattern in the Pacific Ocean is the result of only internal inter-annual and decadal climate variability or if, conversely, there is an anthropogenic fingerprint. Meyssignac et al. (2012), on the basis of Coupled Model Intercomparison Project Phase 3 (CMIP3) models, concluded that over the short altimetry record (17 years at the time of their study) the amplitude of the noise represented by the internal climate variability was so strong in the tropical Pacific that it hindered the detection of an anthropogenic forcing. Hamlington et al. (2014) estimated the contribution of the PDO by an empirical orthogonal function (EOF) analysis of a sea level reconstruction based on 20-year trend patterns from 1950 to 2010. They removed this contribution to the observed sea level and were subsequently able to relate the residual signal to anthropogenic warming of the tropical Indian Ocean (Han et al. 2013), concluding that the anthropogenic fingerprint was already visible in the Pacific Ocean regional sea level trend pattern over the altimetry era, in contrast to the conclusions by Meyssiganc et al. (2012). In a more recent study, Palanisamy et al. (2015a) estimated the contribution of decadal IPO climate variability by linearly regressing the Interdecadal Pacific Oscillation (IPO) climate index on the observed altimetry-based Pacific Ocean sea level signal.

The contribution of the IPO was in good agreement with the decadal sea level fingerprint of Zhang and Church (2012) as well as with that of Hamlington et al. (2014). However, they found that the residual signal contained nonlinear ENSO-related internal variability, indicating that attempts to separate/remove both decadal and inter-annual climate modes from observed altimetry-based sea level signal through the method of linear regression (as in Palanisamy et al. 2015a) or the methodology of Hamlington et al. (2014) do not, in fact, totally eliminate the internal sea level variability. In addition to this major conclusion, based on CMIP5 ensemble mean dynamic sea level data, Palanisamy et al. (2015a) also showed that the anthropogenic sea level fingerprint observed in the tropical Pacific is not comparable to that of the residual signal from altimetry without IPO.

It must be remarked that an important factor that cannot be ignored in these studies is the role of external anthropogenic forcing on internal climate modes. This will be further discussed in a dedicated section in Han et al. (this issue).

5 Anthropogenic Signal in Local Vertical Land Motion

Relative sea level change at a given location depends on the rates of sea level rise but also on the vertical movement of the Earth's crust. Potentially hazardous effects of sea level rise can thus be exacerbated by local ground motions. These can be of natural and/or anthropogenic origin; natural causes include glacial isostatic adjustment, co-seismic, post-seismic, and volcanic deformations, while the anthropogenic land motion is mainly due to gas and fluid extraction (subsidence) or ice mass loss (uplift). High rates of subsidence, combined with current and future rising sea levels, make several coastal regions highly vulnerable to negative impacts of climate change. Land subsidence is the lowering of the land-surface elevation as consequence of changes taking place underground. Currently, observations of land motion are available using space-based and in situ techniques, the most popular being the Global Positioning System (GPS). Alternatively, when GPS data are not available, a combination of altimetry and tide gauge sea level observations can be used to derive the vertical land motion (Fenoglio-Marc et al. 2004; Kuo et al. 2008; Santamaría-Gómez et al. 2014; Wöppelmann and Marcos 2012; Pfeffer and Allemand 2016). Differences between absolute sea level measured by altimetry and relative sea level measured by tide gauges are, in many cases, a good approximation to the vertical motion of the land to which the tide gauge is grounded. However, given the relatively short length of the altimetry period some limitations apply and uncertainties must be carefully estimated. The reader is referred to Wöppelmann and Marcos (2016) for a recent and complete review on geodetic measurements applied to sea level monitoring.

Land subsidence resulting from excessive extraction of groundwater occurs in several parts of the USA and Europe and in many large cities in South East Asia. In Europe, for example, the upper Adriatic coastland experienced significant changes due to the pronounced morphodynamics of the associated environment. In Marina di Ravenna, the land subsidence is about 10 mm/year, while sea level in the Adriatic Sea rises by 2–3 mm/year during the altimetry era (Fenoglio-Marc et al. 2011). Also in the Mediterranean Sea, Wöppelmann and Marcos (2012) identified the highly populated and industrialized city of Thessaloniki (northern Greece) as one of the most exposed to sea level hazards in this region due to large subsidence rates. These were partly attributed to pumping of groundwater for industrial usage (Raucoules et al. 2008). They also identified the iconic site of Venice as suffering from subsidence, a fact well known from previous investigations and already associated with pumping of water wells between about 1940 and 1975 (e.g., Pirazzoli 1987). In the Gulf of Mexico, rates of relative sea level rise range from 1 to

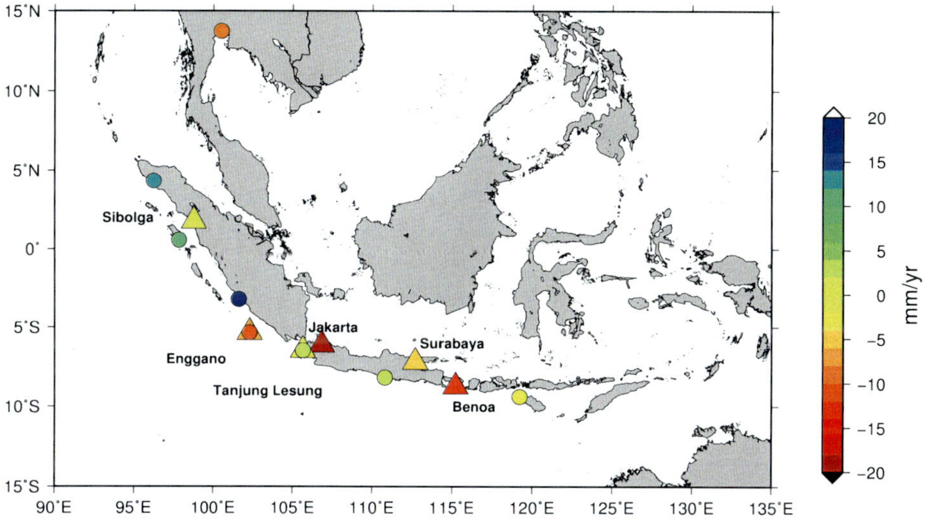

Fig. 5 Land vertical motion from altimetry minus tide gauges (*triangle*) and from GPS (*circle*). Adapted from Fenoglio-Marc et al. (2012)

10 mm/year, with the higher values associated with land subsidence in the northern area (Douglas 2005). Letetrel et al. (2015) estimated vertical land motion rates in excess of −7 mm/year in the Mississippi Delta and up to −4 mm/year along the coast of Texas. Such large subsidence rates were consequence of fluid (oil and gas) withdrawals, in combination with multiple physical processes (e.g., sediment loading).

The region of South East Asia is of particular relevance because, as pointed out in the previous section, it is located in the area where higher rates of sea level rise have been observed during the altimetry era. It is also one of the most populated regions in the world, with more than 10 % of the total world population, long maritime coastlines, and many coastal megacities (e.g., Bangkok, Jakarta, Singapore, Ho Chi Minh, Hong Kong, Kuala Lumpur, Manila, Guangzhou). This area is locally affected by land subsidence, mainly due to ground water depletion in some megacities. As a consequence, sinking rates have been found to be in excess of 10 mm/year in Jakarta (Abidin et al. 2001) and Bangkok (Saramul and Ezer 2014). GPS-controlled tide gauges exist in Thailand (Trisirisatayawong et al. 2011), Indonesia (Schöne et al. 2011), and Malaysia (Simons et al. 2007). The length of the records is however still very short, and data are highly affected by post-seismic motion (Trisirisatayawong et al. 2011). Fenoglio-Marc et al. (2012) found that the land subsidence derived from combining altimetry and tide gauge data is in good agreement with the GPS-derived rates (see also Fig. 5). Moreover, geodetic-based results from the InSAR technique show that land subsidence rates in Jakarta (19 mm/year in Fig. 5) exhibit spatial and temporal variations, with typical rates of 20–100 mm/year (Abidin et al. 2015).

6 Concluding Remarks

Evidence from palaeo-sea level studies indicates that sea level has not changed significantly during the last two millennia (Kemp et al. 2011) and that the departure from Holocene sea level rates can be detected in the early twentieth century (Gehrels and

Woodworth 2013). In spite of this, studies addressing D&A in sea level are scarce due to the limited observations of sea level and its major contributors.

In the last few years, different works have used statistical methodologies to conclude that the natural variability in sea level, i.e. decadal to multi-decadal changes associated with climate variations, cannot explain the observed linear global sea level trend during the past century (Bos et al. 2014; Becker et al. 2014; Dangendorf et al. 2015). Likewise, studies based on climate models and semi-empirical models also separated a significant and increasing contribution of the anthropogenic forcing in global sea level driven by GHG concentrations (Jevrejeva et al. 2009; Kopp et al. 2016; Slangen et al. 2016). There exists therefore an irrefutable fingerprint of an anthropogenic forcing in this key climate change indicator. Nevertheless, due to the fact that local and regional sea level variations have a low signal-to-noise ratio, this response is not always detectable at these smaller spatial scales (Richter and Marzeion 2014; Palanisamy et al. 2015a).

Although not a consequence of higher GHG concentrations, human activities may also influence directly sea level changes. In particular, the impact of dam building has been quantified in −0.55 mm/year on global sea level rise since 1950 (Chao et al. 2008). On the opposite side, continental groundwater extraction contributed with 25 mm of globally averaged sea level change during the twentieth century (Veit and Conrad 2016). In addition to the water exchanges between land and ocean, this mass loading/unloading also induces vertical crustal motion as well as geoid changes that may impact relative sea levels locally. On the other hand, the mechanisms by which sea level is responding to the increased concentration of GHGs and aerosols in the atmosphere are various. The two major contributors to global sea level rise during the twentieth century are ocean thermal expansion and glaciers mass loss. It has been shown that these two processes separately have a significant human fingerprint on top of their corresponding natural variability at least during the last few decades (Marcos and Amores 2014; Slangen et al. 2014; Marzeion et al. 2014). It is likely that, as the number of observations grows, there will be more studies addressing and quantifying the role of each contributor, including the ice sheets for which uncertainties in the past are large.

Acknowledgments M. Marcos acknowledges a "Ramon y Cajal" contract funded by the Spanish Ministry of Economy. This work was supported by the Research Project CLIMPACT (CGL2014-54246-C2-1-R) funded by the Spanish Ministry of Economy. S. Dangendorf acknowledges a Visiting Research Grant from the University of the Balearic Islands. A. Slangen is supported by a CSIRO Office of the Chief Executive Fellowship. H. Palanisamy is supported by a CNES/CLS Ph.D. fund. L. Fenoglio acknowledges the support by DFG in the frame of the COSELE Project. This paper is an outcome of the ISSI workshop on "Integrative study of sea level".

References

Abidin HZ, Djaja R, Darmawan D, Hadi S, Akbar A, Rajiyowiryono H, Sudibyo Y, Meilano I, Kasuma MA, Kahar J, Subarya C (2001) Land subsidence of Jakarta (Indonesia) and its geodetic monitoring. Nat Hazards 23:365–387

Abidin HZ, Andreas H, Gumilar I, Wibowo IRR (2015). On correlation between urban development, land subsidence and flooding phenomena in Jakarta. In: Proceedings of IAHS, 370, 15–20, 2015 proc-iahs.net/370/15/2015/. doi:10.5194/piahs-370-15-2015

Agnew DC (1992) The time-domain behaviour of power-law noises. Geophys Res Lett 19(4):333–336

Allen MR, Tett SFB (1999) Checking for model consistency in optimal fingerprinting. Clim Dyn 15:419–434

Barbosa SM, Silva ME, Fernandes MJ (2008) Time series analysis of sea-level records: characterising long-term variability. In: Donner RV, Barbosa SM (eds) Nonlinear time series analysis in the geosciences. Springer, New York, pp 157–173
Barnett TP, Pierce DW, Schnur R (2001) Detection of anthropogenic climate change in the world's oceans. Science 292:270
Barnett TP, Pierce DW, AchutaRao KM, Gleckler PJ, Santer BD, Gregory JM, Washington WM (2005) Penetration of human induced warming into the world's oceans. Science 309:284–287
Becker M, Karpytchev M, Lennartz-Sassinek S (2014) Long-term sea level trends: natural or anthropogenic? Geophys Res Lett 41:5571–5580. doi:10.1002/2014GL061027
Bilbao RAF, Gregory JM, Bouttes N (2015) Analysis of the regional pattern of sea level change due to ocean dynamics and density change for 1993–2099 in observations and CMIP5 AOGCMs. Clim Dyn. doi:10.1007/s00382-015-2499-z
Bindoff NL, Willebrand J, Artale V, Cazenave A, Gregory J et al (2007) Observations: oceanic climate change and sea level. In: Solomon S, Qin D, Manning M et al (eds) Climate change 2007: the physical science basis, contribution of working group I to the fourth assessment report of the intergovernmental panel on climate change. Cambridge University Press, Cambridge, pp 385–432
Bindoff NL et al (2013) Detection and attribution of climate change: from global to regional. In: Stocker TF, Qin D, Plattner G-K, Tignor M, Allen SK, Boschung J, Nauels A, Xia Y, Bex V, Midgley PM (eds) Climate change 2013: the physical science basis. Contribution of working group I to the fifth assessment report of the intergovernmental panel on climate change. Cambridge University Press, Cambridge
Bordbar MH, Martin T, Latif M, Park W (2015) Effects of long-term variability on projections of twenty-first century dynamic sea level. Nat Clim Change 5:343–347. doi:10.1038/nclimate2569
Bos MS, Williams SDP, Araujo IB, Bastos L (2014) The effect of temporal correlated noise on the sea level rate and acceleration uncertainty. Geophys J Int 196:1423–1430
Bouttes N, Gregory JM (2014) Attribution of the spatial pattern of CO_2-forced sea level change to ocean surface flux changes. Environ Res Lett 9:034004. doi:10.1088/1748-9326/9/3/034004
Bouttes N, Gregory JM, Kuhlbrod T, Smith RS (2014) The drivers of projected North Atlantic sea level change. Clim Dyn 43:1531–1544. doi:10.1007/s00382-013-1973-8
Burke EE, Roe GH (2014) The absence of memory in the climatic forcing of glaciers. Clim Dyn 42:1335–1346
Calafat FM, Chambers DP (2013) Quantifying recent acceleration in sea level unrelated to internal climate variability. Geophys Res Lett 40:3661–3666. doi:10.1002/grl.50731
Carton JA, Giese BS, Grodsky SA (2005) Sea level rise and the warming of the oceans in the Simple Ocean Data Assimilation (SODA) ocean reanalysis. J Geophys Res Oceans 110(C9). doi:10.1029/2004JC002817
Chao BF, Wu YH, Li YS (2008) Impact of artificial reservoir water impoundment on global sea level. Science 320:212–214
Church JA, White NJ (2011) Sea-level rise from the late 19th to the early 21st century. Surv Geophys 32:585–602
Church JA, Clark PU, Cazenave A, Gregory JM, Jevrejeva S, Levermann A, Merrifield MA, Milne GA, Nerem RS, Nunn PD, Payne AJ, PfefferWT, Stammer D, Unnikrishnan AS (2013) Sea level change. In: Stocker TF, Qin D, Plattner G-K, Tignor M, Allen SK, Boschung J, Nauels A, Xia Y, Bex V, Midgley PM(eds) Climate change 2013: the physical science basis. Contribution of working group I to the fifth assessment report of the intergovernmental panel on climate change. Cambridge University Press, Cambridge
Cogley JG (2009) Geodetic and direct mass-balance measurements: comparison and joint analysis. Ann Glaciol 50:96–100
Dangendorf S, Rybski D, Mudersbach C, Müller A, Kaufmann E, Zorita E, Jensen J (2014a) Evidence for long-term memory in sea level. Geophys Res Lett 41:5530–5537. doi:10.1002/2014GL060538
Dangendorf S, Calafat FM, Arns A, Wahl T, Haigh ID, Jensen J (2014b) Mean sea level variability in the North Sea: processes and implications. J. Geophys Res Oceans 119:6820–6841. doi:10.1002/2014JC009901
Dangendorf S, Marcos M, Muller A, Zorita E, Riva R, Berk K, Jensen J (2015) Detecting anthropogenic footprints in sea level rise. Nat Commun. doi:10.1038/ncomms8849
Douglas BC (2005) Gulf of Mexico and Atlantic coast sea level change. In: Sturges W, Lugo-Fernández A (eds) Circulation in the Gulf of Mexico: observations and models. Geophysical monograph series, vol 161. American Geophysical Union, Washington, DC, pp 111–121

England MH, McGregor S, Spence P, Meehl GA, Timmermann A, Cai W, Gupta AS, McPhaden MJ, Purich A, Santoso A (2014) Recent intensification of wind-driven circulation in the Pacific and the ongoing warming hiatus. Nat Clim Change 4(3):222–227. doi:10.1038/nclimate2106

Fenoglio-Marc L, Groten E, Dietz C (2004) Vertical land motion in the Mediterranean Sea from altimetry and tide gauge stations. Mar Geodesy 27(3–4):683–701

Fenoglio-Marc L, Braitenberg C, Tunini L (2011) Sea level variability and trends in the Adriatic Sea in 1993–2008 from tide gauges and satellite altimetry. Phys Chem Earth. doi:10.1016/j.pce.2011.05.014

Fenoglio-Marc L, Schöne T, Illigner J, Becker M, Manurung P, Khafid (2012) Sea level change and vertical motion from satellite altimetry, tide gauges and GPS in the Indonesian Region. Mar Geodesy 35(sup1):137–150. doi:10.1080/01490419.2012.718682

Frankcombe LM, McGregor S, England MH (2014) Robustness of the modes of Indo-Pacific sea level variability. Clim Dyn. doi:10.1007/s00382-014-2377-0

Gardner AS, Moholdt G, Cogley JG, Wouters B, Arendt AA, Wahr J, Berthier E, Hock R, Pfeffer WT, Kaser G, Ligtenberg SRM, Bolch T, Sharp MJ, Hagen JO, van den Broeke MR, Paul F (2013) A reconciled estimate of glacier contributions to sea level rise: 2003 to 2009. Science 340:852–857

Gehrels WR, Woodworth PL (2013) When did modern rates of sea-level rise start? Global Planet Change 100:263–277

Gleckler PJ, Santer BD, Domingues CM, Pierce DW, Barnett TP, Church JA, Taylor KE, AchutaRao KM, Boyer TP, Caldwell PM (2012) Human-induced global ocean warming on multidecadal timescales. Nat Clim Change 2:524–529

Hamlington BD, Strassburg MW, Leben RR, Han W, Nerem RS, Kim K-Y (2014) Uncovering an anthropogenic sea-level rise signal in the Pacific Ocean. Nat Clim Change 4(9):782–785. doi:10.1038/nclimate2307

Han W et al (2013) Intensification of decadal and multi-decadal sea level variability in the western tropical Pacific during recent decades. Clim Dyn. doi:10.1007/s00382-013-1951-1

Hay CC, Morrow E, Kopp RE, Mitrovica JX (2015) Probabilistic reanalysis of twentieth-century sea-level rise. Nature 517:481–484

Hegerl GC et al (2010) Good practice guidance paper on detection and attribution related to anthropogenic climate change. In: Stocker TF et al (ed) Meeting report of the intergovernmental panel on climate change expert meeting on detection and attribution of anthropogenic climate change. IPCC working group I technical support unit, University of Bern, Bern, Switzerland, 8 pp

Hünicke B, Zorita E, Soomere T, Madsen KS, Johansson M, Suursaar Ü (2015) Recent change—sea level and wind waves. Second assessment of climate change for the Baltic Sea Basin. pp 155–185. doi:10.1007/978-3-319-16006-1_9. Print ISBN 978-3-319-16005-4, Online ISBN 978-3-319-16006-1

Ishii M, Kimoto M (2009) Reevaluation of historical ocean heat content variations with time-varying XBT and MBT depth bias corrections. J Oceanogr 65:287–299

Jevrejeva S, Grinsted A, Moore JC, Holgate S (2006) Nonlinear trends and multiyear cycles in sea level records. J Geophys Res 111:C09012. doi:10.1029/2005JC003229

Jevrejeva S, Moore JC, Grinsted A, Woodworth PL (2008) Recent global sea level acceleration started over 200 years ago? Geophys Res Lett 35:L08715

Jevrejeva S, Grinsted A, Moore JC (2009) Anthropogenic forcing dominates sea level rise since 1850. Geophys Res Lett 36:L20706. doi:10.1029/2009GL040216

Jordà G (2014) Detection time for global and regional sea level trends and accelerations. J Geophys Res Oceans 119(10):7164–7174. doi:10.1002/2014JC010005

Kantelhardt JW, Koscielny-Bunde E, Rego HH, Havlin S, Bunde A (2001) Detecting long-range correlations with detrended fluctuation analysis. Phys A 295(3):441–454

Kemp AC, Horton BP, Donelly JP, Mann ME, Vermeer M, Rahmstorf S (2011) Climate related sea-level variations over the past two millennia. Proc Natl Acad Sci USA 105:13252–13257

Köhl A, Stammer D, Cornuelle B (2007) Interannual to Decadal Changes in the ECCO Global Synthesis. J Phys Oceanogr 37(2):313–337. doi:10.1175/JPO3014.1

Kopp RE, Kemp AC, Bittermann K, Horton BP, Donnelly JP, Gehrels WR, Hay CC, Mitrovica JX, Morrow ED, Rahmstorf S (2016) Temperature-driven global sea-level variability in the Common Era. PNAS 113(11):E1434–E1441. doi:10.1073/pnas.1517056113

Kuo CY, Shum CK, Braun A, Cheng KC, Yi Y (2008) Vertical motion determined using satellite altimetry and tide gauges. Terr Atmos Ocean Sci 19(1–2):21–35

Leclercq PW, Oerlemans J, Basagic HJ, Bushueva I, Cook AJ, Le Bris R (2014) A data set of worldwide glacier length fluctuations. Cryosphere 8:659–672

Letetrel C, Karpytchev M, Bouin M-N, Marcos M, Santamaría-Gómez A, Wöppelmann G (2015) Estimation of vertical land movement rates along the coasts of the Gulf of Mexico over the past decades. Cont Shelf Res 111:42–51

Levermann A Peter, Clark U, Marzeion Ben, Milne Glenn A, Pollard David, Radic Valentina, Robinsonh Alexander (2013) The multimillennial sea-level commitment of global warming. PNAS 110(34):13745–13750. doi:10.1073/pnas.1219414110

Levitus S, Antonov JI, Boyer TP, Locarnini RA, Garcia HE, Mishonov AV (2009) Global ocean heat content 1955–2008 in light of recently revealed instrumentation problems, Geophys Res Lett 36(7). doi:10.1029/2008GL037155

Levitus S et al (2012) World ocean heat content and thermosteric sea level change (0–2000 m), 1955–2010. Geophys Res Lett 39:L10603. doi:10.1029/2012GL051106

Lyman JM, Johnson Gregory C (2014) Estimating global ocean heat content changes in the upper 1800 m since 1950 and the influence of climatology choice. J Clim 27:1945–1957. doi:10.1175/JCLI-D-12-00752.1

Lyu K, Zhang X, Church JA, Slangen ABA, Hu J (2014) Time of emergence for regional sea-level change. Nat Clim Change 4(11):1006–1010. doi:10.1038/nclimate2397

Marcos M, Amores A (2014) Quantifying anthropogenic and natural contributions to thermosteric sea level rise. Res Lett, Geophys. doi:10.1002/2014GL059766

Marcos M, Tsimplis MN (2007) Forcing of coastal sea level rise patterns in the North Atlantic and the Mediterranean Sea. Geophys Res Lett 34:L18604. doi:10.1029/2007GL030641

Marzeion B, Cogley JG, Richter K, Parkes D (2014) Attribution of global glacier mass loss to anthropogenic and natural causes. Science 345:919–921

McGregor S, Gupta AS, England MH (2012) Constraining wind stress products with sea surface height observations and implications for Pacific Ocean sea level trend attribution*. J Clim 25(23):8164–8176. doi:10.1175/JCLI-D-12-00105.1

Merrifield MA (2011) A shift in western tropical Pacific sea level trends during the 1990s. J Clim 24(15):4126–4138. doi:10.1175/2011JCLI3932.1

Merrifield MA, Maltrud ME (2011) Regional sea level trends due to a Pacific trade wind intensification. Geophys Res Lett. doi:10.1029/2011GL049576

Merrifield MA, Thompson PR, Lander M (2012) Multidecadal sea level anomalies and trends in the western tropical Pacific. Geophys Res Lett. doi:10.1029/2012GL052032

Meyssignac B, Salas D, Melia Y, Becker M, Llovel W, Cazenave A (2012) Tropical Pacific spatial trend patterns in observed sea level: internal variability and/or anthropogenic signature? Clim Past 8(2):787–802. doi:10.5194/cp-8-787-2012

Monselesan DP, O'Kane TJ, Risbey JS, Church J (2015) Internal climate memory in observations and models. Geophys Res Lett 42:1232–1242. doi:10.1002/2014GL062765

Nidheesh AG, Lengaigne M, Vialard J, Unnikrishnan AS, Dayan H (2013) Decadal and long-term sea level variability in the tropical Indo-Pacific Ocean. Clim Dyn 41(2):381–402. doi:10.1007/s00382-012-1463-4

Oerlemans J (1988) Simulation of historic glacier variations with a simple climate-glacier model. J Glaciol 34:333–341

Palanisamy H, Meyssignac B, Cazenave A, Delcroix T (2015a) Is anthropogenic sea level fingerprint already detectable in the Pacific Ocean? Environ Res Lett 10(8):084024. doi:10.1088/1748-9326/10/8/084024

Palanisamy H, Cazenave A, Delcroix T, Meyssignac B (2015b) Spatial trend patterns in the Pacific Ocean sea level during the altimetry era: the contribution of thermocline depth change and internal climate variability. Ocean Dyn. doi:10.1007/s10236-014-0805-7

Pfeffer J, Allemand P (2016) The key role of vertical land motions in coastal sea level variations: a global synthesis of multisatellite altimetry, tide gauge data and GPS measurements. Earth Planet Sci Lett 439:39–47. doi:10.1016/j.epsl.2016.01.027

Piecuch CG, Ponte RM (2015) Inverted barometer contributions to recent sea level changes along the northeast coast of North America. Geo Phys Res Lett 42:5918–5925. doi:10.1002/2015GL064580

Pirazzoli PA (1987) Recent sea-level changes and related engineering problems in the lagoon of Venice (Italy). Prog Oceanogr 18:323–346. doi:10.1016/0079-6611(87)90038-3

Raucoules D et al (2008) Ground deformation detection of the greater area of Thessaloniki (northern Greece) using radar interferometry techniques. Nat Hazards Earth Syst Sci 8:779–788. doi:10.5194/nhess-8-779-2008

Reichert BK, Bengtsson L, Oerlemans J (2002) Recent glacier retreat exceeds internal variability. J Clim 15:3069–3081

Richter K, Marzeion B (2014) Earliest local emergence of forced dynamic and steric sea-level trends in climate models. Environ Res Lett 9(11):114009. doi:10.1088/1748-9326/9/11/114009

Roe GH, Baker MB (2014) Glacier response to climate perturbations: an accurate linear geometric model. J Glaciol 60:670–684

Santamaría-Gómez A, Gravelle M, Wöppelmann G (2014) Long-term vertical land motion from double-differenced tide gauge and satellite altimetry data. J Geodesy 88:207–222. doi:10.1007/s00190-013-0677-5

Saramul S, Ezer T (2014) Spatial variations of sea level along the coast of Thailand: impacts of extreme land subsidence, earthquakes and the seasonal monsoon. Global Planet Changes 122:70–81. doi:10.1016/j.gloplacha.2014.08.012

Schöne T, Illigner J, Manurung P, Subarya C, Khafid Zech C, Galas R (2011) GPS-controlled tide gauges in Indonesia a German contribution to Indonesia's Tsunami Early Warning System. Nat Hazards Earth Syst Sci 11:731–741

Simons WJF, Socquet A, Vigny A, Ambrosius BAC, Abu SH, Promthong C, Subarya C, Sarsito DA, Matheussen S, Morgan P, Spakman W (2007) A decade of GPS in Southeast Asia: resolving Sundaland motion and boundaries. J Geophys Res 112(B06420). doi:10.1029/2005JB003868

Slangen ABA, Church JA, Zhang X, Monselesan D (2014) Detection and attribution of global mean thermosteric sea-level change. Geophys Res Lett 41(16):5951–5959. doi:10.1002/2014GL061356

Slangen ABA, Church JA, Zhang X, Monselesan D (2015) The sea-level response to external forcings in CMIP5 climate models. J Clim. doi:10.1175/JCLI-D-15-0376.1

Slangen ABA, Church JA, Agosta C, Fettweis X, Marzeion B, Richter K (2016) Anthropogenic forcing dominates global mean sea-level rise since 1970 Nat Climate Change. doi:10.1038/nclimate2991

Stammer D, Cazenave A, Ponte RM, Tamisiea ME (2013) Causes for contemporary regional sea level changes. Annu Rev Mar Sci. doi:10.1146/annurev-marine-121211-172406

Thompson PR, Merrifield Mark A, Wells Judith R, Chang Chantel M (2014) Wind-driven coastal sea level variability in the northeast pacific. J Clim 27:4733–4751. doi:10.1175/JCLI-D-13-00225.1

Timmermann A, McGregor S, Jin F-F (2010) Wind effects on past and future regional sea level trends in the southern Indo-Pacific*. J Clim 23(16):4429–4437. doi:10.1175/2010JCLI3519.1

Trisirisatayawong I, Naeije M, Simons W, Fenoglio-Marc L (2011) Sea level change in the Gulf of Thailand from GPS-corrected tide gauge data and multi-satellite altimetry. Global Planet Change 76:137–151

Veit E, Conrad CP (2016) The impact of groundwater depletion on spatial variations in sea level change during the past century. Geophys Res Lett 43:3351–3359. doi:10.1002/2016GL068118

Venzke S, Allen MR, Sutton RT, Rowell DP (1999) The atmospheric response over the North Atlantic to decadal changes in sea surface temperature. J Clim 12:2562–2584

Wakelin SL, Woodworth PL, Flather RA, Williams JA (2003) Sea-level dependence on the NAO over the NW European Continental Shelf. Geophys Res Lett 30(7):1403. doi:10.1029/2003GL017041

Wöppelmann G, Marcos M (2012) Coastal sea level rise in southern Europe and the nonclimate contribution of vertical land motion. J Geophys Res 117:C01007. doi:10.1029/2011JC007469

Wöppelmann G, Marcos M (2016) Vertical land motion as a key to understanding sea level change and variability. Rev Geophys. doi:10.1002/2015RG000502

Wöppelmann G, Marcos M, Santamaría-Gómez A, Martín-Míguez B, Bouin M-N, Gravelle M (2014) Evidence for a differential sea level rise between hemispheres over the twentieth century. Geophys Res Lett 41:1639–1643. doi:10.1002/2013GL059039

Wunsch Carl, Heimbach Patrick (2014) Bidecadal thermal changes in the abyssal ocean. J Phys Oceanogr 44:2013–2030. doi:10.1175/JPO-D-13-096.1

Zhang X, Church JA (2012) Sea level trends, interannual and decadal variability in the Pacific Ocean. Geophys Res Lett. doi:10.1029/2012GL053240

Part V
From Observations to Sea Level Projections

Surv Geophys (2017) 38:349–383
DOI 10.1007/s10712-016-9401-3

Testing the Quality of Sea-Level Data Using the GECCO Adjoint Assimilation Approach

Martin G. Scharffenberg[1] · Armin Köhl[1] · Detlef Stammer[1]

Received: 6 June 2016 / Accepted: 2 November 2016 / Published online: 26 November 2016

Abstract Besides providing an estimate of the changing ocean state, an important result of the dynamically consistent estimating the circulation and climate of the ocean (ECCO) state estimate approach is the provision of a posterior model–data residuals which contain important information about elements in the assimilated observations that are inconsistent with the model dynamics or with the information present in other ocean data sets that are being used as constraints in the assimilation procedure. Based on decreased GECCO2 model–data residuals, upon using the altimeter data through the ESA climate change initiative (cci) sea-level (SL) project, we show here that the recently reprocessed ESA SL_cci altimeter data set (*SL*1) has been improved relative to the earlier AVISO altimetry data set and is now more consistent with the GECCO2 estimate and with the information about the changing ocean state embedded in other ocean data sets. The improvement can be shown to exist separately for both TOPEX/POSEIDON and ERS data sets. The study reveals that especially in regions characterized by small sea surface height (SSH) variability and small signal-to-noise ratio in the SSH data, improvements can be on the order of 30% of previously existing model–data residuals. However, in some regions we can find degradations, particulary in those where GECCO2 has little skill in representing the altimeter data and where evaluation of the products with GECCO2 is thus not advisable. Upon the assimilation of the new *SL*1 data set, the GECCO2 synthesis was further improved. However, adding the sea surface temperature (SST) from the SST_cci project as additional constrain, no further impact can be identified.

GECCO: German partner of the Estimating the Circulation and Climate of the Ocean effort

✉ Martin G. Scharffenberg
martin.scharffenberg@uni-hamburg.de

Armin Köhl
armin.koehl@uni-hamburg.de

Detlef Stammer
detlef.stammer@uni-hamburg.de

[1] Universität Hamburg, Bundesstrae 53, 24106 Hamburg, Germany

Keywords Altimeter observations · Data assimilation · Ocean state estimate · Ocean Modeling

1 Introduction

Over the last decades, coupled climate general circulation models have become a firm part of climate research, and by now are an essential tool to simulate the entire climate system as well as the interaction between its various components. As such, they are important for the analysis of climate processes, and climate sensitivity and feedback mechanisms; at the same time, they are essential to interpret climate observations and underlying processes. When it comes to future climate change, coupled climate models are the primary means to predict and project changes in the entire climate system. Respective models are now being used in the context of the World Climate Research Program (WCRP) climate model intercomparison project (CMIP) to provide multi-model productions and projections forming the basis for the Intergovernmental Panel on Climate Change (IPCC) assessment reports.

In a similar vein, ocean-only models have also become an essential tool to investigate important processes shaping and changing the ocean circulation, including mixing, water mass formation, transport processes, and changing circulation as well as air–sea feedbacks. In contrast to coupled climate models, ocean hind-casts driven by atmospheric forcing fields aim to provide simulations of the past and ongoing changes in the ocean and to identify underlying processes. But before results from ocean models can be used to identify real-world climate processes, the realism of them needs to be tested against ocean and climate observations. Respective tests have been going on for quite some time (e.g., Stammer et al. 1996); they now have been extended to encompass entire climate models using the full suit of climate observations (e.g., Flato et al. 2013). In this context, data from global networks have become indispensable for climate model quality tests, such as the Argo float program in the ocean. However, it is especially the capability of satellite data to observe the global ocean and its variability which allow to test ocean models in great detail jointly with information available from in situ data sets. The same holds for coupled climate models. Respective comparison studies serve to enhance our confidence in climate models, and also help to improve them further.

When it comes to the improvement of the model skill in simulating the observed ocean state, observations can be used directly as constraints during a data assimilation procedure. Stammer et al. (2016) provide a review of assimilation approaches and their application in climate science. Using dynamically consistent smoother approaches, uncertain model parameters are being adjusted in this process such that the model simulates the observed ocean state, including its variability. In the context of ocean state estimation, sea surface height (SSH) fields are being assimilated on a regular basis jointly with other satellite data and all available in situ data sets. The importance of SSH fields in this context originates from their significance for constraining the oceans flow field. This also holds for hydrographic data sets but is less obvious for essential climate variables (ECVs) over the ocean such as sea surface temperature (SST) or ocean color. Nevertheless, existing ocean state estimation efforts attempt to use all available data sets as constraints, including also surface momentum and buoyancy fluxes and ocean velocity and transport data sets.

Analyses of sea-level variability including secular trends on the basis of such an ocean synthesis are provided by Köhl and Stammer (2008). These consistent estimates of the ocean circulation enable the identification of the causes of trends; they provided, for instance, the result that much of the regional sea-level trends over the last decades are related to changes in wind forcing. An important output of assimilation procedures is information about model–data residuals, which, depending on the details of the assimilation approach, can be used to identify problems in the data. An example is shown by Stammer et al. (2007) who were able to show that, based on the GECCO estimate (German partner of the Estimating the Circulation and Climate of the Ocean effort, ECCO), model–data residuals of the time-mean dynamic topography can be interpreted as errors in the EGM96 geoid.

In this present study, the recent and higher resolution GECCO2 ocean synthesis framework is used to assess the quality of satellite altimeter data emerging from the ESA climate change initiative (cci) data improvement project of SSH data. As we will show below, the GECCO procedure does identify a significant level of improvement in the data sets. The improved SSH fields represent the ECV of sea level (SL) computed in the framework of the ESA SL_cci (Ablain et al. 2015). Encouraged by these results, the new data sets are then being assimilated in a separate assimilation run and the impact of the SSH improvement on the estimated state is then further analyzed. Finally, we test in a third assimilation run the additional information content in the new ESA SST_cci data sets by including this also into the state estimation.

The remaining paper is structured as follows: Sect. 2 explains the used altimeter data sets, the GECCO2 ocean model that is used for the assimilation runs as well as the methods applied for the analysis. Section 3 discusses the quality assessment of the improved ESA SL_cci data while Sect. 4 analyzes the influence of the assimilation of the improved SL data into the GECCO2 synthesis. In Sect. 5, the changes in regional sea-level trends are presented, and Sect. 6 discusses the influence on the annual SL signal. Concluding remarks are provided in Sect. 7.

2 Data, Model, and Methodology

2.1 SSH Data

Within the SL_cci project, the SSH product previously provided by AVISO and now by Copernicus Marine Environment Monitoring Service (CMEMS; http://marine.copernicus.eu/services-portfolio/access-to-products/; Global Ocean Along-Track Sea Level Anomalies Reprocessed (1993-ongoing); Product ID: SEALEVEL_GLO_SLA_L3_REP_OBSERVATIONS_008_018; henceforth called *SL0*) has been further improved and updated individually for each satellite series (Ablain et al. 2015). The main improvements include the reduction in orbit errors and wet/dry atmospheric correction errors, reduction in instrumental drifts and bias, intercalibration biases, intercalibration between missions and combination of the different sea-level data sets, and an improvement of the mean sea surface. This effort leads to homogenized corrections between all the missions in order to reduce the sources of discrepancies. From this effort, two updated SSH products are now available (SL1, SL1.1, Cazenave et al. 2014), the first covering the 18-year time period from January 1993 to October 2010. A second product covers the extended 21-year-long period from January 1993 to December 2013. Both data products consist of SSH anomalies relative to the DTU2010 mean sea surface (Anderson et al. 2015), which have been reprocessed using new updated geophysical standards and multi-mission cross calibration.

In this study, the extended product SL1.1 is used which in the following will be referred to as *SL*1. Two different versions of this product are available, consisting of (1) gridded SSH anomaly fields combining all altimetric satellite measurements, with a regular spatial resolution of 0.25° × 0.25° and with a monthly temporal resolution from 1993 to 2013, and (2) along-track (AT) SSH anomalies covering the period 1992/93–2013 and provided separately for each altimetric mission. In the following, only the along-track SSH anomalies from the ERS- and from the TOPEX/POSEIDON (T/P) satellite series will be used during various analyses and assimilation runs. The ERS-1/2 and ENVISAT satellite missions (https://earth.esa.int) are orbiting at an altitude of about 780 km in a Sun-synchronous polar orbit with 98.52° inclination and a repeat cycle of 35 days. ERS-1 covers the period 1993–1995 (repeat cycle 17–43), followed by ERS-2 covering the years 1995–2002 (repeat cycle 1–78), and ENVISAT for the period 2002–2010/11 (repeat cycle 10–94). The T/P and Jason-1/2 satellite missions (http://sealevel.jpl.nasa.gov/missions) are flying in a circular orbit with 66° inclination at an altitude of 1336 km, and a repeat cycle of 9.91 days, covering 95% of the ice-free oceans. The T/P satellite was flying in this orbit from 1992 to 2002 (repeat cycle 10–353), Jason-1 covered the period 2002–2008 (repeat cycle 11–249), and Jason-2 followed on from 2008 to 2010/2013 (repeat cycle 11–91).

The along-track SSH level 3 products were filtered previously by the SL_cci project with a filter length scale that varied by latitude and were interpolated onto latitudinally varying reference positions prior to delivery. The applied filter length scales are 250 km (0°–10°), 200 km (10°–20°), 150 km (20°–30°), 100 km (30°–40°) and 65 km (40°–90°). The original along-track resolution of about 6.2 km has been interpolated onto reference positions of 18.6 km (90°–30°), 24.9 km (30°–20°), 31.8 km (20°), 37.4 km (20°–10°), and 43.7 km (10° to the equator). The details are summarized in Table 1 (column 1–3).

2.2 The GECCO2 Synthesis

In general terms, data assimilation methods can be separated into filtering methods and smoothing methods. Within these two categories, some methods are relatively simple and computationally efficient (e.g., optimal interpolation), whereas others are more rigorous yet

Table 1 SSH along-track filter length scales that have been applied during the generation of the level 3 (L3) *SL*0 and *SL*1 data sets (column 2) and their interpolated along-track resolutions (column 3) for TP (ERS)

Latitude range	SL_cci filter length scale (km)	SL_cci AT resolution (km) TP (ERS)	Filter length scale (km) TP (ERS) to match GECCO2 resolution		
			*f*7	*f*9	*f*11
90°–40°	65				
40°–30°	100	18.6 (19.7)	130.2 (137.9)	167.4 (177.3)	204.6 (216.7)
30°–20°	150	24.9 (26.4)	174.3 (184.8)	224.1 (237.6)	273.9 (290.4)
20°		31.2 (33.0)	218.4 (231.0)	280.8 (297.0)	343.2 (363.0)
20°–10°	200	37.4 (39.7)	261.8 (277.9)	336.6 (357.3)	411.4 (436.7)
10°–0°	250	43.7 (46.4)	305.9 (324.8)	393.3 (417.6)	480.7 (510.4)

As the GECCO2 model is smoother than the assimilated data sets *SL*0 and *SL*1, column 4–6 indicate the filter length scales of the *SL*0 and *SL*1 data sets averaged over 7, 9 or 11 AT points to best match the GECCO2 resolution computationally intensive (e.g., Kalman filters, 4D variational/adjoints). Available ocean data assimilation schemes vary in the way the individual data assimilation components are defined and the extent to which optimal values are subject to constraints (e.g., Stammer et al. 1996). Obtaining dynamically self-consistent estimates of the ocean circulation requires the use of mathematically consistent "smoother" data assimilation approaches as realized by adjoint methods. Any smoother assimilation approach adjusts uncertain model parameters to bring the model into consistency with ocean observations. See Stammer et al. (2016) for details on the various methods used during assimilation efforts.

The adjoint method used here is a variational method that seeks to minimize an objective function of weighted model–data differences that describes a conditional likelihood of the obtained solution given the data and their uncertainties. The optimization of this function involves an iterative process in which the adjoint to the tangent linear model is used to calculate the gradient of the objective function with respect to a number of parameters. In the case of the GECCO2 estimate (Köhl 2015), parameters that are optimized to bring the model into consistency with the data include the initial temperature and salinity conditions and the atmospheric state variables from which surface forcing fields are derived via bulk formulae. In this way, the ocean state estimation ultimately leads to estimates of the surface forcing fields that are required to simulate the observed ocean in a best possible way (given the model resolution and the model physics). Stammer et al. (2004) have shown that outside of western boundary current regions, the resulting surface flux fields show improved skill and that large biases present in the NCEP fields seemed to be remedied (see also Menemenlis et al. 2005, who came to a similar conclusion).

All assimilation methods rely on the assumption of a perfect model and in the adjoint only the to be optimized parameters are regarded as uncertain. In reality, models have substantial biases associated with unresolved and insufficiently parameterized processes. These biases lead to posterior model–data differences that may be larger than the observational error. It is therefore not guaranteed that, after assimilation the model result is closer to the truth than the data. In particular, the coarse resolution model employed here does not resolve the mesoscale signal observed by the altimeter data which leads to a substantial representation error, which causes the posterior error to be significantly larger than the error of the altimeter data in most regions. Although one might assume that in this case the evaluation of the posterior error will not be useful for the evaluation of observational errors, model biases can be assumed to be largely uncorrelated with the observational error. Since the posterior model–data residuals depend on both the model and the data error, changes in one of the error contributions will still be revealed by a posterior error evaluation and provide useful information on data error.

This present study is based on a previous GECCO2 solution covering the period from 1948 to 2011 that had been optimized over 23 iterations. See Köhl (2015) for a description of the GECCO2 ocean state estimate and the data sets used as constraints. Starting from this already optimized state, three additional assimilation runs were performed as part of this study, all starting from iteration 23, and carrying out five additional iterations. The only differences between these three assimilation runs are the assimilated SSH and SST data sets used as constraints. As before, the GECCO2 model continued to assimilate in situ temperature and salinity data, from the EN3 data base (Ingleby and Huddleston 2007), SST data, and a mean dynamic topography. For each experiment, 5 iterations were added to the previous solution to have a consistent set of solutions.

The first assimilation run is a continuation of the original GECCO2 synthesis which assimilates the AVISO SSH fields *SL0*. In contrast, the second assimilation run uses the updated SL product *SL*1 as SSH constraint. The small number of just 5 additional iterations is justified by the small differences between the different SL products *SL*0 and *SL*1 which do not warrant a complete new synthesis. Over the course of the five additional iterations, the cost function contribution (sum of weighted squared differences to the data) representing the altimeter data misfit reduced typically by 20%; however, during the last iterations, the initial steep improvement flattened out, suggesting that the additional optimization found a new minimum. Because of the numerical model formulation, the global mean sea level (GMSL) of the GECCO2 model is zero at each time step; thus, GECCO2 does not have a GMSL rise. Therefore, before assimilation, the GMSL computed from mapped altimeter data was subtracted from the along-track data (*SL*0 and *SL*1) at each time.

A third assimilation run assimilated *SL*1 and additionally an updated sea surface temperature product from the ESA SST_cci. In the following, the three resulting GECCO2 outputs will be referred to as *G*0, *G*1, and *G*1sst, respectively.

2.3 Methodology

It is important to note that the GECCO2 synthesis adapts to the assimilated *SL*0 and *SL*1 data products, respectively, and also to additional data sources such that a cost function, measuring the weighted quadratic differences between model and data, is being minimized. Due to the assimilation of additional data, the assimilation procedure is able to compensate for observational errors in one data source. As was shown before, GECCO2 rejects components in the assimilated altimeter data that are dynamically incompatible with the model solution and the other data sets (within the prescribed errors); by comparing both SL data sets (*SL*0 and *SL*1) against the synthesis, errors of the data can then be revealed. These capabilities have been demonstrated for the mean dynamic topography (Stammer et al. 2007). Although the model suffers from large biases and its resulting sea-level fields cannot be assumed free of error or even of smaller error than the assimilated data, the improvement of the data does show up as reduction in model–data residuals since the improved data will be in better agreement with additionally assimilated data because the error of different data sources can be assumed to be uncorrelated.

To illustrate the variability present in the SL products as compared to the GECCO2 synthesis, Fig. 1 compares the standard deviation (STD) of the SL product *SL*1 with that obtained from the GECCO2 synthesis product *G*1. The GECCO2 synthesis (*G*1) can reproduce the general large-scale variability seen in the data (*SL*1); however, as the model is in zonal direction setup on a 1° grid and in meridional direction on a 1°/3° grid (for details, see Köhl 2015), it is not able to resolve eddies and lacks variability in regions of high mesoscale activity such as the boundary currents or along the Antarctic Circumpolar Current (ACC) in the Southern Ocean.

To assess the improvement of the *SL*1 data set, the daily output fields of SL of both GECCO2 solutions (*G*0 and *G*1) are compared to the two along-track SL products *SL*0 and *SL*1 itself, using two different approaches:

1. The first approach uses the GECCO2 model grid (M) as the basis for the analysis of the root mean squared (RMS) differences. Here, the along-track data sets (SL) have been interpolated onto the model grid, whereas
2. the second approach uses the full resolution of the SL data sets and performs the same analysis for each along-track satellite position (AT) individually, while interpolating the GECCO2 solution onto the respective satellite positions.

As described in Sect. 2.1, the along-track data sets (*SL*0 and *SL*1) have been interpolated onto reference positions along the satellite tracks prior to delivery by AVISO and SL_cci.

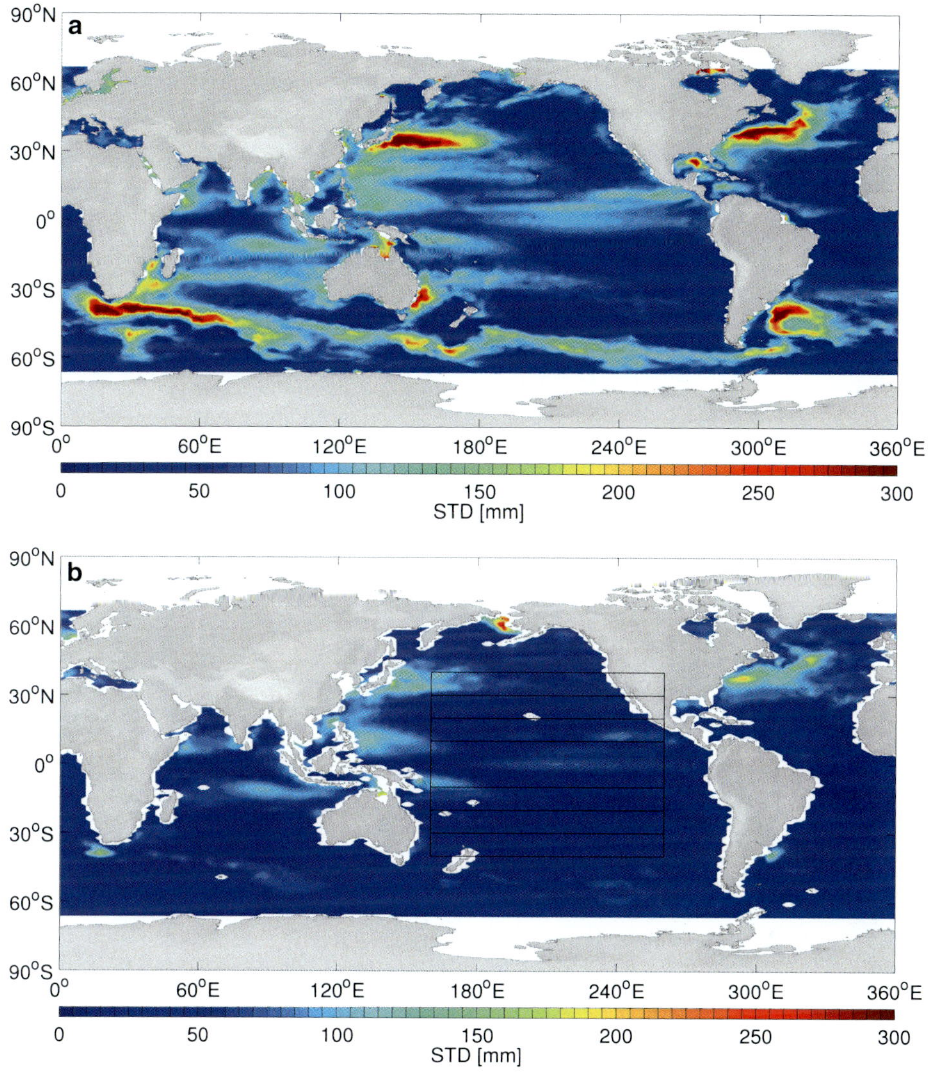

Fig. 1 STD on the TP tracks for the *SL*1 product (**a**) and for *G*1 (**b**). The *boxes* in **b** indicate the regions for which zonal wave number spectra are calculated in Sect. 4

Therefore, for the second approach, the daily GECCO2 synthesis results (*G*0 and *G*1) were interpolated onto the satellite tracks that matched the respective days for the respective along-track positions. For display purpose only, the data (*SL*0, *SL*1) and the synthesis results (*G*0, *G*1, *G*1sst) have been gridded on a $1^\circ \times 1^\circ$ grid for the ERS time series and on a $2^\circ \times 1^\circ$ grid for the TP time series. 190 and 664 repeat cycles were used from the ERS- and from the TP time series, respectively. As both approaches have comparable results, only the AT analysis is discussed. A comparison between the analysis at model and along-track resolution is given in Table 2.

The first step was to compare the *G*0 solution to *SL*0 and to the updated SL product *SL*1. In the second step, the updated SL product *SL*1 has been assimilated into the GECCO2

Table 2 Global mean (GM) ratios of RMS differences for TP and ERS time series as percentages of improvement, computed on the model grid (M), on the along-track points (AT) as well as on the along-track points using a running filter of 7 (*f*7)–11 (*f*11) points, for latitudes between 66°S and 66°N

RMS ratios	TP time series					ERS time series				
	M	AT (18.6–43.7 km)	*f*7 (130.2–305.9 km)	*f*9 (167.4–393.3 km)	*f*11 (204.6–480.7 km)	M	AT (19.7–46.4 km)	*f*7 (137.9–324.8 km)	*f*9 (177.3–417.6 km)	*f*11 (216.7–510.4 km)
$\frac{G0SL0}{G0SL1}$	2.53	1.97	2.71	2.97	3.21	2.99	2.35	3.08	3.30	3.51
$\frac{G1SL0}{G1SL1}$	3.53	2.50	3.44	3.77	4.08	3.56	2.62	3.48	3.76	4.01
$\frac{G1sstSL0}{G1sstSL1}$	3.84	2.43	3.30	3.59	3.87	3.87	2.58	3.41	3.67	3.91
$\frac{G0SL0}{G1SL1}$	3.80	3.25	4.37	4.75	5.09	4.07	3.32	4.40	4.74	5.03
$\frac{G0SL1}{G1SL1}$	1.37	1.26	1.62	1.73	1.82	1.14	0.95	1.29	1.39	1.47

The upper 3 rows illustrate the RMS differences of the individual assimilation runs $G0 - \frac{G0SL0}{G0SL1}$, $G1 - \frac{G1SL0}{G1SL1}$, and $G1sst - \frac{G1sstSL0}{G1sstSL1}$. The lower rows give the cross comparison of $G0$ compared to $SL0$ and $G1$ as compared to $SL1$ ($\frac{G0SL0}{G1SL1}$). Further, $\frac{G0SL1}{G1SL1}$ illustrates the improvement of the GECCO2 model due to the improved assimilated ECV $SL1$

model producing the $G1$ solution, which in turn has been compared to the SL products $SL0$ and $SL1$, as respective analysis of the model–data residuals that emerge in all cases.

The assessment of data products has been done separately for the ERS- and the T/P-satellite data sets. In each case, residuals were investigated globally and regionally, as RMS differences. Smaller residuals were expected for the updated SL product $SL1$, but strictly speaking this is not the case everywhere. The RMS differences are calculated as a measure to determine the difference in the SL products $SL0$ and $SL1$ compared to both GECCO2 solutions. The RMS difference-based skill score has been calculated as:

$$G0SL0 = \frac{\sigma_{(G0-SL0)}}{\sqrt{\left(\sigma^2_{G0} + \sigma^2_{SL0}\right)}}$$

$$G0SL1 = \frac{\sigma_{(G0-SL1)}}{\sqrt{\left(\sigma^2_{G0} + \sigma^2_{SL0}\right)}}$$

$$G1SL0 = \frac{\sigma_{(G1-SL0)}}{\sqrt{\left(\sigma^2_{G1} + \sigma^2_{SL0}\right)}}$$

$$G1SL1 = \frac{\sigma_{(G1-SL1)}}{\sqrt{\left(\sigma^2_{G1} + \sigma^2_{SL0}\right)}}$$

with σ being the STD of the time series and $GXSLX$ are the weighted RMS residuals. Note that the notation RMS differences in anomalies is trivially the same as the STD of anomalies, when referenced to the same averaging period. Again, $G0$ and $G1$ denote the GECCO2 model runs assimilated $SL0$ and $SL1$, respectively. Further, $SL0$ and $SL1$ are the SL data sets itself that have been used for the comparison, respectively. Smaller residuals indicate an improvement of $SL1$ compared to $SL0$. The skill score is designed such that zero indicates perfect skill while a value of one is yielded by two white noise time series and consequently corresponds to no skill.

Of the possible RMS differences, $G0SL0$ is shown exemplarily in Fig. 2, illustrating that the RMS differences vary regionally. In the tropical regions, GECCO2 and the SL product are very close together (blue), indicating a very good representation of the GECCO2 model solution of the assimilated SL data, whereas especially in energetic regions both clearly differ from each other (red). The analysis shows where the GECCO2 solution represents the assimilated SL product well. As has been shown in Fig. 1 for the STD, GECCO2 is missing energy in energetic regions. The model is not able to adjust to mesoscale features mainly because they are not represented by the limited resolution of the model. However, comparing Figs. 1 and 2 reveals that not all regions with large discrepancies are associated with high variability, particularly in the midlatitude eastern Pacific low skill exists in quiet regions. The overall RMS differences for both the T/P and the ERS data sets are the same; however, as a result of the unequal sampling of both satellite observations, the small-scale structures differ from each other. The following sections discuss the disparities of the individual RMS differences.

3 Sea-Level Quality Assessment

To test whether the $SL1$ data set has been improved as compared to the previous product $SL0$, the RMS differences ($G0SL0$, $G0SL1$, $G1SL0$ and $G1SL1$) of both SL data sets to the GECCO2 synthesis results are compared. In the first step, the synthesis result of the GECCO2 model ($G0$) that assimilated the original SL product $SL0$ is compared to both SL

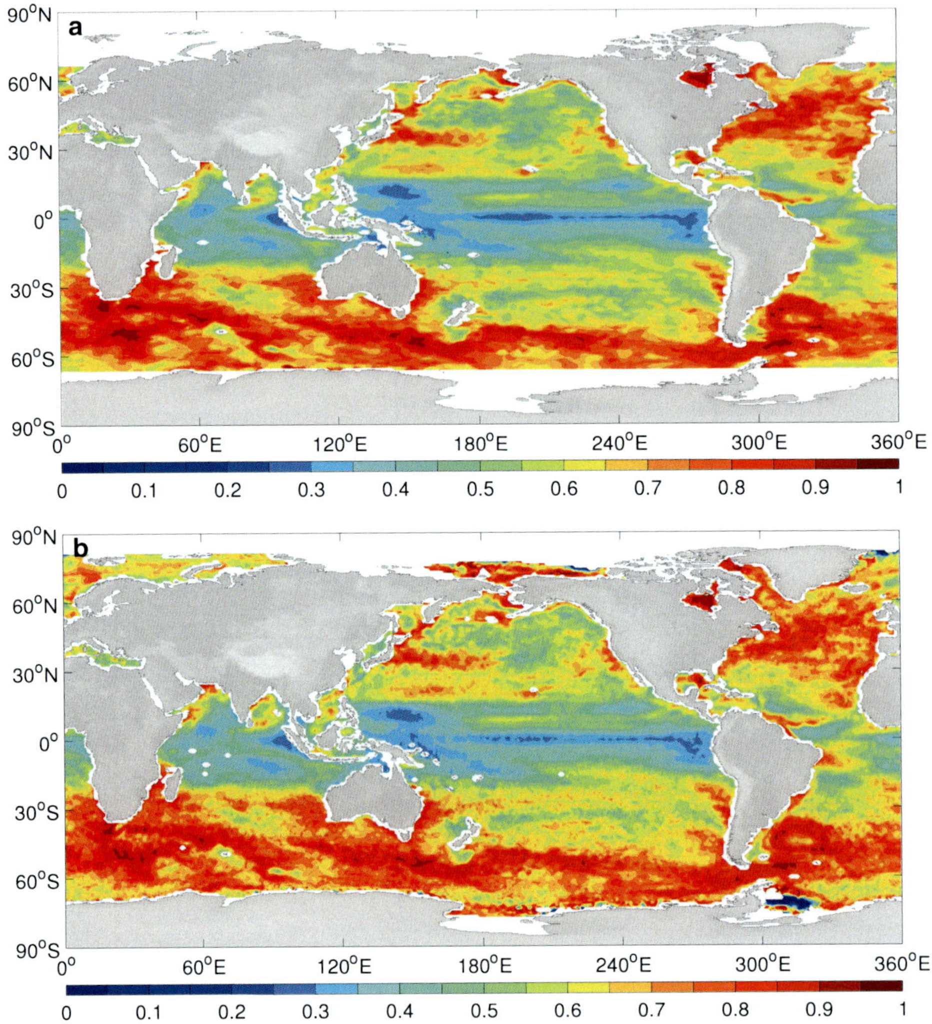

Fig. 2 RMS differences in $G0SL1$ exemplarily, as differences between the GECCO2 model, and the satellite time series for TOPEX/POSEIDON, Jason-1 and Jason-2 (**a**) and for ERS-1, ERS-2, and ENVISAT (**b**)

data sets. For that purpose, the ratio of the RMS differences $\frac{G0SL0}{G0SL1}$ is calculated and shown in Fig. 3 as percentages of improvement for both satellite data sets.

Smaller RMS differences between the model truth $G0$ and the $SL1$ data ($G0SL1$) as compared to $G0SL0$ are interpreted as improvement of the $SL1$ data set, in the way that it is in closer agreement with the model truth. The smaller residuals of $G0SL1$ lead to positive RMS ratios in Fig. 3, whereas increased residuals lead to negative RMS ratios. Therefore, red indicates an improvement of the $SL1$ product, whereas blue indicates a degradation of the data set. Remarkable improvements of up to 30% can be seen in all equatorial regions, on the Argentine shelf and in parts of the ACC. In most other parts of the ocean, improvements of up to 10% are evident. A few regions also show a degradation of up to

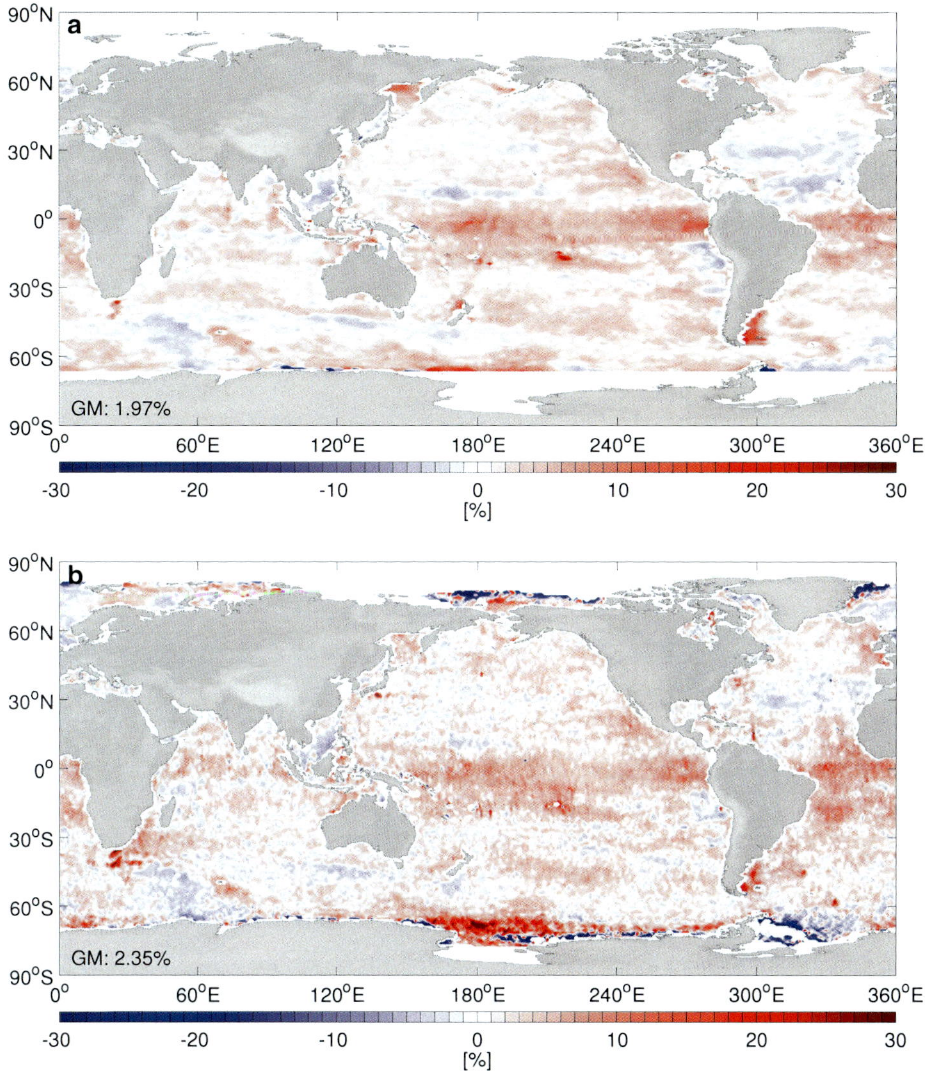

Fig. 3 Ratio of RMS differences, for $\frac{G0SL0}{G0SL1}$, shown as percentages of improvement, for TP time series (**a**) and ERS time series (**b**). The global mean (GM) percentage of improvement is given for latitudes between 66°S and 66°N

10%. In these regions, *G*0 shows little skill when compared to *SL*0 despite the fact that the STDs (Fig. 1) in these regions are small, implying that the assumption of the model serves as truth breaks down.

The global mean (GM) percentages of improvement between latitudes of 66°S and 66°N are on the order of 1.97% (TP) and 2.35% (ERS) and are given in the lower left corner of Fig. 3. The improvements of *SL*1 are remarkable despite the fact that the GECCO2 synthesis was adjusted to minimize the difference to the assimilated *SL*0 data set and, therefore, should be in best agreement with *SL*0. However, since many other climate data have been assimilated as well the final state may reject the assimilated SSH data partly if it is inconsistent with information embedded in other data sets. Therefore, the GECCO2 ocean synthesis may filter data errors even though large biases to the data remain in GECCO2 and GECCO2 cannot be regarded as the best product with the smallest error. Since GECCO2 reveals a closer agreement with $SL1$, it is hypothesized that $SL1$ is in better agreement with the other data than $SL0$.

Overall, the improvements of the satellite time series of T/P and ERS are consistent with each other. However, since both satellite time series have been improved individually (Ablain et al. 2015), resulting improvements show different spatial characteristics; they also differ due to different spatial resolutions and the different temporal sampling of 9.91 days for the TP time series and of 35 days for the ERS time series. Furthermore, the T/P tracks have a wider track spacing making it necessary to display the results at a $2^\circ \times 1^\circ$ grid, whereas the ERS tracks have a smaller track spacing that allows for a display at a $1^\circ \times 1^\circ$ grid.

4 Influence of Assimilation of Improved Sea Level

The significant improvement of $SL1$ within the ESA SL_cci project (Ablain et al. 2015) as compared to $SL0$ has been demonstrated in the previous section by comparing both SL data sets to the GECCO2 ocean synthesis. This section will now investigate the impact on the GECCO2 model by assimilating the improved $SL1$ data set. For that purpose, $SL1$ has been assimilated in the GECCO2 synthesis for iterations 23–28, replacing $SL0$. The GECCO2 synthesis results from this assimilation run are referred to as $G1$. As in the previous section, the ratios of both RMS differences can be calculated for $G1$ as $\frac{G1SL0}{G1SL1}$. It is expected that the assimilation of $SL1$ in the GECCO2 synthesis leads to even smaller RMS differences for $G1SL1$, as in $G1$ the differences to $SL1$ are minimized and since the constraint to be close to $SL0$ no longer exists, it should deviate more from $SL0$.

In order to give an overview and to allow for an inter-comparison of the various ratios of RMS differences within this paper, global mean ratios of RMS differences are presented in Table 2. The global mean ratios of RMS differences are also indicated in the lower left corner of each plot for the ratios of RMS differences.

In Sect. 3, the analysis at the AT positions has been shown for $\frac{G0SL0}{G0SL1}$. As can be seen in Table 2, the global mean ratio of RMS differences deviates between the analysis at AT resolution (1.97% for TP and 2.35% for ERS time series) and at model resolution (2.53% for TP and 2.99% for ERS time series). The disparity is expected, as the model was adjusted to the data on the model grid. The model resolution of about 1° allows for structures on the order of 400 km to be resolved and bringing the model on the satellite tracks by interpolation will lead to even smoother model results. To identify the length scales the GECCO2 model is able to resolve, and for which a meaningful inter-comparison between model and data is possible, zonal wave number spectra have been calculated in the Pacific as this region is not affected by continental boundaries. The regions are marked in Fig. 1b.

At each latitude and for each daily time step (1992–2012) of the $G0$ solution, a zonal wave number spectrum has been calculated in the longitude range of 160°E–260°E using a Hamming window over the whole longitude range while interpolating missing values linearly in between and zero-padding missing values at the ends. The spatial and temporal mean zonal wave number spectra are shown in Fig. 4 combined for the northern and southern latitude ranges of 10°S–10°N, 10°–20°N/S, 20°–30°N/S and 30°–40°N/S. The

Fig. 4 Meridional GECCO2 ($G0$) wave number spectra for the 20-year period in the Pacific, for the latitude bands of 10°S–10°N, 10°–20°NS, 20°–30°NS and 30°–40°NS. The regions are marked in Fig. 1b, and the northern and southern latitude bands are combined. The indicated slope is k^{-6}

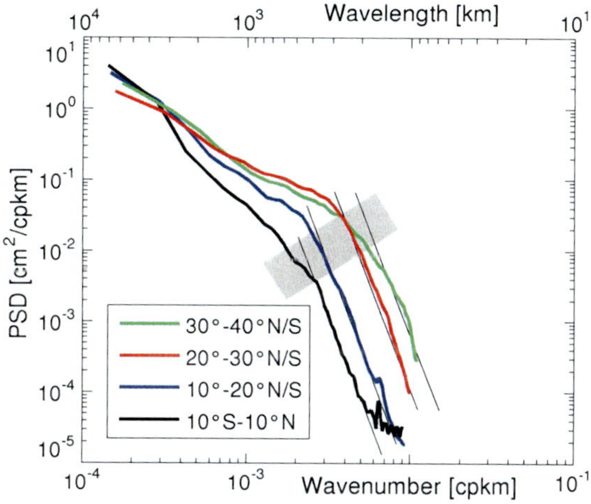

model cutoff length scale depends mainly on the prescribed viscosity, but implicit (numerical) viscosity and diffusivity also play a role. To first order, it is expected that the high wave number end of the spectrum is governed by the action of the harmonic viscosity and therefore will show a slope of k^{-4} for velocities and a slope of k^{-6} for SL. Slopes of k^{-6} are added to Fig. 4 and are thus utilized to determine the cut off. Since a smooth transition into the k^{-6} slope regime exists, the determination of the cutoff bears some vagueness indicated by the gray box, which ultimately causes some arbitrariness in the determination of the scale.

From the figure, it appears that the GECCO2 model cutoff is found to be latitudinal dependent and on the order of 400 km (5°), 360 km (15°), 250 km (25°), and 170 km (35°). As can be seen in Table 1 (columns 4–6), the above model cutoff length scales correspond to the AT running mean filter length scales that take 7, 9, and 11 AT values into account. In order to generate the AT ratios of RMS differences that are comparable for AT and M resolution, the ratios of RMS differences are shown in the following for the *f*9-filter, as it is closest to the cutoff seen in the zonal wave number spectra of about 200–400 km, that the model is actually able to resolve. Therefore, the ratios of RMS differences from Fig. 3 are shown again, but this time low-pass filtered with *f*9, in Fig. 5a, d to illustrate the rate of improvement of the resolved features.

The global mean ratios of RMS differences for the different assimilation runs $G0$, $G1$, and $G1$sst, as displayed in Table 2 (first 3 rows), clarify the improvement of $SL1$ over $SL0$, as all the ratios are positive, indicating a smaller RMS difference between each GECCO2 synthesis and $SL1$ and a larger for $SL0$. The global mean ratios of RMS differences increase further, when assimilating the updated $SL1$ in the GECCO2 synthesis ($\frac{G1SL0}{G1SL1}$). When additionally to $SL1$, assimilating an updated version of SST from the ESA SST_cci (Merchant et al. 2014), the global mean ratios of RMS differences ($\frac{G1\mathrm{sst}SL0}{G1\mathrm{sst}SL1}$) are comparable to those of the assimilation run with the previous SST estimates $G1$. The changed SST product therefore does not alter the GECCO2 synthesis with respect to SL.

The total improvement due to the updated satellite data $SL1$ and its assimilation into the GECCO2 synthesis can be revealed by the ratio of the differences in $G0$ and $SL0$ by using only the previous data set $SL0$, and the differences in $G1$ and $SL1$ by using the updated SL data set only. This ratio $\frac{G0SL0}{G1SL1}$ as shown in Fig. 5b, e highlights the reduction in the RMS

Fig. 5 Ratio of RMS differences in low-pass (*f*9)-filtered data, for $\frac{G0SL0}{G0SL1}$ (**a**, **d**), for $\frac{G0SL0}{G1SL1}$ (**b**, **e**) and for $\frac{G0SL1}{G1SL1}$ (**c**, **f**), shown as percentages of improvement, for TP time series (**a–c**) and ERS time series (**d–f**). The global mean (GM) percentage of improvement is given for latitudes between 66°S and 66°N

differences in most regions of the world oceans and leading to an improvement of more than 30% in many regions. Hence, additionally to the equatorial regions, the Argentine shelf and parts of the ACC, improvements of more than 30% can now be seen in the northern Indian Ocean, the north Pacific, subtropical regions, and large regions south of the ACC as well. As for the ratio $\frac{G0SL0}{G0SL1}$ (Fig. 5a, d), degradations of *SL*1 exist in isolated regions, where the GECCO2 synthesis adapts less well to the assimilated *SL*1 product. The regions showing a degradation match with regions of small STD (Fig. 1) implying that the assumption of model serves as truth breaks down. As for the improvements, the degradations are amplified as well. The global mean improvement sums up to 4.75 and 4.74% for the TP and ERS data sets, respectively. While Fig. 5b, e gives the combined improvement due to the assimilation (*G*1) and the updated product (*SL*1), Fig. 5c, f answers the question about the improvement of the GECCO2 synthesis by assimilating the updated *SL*1 data set alone. Here, the ratio $\frac{G0SL1}{G1SL1}$ compares the different assimilation runs *G*0 and *G*1 while calculating the RMS differences to the same updated *SL*1 data set.

The GECCO2 synthesis takes most profit from the updated *SL*1 data set in the northern Indian Ocean, the ACC, and the north Pacific. Degradations are evident in the north Atlantic, the eastern north Pacific as well as scattered regions around the globe, however, leading to an overall global mean improvement of the GECCO2 synthesis of 1.73 and 1.39% as seen from the TP and ERS satellite time series, indicating that the model physics accepted the assimilated *SL*1 better as compared to *SL*0 and therefore adapted closer to the *SL*1 data set. Although degradations seem surprising given that the synthesis was constrained to the data, the problem is far to complex to expect a convergence to a minimum within only a few iterations. Not all changes can thus be expected to causally relate to the change in data.

Nevertheless, the GECCO2 synthesis benefits from the updated *SL*1 data set at least with respect to SL. Consequently, the question arises to what degree the GECCO2 synthesis improves in general and in particular to those additionally assimilated data. To tackle this question, Fig. 6 displays the change in the weighted global mean RMS differences in the GECCO2 synthesis after assimilating *SL*1 instead of *SL*0. The reduction in RMS differences indicates an improvement of 2.4% for SSH, in the order of 1% for SST of AMSRE and EN3 and of about 2.5% for EN3 salinities, while having only a minor increase on the order of 0.2% for other variables.

So far, the effect of applying the low-pass filter (*f*7, *f*9, and *f*11) to the AT data has revealed larger improvements for larger filtering length scales (Fig. 5; Table 2). The main improvements seem to exist on length scales above several hundred km, and the question appears whether an effect on the smaller length scales exists, although for the smallest scales no significant effect can be expected due to the resolution of the GECCO2 model. The high-pass filter is represented as the difference between the individual SL data set and its filtered (*f*9) product. The remaining ratios of RMS differences for the high-pass filtered data set are shown exemplarily for $\frac{G0SL0}{G0SL1}$ in Fig. 7. For both satellite time series (TP and ERS), no clear influence can be seen at high frequencies.

A patchy structure is evident for both satellite time series. However, the ratios of RMS differences in the high-pass-filtered data depict a different behavior for both satellite time series indicating the different changes that were applied to improve the two different SL

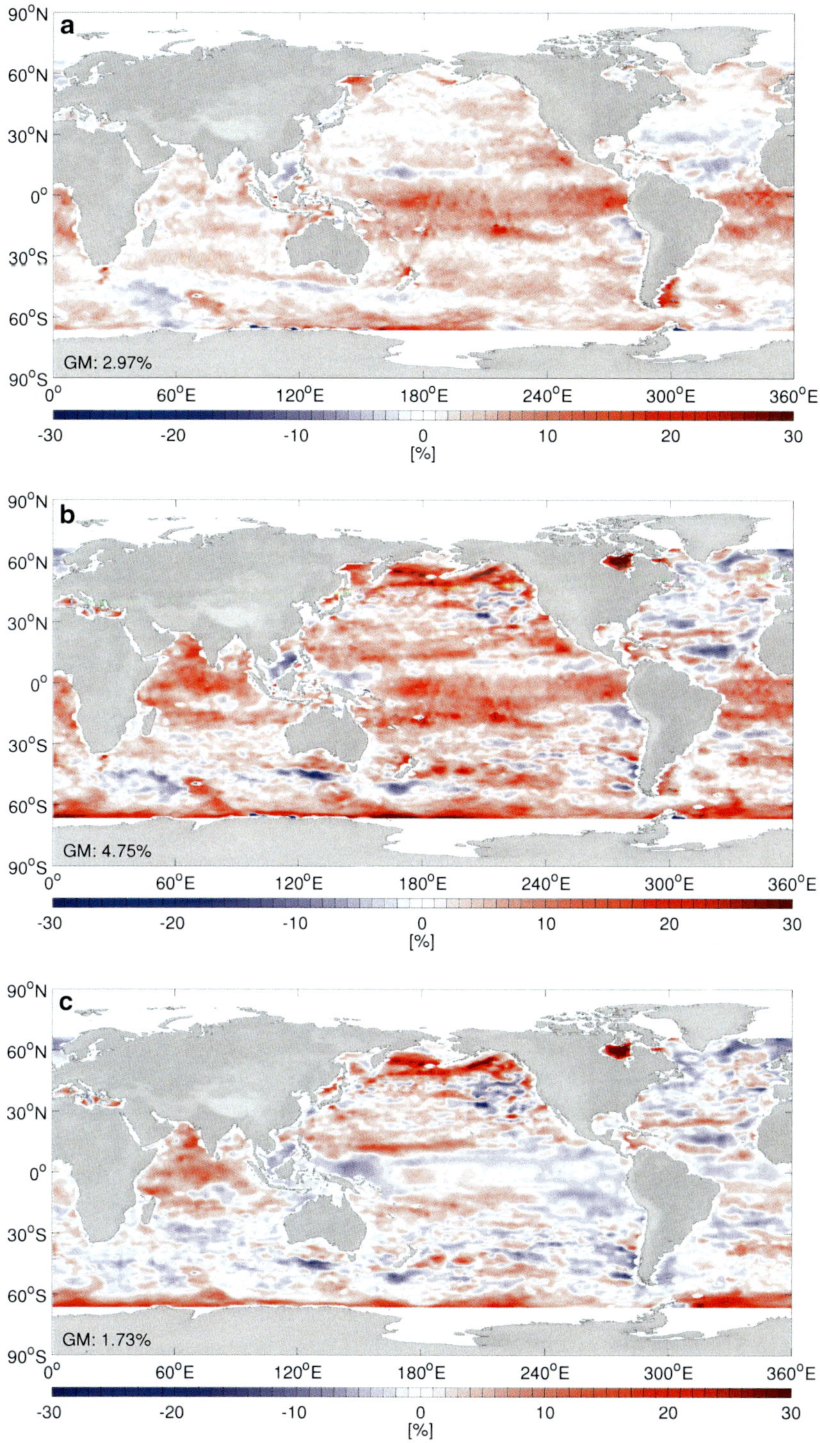
a
GM: 2.97%
b
GM: 4.75%
c
GM: 1.73%
90°N
60°N
30°N
0°
30°S
60°S
90°S
0°
60°E
120°E
180°E
240°E
300°E
360°E
-30
-20
-10
0
10
20
30
[%]

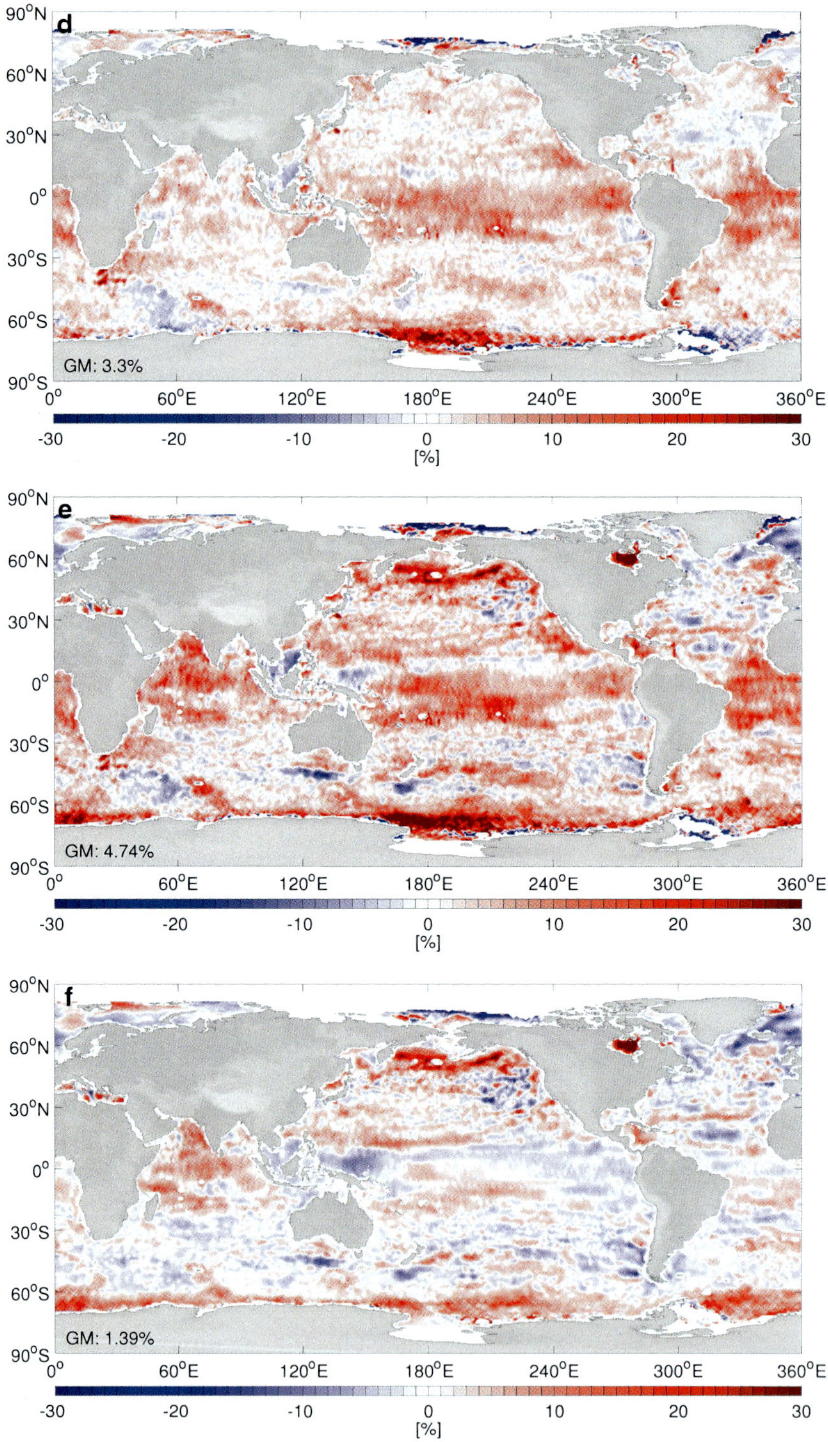

Fig. 5 continued

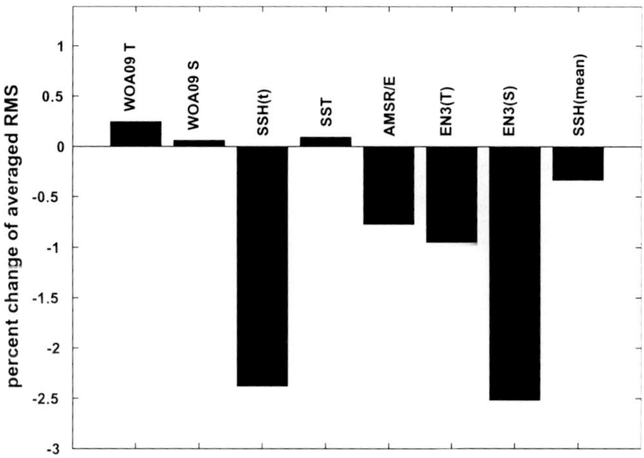

Fig. 6 Global mean weighted RMS differences in GECCO2 for assimilated data sources after assimilation of *SL*1 in percent. The weighting derives from the prior model–data errors that enter the cost function during the assimilation procedure. For absolute values, see Fig. 1 in Köhl (2015)

time series (Ablain et al. 2015). The global mean of −0.28 (TP time series) and 0.16% (ERS time series) cannot be assumed to be significant due to the patchy structure. As expected, the GECCO2 synthesis clearly shows no improvement on the small scales.

5 Regional Sea-Level Trends

The ratios of RMS differences proved an advancement of the SL data set through the ESA SL_cci effort. The accuracy of the altimeter-based regional sea-level trends (Ablain et al. 2015) has been another aim of the ESA SL_cci effort. Therefore, the overall performance of the trends will be examined, to clarify the changes due to the update of the SL data set *SL*1.

As described in Sect. 2.2, the global mean sea level (GMSL) of the GECCO2 synthesis is zero at each time step, and therefore, in order to treat the SL data sets *SL*0 and *SL*1 in an analogous manner, the GMSL time series had to be subtracted from the SL data sets accordingly, leaving the regional trends untouched. Thus, the GECCO2 synthesis can be used to assess the changes in the regional trend pattern of the SL data sets.

Exemplarily, the regional trends of *SL*1 are shown in Fig. 8a, d for the TP and ERS time series. Both indicate the strongest positive regional trends in the equatorial western Pacific and Indian Ocean. Further positive regional trends are evident in the tropical Atlantic, western Pacific as well as in parts of the ACC. Regions of negative regional trends are the eastern Pacific, the northern Indian Ocean, and parts of the subtropical Atlantic.

The corresponding regional trends of the GECCO2 synthesis *G*1 sampled at the TP and ERS along-track positions and times are shown in Fig. 8b, e, respectively. While the overall regional trend pattern is depicted in *G*1 as well, as the large trends of about 15 mm/year in the western equatorial Pacific for *SL*1, *G*1 has large positive trends in regions where *SL*1 does not, for example in parts of the ACC and in the north Atlantic. In order to accentuate the differences between the regional trends of data and model, Fig. 8c, f shows the regional trend differences for $G1 - SL1$, hence the closer to zero the better the

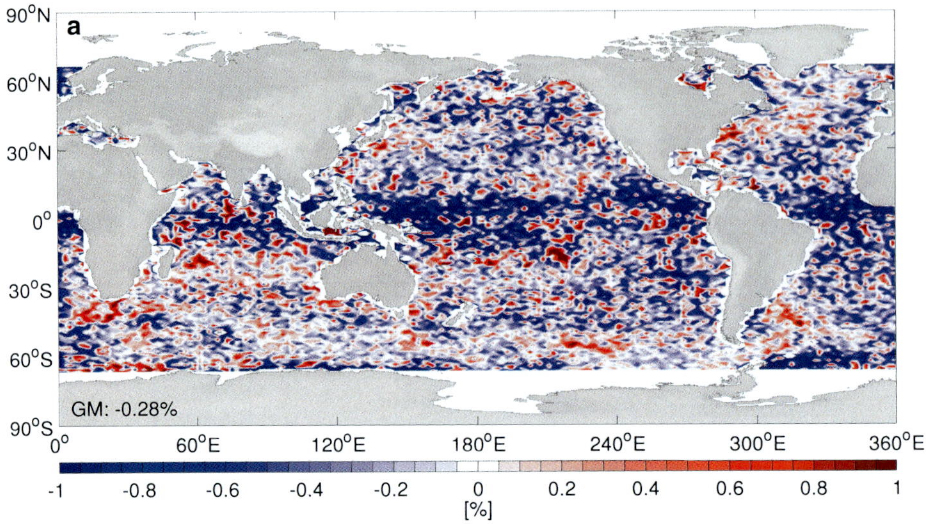

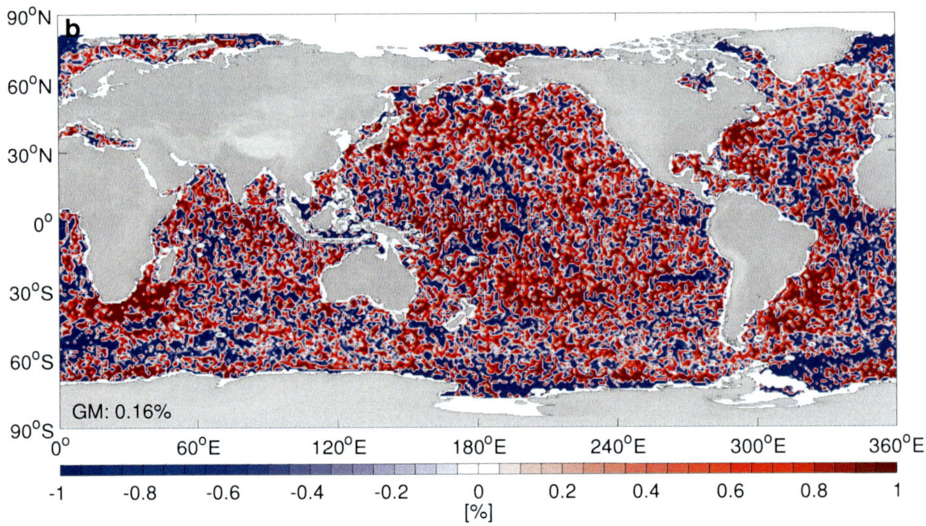

Fig. 7 Ratio of RMS differences in high-pass-filtered data, for $\frac{GOSL0}{GOSL1}$, shown as percentages of improvement, for TP time series (**a**) and ERS time series (**b**). The figure is truncated at $\pm 1\%$. The global mean (GM) percentage of improvement is given for latitudes between 66°S and 66°N

agreement between the regional trends of $G1$ and $SL1$. Especially in the North Atlantic, the South China Sea, and the ACC, the regional trends disagree with each other, indicating that the model physics of the GECCO2 synthesis provides a solution different from the $SL1$ data set. However, in most regions there is a close agreement.

Fig. 8 Regional trends of $SL1$ (**a**, **d**), of $G1$ (**b**, **e**), and of the respective trend differences $G1 - SL1$ ▶ (D_G1SL1, **c**, **f**), for TP time series (**a–c**) and ERS time series (**d–f**). The absolute global mean (|GM|) percentage of the trend differences (**c**, **f**) is given for latitudes between 66°S and 66°N

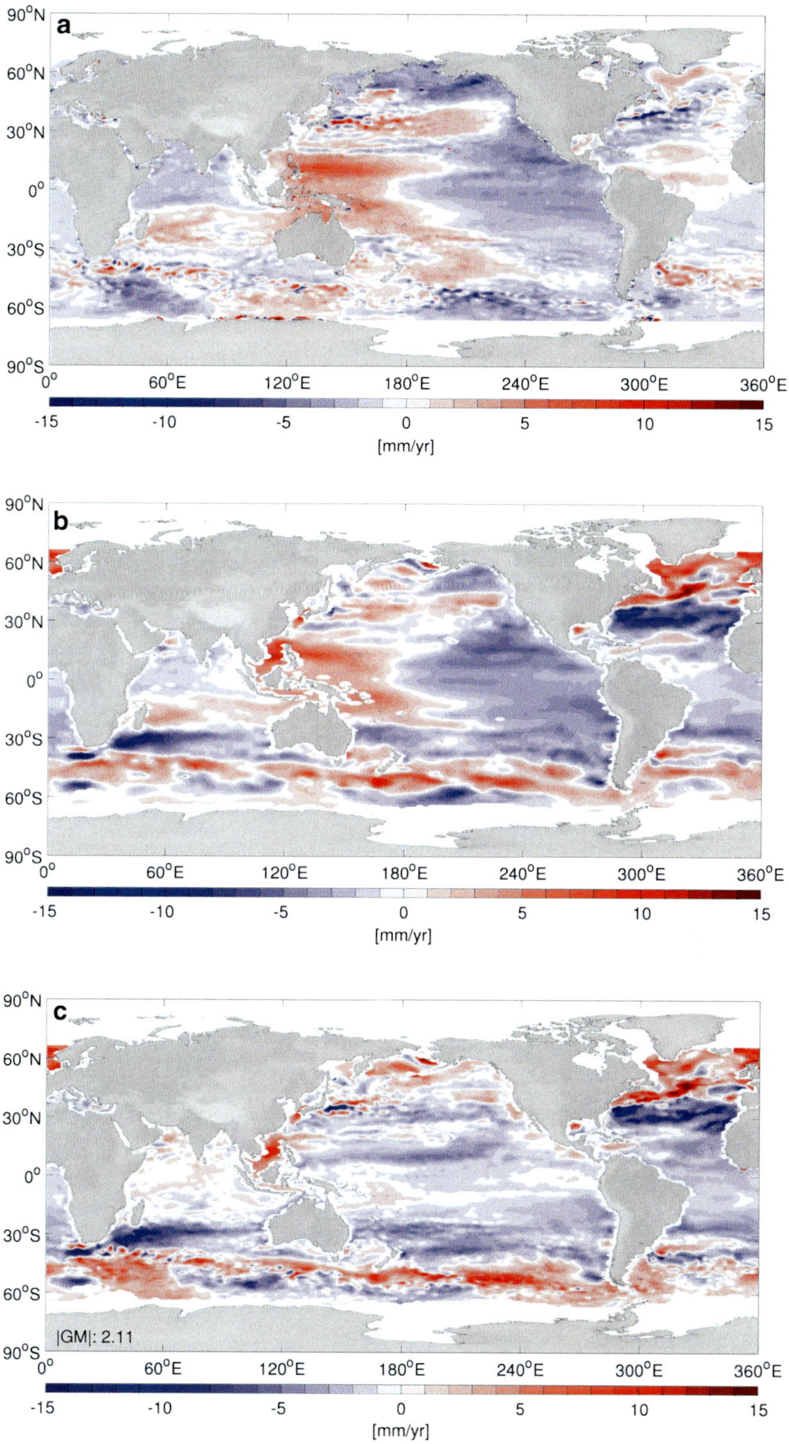
a
90°N
60°N
30°N
0°
30°S
60°S
90°S
0°
60°E
120°E
180°E
240°E
300°E
360°E
-15
-10
-5
0
5
10
15
[mm/yr]
b
90°N
60°N
30°N
0°
30°S
60°S
90°S
0°
60°E
120°E
180°E
240°E
300°E
360°E
-15
-10
-5
0
5
10
15
[mm/yr]
c
90°N
60°N
30°N
0°
30°S
60°S
90°S
|GM|: 2.11
0°
60°E
120°E
180°E
240°E
300°E
360°E
-15
-10
-5
0
5
10
15
[mm/yr]

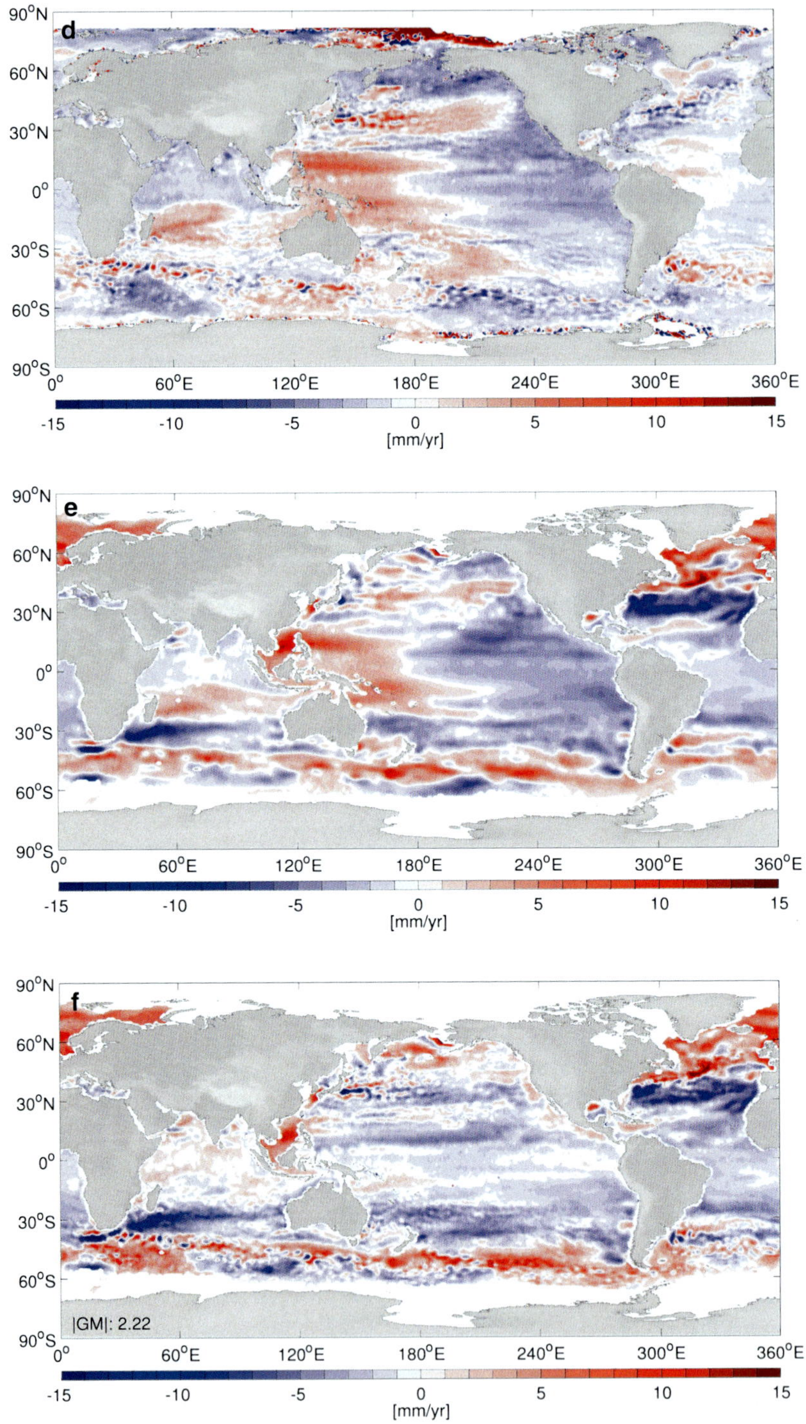

Fig. 8 continued

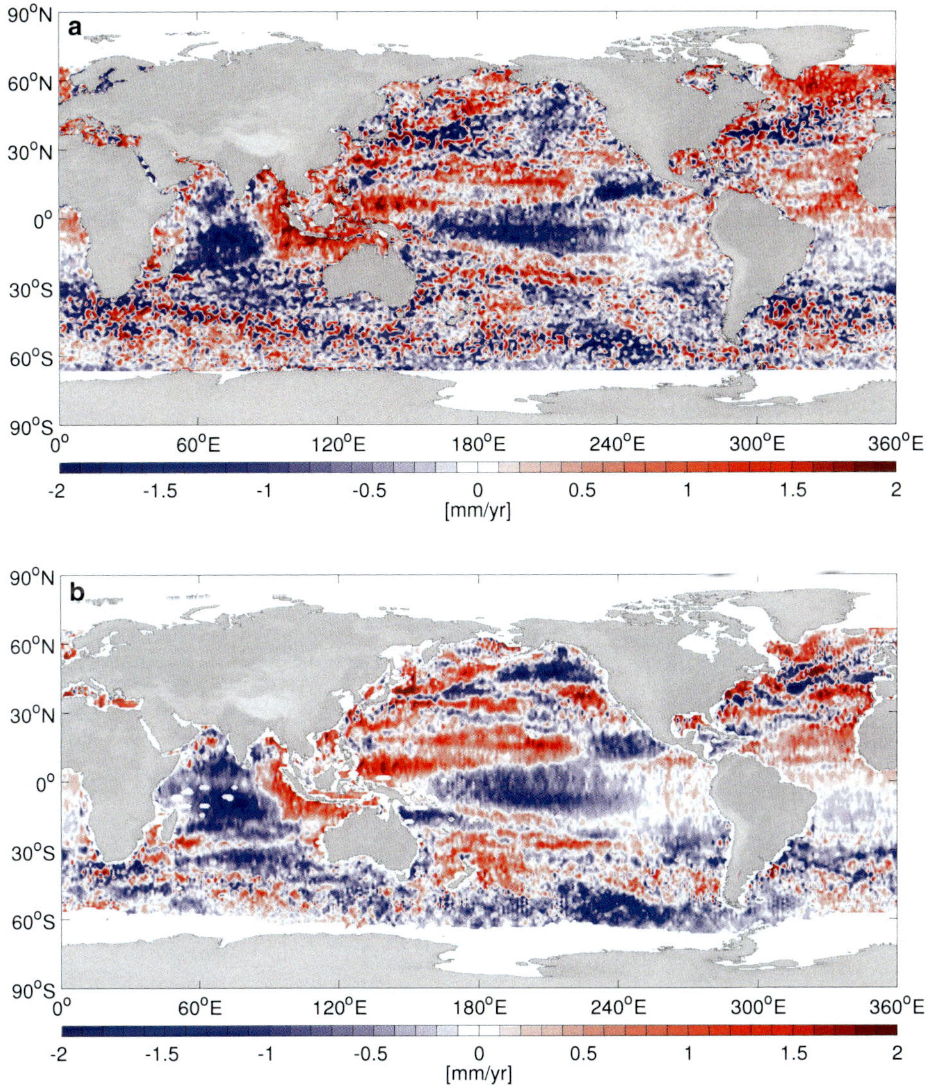

Fig. 9 Regional trend differences in TP−ERS time series for *SL*1 (**a**) and for *G*1 (**b**)

The analysis has been conducted along the TP as well as along the ERS ground tracks, with their unequal temporal and spacial sampling patterns. When analyzing the regional trend differences in *SL*1 in Fig. 8a, d and in *G*1 Fig. 8b, e, differences between the sampling pattern of TP and ERS can be seen. These differences in regional trend differences are shown in Fig. 9 for *SL*1 (a) and for *G*1 (b). The differences resulting from sampling the GECCO2 synthesis along the different satellite ground tracks only (Fig. 9b) have a distinct pattern that is clearly depicted by the regional trend differences between the TP and ERS altimeter results of *SL*1 as well, as can be seen in Fig. 9a. The differences in regional trend differences between the TP and the ERS sampling are on the order of ±2 mm/year, which is about an order of magnitude smaller than the actual signal.

To identify the changes among the data sets and the GECCO2 synthesis results, respectively, the changes in regional trends between $SL0$ and $SL1$ as well as between $G0$ and $G1$ are addressed in Fig. 10 by means of their respective differences. The large correlated structures of regional trend differences (Ablain et al. 2015) are here shown for the TP and ERS time series separately (Fig. 10a, c), distinguishing again the different corrections applied for both satellite time series, as well as their sampling differences. The

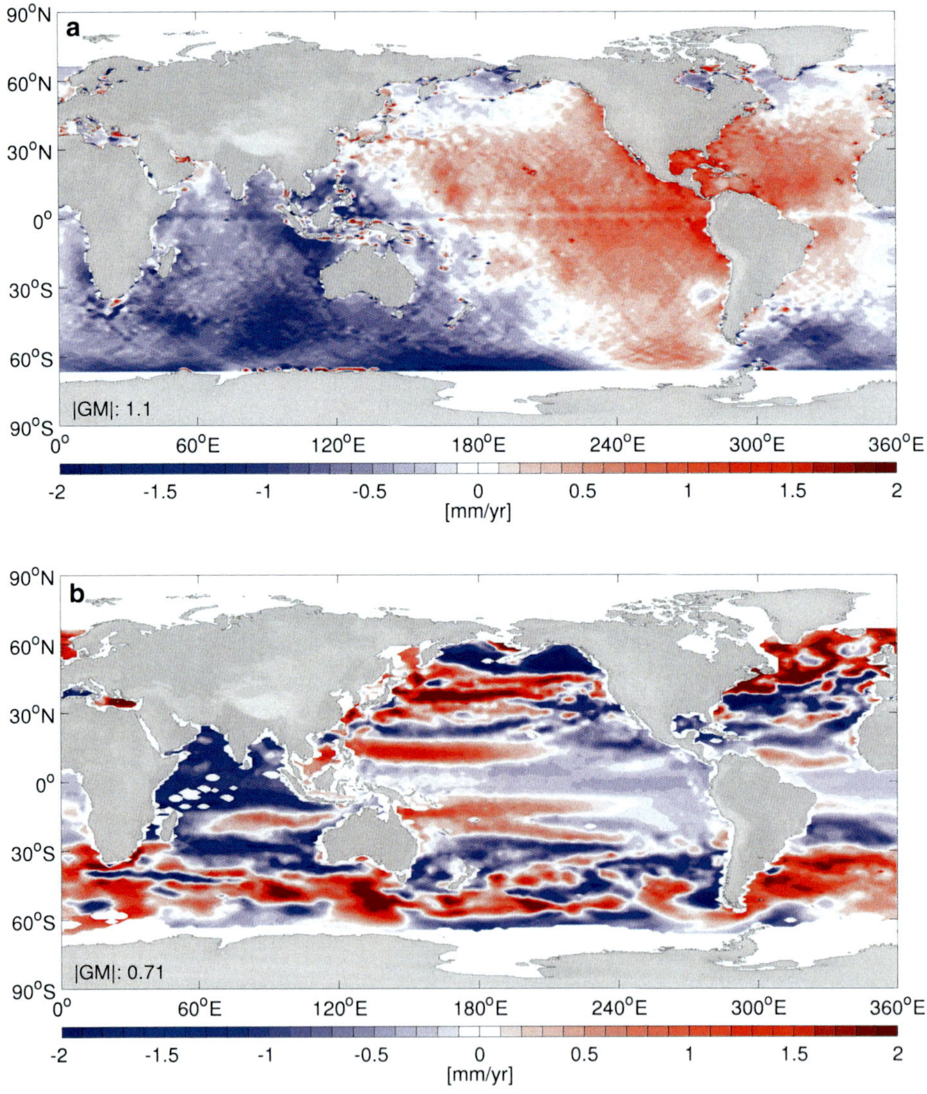

Fig. 10 Regional trend differences of the two data sets $SL1 - SL0$ (**a**, **c**), and the two GECCO2 assimilation runs $G1 - G0$ (**b**, **d**), for TP series (**a**, **b**) and ERS series (**c**, **d**). The absolute of global mean (|GM|) percentages of regional trend differences is given for latitudes between 66°S and 66°N

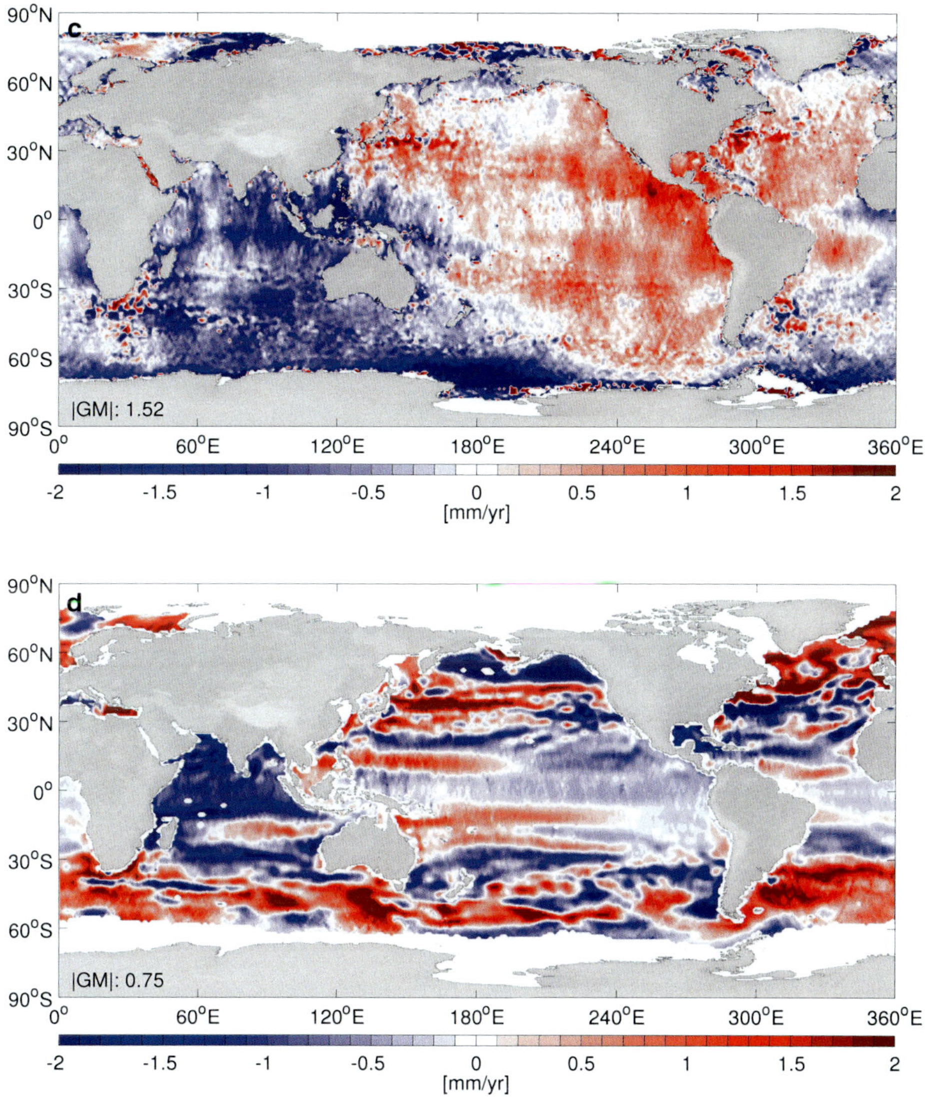

Fig. 10 continued

differences primarily result from the new orbit solutions, the new atmospheric fields, the new wet tropospheric corrections, and the geographical biases when linking altimetry missions together (Ablain et al. 2015). For both satellite time series, the large-scale pattern gives a positive SL trend centered in the eastern equatorial Pacific and Gulf of Mexico, whereas a pattern of negative SL trend is centered in the ACC in the Indian Ocean. The global mean of the absolute regional trends (|GM|) is given in the lower left corner of the plots and additionally is given in Table 3 for comparison. According to the absolute global mean values, the regional trends changed by as much as 1.11 and 1.52 mm/year for the TP and ERS time series, respectively.

Fig. 11 Differences in absolute regional trend differences of $|D\_G1SL1| - |D\_G1SL0|$ (**a**, **d**), of $|D\_G1SL1| - |D\_G0SL1|$ (**b**, **e**) and of $|D\_G1SL1| - |D\_G0SL0|$ (**c**, **f**), for TP series (**a–c**) and ERS series (**d–f**). The global mean (GM) percentages of absolute regional trend differences are given for latitudes between 66°S and 66°N

While $G0$ and $G1$ tried to adapt to the assimilated data sets $SL0$ and $SL1$, respectively, the differences between the regional trends of $G0$ and $G1$ (Fig. 10b, d) are expected to provide a version of the trend difference between $SL0$ and $SL1$ filtered by the assimilation procedure for both satellite time series. Apparently, the regional trend differences in the GECCO2 synthesis $G1 - G0$ do not reflect the large-scale pattern of regional trend changes as seen for the SL data sets (Fig. 10a, c). The reason can be the smallness of the trend difference. As indicated above, the solution cannot be considered converged and some changes may be not causally connected to the changes in the input data. However, the differences rather reflect the large-scale circulation pattern. An increase in the $G1$ trends can be found in the north and south equatorial currents (NEC, SEC) in the Atlantic and Pacific as well as the SEC in the Indian Ocean, many parts of the ACC, the Kuroshio and North Atlantic Current as well as large parts of the North Atlantic, whereas a decrease in the trends can be seen, for example, along the Equatorial Counter Currents, the eastern tropical and northern Pacific.

Using $G0$, $G1$, and $G1$sst as measures to evaluate the regional trends of $SL0$ and $SL1$, the absolute global mean regional trend differences between all possible data–model combinations are shown in Table 3 as $GX - SLX$. The smaller absolute regional trend differences indicate a closer agreement of model and data in terms of their regional trends. For $G0$, the global mean regional trend differences $G0 - SL0$ and $G0 - SL1$ are of the same order of magnitude, and only slightly larger for $SL1$, indicating neither an improvement nor a

Table 3 Global mean ratios of trend differences for TP and ERS time series as differences in mm/year, computed on the along-track points

TREND differences [mm/year]	TP time series	ERS time series
$\|SL1 - SL0\|$	1.11	1.52
$\|G1 - G0\|$	0.71	0.75
$\|G0 - SL0\|$	2.02	2.15
$\|G0 - SL1\|$	2.07	2.18
$\|G1 - SL0\|$	2.04	2.2
$\|G1 - SL1\|$	2.11	2.22
$\|G1sst - SL0\|$	2.28	2.47
$\|G1sst - SL1\|$	2.35	2.48
$D\_G1SL1 - D\_G0SL0$	0.92	0.98
$D\_G1SL1 - D\_G0SL1$	0.71	0.74
$\|D\_G1SL1\| - \|D\_G1SL0\|$	0.06	0.02
$\|D\_G1SL1\| - \|D\_G0SL1\|$	0.03	0.04
$\|D\_G1SL1\| - \|D\_G0SL0\|$	0.08	0.07

The upper 2 rows illustrate the trend differences between the SL products $SL1 - SL0$ and the different GECCO2 assimilation runs $G1 - G0$. Further, the middle rows give the trend differences between the model runs ($G0$ and $G1$) and the SL products ($SL0$ and $SL1$). The lower 5 rows give the differences between the model-data differences and as differences in the absolute differences

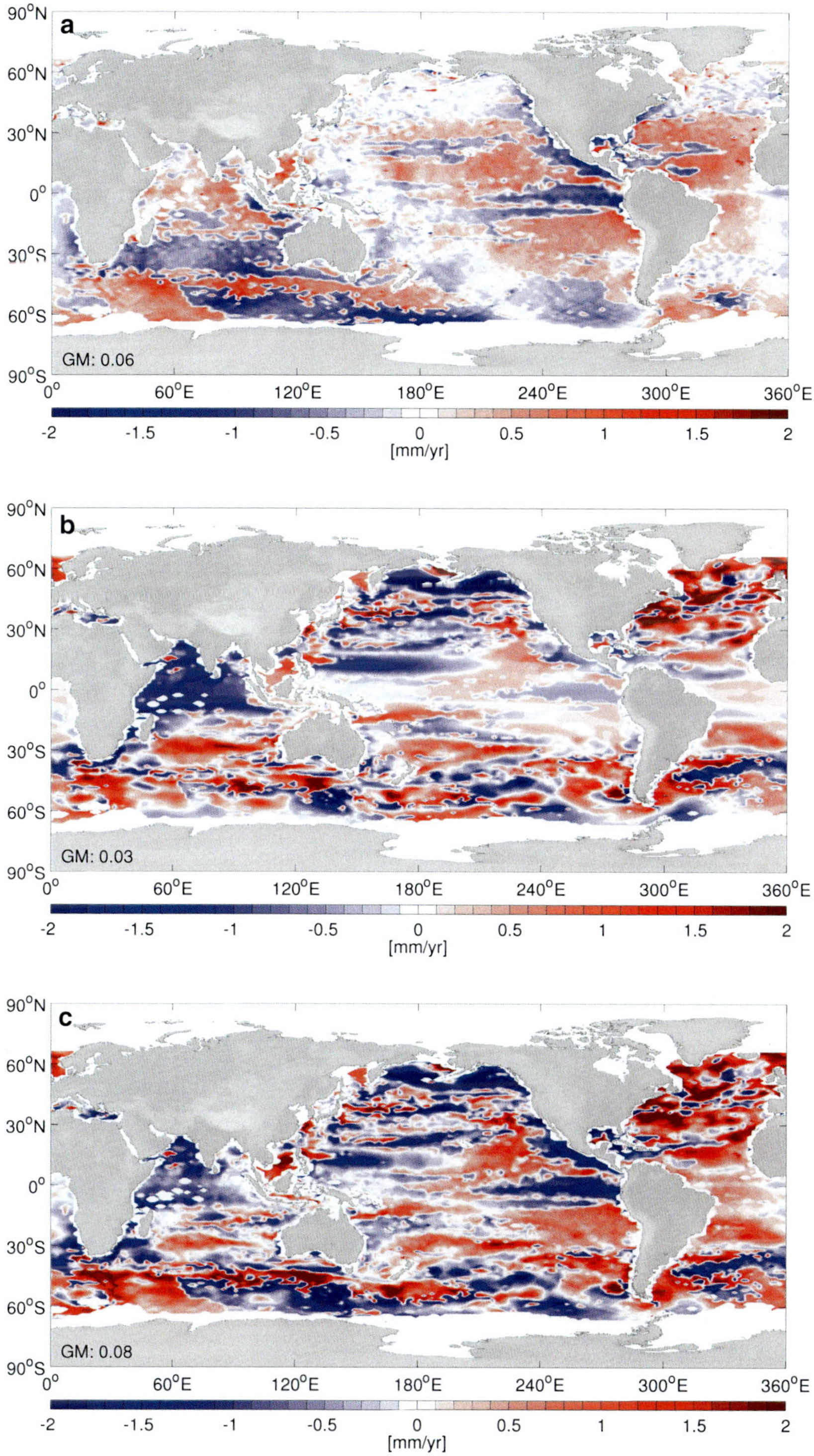
a
GM: 0.06
b
GM: 0.03
c
GM: 0.08
90°N
60°N
30°N
0°
30°S
60°S
90°S
0°
60°E
120°E
180°E
240°E
300°E
360°E
-2
-1.5
-1
-0.5
0
0.5
1
1.5
2
[mm/yr]

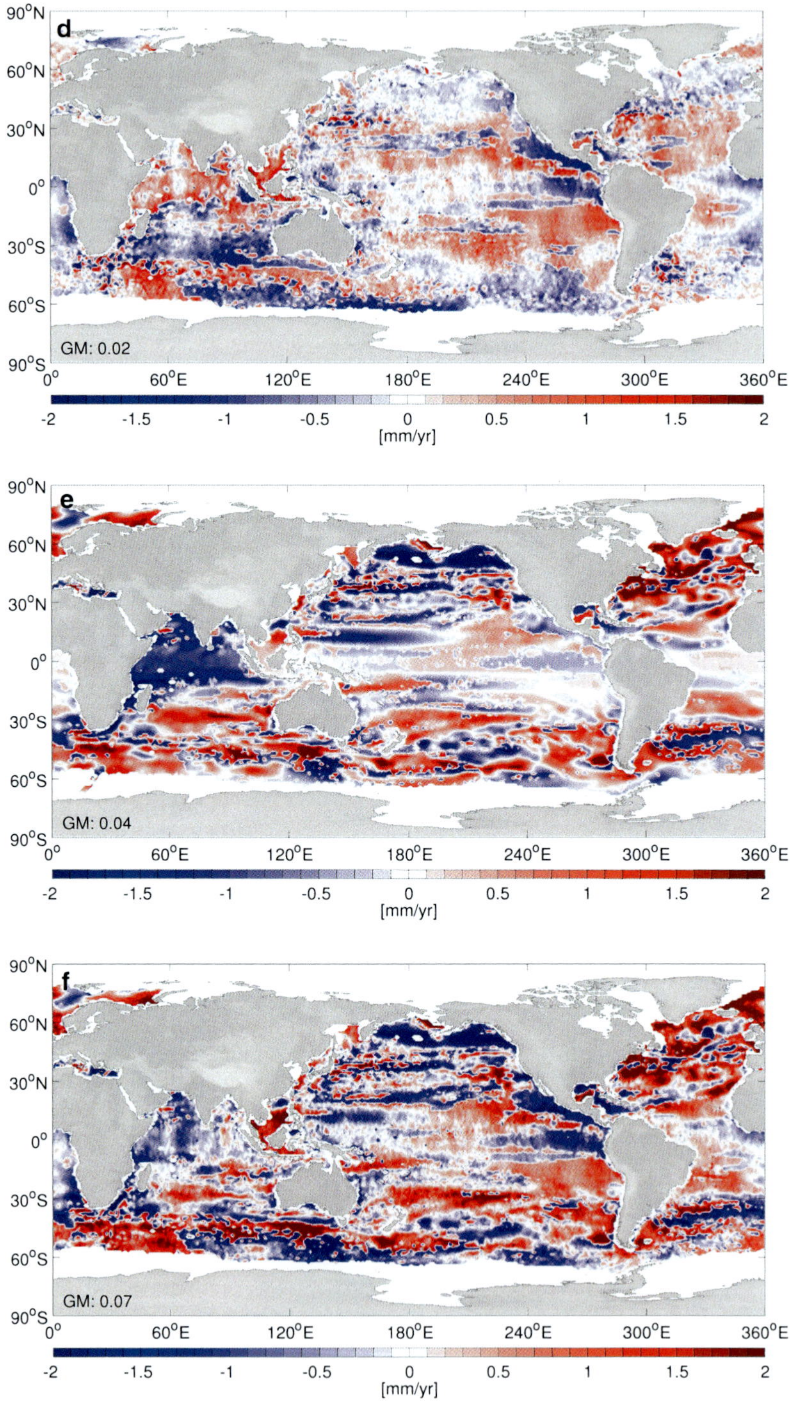

Fig. 11 continued

degradation of the $SL1$ data set in terms of regional trends as compared to the GECCO2 synthesis. The same holds true for the global mean trend differences in $G1$ as those global mean regional trend differences are nearly identical to the ones of $G0$. The additional assimilation of the updated SST ($G1$sst), however, leads to larger regional trend differences as compared to $G0$ and $G1$, indicating a degradation of regional trend differences as compared to $G0$ and $G1$. Therefore, no further assessment is undertaken for the GECCO2 synthesis $G1$sst.

To assess the changes in the regional trend differences between data and model directly, the absolute values of their differences (D_GXSLX) are subtracted, respectively, as shown in the lower three rows of Table 3 for the global mean and in Fig. 11 for the regional distribution. Using this measure, the expected better data $SL1$ and the expected better model synthesis $G1$ are anticipated to have the smaller absolute global mean differences ($|D\_G1SL1|$) in terms of regional trends (see Table 3). Therefore, negative values indicate a closer agreement between the first model and data, whereas positive values indicate a closer agreement between the second.

However, for the comparison of $|D\_G1SL1| - |D\_G1SL0|$, the resulting positive differences (0.06 mm/year for TP and 0.02 for ERS) of the global mean differences in the regional trends indicate a closer agreement of $G1$ to $SL0$ as to $SL1$. When comparing regional trend differences in the GECCO2 synthesis $G0$ and $G1$ against $SL1$ ($|D\_G1SL1| - |D\_G0SL1|$), $G0$ seems to be in closer agreement with SL1 than $G1$. Even more when comparing the differences $D\_G1SL1 - D\_G0SL0$, the closer agreement is found for the $SL0$ data set and $G0$.

The regional distribution of trend distribution differences is given in Fig. 11. The comparison of $SL0$ and $SL1$ ($|D\_G1SL1| - |D\_G1SL0|$, Fig. 11a, d) shows an improvement (blue) of the SL1 data set in the north Atlantic. Yet, in the same region, the comparison of $G0$ and $G1$ ($|D\_G1SL1| - |D\_G0SL1|$, Fig. 11b) shows a degradation of $G1$ (red). The combined effect is shown in Fig. 11c, f.

We note, however that, because the global mean differences of regional trend differences are very small, the comparisons of regional trends of the SL products to the GECCO2 synthesis results are not able to detect an improvement or degradation of the regional trend pattern. The different regional trends of the data sets and the GECCO2 synthesis are the reason for the inability to clearly identify the regional trend changes from $SL0$ to $SL1$ as improvement or degradation.

6 Annual Sea-Level Signal

Sea level has a significant annual signal which is an important component of the overall SL signal. In this section, the influence of the changes from $SL0$ to $SL1$ on the annual signal is evaluated. For clarity, the evaluation is shown for the TP data set only. The annual SL signal has amplitudes of more than 150 mm in the equatorial Pacific and Indian Ocean as well as in the Kuroshio and Gulf Stream regions, and is shown in Fig. 12a exemplarily for $SL1$. In most other parts of the world's oceans, the annual signal does not exceed 50 mm. The phase pattern (Fig. 12d) shows a clear difference between the northern and southern hemisphere reflecting in part the annual changes in incoming solar radiation. The corresponding annual SL pattern for the GECCO2 synthesis $G1$ (Fig. 12b, e) shows similar structures as the SL data, for the amplitudes as well as for the phase. The differences

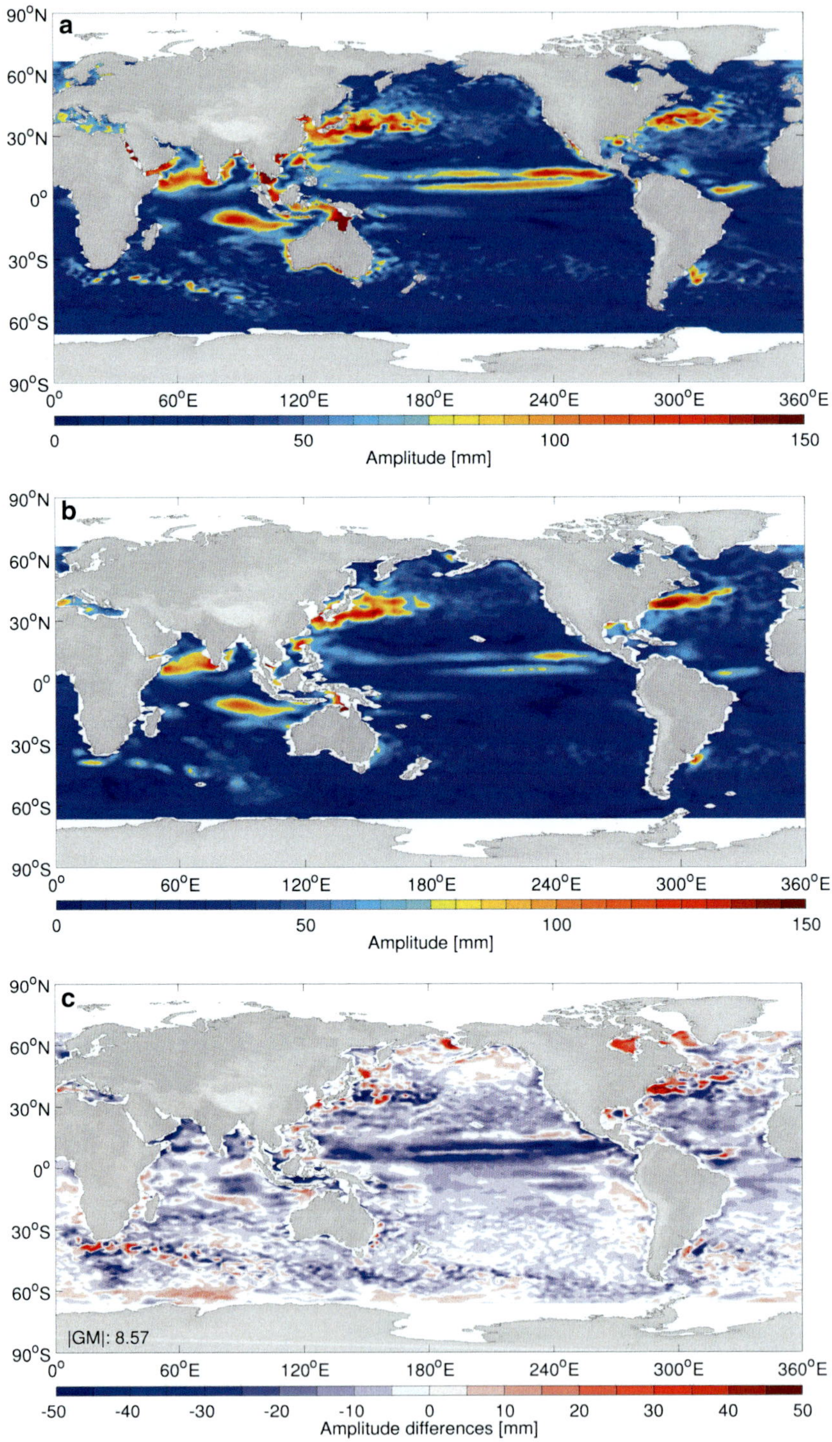
a
90°N
60°N
30°N
0°
30°S
60°S
90°S
0° 60°E 120°E 180°E 240°E 300°E 360°E
0 50 100 150
Amplitude [mm]
b
90°N
60°N
30°N
0°
30°S
60°S
90°S
0° 60°E 120°E 180°E 240°E 300°E 360°E
0 50 100 150
Amplitude [mm]
c
90°N
60°N
30°N
0°
30°S
60°S
90°S
|GM|: 8.57
0° 60°E 120°E 180°E 240°E 300°E 360°E
-50 -40 -30 -20 -10 0 10 20 30 40 50
Amplitude differences [mm]

◀**Fig. 12** Annual amplitude (**a–c**) and phase (**d–f**) of *SL*1 (**a, d**), of *G*1 (**b, e**), and of annual amplitude and phase differences as $G1 - SL1$(**c, f**), for TP series. The absolute global mean (|GM|) percentages of annual amplitude and phase differences are given for latitudes between 66°S and 66°N

between *G*1 and *SL*1 are displayed in Fig. 12c, f. The largest differences in amplitude occur in the Gulf Stream region, the equatorial eastern Pacific, and in parts of the ACC. The phase differences are patchy and are in the range of ±50 days.

Next to the differences between *G*1 and *SL*1, the changes between the two data sets *SL*0 and *SL*1 (Fig. 13a, c) as well as between *G*0 and *G*1 (Fig. 13b, d) are equally important. Large-scale patterns can be seen for the differences in the SL data sets (Fig. 13a). In contrast, the differences in the GECCO2 synthesis results *G*0 and *G*1 (Fig. 13b) not only have a different structure but also are larger as those between the SL data sets. This also holds true for the changes in the phase pattern (Fig. 13d). The absolute global mean amplitude and phase changes sum up to 1 mm/year and 4.04 days for the SL data sets and 1.88 mm/year and 6.17 days for the GECCO2 results.

To evaluate whether the updated *SL*1 data set has an improved annual signal when compared to the GECCO2 synthesis, Fig. 14a, c shows the changes in the SL data sets as absolute changes in the differences between model and data as $|D\_G1SL1| - |D\_G1SL0|$. A very patchy structure is evident that sums up to an amplitude change of 0.04 mm/year and a phase change in −0.3 days. As in Sect. 5 for the trends (Fig. 11), blue indicates regions where the data set *SL*1 is closer to *G*1 as *SL*0 was to *G*1. When further taking not only the improvement of the SL data set into account, but additionally the influence of the assimilation of *SL*1 into the GECCO2 synthesis, even larger changes become evident as shown in Fig. 14b, d for $|D\_G1SL1| - |D\_G0SL0|$. However, as the global mean increases, it indicates that the annual signal of *SL*0 is in better agreement with the annual signal of *G*0 as compared to the difference between *SL*1 and *G*1. The annual phase seems to be in better agreement between *SL*1 and *G*1 as compared to both *D_G*1*SL*0 and *D_G*0*SL*0. Overall, the influence of the updated data set *SL*1 on the annual signal is very small when compared to the GECCO2 synthesis results.

7 Conclusions

In this study, we were able to demonstrate that the new *SL*1 data set, generated by the ESA SL_cci project, is in better agreement with the GECCO2 synthesis and the various global oceanographic data sets assimilated therein (Köhl 2015) than was the case with an earlier version of the same data (*SL*0). The improvement can be shown to exist separately for both TOPEX/POSEIDON and ERS data sets. The study reveals that especially in regions characterized by small SSH variability, and thereby small signal-to-noise ratio in the SSH data, improvements can be on the order of 30% of previously existing model–data residuals. Geographically, those improvements are especially obvious in all equatorial regions, on the Argentine shelf as well as in large parts of the ACC.

However, we note that in some regions we can find degradations, i.e., the residuals do become larger upon comparing to the *SL*1 data set, particularly in regions where the previous GECCO2 synthesis has little skill in representing the altimeter data. In those regions, for instance in parts of the North Atlantic and the Southern Ocean, changes are therefore likely to be insignificant. As can be expected from this, the GECCO2 synthesis was further improved upon the assimilation of the new *SL*1 data set, However, this does not

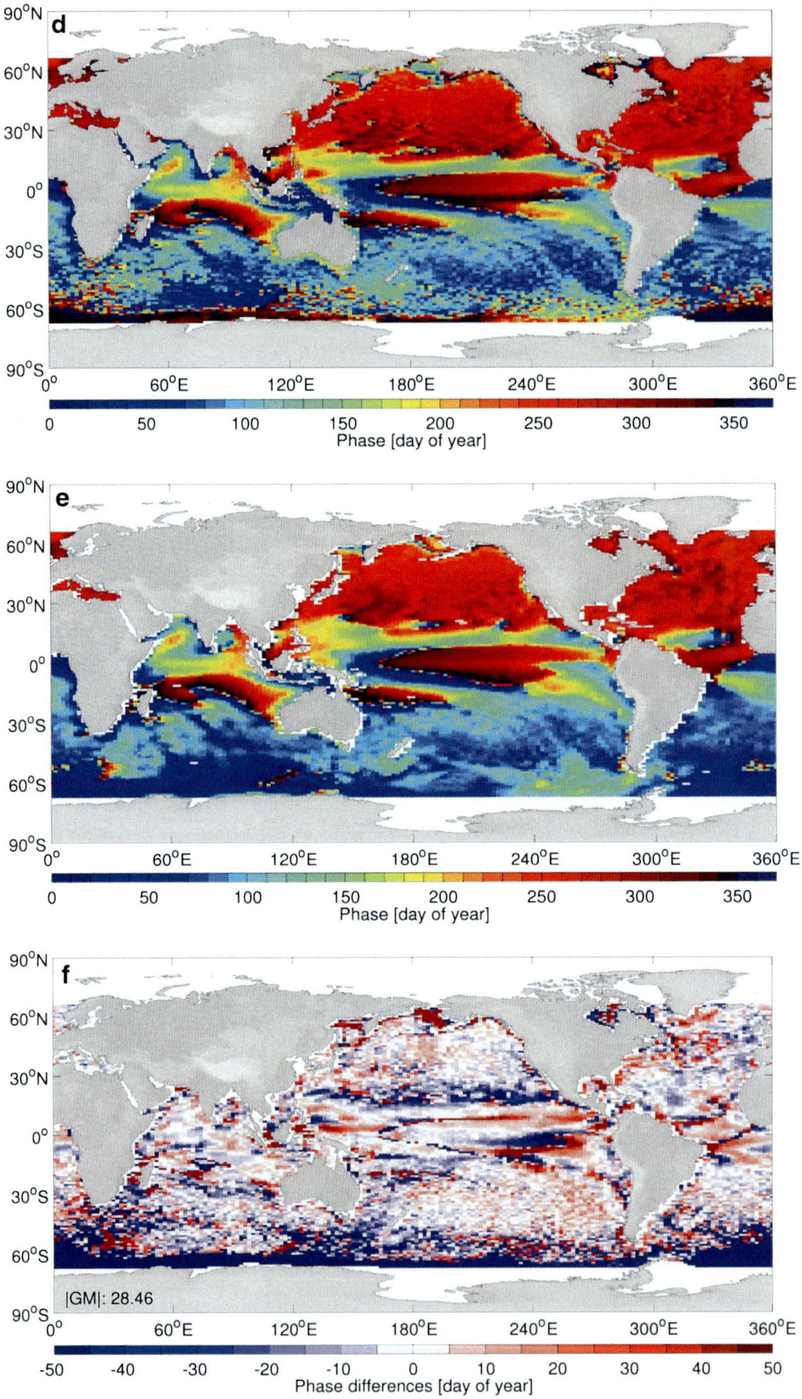

Fig. 12 continued

hold for all regions suggesting incompatibility of the model physics with the information content of SSH data in those regions.

Nevertheless, we are able to show that through the assimilation of the *SL*1 data, other model–data residuals are also reduced such as differences in in situ T and S profiles, indicating that the *SL*1 data set is in better agreement with the model dynamics and with the information content of other ocean data sets. However, these changes are naturally small and over large parts of the world ocean statistically insignificant, since the overall improvement in agreement between GECCO2 and the SL products is only a

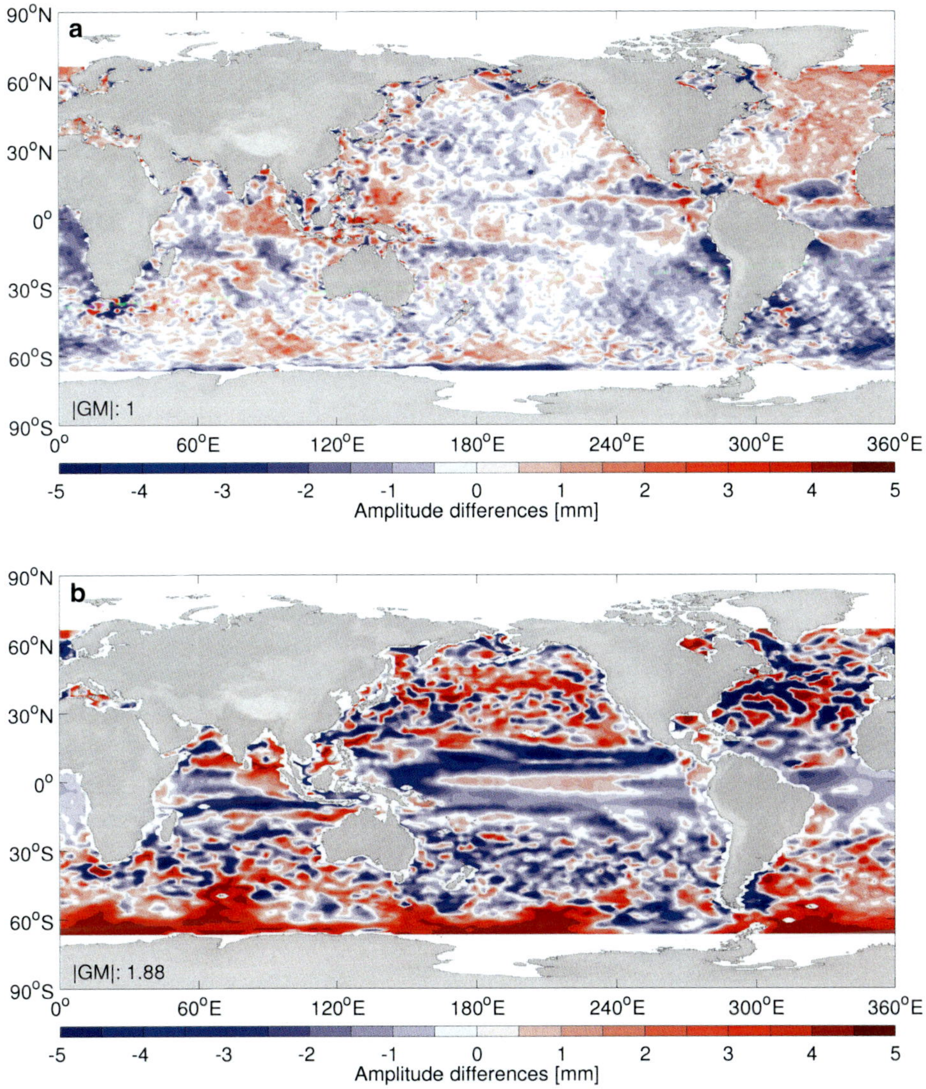

Fig. 13 Annual amplitude differences (**a**, **b**) and phase differences (**c**, **d**) of the two data sets *SL*1 − *SL*0 (**a**, **c**), and the two GECCO2 assimilation runs *G*1 − *G*0 (**b**, **d**), for TP series

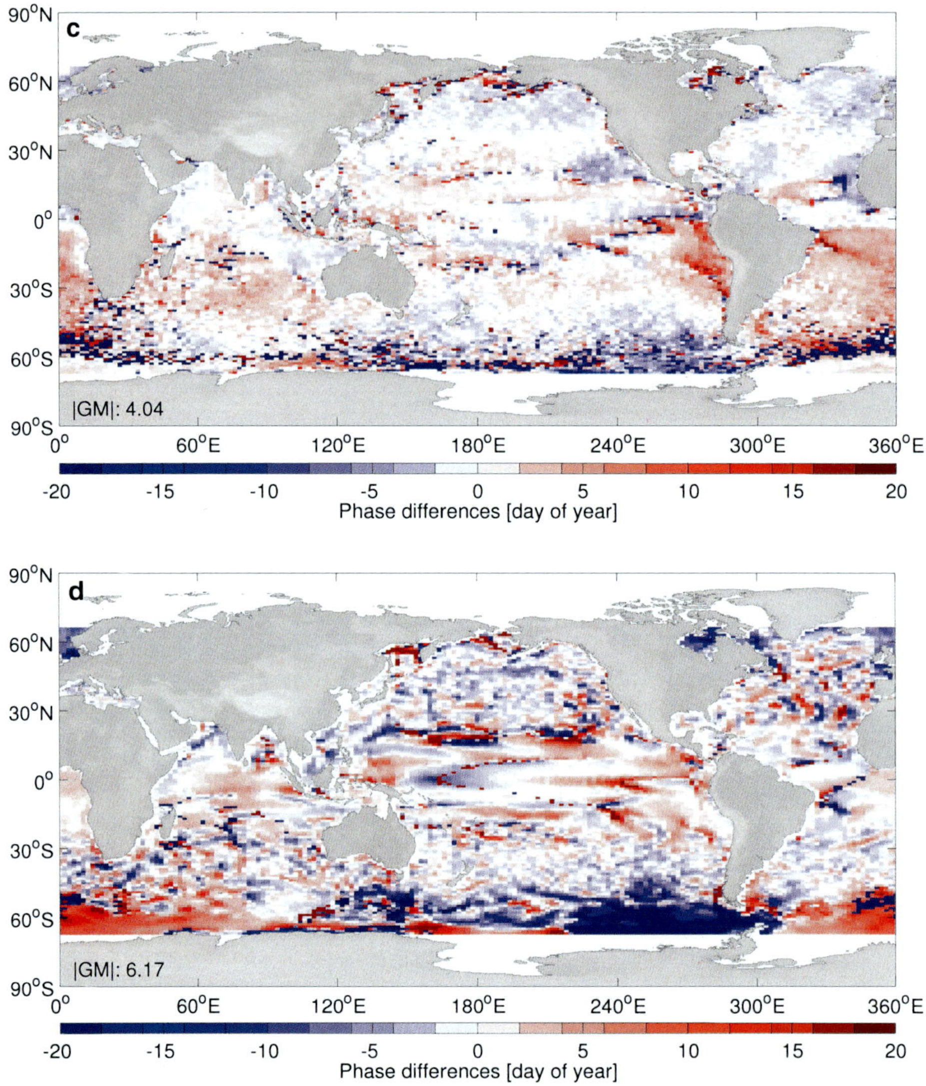

Fig. 13 continued

few percent. In contrast, the impact on sea-level trends, its seasonal amplitude and phase remain inconsistent with no clear connection to the pattern of the changes between *SL*0 and *SL*1. The additional assimilation of the ESA SST_cci (*G*1sst) did not have a strong influence.

Our work suggests that using gridded altimeter products shows smaller residuals relative to AT data. We were able to show that this apparent difference results simply from the more heavily smoothed gridded products. Smoothing the AT data in a similar manner does lead to smaller residuals. We nevertheless believe that the use of AT data over gridded fields has the strong advantage of constraining the barotropic fast movements in the model.

Respective signals are being filtered out in gridded fields. Nevertheless, our study also suggests that in future GECCO approaches, we should either use a more sophisticated error covariance allowing to more effectively down-weight eddy signal in the data, thereby constraining more the large-scale signal of the models or, alternatively, the along-track data can be filtered to remove the eddy component. In that case, the data error information used during the assimilation can substantially be decreased allowing to more effectively feel the large-scale altimetry. Both approaches need to be tested in future

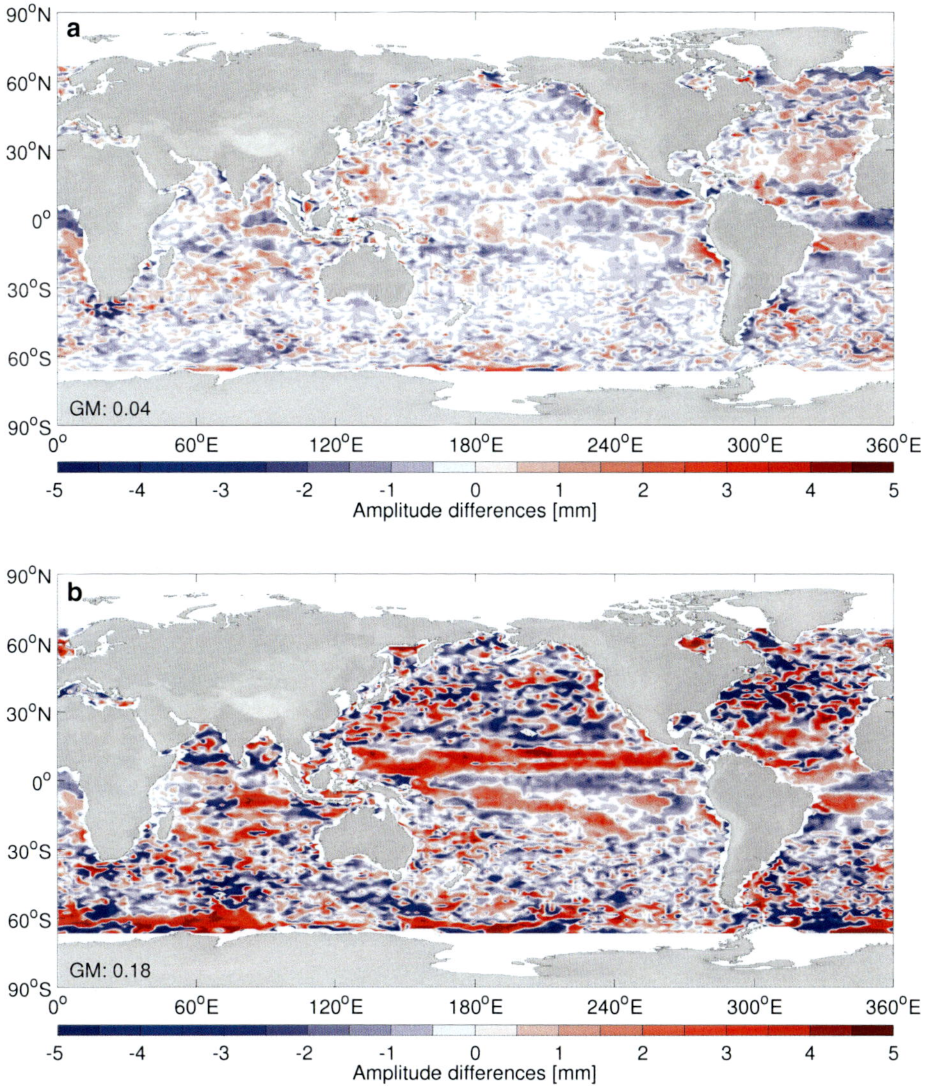

Fig. 14 Absolute annual amplitude (**a**, **b**) and phase changes (**c**, **d**) in absolute amplitude and phase differences of $|D\_G1SL1| - |D\_G1SL0|$ (**a**, **c**), and $|D\_G1SL1| - |D\_G0SL0|$ (**b**, **d**), for TP series

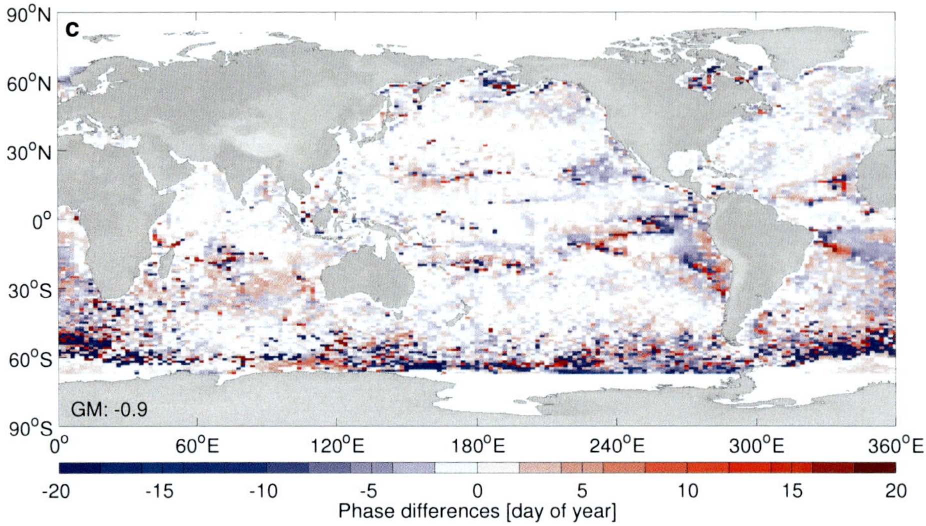

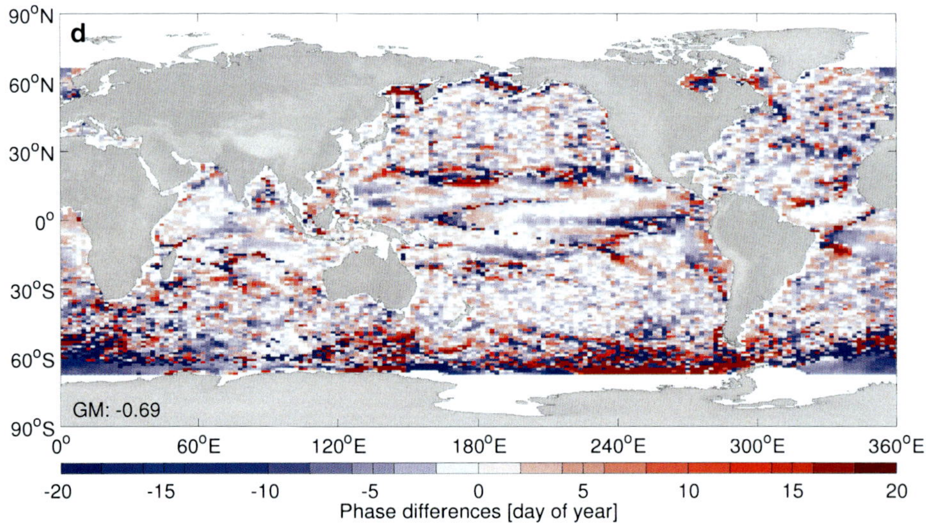

Fig. 14 continued

assimilation runs. They will likely lead to assimilation results that are better constrained by altimeter data sets.

Acknowledgements The paper is an outcome of the ISSI workshop on Integrative study of sea level. The study was funded in part through the ESA CCI Sea Level Project (4000109872/13/I-NB SL-CCI-subcontract CLS/UOHCLSDOS-14.002-Sea Level CCI) and the BMBF-funded RACE project (FZ 03F0651A). The helpful comments of Nicolas Champollion and an anonymous referee are appreciated.

References

Ablain M, Cazenave A, Larnicol G, Balmaseda M, Cipollini P, Faugère Y, Fernandes MJ, Henry O, Johannessen JA, Knudsen P, Andersen O, Legeais J, Meyssignac B, Picot N, Roca M, Rudenko S, Scharffenberg MG, Stammer D, Timms G, Benveniste J (2015) Improved sea level record over the satellite altimetry era (1993–2010) from the climate change initiative project. Ocean Sci 11:67–82. doi:10.5194/os-11-67-2015

Andersen OB, Knudsen P, Stenseng L (2015) The DTU13 MSS (mean sea surface) and MDT (mean dynamic topography) from 20 years of satellite altimetry. International Association of Geodesy Symposia. Springer, Berlin, pp 1–10. doi:10.1007/1345_2015_182

Cazenave A, Larnicol G, Legeais J-F, Ablain M, Faugère Y, Mbajon Njiche S, Timms G, Meyssignac B, Balmaseda M, Zuo H, Johannessen J, Scharffenberg MG, Stammer D, Andersen O, Knudsen P, Zawadzki L, Thibaut P, Poisson J-C, Picard B, Carrre L, Mertz F, Lauret O, Rudenko S, Farquhar C, Pechorro E, Roca M, Nilo P, Cipollini P, Calafat F, Fernandes J, Lazaro C, Quartly G, Kurekin A, Nencioli F, Fenoglio-Marc L, Benveniste J, Lucas B, Dinardo S, Ambrozio A (2014) ESA sea level climate change initiative (ESA SL_cci): sea level essential climate variable products, version 1.1., December 2014. doi:10.5270/esa-sea_level_cci-1993_2013-v_1.1-201412

Flato G, Marotzke J, Abiodun B, Braconnot P, Chou SC, Collins W, Cox P, Driouech F, Emori S, Eyring V, Forest Chr, Gleckler P, Guilyardi E, Jakob Chr, Kattsov V, Reason Chr, Rummukainen M (2013) Evaluation of climate models. In: Stocker TF, Qin D, Plattner GK, Tignor M, Allen SK, Boschung J, Nauels A, Xia Y, Bex V, Midgley PM, (eds) Climatic change 2013: The physical science basis. Contribution of working group I to the fifth assessment report of the intergovernmental panel on climate change. Cambridge University Press, Cambridge, United Kingdom and New York, NY, USA

Ingleby B, Huddleston M (2007) Quality control of ocean temperature and salinity profiles—historical and real-time data. J Mar Syst 65(1):158–175. doi:10.1016/j.jmarsys.2005.11.019

Köhl A (2015) Evaluation of the GECCO2 ocean synthesis: transports of volume, heat and freshwater in the Atlantic. Q J R Meteorol Soc 141:166–181. doi:10.1002/qj.2347

Köhl A, Stammer D (2008) Decadal sea level changes in the 50-year GECCO ocean synthesis. J Clim 21(9):1876–1890. doi:10.1175/2007JCLI2081.1

Menemenlis D, Fukumori I, Lee T (2005) Using Greens functions to calibrate an ocean general circulation model. Mon Weather Rev 133:1224–1240

Merchant CJ, Embury O, Roberts-Jones J, Fiedler E, Bulgin CE, Corlett GK, Good S, McLaren A, Rayner N, Morak-Bozzo S, Donlon C (2014) Sea surface temperature datasets for climate applications from phase 1 of the European space agency climate change initiative (SST CCI). Geosci Data J 1:179–191. doi:10.1002/gdj3.20

Stammer D, Tokmakian R, Semtner A, Wunsch C (1996) How well does a 1/4° global circulation model simulate large-scale oceanic observations? J Geophys Res 101(C11):25779. doi:10.1029/96JC01754

Stammer D, Ueyoshi K, Köhl A, Large WG, Josey SA, Wunsch C (2004) Estimating air-sea fluxes of heat, freshwater and momentum through global ocean data assimilation. J Geophys Res 109:C05023. doi:10.1029/2003JC002082

Stammer D, Köhl A, Wunsch C (2007) Impact of accurate geoid fields on estimates of the ocean circulation. J Atmos Ocean Technol 24(8):1464–1478. doi:10.1175/JTECH2044.1

Stammer D, Balmaseda M, Heimbach P, Köhl A, Weaver A (2016) Ocean data assimilation in support of climate applications: status and perspectives. Ann Rev Mar Sci 8:491–518. doi:10.1146/annurev-marine-122414-034113

Surv Geophys (2017) 38:385–406
DOI 10.1007/s10712-016-9374-2

A Review of Recent Updates of Sea-Level Projections at Global and Regional Scales

A. B. A. Slangen[1,2] · F. Adloff[3] · S. Jevrejeva[4] · P. W. Leclercq[5] · B. Marzeion[6] · Y. Wada[7,8,9,10] · R. Winkelmann[11]

Received: 3 January 2016 / Accepted: 18 May 2016 / Published online: 6 June 2016

Abstract Sea-level change (SLC) is a much-studied topic in the area of climate research, integrating a range of climate science disciplines, and is expected to impact coastal communities around the world. As a result, this field is rapidly moving, and the knowledge and understanding of processes contributing to SLC is increasing. Here, we discuss noteworthy recent developments in the projection of SLC contributions and in the global mean and regional sea-level projections. For the Greenland Ice Sheet contribution to SLC, earlier estimates have been confirmed in recent research, but part of the source of this contribution has shifted from dynamics to surface melting. New insights into dynamic discharge processes and the onset of marine ice sheet instability increase the projected range for the Antarctic contribution by the end of the century. The contribution from both ice sheets is projected to increase further in the coming centuries to millennia. Recent updates of the global glacier outline database and new global glacier models have led to slightly lower projections for the glacier contribution to SLC (7–17 cm by 2100), but still

✉ A. B. A. Slangen
aimee.slangen@gmail.com

1 Institute for Marine and Atmospheric research Utrecht (IMAU), Utrecht University, Utrecht, The Netherlands

2 CSIRO Oceans and Atmosphere, CSIRO, Hobart, Australia

3 CNRM-GAME, Météo-France, CNRS, Toulouse, France

4 National Oceanographic Centre, Liverpool, UK

5 Department of Geosciences, University of Oslo, Oslo, Norway

6 Institute of Geography, University of Bremen, Bremen, Germany

7 NASA Goddard Institute for Space Studies, Columbia University, New York, NY, USA

8 Center for Climate Systems Research, Columbia University, New York, NY, USA

9 Department of Physical Geography, Utrecht University, Utrecht, The Netherlands

10 International Institute for Applied Systems Analysis, Laxenburg, Austria

11 Potsdam Institute for Climate Impact Research, University of Potsdam, Potsdam, Germany

project the glaciers to be an important contribution. For global mean sea-level projections, the focus has shifted to better estimating the uncertainty distributions of the projection time series, which may not necessarily follow a normal distribution. Instead, recent studies use skewed distributions with longer tails to higher uncertainties. Regional projections have been used to study regional uncertainty distributions, and regional projections are increasingly being applied to specific regions, countries, and coastal areas.

Keywords Sea-level change · Regional sea-level change · Sea-level projections · Ice sheets · Glaciers · Terrestrial water storage · Mediterranean

1 Introduction

As one of the most well-known consequences of climate change, sea-level change (SLC) is relevant to coastal communities and stakeholders around the world. A large number of the world's population ($\sim$10 %, McGranahan et al. 2007) lives and works near the coast and depends on the ocean as their primary source of food and livelihood. An increase in mean sea level can increase the impacts of storm surges and the risk of flooding events in coastal zones (Wong et al. 2013). To make well-informed decisions about protective or adaptive measures, it is crucial that decision makers are provided with the best possible projections of SLC. Projecting future SLC and understanding the physical processes that contribute to SLC is therefore an important and rapidly evolving research topic.

SLC is a result of changes in many different parts of the climate system and can therefore be seen as an integrative measure of climate change. Over 90 % of the energy that is stored in the climate system ends up in the ocean (Rhein et al. 2013), causing thermal expansion and sea-level rise. In addition, ice sheets and glaciers lose mass due to increasing temperatures (both atmospheric and in the ocean, Vaughan et al. 2013) and reservoirs of water on land change due to human intervention (Church et al. 2013), which not only changes the amount of water in the oceans, but also the Earth's gravitational field. The solid Earth also responds to the redistribution of mass on the Earth surface both for present-day and for distant past (last glacial maximum, $\sim$20,000 years ago) mass variations, changing the height of the ocean floor. In the past century, global mean sea level has already increased by 19 $\pm$ 2 cm (1901–2010, Church et al. 2013), a rise that is expected to continue and accelerate in the coming centuries.

The Intergovernmental Panel on Climate Change Fifth Assessment Report (IPCC AR5) chapter on sea-level rise (Church et al. 2013) presented a comprehensive assessment of papers up to the IPCC working group 1 cut-off date of March 2013. In the chapter, important strides were made compared to the IPCC Fourth Assessment report (AR4) by progress in closing the twentieth century sea-level budget, the addition of an assessment of the ice sheet dynamical contribution to SLC and by making regional sea-level projections for the twenty-first century. However, a lot of research has been completed since IPCC AR5, and the lead authors of the chapter on SLC have recently published an update of their work (Clark et al. 2015). Here, we also focus on work that has been published since AR5 and aim to complement the review by Clark et al. (2015) by including more recent publications for the different contributions where available and by presenting overviews of research on the terrestrial water storage (TWS) contribution and the Mediterranean region, which were not discussed in Clark et al. (2015). The case of the Mediterranean is chosen because it is an area that is vulnerable to SLC due to the high population densities around the basin, and a lot of sea-level research is done specifically for this region.

First, we present an overview of recent work on contributions to SLC due to mass changes in glaciers and ice sheets (Sect. 2) and TWS changes (Sect. 3). Then, we will discuss global mean sea-level projections and new ways to treat the uncertainties thereof (Sect. 4). Recent advances in and uses of regional sea-level projections are presented in Sect. 5. Thermal expansion and dynamical ocean fields are not discussed in a separate section but are included in Sects. 4 and 5, as the most up-to-date projections are based on climate model output which has not changed since IPCC AR5. Finally, Sect. 6 presents research on sea-level projections in the Mediterranean region.

2 Land Ice Mass Change Projections

2.1 Ice Sheet Projections

The ice sheets on Greenland and Antarctica are by far the largest potential source of future SLC, storing approximately 65 m sea-level equivalent (SLE, Vaughan et al. 2013; Clark et al. 2015). Both ice sheets have increasingly lost mass in the past decades (Rignot et al. 2011) and are expected to dominate the sea-level budget on the long term (Church et al. 2013).

2.1.1 Greenland

Mass loss from Greenland is controlled by changes in surface mass balance (SMB) and dynamic discharge, including the effects of basal lubrication and ocean warming. IPCC AR5 (Church et al. 2013) estimated that Greenland would contribute between 0.04 and 0.10 m for the RCP2.6 (Representative Concentration Pathway, Moss et al. 2010) scenario and 0.07–0.21 m for the RCP8.5 scenario by the end of this century. For the lower emission scenarios, surface melting and dynamic discharge were expected to contribute equally to the overall mass loss. For RCP8.5, the mass balance was projected to be dominated by increased melting at the surface.

If a certain threshold is passed, the feedback between the lowering surface elevation and increasing surface melting can lead to additional ice loss and eventually even the complete loss of the Greenland Ice Sheet (Ridley et al. 2010; Robinson et al. 2012). Edwards et al. (2014) found that this positive feedback might be less significant for this century than previously expected. They estimate the surface elevation feedback to account for at most an additional 6.9 % ice loss from Greenland as opposed to the 15 % estimated in AR5. Other recent studies confirm the conclusion from AR5 that basal lubrication will likely not have a significant effect on Greenland mass loss within this century (Shannon et al. 2013).

Based on a higher-order ice sheet model driven by temperature changes from Atmosphere-Ocean General Circulation Models (AOGCMs) results, Fürst et al. (2015) project a Greenland contribution of 0.01–0.17 m to SLC within the twenty-first century in response to both atmospheric and oceanic warming. In contrast to previous studies, they conclude that future ice loss will be dominated by surface melting rather than dynamic discharge because the marine ice margin will retreat over time, reducing the contact area between ice and ocean water and thus limiting dynamic discharge. For the two lower emission scenarios, RCP2.6 and RCP4.5, simulations yield a contribution to SLC between 0.03 and

0.32 m by 2300 (Fig. 1). These new projections thus fall within the AR5 likely ranges but with a higher contribution from surface melting as opposed to dynamic discharge (Clark et al. 2015).

2.1.2 Antarctica

The mass balance of the Antarctic Ice Sheet is determined by changes in the SMB as well as changes in the ice flux across the grounding line resulting from enhanced basal sliding, calving or sub-shelf melting. Since surface melting will remain negligible within the twenty-first century (Vizcaino et al. 2010; Huybrechts et al. 2011), the SMB is predominantly determined by snow accumulation. Evidence from paleodata and projections from global and regional climate models show that snowfall in Antarctica is very likely to increase with future atmospheric warming (Frieler et al. 2015) by 5 ± 1 % per degree warming. The resulting mass gain might however be compensated or even overcompensated by dynamic effects (Winkelmann et al. 2012).

In AR5, the overall contribution of Antarctica to SLC was estimated to range from −0.03 to 0.14 m for RCP2.6 and −0.06 to 0.12 m for RCP8.5 by 2100 compared to 1986–2005 (Church et al. 2013). The SLC arising from rapid dynamics was projected to be −0.01 to 0.16 m, deduced from a combination of model results, expert judgement, and statistical extrapolation of current trends. Due to insufficient understanding of the underlying processes, scenario dependence for rapid dynamics could not be established in IPCC AR5.

Significant progress has been made since IPCC AR5 to understand the dynamic processes and quantify their effect on Antarctic ice loss for the twenty-first century and beyond. Pollard et al. (2015) found that crevasse-induced ice shelf loss can lead to the onset of rapid ice discharge from several Antarctic drainage basins. Based on the results from the SeaRISE model intercomparison project (Nowicki et al. 2013), Levermann et al. (2014) developed a probabilistic approach to estimate the future sea-level contribution from Antarctica, combining uncertainty in the climatic boundary conditions, the oceanic response and the ice sheet response. The results, based on linear response theory, correspond with the recently observed mass loss from the Antarctic Ice Sheet (Shepherd et al.

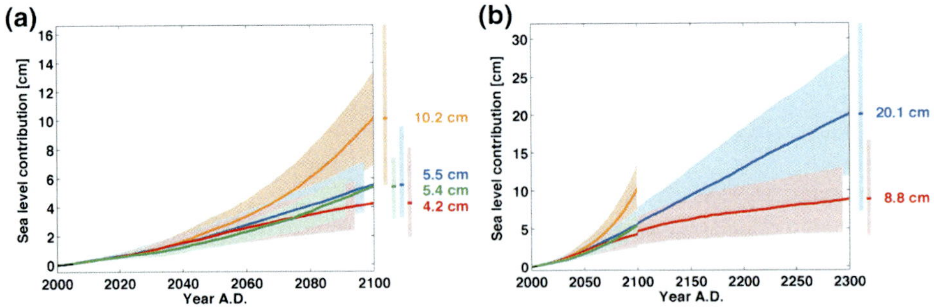

Fig. 1 Projected global mean sea-level contribution (cm) from the Greenland Ice Sheet (surface mass balance and dynamics) using a three-dimensional ice flow model driven by output from 10 atmosphere–ocean general circulation models **a** for four RCP climate scenarios over the twenty-first century and **b** for two RCP climate scenarios until 2300 [reproduced from Fürst et al. (2015)]. The shaded area indicates the ensemble mean $\pm 1\sigma$, while the vertical bars show the spread (all climate models) at the end of 2100 and 2300, respectively

2012). Levermann et al. (2014) find that the 90 % uncertainty associated with the contribution from Antarctica, mainly due to ocean warming, reaches up to 0.23 m (median 0.07 m; 90 % 0.0–0.23 m) by 2100 for RCP2.6 and up to 0.37 m (median 0.09 m; 90 % 0.01–0.37 m) for RCP8.5 (Fig. 2).

IPCC AR5 concluded that the collapse of marine ice sheet basins could cause additional SLC above the likely range of 'up to several tenths of a metre' (Church et al. 2013), but the timing could not be quantified. The mechanism underlying such a potential collapse is the marine ice sheet instability (MISI, Weertman 1974; Mercer 1978): several Antarctic basins are partly grounded below sea level, on bedrock generally sloping downwards towards the interior of the ice sheet. If the grounding line retreats into such an area, it could become unstable.

Shortly after the release of AR5, several studies were published showing with increasing certainty that parts of West Antarctica might in fact already be undergoing unstable grounding line retreat (Favier et al. 2014; Joughin et al. 2014; Rignot et al. 2014). The retreat was most likely caused by warm circumpolar deep water reaching the ice shelf cavities in recent years—whether this process was influenced by anthropogenic climate change is not yet clear.

Using a process-based statistical approach, Ritz et al. (2015) derived probability estimates for exceeding particular thresholds in the marine basins of Antarctica as a function of time if MISI is triggered. Their results suggest that particularly in the Amundsen Sea Sector, large and rapid ice loss due to the marine ice sheet instability could be initiated within this century. By 2100, the total ice loss from such rapid dynamics is estimated to contribute up to 0.3 m to global SLC, quantifying and narrowing down the IPCC AR5 uncertainties, and 0.72 m by 2200 (95 % quantiles). Large uncertainties remain, especially with respect to the effect of basal sliding on the ice flux (Ritz et al. 2015).

These advances made in estimating both the gradual response to oceanic warming and the possibly abrupt onset of self-sustained grounding line retreat can be consolidated into a new uncertainty range for Antarctic ice loss. It contains the IPCC likely range but leads to an overall larger spread for the twenty-first century sea-level projections.

However, a recent paper by Pollard and DeConto (2016) includes a number of processes in their model simulation that were not included in models before, such as the hydrofracturing of Antarctic ice shelves due to atmospheric warming and subsequent ice cliff instabilities. The model is found to be in relatively good agreement with geological

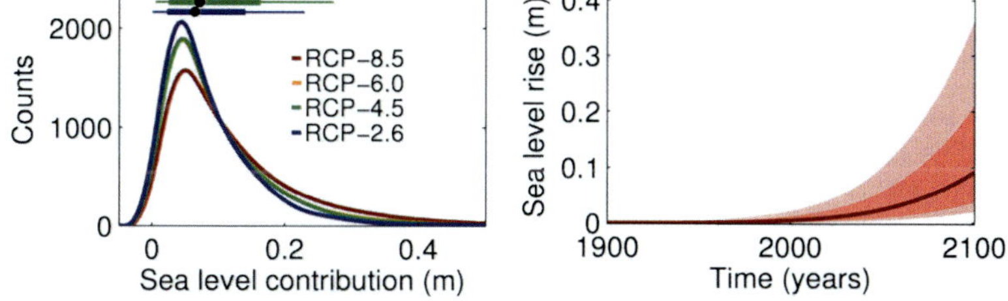

Fig. 2 Projected global mean sea-level contribution (m) from the Antarctic Ice Sheet in the twenty-first century. **a** Uncertainty range including climate, ocean, and ice dynamic uncertainty for the year 2100 (*top: thick line* is 66 % range, *thin line* is 90 % range). *Different colours* represent different climate scenarios used to drive three Antarctic Ice Sheet models. **b** Time series of future SLC from Antarctica (median 66 and 90 % uncertainty ranges) [reproduced from Levermann et al. (2014)]

estimates of the Pliocene (~3 million years ago) and the last interglacial (130,000–115,000 years ago). Depending on the geologic criteria used, they find possible contributions up to 1.05 ± 0.30 m (1σ) by 2100 under the RCP8.5 scenario. This means that the possibility of 1 m SLC from the Antarctic Ice Sheet by 2100 still cannot be excluded.

2.1.3 Long-Term Projections

Sea level will continue to rise well beyond 2100, even under strong mitigation scenarios (Church et al. 2013). Due to the long lifetime of anthropogenic CO_2 in the atmosphere and the consequent slow decline in temperatures, greenhouse gas emissions within this century can induce a sea-level commitment of several metres for the next millennia (Levermann et al. 2013). On these timescales, the Greenland Ice Sheet shows critical threshold behaviour with respect to atmospheric warming due to the surface elevation feedback (Ridley et al. 2010; Robinson et al. 2012).

Long-term projections from different process-based model simulations are now also available for the Antarctic Ice Sheet (Golledge et al. 2015; Winkelmann et al. 2015). Since several ice basins in Antarctica are potentially preconditioned to become subject to MISI, the response of the ice sheet to global warming might also be highly nonlinear. Golledge et al. (2015) find that the irreversible retreat of major Antarctic drainage basins can only be avoided if greenhouse gas emissions do not exceed the RCP2.6 level. Winkelmann et al. (2015) studied the evolution of Antarctica on millennial timescales and showed that the West Antarctic Ice Sheet becomes unstable after 600–800 GtC of additional carbon emissions. They further concluded that, on a multi-millennial timescale, Antarctica could become essentially ice-free for a scenario in which all available fossil carbon resources are combusted (10,000 GtC). These new studies suggest that the rate of SLC for higher emission scenarios could reach values of up to a few metres per century beyond 2100.

2.2 Glacier Projections

Glacier mass loss constituted a large contributor to twentieth century SLC (Gregory et al. 2013). Despite accelerating mass loss of the ice sheets (Shepherd et al. 2012), glacier mass loss continues to be a main component of SLC (Church et al. 2011) and is likely to remain an important factor in the twenty-first century. The AR5 evaluation of projected glacier mass loss in 2081–2100 relative to 1986–2005 ranges from 0.04 to 0.23 m, based on the results of four process-based models across different forcing scenarios (Church et al. 2013).

There are five glacier models operating on a global scale which have published projections of glacier mass change under the RCP scenarios (Marzeion et al. 2012; Hirabayashi et al. 2013; Radić et al. 2014; Slangen et al. 2014; Huss and Hock 2015) and one study which uses the Special Report on Emission Scenarios (SRES) to drive their glacier model (Giesen and Oerlemans 2013). They all combine a glacier surface mass balance model with a model that accounts for the response of glacier geometry to changes in glacier mass. The calculation of both the glacier mass balance and geometry change varies across the different models. All models except Huss and Hock (2015) were used in IPCC AR5, but some have been updated since, as will be detailed below.

Slangen et al. (2012, 2014) calculate the glacier mass balance from the sensitivity of the surface mass balance to temperature change and changes in precipitation. This sensitivity is parameterised by relations that are calibrated on more detailed model studies for 12 individual glaciers (Zuo and Oerlemans 1997). The initial areas of the glaciers are based on WGI-XF (World Glacier Inventory, extended format, Cogley 2009), and the glacier volumes are based on volume–area scaling. The glacier projections are forced by 14 models from the CMIP5 database (Taylor et al. 2012a) for each of the RCP4.5 and RCP8.5 scenarios.

Radić et al. (2014) and Marzeion et al. (2012) both use an approach in which accumulation and ablation are modelled explicitly. Accumulation is calculated by summing the solid precipitation over the glacier characterised by an area distribution over elevation. Ablation is calculated with a temperature-index method in both studies. Following Radić and Hock (2011), Radić et al. (2014) calculate the surface mass balance for each glacier at different elevation bands, whereas Marzeion et al. (2012) calculate melt from the temperature at the glacier tongue elevation only. Both studies use mass balance observations to calibrate the modelled glacier mass balance. In order to account for glacier retreat to higher elevations and thus allow for new equilibrium in a different climate, Radić et al. (2014) remove, or add in case of modelled mass gain, mass in the lowest elevation bins of the modelled glaciers, based on volume–area scaling. Marzeion et al. (2012) combine volume–length scaling with the mean slope of the glacier surface to let the glacier terminus retreat to higher elevations, or advance to lower elevations. They also include a response time between volume changes on the one hand, and length and area changes on the other. Their model is validated against in situ and geodetic mass balance observations of individual glaciers. Marzeion et al. (2012) do not model peripheral glaciers (PGs) in Antarctica explicitly, but apply the global mean specific mass balance rate as a rough approximation.

The results of Radić et al. (2014) shown in Fig. 3 are from projections that are forced by 14 models from the CMIP5 database for each of the RCP4.5 and RCP8.5 scenarios. The results of Marzeion et al. (2012) were updated based on a more recent version of the Randolph Glacier Inventory (RGI). Their projections were forced by 13 CMIP5 models for the RCP2.6 scenario, 15 models for RCP4.5, 11 models for RCP6.0, and 15 models for RCP8.5.

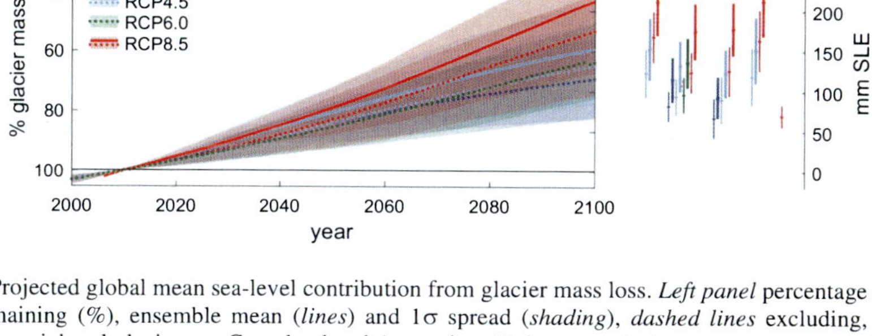

Fig. 3 Projected global mean sea-level contribution from glacier mass loss. *Left panel* percentage of glacier mass remaining (%), ensemble mean (*lines*) and 1σ spread (*shading*), *dashed lines* excluding, *full lines* including peripheral glaciers on Greenland and Antarctica. *Right panel* glacier contribution to SLC by 2100 (mm), ensemble mean and 1 ensemble standard deviation in 2100; *thick lines* including, *thin lines* excluding peripheral glaciers. All numbers are relative to 2010

Huss and Hock (2015) use a temperature-index model to calculate mass changes for every individual glacier, but their model approach is different from the earlier models discussed above. They do not use volume–area or volume–length scaling. Instead, they derive the initial glacier volume following Huss and Farinotti (2012). This method provides ice thickness, and therefore glacier volume and glacier bed elevation, distributed over 10-m elevation intervals for every glacier. Glacier geometry changes due to changes in calculated glacier mass are distributed over the glacier elevation following the parameterisation of Huss et al. (2010). Furthermore, Huss and Hock (2015) explicitly compute mass loss through calving using the simple model of Oerlemans and Nick (2005) that describes the calving rate as a linear function of the height of the calving front. Finally, they subtract the mass loss of glacier ice below sea level, which does not contribute to SLC, from the total of calculated glacier mass loss in their calculation of the glacier contribution to SLC (note that, for comparability, the loss of ice below sea level is also included in their numbers shown in Fig. 3). For calibration, Huss and Hock (2015) assume that mean specific balance rate of each individual glacier should equal the observed region-wide mean specific balance rate (Gardner et al. 2013) within a range of ± 0.1 m w.e.a^{-1} (m of water equivalent per year). The model is validated against in situ and geodetic mass balance observations, as well as observed area changes and calving rates, for individual glaciers. The results of Huss and Hock (2015) used here are from projections that are forced by 12 models from the CMIP5 database for the RCP2.6 scenario and 14 models for the RCP4.5 and RCP8.5 scenarios.

Hirabayashi et al. (2013) use a grid-based approach to modelling glacier mass change. Within each 0.5×0.5 degree grid cell, individual glaciers are lumped together as one glacier, while applying sub-gridscale elevation bands preserves the vertical elevation distribution of the ice area within each grid cell. Their model was calibrated against the observations of Dyurgerov and Meier (2005) and does not cover PGs on Greenland or Antarctica. The projections used here are forced by 10 models from the CMIP5 database for the RCP8.5 scenario only.

In each of the five global studies described above, the mass balance is calculated with a temperature-index model. Giesen and Oerlemans (2013) apply a more complex surface mass balance model that besides the dependence of glacier mass balance on temperature and precipitation also includes incoming solar radiation in the calculation of ablation. They calibrate this model to 89 glaciers with in situ observations of winter and summer mass balance and then upscale the results to all glaciers. Their projections for the twenty-first century are based on an ensemble of CMIP3 model runs for scenario A1B. They find a significant effect of projected decrease in incoming solar radiation in the Arctic region on the projected sea-level contribution. The twenty-first century global glacier mass loss found in Giesen and Oerlemans (2013) is significantly less than in other studies (Marzeion et al. 2012; Radić et al. 2014; Slangen et al. 2014) for comparable RCPs. In a regional study of future surface mass balance with the high-resolution regional climate model MAR, Lang et al. (2015) find significantly less mass loss for Svalbard than Marzeion et al. (2012) and Radić et al. (2014). Suggested explanations for this discrepancy are the coarse resolution of the global climate models that were used to force the global glacier models, and a better representation of the physical processes in the regional climate model compared to the empirical temperature-index mass balance models. Lang et al. (2015) also find a significant reduction in the incident solar radiation due to increased cloudiness, supporting the findings of Giesen and Oerlemans (2013) for the Arctic. Huss and Hock (2015) also find a 16–22 % lower projected glacier mass loss when they include incoming solar radiation (assumed to be constant in time) in a sensitivity experiment with their glacier mass balance model.

Figure 3 shows the projected glacier mass loss from the five global studies under RCP scenarios. They all show a large spread in the projected global glacier mass loss within the ensemble of different climate model runs for the same scenario. The ensemble standard deviation within each scenario is comparable to the differences between the ensemble means of different scenarios. Also, the differences between the different glacier models, but identical scenarios, are of comparable magnitude. The exception is the projection of Hirabayashi et al. (2013), which for the RCP8.5 scenario projects glacier mass loss comparable to the other models' projections for RCP2.6.

Updates of existing projections (Marzeion et al. 2012) and new models (Huss and Hock 2015) published after the IPCC AR5 have generally lead to slightly lower projected mass losses (Table 1). For the RCP8.5 scenario for instance, IPCC AR5 projected a contribution of 16 ± 7 cm, while Huss and Hock (2015) and the updated Marzeion et al. (2012) present projections around 12.5 cm for the same scenario. In the case of Marzeion et al. (2012), this is attributable to updates of the RGI; it is unclear for Huss and Hock (2015) since no previous estimate existed. On the other hand, the results of Slangen et al. (2012, 2014) are very similar to those of Radić et al. (2014). The results of Giesen and Oerlemans (2013) and Lang et al. (2015) suggest that a projected decrease in Arctic incoming solar radiation could lead to a lower projected mass loss than is given by the temperature-index models. However, a direct comparison of the individual studies is complicated through the differing compositions of the ensembles used for forcing the glacier models. Therefore, a coordinated glacier model intercomparison is currently underway to better understand the causes of the model and ensemble spread.

3 Terrestrial Water Storage Change Projections

Terrestrial water storage (TWS) change can result in a positive contribution to SLC due to a net transfer of water from long-term groundwater storage to the active hydrological cycle and eventually to ocean storage (Gornitz 1995; Taylor et al. 2012b). Other terrestrial components potentially influencing SLC include water impoundment behind dams (which can cause sea-level fall), drainage of endorheic lakes (mostly from the Aral Sea) and wetlands, deforestation, and changes in natural water storage (soil moisture, groundwater,

Table 1 Projected glacier contributions to SLC for 2010–2100 (mm, ensemble mean ± 1σ), for four different RCP scenarios, peripheral glaciers excluded (numbers in brackets include peripheral glaciers)

Study	RCP2.6	RCP4.5	RCP6.0	RCP8.5
Hirabayashi et al. (2013)	–	–	–	73 ± 14
Huss and Hock (2015)	67 ± 25 (93 ± 26)	90 ± 29 (123 ± 30)	–	126 ± 31 (178 ± 33)
Marzeion et al. (2012, updated)	82 ± 19 (115 ± 28)	94 ± 23 (132 ± 32)	96 ± 22 (136 ± 31)	124 ± 25 (175 ± 35)
Radić et al. (2014)	–	122 ± 36 (155 ± 41)	–	167 ± 38 (216 ± 44)
Slangen and van de Wal (2011, updated)	–	123 ± 30 (153 ± 39)	–	168 ± 32 (212 ± 42)

permafrost, and snow). Natural TWS change mostly varies with decadal climate variation with no significant trend.

Chao et al. (2008) found that the volume of water accumulated in dams up to 2010 was equivalent to a sea-level fall of ~30 mm. However, Lettenmaier and Milly (2009) indicated that the volume of silt accumulated in dams should be removed from the estimate, which is equal to ~4 mm less sea-level fall. Indeed, silting-up of existing dams may already be, or in coming decades may become, a larger effect on impoundment than construction of new dam capacity (Wisser et al. 2013).

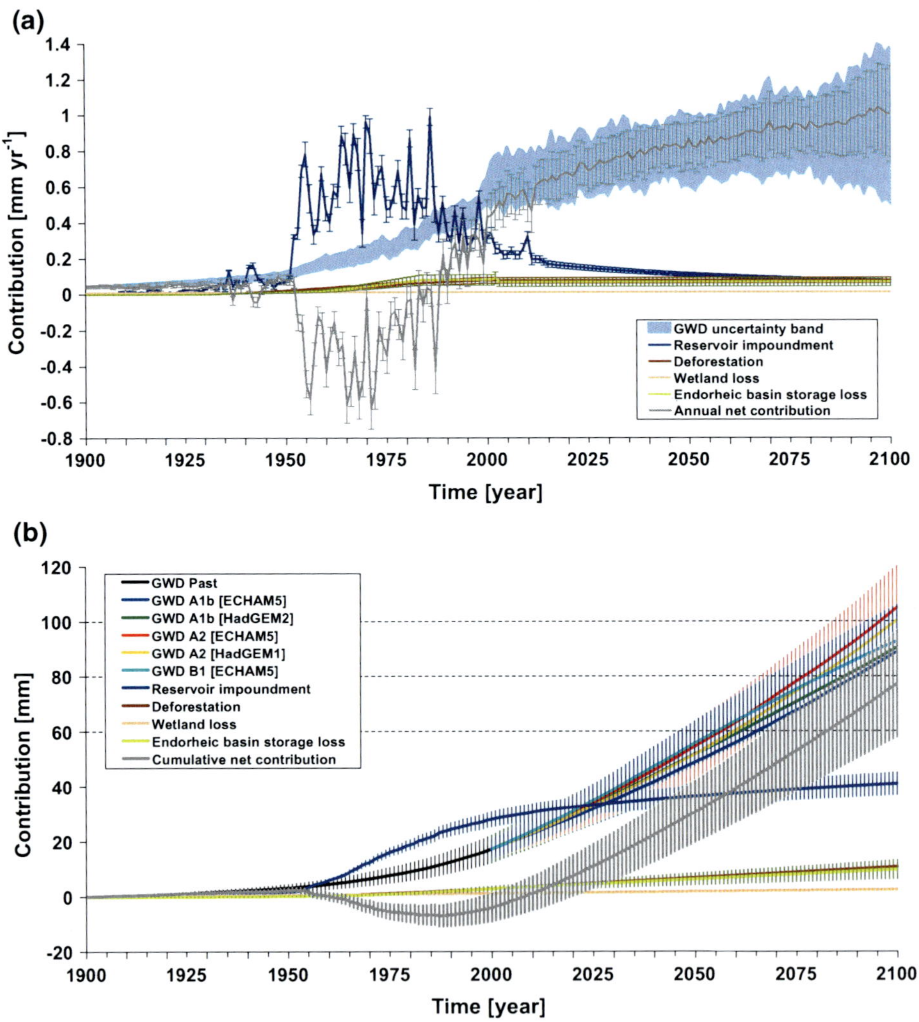

Fig. 4 Historical and projected terrestrial water contributions to SLC for a range of processes. **a** Yearly rates for 1900–2100 (mm year^{-1}) and **b** cumulative contribution to SLC since 1900 (mm). *Bars* indicate 1σ standard deviation. *Blue band* in (**a**) is based on 10,000 Monte Carlo realisations from five future projections of groundwater depletion, individual projections, and uncertainties shown in (**b**) [from Wada et al. (2012)]

Using a global hydrological model, Wada et al. (2012) estimated that the contribution of groundwater depletion (GWD) to SLC increased from 0.035 ± 0.009 mm year^{-1} in 1900 to 0.57 ± 0.09 mm year^{-1} in 2000 (Fig. 4). These figures were recently revised to lower values in Wada et al. (2016), who found a sea-level contribution of 0.12 ± 0.04 mm year^{-1} for the period 1993–2010 using a coupled climate–hydrological model. A volume-based study by Konikow (2011) also found slightly lower values than Wada et al. (2012) using direct groundwater observations, calibrated groundwater modelling, GRACE satellite data, and partly extrapolation for some regions. Also combining hydrological modelling with information from well observations and GRACE satellites, Döll et al. (2014) estimated the SLC contribution of GWD was 0.31 mm year^{-1} during 2000–2009. Another study (Pokhrel et al. 2012) used an integrated water resources assessment model to estimate all changes in TWS. However, their estimate is likely to overestimate the GWD contribution, because the model did not account for any physical constraints on the amount of groundwater pumping.

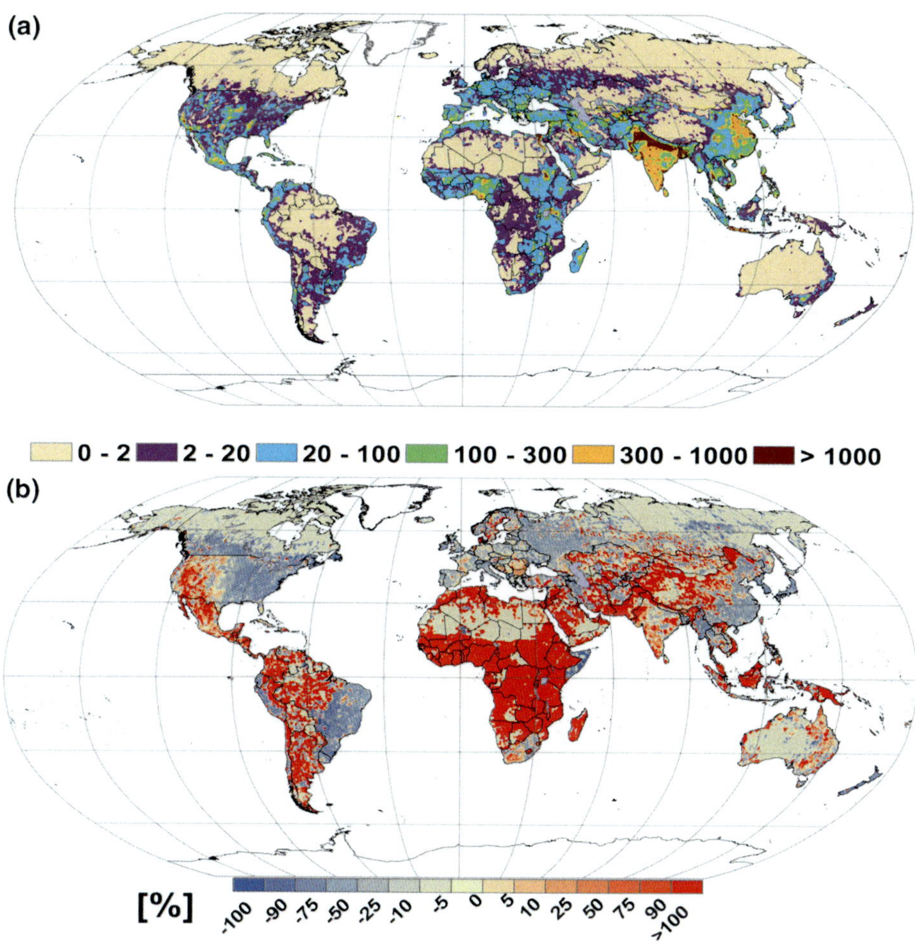

Fig. 5 **a** Projected global human water consumption in 2099 (million m^3 year^{-1}) and **b** the relative change (%) between 2010 and 2099 [from Wada and Bierkens (2014)]

Satellite observations have opened a path to monitor groundwater storage changes in data scarce regions (Famiglietti 2014). Since its launch in 2002, the GRACE satellite has been increasingly employed to quantify GWD at regional scales (Rodell et al. 2009; Famiglietti et al. 2011). GWD can be assessed after subtracting remaining TWS changes from GRACE-derived total TWS changes. However, coarse spatial resolution and noise contamination inherent in GRACE data hinder their global application in estimating GWD (Longuevergne et al. 2010).

Future projections of the GWD contribution to SLC are subject to large uncertainties due to the combination of climate projections from Atmosphere-Ocean General Circulation Models (AOGCMs) with future socio-economic and land-use scenarios that are inherently uncertain. The TWS contribution to SLC is projected to be 38.7 ± 12.9 mm, based on CMIP3 climate model output (Wada et al. 2012; Church et al. 2013). Since IPCC AR5, the groundwater model simulation has been updated, based on the latest CMIP5 climate and IPCC AR5 socio-economic data sets [see Fig. 5 for the latest projection of human water consumption from Wada and Bierkens (2014)], but does not provide the GWD contribution to SLC yet. The existing twenty-first century projections indicate increasing GWD caused by (1) increasing water demand due to population growth, and (2) an increased evaporation projected in irrigated areas due to changes in precipitation variability and higher temperatures. Groundwater depletion will be limited by decreasing surface water availability and groundwater recharge, which may cause groundwater resources to become exhausted at some time in the coming century (Gleeson et al. 2015).

4 Global Mean Sea-Level Projections

Before we discuss total global mean sea-level projections, we briefly discuss thermal expansion, as this is one of the most important contributors to global mean sea-level change. The majority of the net energy increase in the Earth's climate system is stored in the ocean, increasing the ocean heat content, which leads to warming and expansion of the ocean water. The resulting global mean thermosteric SLC by 2100 is projected to be 0.14 m (±0.04 m) for the RCP2.6 scenario, up to 0.27 m (±0.06 m) for the RCP8.5 scenario in IPCC AR5 (Church et al. 2013). New results are expected when the output of the sixth Climate Modelling Intercomparison project is released from 2017 onwards.

Although the focus in sea-level science is gradually moving towards regional SLC projections, as this is more relevant for coastal adaptation, there are still lessons to be learnt from the global mean SLC. The signal-to-noise ratio is smaller in the global mean, allowing a focus on long-term changes rather than local, short-term variability. As a result, it can be used to focus on narrowing uncertainties in the projections.

A notable development in global mean sea-level projections since IPCC AR5 is the use of a probabilistic approach to explore uncertainties in sea-level projections beyond the likely range (Jevrejeva et al. 2014; Kopp et al. 2014; Grinsted et al. 2015). In this approach, the projections (as presented in IPCC AR5) are blended with expert assessments of the Greenland and Antarctic Ice Sheet contributions (Bamber and Aspinall 2013) or expert assessments of total SLC (Horton et al. 2014). Expert assessments of, for instance, the potential contribution from ice sheets can be a useful tool to assess the uncertainty ranges, because the ice sheet experts know which particular physical processes (e.g., calving, ice sheet–ocean interaction) are insufficiently represented in their ice sheet models. One should keep in mind, however, that the current changes in the climate system are unprecedented, and estimates based on intuition, such as expert assessments, should therefore be used with care.

Figure 6 demonstrates the difference between the conventional and probabilistic approaches for global mean sea-level projections. Probabilistic projections allow the selection of specific probability levels to estimate low-probability/high-risk SLC projections, which by definition are unlikely to be reached, but cannot be ruled out given paleoclimate proxy information and the limitations in process-based modelling (Jevrejeva et al. 2014). They also allow for the use of probability distributions that do not follow a Gaussian distribution, such as skewed probability distributions with a longer tail to high SLC projections (Fig. 6).

In addition to the studies focusing on uncertainties in the global mean, a new application of the semi-empirical approach was published recently by Mengel et al. (2016). Semi-empirical models were developed after IPCC AR4 to offer an alternative to more complicated physical models of SLC. They are based on the assumption that sea level in the future will respond to imposed climate forcing as it has in the past, which may not hold if potentially nonlinear physical processes, such as marine ice sheet instability or thermal expansion, do not scale in the future as they have in the past. Mengel et al. (2016) calibrate the semi-empirical model for each contribution separately, such that the timescales of each contribution are considered in the calibration of the model. Their projected global mean SLC by 2100 is 84.5 cm (57.4–131.2 cm; median, 5th and 95th percentile) for the RCP8.5 scenario. This brings the semi-empirical models closer to the process-based IPCC AR5 estimates of 74 cm (52–98 cm) than other, larger, semi-empirical estimates at the time of IPCC AR5 (Church et al. 2013, Table 13.6).

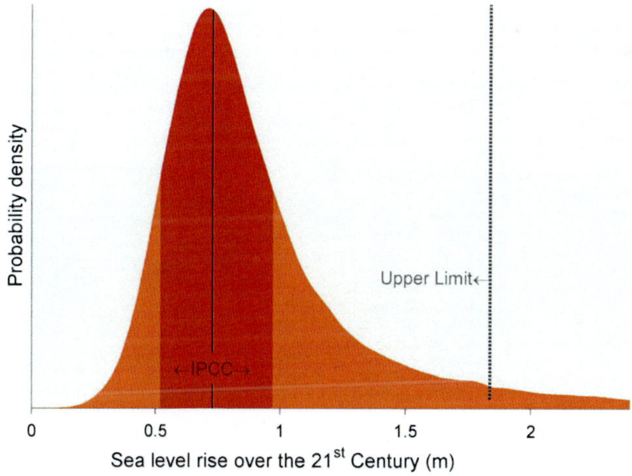

Fig. 6 Projected global mean sea-level rise by 2100 relative to 2000 for the RCP8.5 scenario and uncertainty (m). *Dark orange* represents the mean (*black line*) and likely range from IPCC AR5 (Church et al. 2013), *light orange* represents the probabilistic uncertainties from Jevrejeva et al. (2014). The vertical dotted black line represents the 95 % probability estimate of sea-level rise in 2100 (1.8 m) [from Jevrejeva et al. (2014)]

5 Regional Sea-Level Projections

Regional SLC can deviate substantially from the global mean due to a number of processes. Firstly, oceanic and atmospheric circulation changes and heat and salt redistribution in the ocean change the density of the water as well as redistribute mass within the oceans (Yin et al. 2010; Yin 2012). Secondly, any redistribution of mass between ocean and land, such as land ice mass change or TWS, affects the gravitational field of the Earth and causes viscoelastic deformation of the Earth's crust, the combination of which results in distinct sea-level patterns referred to as 'fingerprints' (Farrell and Clark 1976; Mitrovica et al. 2001). Thirdly, regional sea level can be influenced by vertical land motion, such as tectonic activity or glacial isostatic adjustment (GIA). GIA is the present-day viscous deformation of the Earth's crust as a result of ice melt after the last glacial maximum, which in turn also affects the gravitational field (Peltier 2004). GIA can have large local effects, while on a global mean scale the effect is negligible.

IPCC AR5 (Church et al. 2013) adopted the approach from Slangen et al. (2012, 2014) to compute regional sea-level projections by combining climate model results for thermal expansion and circulation changes with offline models to compute gravitational fingerprints as a result of mass change and GIA. Using this approach, both IPCC AR5 and Slangen et al. (2014) project regional sea-level values up to 20 % larger than the global mean in equatorial regions (Fig. 7), while close to regions of ice mass loss the values can be as small as 50 % of the global mean, mainly as a result of the gravitational effect. The meridional dipole in the Southern Ocean and the dipole in the North Atlantic are associated with the response of dynamic sea level (DSL) to increasing greenhouse gas forcing (Bilbao et al. 2015; Slangen et al. 2015), through wind stress and surface heat flux changes (Bouttes and Gregory 2014).

Carson et al. (2015) used the regional projections from Slangen et al. (2014) to study coastal SLC and found that coastal deviations from the global mean by 2100 can be up to 20 cm. The same regional sea-level projections were also used for a number of national assessments, such as Simpson et al. (2014) in Norway and Han et al. (2014, 2015) in Canada, where the global GIA model estimates were corrected or replaced by more accurate local GIA models or GPS measurements. Other regional assessments were done in Australia (CSIRO and Bureau of Meteorology 2015; Mcinnes et al. 2015) and the Netherlands (de

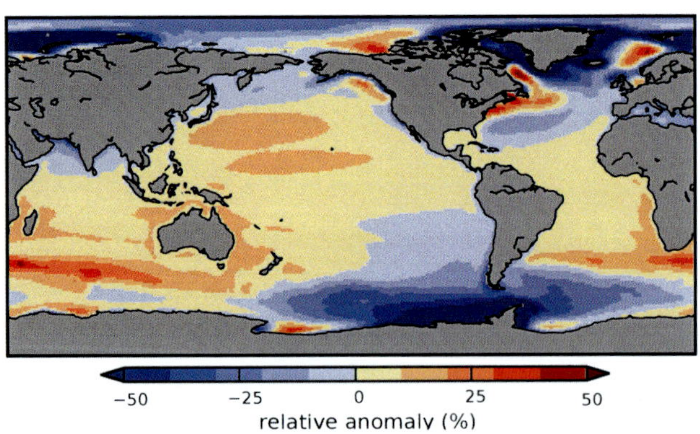

Fig. 7 Relative regional sea-level anomaly from the global mean change (between 1986–2005 and 2081–2100, %), based on the CMIP5-RCP4.5 scenario [from Slangen et al. (2014)]

Vries et al. 2014), which build on the IPCC-type regional sea-level projections. However, to really make a step forward in these national assessments, finer grid resolutions will be required to improve the model representation of ocean dynamic processes.

Using a probabilistic approach, Kopp et al. (2014) combined climate model information with an expert elicitation of the ice sheet contributions (Bamber and Aspinall 2013) to provide complete probability distributions of regional SLC projections. While the mean SLC is similar to IPCC AR5, Kopp et al. (2014) present high-end estimates which can be of particular interest and relevance for coastal management purposes. Following onto this, Little et al. (2015a) combined probability distributions with statistical models to estimate coastal flooding risk due to storm surges and SLC. They found that the risk of floods at the US East coast substantially increases as a result of SLC and changes in the frequency and intensity of tropical cyclones. However, these results were based on SLC from climate models only and do not include the SLC as a result of land-ice melt or TWS, which could lead to even larger flood risks.

To study the sources of uncertainty in sea level from climate models, Little et al. (2015b) decomposed the uncertainty into several components: model uncertainty, internal variability, scenario uncertainty, and a model–scenario interaction component. They found that, in the global mean, model uncertainty is the dominant term in the variance, whereas the variance due to scenario uncertainty increases in the twenty-first century and variance due to internal variability is initially large but decreases quickly. Locally, the contribution of each source of uncertainty can be very different, depending on the local magnitude of internal variability versus the response to external climate forcings. Both Hu and Deser (2013) and Bordbar et al. (2015) showed that internal variability in some locations can even be sufficiently large to be the main source of uncertainty all through the twenty-first century. As a result of the large internal variability, the time of emergence of SLC for DSL only (Lyu et al. 2014, their Fig. 2a) is beyond 2100 in over 80 % of the ocean area. The area with an emerging signal increases significantly (to almost 100 % by 2080) when thermal expansion, land ice, GIA, and GWD are included. For a further discussion of the literature on the effect of unforced variability on sea level and detection and attribution of SLC, see Han et al. and Marcos et al. in this issue, respectively.

The effect of freshwater input into the ocean as a result of land ice mass loss has been discussed in a number of studies, which have produced climate projections with integrated realistic estimates for glacier and ice sheet melt water run-off (Howard et al. 2014; Agarwal et al. 2015; Lenaerts et al. 2015). The first two studies focus on the impact of the freshwater forcing on DSL and find, using different models and different scenarios, that the impact is small (in the order of several cm) compared to the total SLC projected for the twenty-first century. However, both Howard et al. (2014) and Lenaerts et al. (2015) find that adding ice sheet freshwater forcings leads to a slight weakening of the Atlantic Meridional Overturning Circulation, indicating that it is important to include the freshwater forcing in climate models.

6 Mediterranean Sea-Level Projections

The Mediterranean is a semi-enclosed basin, linked to the open ocean through the Strait of Gibraltar. The high population density at the coast makes this basin particularly vulnerable to future SLC. Mediterranean sea level is influenced by various complex processes such as mass fluctuations (e.g., additional water input), variation in the density structure (steric effect), changes in circulation, waves, atmospheric pressure variations, and changes in the hydrographic conditions of incoming Atlantic water. These different components contribute to SLC at different timescales, from daily to interdecadal.

So far, global climate modelling attempts to assess future SLC in the Mediterranean did not deliver a consistent signal. Marcos and Tsimplis (2008) used projections from IPCC models to assess the interannual variation in steric sea level averaged for the Mediterranean, under the SRES A1B scenario, and found that global models did not agree on a trend. Indeed, their coarse resolution does not enable an accurate representation of important small-scale processes acting in the Mediterranean region, which are important to represent the water masses of the basin accurately. Additionally, AOGCMs have difficulties to simulate a reasonable water exchange at Gibraltar, which strongly influences the circulation and the changes in sea level in the Mediterranean Sea.

High-resolution regional climate modelling is thus needed to answer the question of ongoing Mediterranean SLC (Calafat et al. 2012). In addition to the thermosteric component, the contribution from changes in salinity has to be taken into account for the Mediterranean, since climate projections predict that the basin will become saltier in the future. Jordà and Gomis (2013) underlined that the saltening of the Mediterranean has two counteracting effects on sea level. Firstly, the halosteric effect leads to contraction of the water and thus a sea-level fall (-0.10 mm year^{-1} for 1960–2000). In contrast, the addition of salt to the basin in terms of mass leads to a sea-level rise ($+0.12$ mm year^{-1} for 1960–2000). As a simplification, these two contradicting effects can be neglected and Mediterranean mean SLC can be restricted to its thermosteric component.

Two recent studies have analysed Mediterranean SLC in future scenarios with regional models. Carillo et al. (2012) projected a thermosteric sea-level rise from 5 to 7 cm by 2050 (vs. 1951–2000) for the A1B scenario. With a six-member ensemble of scenario simulations, Adloff et al. (2015) found a larger sea-level rise of 10–20 cm in 2050 and 45–60 cm in 2099 (with respect to 1961–1990). In both studies, a large source of uncertainty is attributed to the hydrographic characteristics of the Atlantic boundary conditions prescribed in the Mediterranean model. Using the ensemble of Adloff et al. (2015), Fig. 8 shows the comparison of the spread of thermosteric sea-level response of the Mediterranean linked to (1) the choice of hydrographic conditions of Atlantic waters prescribed at the western boundary of the Mediterranean, and (2) the choice of the socio-economic

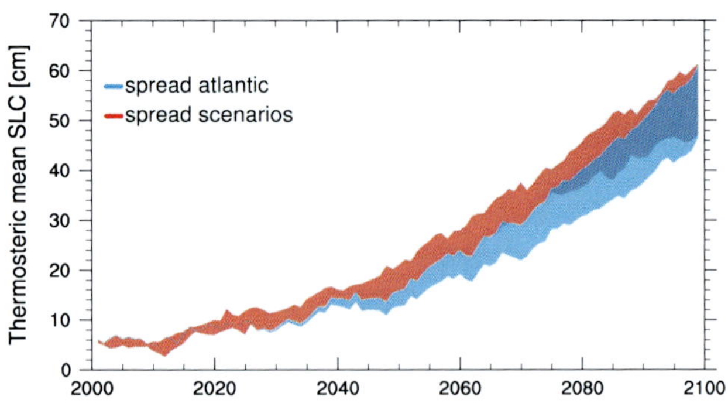

Fig. 8 Cumulative thermosteric sea-level change w.r.t. 1961–1990 (cm), averaged over the Mediterranean Sea from the six-member ensemble scenario simulations from Adloff et al. (2015). In *blue*, the uncertainties linked to the choice of the prescribed hydrographic conditions of Atlantic waters west of Gibraltar, and in *red*, the uncertainties linked to the choice of the socio-economic scenario

scenario. These results confirm how much the Mediterranean response in the future is driven by the Atlantic behaviour and raises the importance of the data set (mostly AOGCMs-derived) used to force the regional model at the boundary with the open ocean. Keeping in mind that the range of changes in near-Atlantic hydrography explored in the study by Adloff et al. (2015) is much smaller than the spread among CMIP models, it only gives a lower bound for the range of uncertainties in Mediterranean sea-level projections.

In comparison with the significant progress at the global scale, the advances at the Mediterranean scale remain small in terms of sea-level representation in regional ocean models. There is a significant lack of regional studies dealing with Mediterranean sea level, for hindcast periods as well as for projections, and none of them accounts for a proper Atlantic sea-level signal. The next step would be to include this missing feature and prescribe the complete sea-level signal at the Atlantic western boundary of Mediterranean regional models. This would allow the correct evolution of the Atlantic Ocean, which determines a part of the Mediterranean's behaviour, to be accounted for.

7 Synthesis

The field of sea-level research and all of its contributions is moving quickly, and a lot of work has been done since IPCC AR5. Here, we have reviewed the recent literature of projected sea-level contributions of ice sheets, glaciers, and terrestrial water storage to sea-level change. Furthermore, we discussed recent advances in global, regional, and Mediterranean sea-level projections. We did not discuss contributions that have seen little progress since IPCC AR5, most notably the thermal expansion and ocean dynamics components. However, these components are expected to be updated once the new model runs of the sixth phase of the Climate Model Intercomparison Project (CMIP6) become available.

The most recent sea-level projections for the Greenland Ice Sheet of 0.01–0.17 m by 2100 largely fall within the IPCC AR5 likely range for the twenty-first century. However, the contribution of surface melting is larger and the contribution of dynamic discharge is smaller than in IPCC AR5. Most projections for the Antarctic Ice Sheet since IPCC AR5 limit the sea-level contribution as a result of dynamic discharge and the potential onset of the marine ice sheet instability to 0.3 m by the end of this century. From the response to ocean warming, which is likely to dominate the dynamic Antarctic contribution, the 90 % uncertainty reaches up to 0.37 m by 2100 under the RCP8.5 scenario. However, Pollard and DeConto (2016) challenge this and project changes of well over 1 m by 2100 under the RCP8.5 scenario. All publications project that the bulk of SLC from Greenland and Antarctica will, however, occur after 2100 and might surpass several metres within the next centuries to millennia.

Glacier mass loss has been one of the main contributors to sea-level rise in the twentieth century and is expected to remain an important contributor in the next century. The latest findings, based on updates of glacier outlines used in existing projections and also new glacier models, project slightly lower contributions to sea-level rise from glaciers compared to IPCC AR5: from projections around $\sim$0.16 m in IPCC AR5 to the order of $\sim$0.12 m for the RCP8.5 scenario in more recent publications.

The sea-level contribution of changes in terrestrial water storage (TWS) has been difficult to estimate from observations in the past, but satellite observations now allow for better monitoring of changes in land water storage. Groundwater depletion is projected to increase due to growing water demand as a result of population growth and increasing evaporation. The projected contribution of TWS is 38.7 ± 12.9 mm for the period 2010–2100 (ensemble mean ± 1σ).

In projecting global mean SLC, the focus has turned towards providing better uncertainty estimates by using probabilistic methods and skewed uncertainty distributions. This may lead to better estimates of the low-probability/high-risk events in a changing climate. So far, these improved uncertainty distributions are based on expert elicitations, but as models evolve hopefully the uncertainty estimates will be based on modelling of physical processes in the near future.

Although significant advances have been made in recent years in projecting regional SLC, there are still a number of challenges that remain. The modelling and understanding of the ocean dynamical processes and incorporation of freshwater forcing as a result of ice sheet melt in climate models is an ongoing process. Ideally, the surface mass balance modelling of the ice sheets and glaciers would become part of the AOGCMs to obtain consistent results and include feedbacks between the ice sheets and glaciers with the rest of the climate system.

Ideally, sea-level change should be estimated on a national level, which is what coastal planners are interested in, but the spatial resolution of the current sea-level projections is still relatively coarse. To provide decision makers with better local estimates, models will need to use finer grid resolutions to account for local effects, such as coastal evolution and sediment transport. The increasing number of GPS measurements is also useful for local cases, as they give better estimates of vertical land motion, which can be large locally. In addition, new approaches now offer possibilities to link changes in flood risk and sea-level extremes to regional SLC.

Recent regional modelling studies in the Mediterranean have pointed out the relevance of the Atlantic signal, which largely contributes to the Mediterranean sea-level variability and represents one of the main sources of uncertainty in sea-level projections of the basin. Ongoing regional simulations are starting to tackle this issue and show that the prescription of sea-level information from the near-Atlantic at the lateral boundary significantly improves the Mediterranean sea-level representation at basin scale for hindcast periods. This will be added in future scenario simulations of the Mediterranean Sea.

Acknowledgments This paper is a result of the ISSI Workshop on Integrative Study of Sea Level, held in Bern, Switzerland, 2–6 February 2015. A.S. was supported by a CSIRO Australia Office of the Chief Executive Fellowship and the NWO-Netherlands Polar Programme. P.L. was funded by the European Research Council under the European Union's Seventh Framework Programme (FP/2007–2013)/ERC Grant Agreement No. 320816. Y.W. was supported by a Japan Society for the Promotion of Science (JSPS) Overseas Research Fellowship (Grant No. JSPS-2014-878). F.A. was supported by an AXA postdoctoral fellowship.

References

Adloff F et al (2015) Mediterranean Sea response to climate change in an ensemble of twenty first century scenarios. Clim Dyn. doi:10.1007/s00382-015-2507-3

Agarwal N, Jungclaus JH, Köhl A, Mechoso CR, Stammer D (2015) Additional contributions to CMIP5 regional sea level projections resulting from Greenland and Antarctic ice mass loss. Environ Res Lett 10(7):074008. doi:10.1088/1748-9326/10/7/074008

Bamber JL, Aspinall WP (2013) An expert judgement assessment of future sea level rise from the ice sheets. Nat Clim Chang 3(4):424–427. doi:10.1038/nclimate1778

Bilbao RAF, Gregory JM, Bouttes N (2015) Analysis of the regional pattern of sea level change due to ocean dynamics and density change for 1993–2099 in observations and CMIP5 AOGCM's. Clim Dyn 45(9):2647–2666. doi:10.1007/s00382-015-2499-z

Bordbar MH, Martin T, Latif M, Park W (2015) Effects of long-term variability on projections of twenty-first century dynamic sea level. Nat Clim Chang 5:343–347. doi:10.1038/nclimate2569

Bouttes N, Gregory JM (2014) Attribution of the spatial pattern of CO_2-forced sea level change to ocean surface flux changes. Environ Res Lett 9(3):034004. doi:10.1088/1748-9326/9/3/034004

Calafat FM, Jordà G, Marcos M, Gomis D (2012) Comparison of Mediterranean sea level variability as given by three baroclinic models. J Geophys Res 117(C2):C02009. doi:10.1029/2011JC007277

Carillo A, Sannino G, Artale V, Ruti PM, Calmanti S, Dell'Aquila A (2012) Steric sea level rise over the Mediterranean Sea: present climate and scenario simulations. Clim Dyn 39(9–10):2167–2184. doi:10.1007/s00382-012-1369-1

Carson M, Köhl A, Stammer D, Slangen ABA, Katsman CA, van de Wal RSW, Church JA, White NJ (2015) Coastal sea level changes, observed and projected during the 20th and 21st Century. Clim Change 134(1):269–281

Chao BF, Wu YH, Li YS (2008) Impact of artificial reservoir water impoundment on global sea level. Science 320(5873):212–214. doi:10.1126/science.1154580

Church JA, White NJ, Konikow LF, Domingues CM, Cogley JG, Rignot E, Gregory JM, van den Broeke MR, Monaghan AJ, Velicogna I (2011) Revisiting the Earth's sea-level and energy budgets from 1961 to 2008. Geophys Res Lett. doi:10.1029/2011GL048794

Church JA et al (2013) Sea level change. In: Stocker TF, Qin D, Plattner G-K, Tignor M, Allen SK, Boschung J, Nauels A, Xia Y, Bex V, Midgley PM (eds) Climate change 2013: the physical science basis, contribution of working group I to the fifth assessment report of the intergovernmental panel on climate change. Cambridge University Press, Cambridge, and New York

Clark PU, Church JA, Gregory JM, Payne AJ (2015) Recent progress in understanding and projecting regional and global mean sea level change. Curr Clim Chang Rep 1(4):224–246. doi:10.1007/s40641-015-0024-4

Cogley JG (2009) A more complete version of the World Glacier Inventory. Ann Glaciol 50:32–38

CSIRO and Bureau of Meteorology (2015) Climate change in Australia Information for Australia's Natural Resourche Management Regions: technical report

de Vries H, Katsman C, Drijfhout S (2014) Constructing scenarios of regional sea level change using global temperature pathways. Environ. Res. Lett. 9(11):115007. doi:10.1088/1748-9326/9/11/115007

Döll P, Müller Schmied H, Schuh C, Portmann FT, Eicker A (2014) Global-scale assessment of groundwater depletion and related groundwater abstractions: combining hydrological modeling with information from well observations and GRACE satellites. Water Resour Res 50(7):5698–5720. doi:10.1002/2014WR015595

Dyurgerov MB, Meier MF (2005) Glaciers and the changing earth system: a 2004 snapshot. Technical report, Institute of Arctic and Alpine Research, University of Colorado, Boulder, Occasional Paper No 58

Edwards TL et al (2014) Effect of uncertainty in surface mass balance–elevation feedback on projections of the future sea level contribution of the Greenland ice sheet. Cryosphere 8(1):195–208. doi:10.5194/tc-8-195-2014

Famiglietti JS (2014) The global groundwater crisis. Nat Clim Chang 4(11):945–948. doi:10.1038/nclimate2425

Famiglietti JS, Lo M, Ho SL, Bethune J, Anderson KJ, Syed TH, Swenson SC, de Linage CR, Rodell M (2011) Satellites measure recent rates of groundwater depletion in California's Central Valley. Geophys Res Lett, Geophys. doi:10.1029/2010GL046442

Farrell WE, Clark JA (1976) On Postglacial Sea Level. Geophys JR Astron Soc 46:647–667

Favier L, Durand G, Cornford SL, Gudmundsson GH, Gagliardini O, Gillet-Chaulet F, Zwinger T, Payne AJ, Le Brocq AM (2014) Retreat of Pine Island Glacier controlled by marine ice-sheet instability. Nat Clim Chang 4(2):117–121. doi:10.1038/nclimate2094

Frieler K, Clark PU, He F, Ligtenberg SRM, van den Broeke MR, Winkelmann R, Reese R, Levermann A (2015) Consistent evidence of increasing antarctic accumulation with warming. Nat Clim Chang 5:348–352

Fürst JJ, Goelzer H, Huybrechts P (2015) Ice-dynamic projections of the Greenland ice sheet in response to atmospheric and oceanic warming. Cryosphere 9(3):1039–1062. doi:10.5194/tc-9-1039-2015

Gardner AS et al (2013) A reconciled estimate of glacier contributions to sea level rise: 2003 to 2009. Science 340(6134):852–857. doi:10.1126/science.1234532

Giesen RH, Oerlemans J (2013) Climate-model induced differences in the 21st century global and regional glacier contributions to sea-level rise. Clim Dyn 41(11–12):3283–3300. doi:10.1007/s00382-013-1743-7

Gleeson T, Befus KM, Jasechko S, Luijendijk E, Cardenas MB (2015) The global volume and distribution of modern groundwater. Nat Geosci 9:161–167. doi:10.1038/ngeo2590

Golledge NR, Kowalewski DE, Naish TR, Levy RH, Fogwill CJ, Gasson EGW (2015) The multi-millennial Antarctic commitment to future sea-level rise. Nature 526(7573):421–425. doi:10.1038/nature15706

Gornitz V (1995) Sea-level rise: a review of recent past and near-future trends. Earth Surf Process Landforms 20(1):7–20. doi:10.1002/esp.3290200103

Gregory JM et al (2013) Twentieth-century global-mean sea level rise: Is the whole greater than the sum of the parts? J Clim 26(13):4476–4499. doi:10.1175/JCLI-D-12-00319.1

Grinsted A, Jevrejeva S, Riva REM, Dahl-Jensen D (2015) Sea level rise projections for northern Europe under RCP8.5. Clim Res 64:15–23. doi:10.3354/cr01309

Han G, Ma Z, Bao H, Slangen A (2014) Regional differences of relative sea level changes in the Northwest Atlantic: historical trends and future projections. J Geophys Res Ocean 119(1):156–164. doi:10.1002/2013JC009454

Han G, Ma Z, Chen N, Thomson R, Slangen A (2015) Changes in mean relative sea level around Canada in the twentieth and twenty-first centuries. Atmos Ocean. doi:10.1080/07055900.2015.1057100

Hirabayashi Y, Zang Y, Watanabe S (2013) Projection of glacier mass changes under a high-emission climate scenario using the global glacier model HYOGA2. Hydrol Res 7(1):6–11. doi:10.3178/HRL.7.6

Horton BP, Rahmstorf S, Engelhart SE, Kemp AC (2014) Expert assessment of sea-level rise by AD 2100 and AD 2300. Quat Sci Rev 84:1–6. doi:10.1016/j.quascirev.2013.11.002

Howard T, Ridley J, Pardaens AK, Hurkmans RTWL, Payne AJ, Giesen RH, Lowe JA, Bamber JL, Edwards TL, Oerlemans J (2014) The land-ice contribution to 21st-century dynamic sea level rise. Ocean Sci 10(3):485–500. doi:10.5194/os-10-485-2014

Hu A, Deser C (2013) Uncertainty in future regional sea level rise due to internal climate variability. Geophys Res Lett 40:2768–2772. doi:10.1002/grl.50531

Huss M, Farinotti D (2012) Distributed ice thickness and volume of all glaciers around the globe. J Geophys Res Earth Surf 117(F4):F04010. doi:10.1029/2012JF002523

Huss M, Hock R (2015) A new model for global glacier change and sea-level rise. Front Earth Sci 3(September):1–22. doi:10.3389/feart.2015.00054

Huss M, Jouvet G, Farinotti D, Bauder A (2010) Future high-mountain hydrology: a new parameterization of glacier retreat. Hydrol Earth Syst Sci 14(5):815–829. doi:10.5194/hessd-7-345-2010

Huybrechts P, Goelzer H, Janssens I, Driesschaert E, Fichefet T, Goosse H, Loutre M-F (2011) Response of the Greenland and Antarctic Ice Sheets to multi-millennial greenhouse warming in the earth system model of intermediate complexity LOVECLIM. Surv Geophys 32(4):397–416

Jevrejeva S, Grinsted A, Moore JC (2014) Upper limit for sea level projections by 2100. Environ Res Lett 9(10):104008. doi:10.1088/1748-9326/9/10/104008

Jordà G, Gomis D (2013) On the interpretation of the steric and mass components of sea level variability: the case of the Mediterranean basin. J Geophys Res Ocean 118(2):953–963. doi:10.1002/jgrc.20060

Joughin I, Smith BE, Medley B (2014) Marine ice sheet collapse potentially under way for the Thwaites Glacier Basin, West Antarctica. Science 344(6185):735–738

Konikow LF (2011) Contribution of global groundwater depletion since 1900 to sea-level rise. Geophys Res Lett. doi:10.1029/2011GL048604

Kopp RE, Horton RM, Little CM, Jerry X, Oppenheimer M, Rasmussen DJ, Strauss BH, Tebaldi C (2014) Probabilistic 21st and 22nd century sea-level projections at a global network of tide gauge sites. Earth's Future. doi:10.1002/2014EF000239

Lang C, Fettweis X, Erpicum M (2015) Future climate and surface mass balance of Svalbard glaciers in an RCP85 climate scenario: a study with the regional climate model MAR forced by MIROC5. Cryosphere 9(3):945–956. doi:10.5194/tc-9-945-2015

Lenaerts JTM, Le Bars D, van Kampenhout L, Vizcaino M, Enderlin EM, van den Broeke MR (2015) Representing Greenland ice sheet freshwater fluxes in climate models. Geophys Res Lett 42(15):6373–6381. doi:10.1002/2015GL064738

Lettenmaier DP, Milly PCD (2009) Land waters and sea level. Nat Geosci 2(7):452–454. doi:10.1038/ngeo567

Levermann A, Clark PU, Marzeion B, Milne GA, Pollard D, Radic V, Robinson A (2013) The multimillennial sea-level commitment of global warming. Proc Natl Acad Sci USA 110(34):13745–13750. doi:10.1073/pnas.1219414110

Levermann A et al (2014) Projecting Antarctic ice discharge using response functions from SeaRISE ice-sheet models. Earth Syst Dyn 5(2):271–293. doi:10.5194/esd-5-271-2014

Little CM, Horton RM, Kopp RE, Oppenheimer M, Vecchi GA, Villarini G (2015a) Joint projections of US East Coast sea level and storm surge. Nat Clim Chang 5:1114–1120. doi:10.1038/nclimate2801

Little CM, Horton RM, Kopp RE, Oppenheimer M, Yip S (2015b) Uncertainty in twenty-first-century CMIP5 sea level projections. J Clim 28(2):838–852. doi:10.1175/JCLI-D-14-00453.1
Longuevergne L, Scanlon BR, Wilson CR (2010) GRACE Hydrological estimates for small basins: evaluating processing approaches on the High Plains Aquifer, USA. Water Resour Res. doi:10.1029/2009WR008564
Lyu K, Zhang X, Church JA, Slangen ABA, Hu J (2014) Time of emergence for regional sea-level change. Nat Clim Chang 4:1006–1010. doi:10.1038/nclimate2397
Marcos M, Tsimplis MN (2008) Coastal sea level trends in Southern Europe. Geophys J Int 175:70–82
Marzeion B, Jarosch AH, Hofer M (2012) Past and future sea-level change from the surface mass balance of glaciers. Cryosphere 6:1295–1322. doi:10.5194/tc-6-1295-2012
McGranahan G, Balk D, Anderson B (2007) The rising tide: assessing the risks of climate change and human settlements in low elevation coastal zones. Environ Urban 19(1):17–37
Mcinnes KL, Church JA, Monselesan D, Hunter JR, O'Grady JG, Haigh ID, Zhang X (2015) Sea-level rise projections for Australia: information for impact and adaptation planning. Meteorol Oceanogr J, Aust
Mengel M, Levermann A, Frieler K, Robinson A, Marzeion B, Winkelmann R (2016) Future sea level rise constrained by observations and long-term commitment. Proc Natl Acad Sci. doi:10.1073/pnas.1500515113
Mercer JH (1978) West Antarctic ice sheet and CO2 greenhouse effect: a threat of disaster. Nature 271:321–325
Mitrovica JX, Tamisiea ME, Davis JL, Milne GA (2001) Recent mass balance of polar ice sheets inferred from patterns of global sea-level change. Nature 409:1026–1029
Moss RH et al (2010) The next generation of scenarios for climate change research and assessment. Nature 463(747–756):747–756. doi:10.1038/nature08823
Nowicki S et al (2013) Insights into spatial sensitivities of ice mass response to environmental change from the SeaRISE ice sheet modeling project I: Antarctica. J Geophys Res Earth Surf 118(2):1002–1024. doi:10.1002/jgrf.20081
Oerlemans J, Nick FM (2005) A minimal model of a tidewater glacier. Ann Glaciol 42:1–6
Peltier WR (2004) Global glacial isostasy and the surface of the Ice-Age Earth: the ICE-5G (VM2) model and GRACE. Annu Rev Earth Planet Sci 32:111–149
Pokhrel YN, Hanasaki N, Yeh PJ-F, Yamada TJ, Kanae S, Oki T (2012) Model estimates of sea-level change due to anthropogenic impacts on terrestrial water storage. Nat Geosci 5(6):389–392. doi:10.1038/ngeo1476
Pollard D, DeConto RM (2016) Contribution of Antarctica to past and future sea-level rise. Nature 531(7596):591–597. doi:10.1038/nature17145
Pollard D, DeConto RM, Alley RB (2015) Potential Antarctic Ice Sheet retreat driven by hydrofracturing and ice cliff failure. Earth Planet Sci Lett 412:112–121. doi:10.1016/j.epsl.2014.12.035
Radić V, Hock R (2011) Regionally differentiated contribution of mountain glaciers and ice caps to future sea-level rise. Nat Geosci. doi:10.1038/NGEO1052
Radić V, Bliss A, Cody AB, Hock R, Miles E, Cogley JG (2014) Regional and global projections of twenty-first century glacier mass changes in response to climate scenarios from global climate models. Clim Dyn 42:37–58. doi:10.1007/s00382-013-1719-7
Rhein M et al (2013) Observations: ocean. In: Stocker TF, Qin D, Plattner G-K, Tignor M, Allen SK, Boschung J, Nauels A, Xia Y, Bex V, Midgley PM (eds) Climate change 2013: the physical science basis. Contribution of Working Group I to the fifth assessment report of the intergovernmental panel on climate change. Cambridge University Press, Cambridge and New York
Ridley J, Gregory JM, Huybrechts P, Lowe JA (2010) Thresholds for irreversible decline of the Greenland ice sheet. Clim Dyn 35:1049–1057. doi:10.1007/s00382-009-0646-0
Rignot E, Velicogna I, van den Broeke MR, Monaghan A, Lenaerts J (2011) Acceleration of the contribution of the Greenland and Antarctic ice sheets to sea level rise. Geophys Res Lett. doi:10.1029/2011GL046583
Rignot E, Mouginot J, Morlighem M, Seroussi H, Scheuchl B (2014) Widespread, rapid grounding line retreat of Pine Island, Thwaites, Smith, and Kohler glaciers, West Antarctica, from 1992 to 2011. Geophys Res Lett 41(10):3502–3509. doi:10.1002/2014GL060140
Ritz C, Edwards TL, Durand G, Payne AJ, Peyaud V, Hindmarsh RCA (2015) Potential sea-level rise from Antarctic ice-sheet instability constrained by observations. Nature. doi:10.1038/nature16147
Robinson A, Calov R, Ganopolski A (2012) Multistability and critical thresholds of the Greenland ice sheet. Nat. Clim. Chang. 2(6):429–432. doi:10.1038/nclimate1449
Rodell M, Velicogna I, Famiglietti JS (2009) Satellite-based estimates of groundwater depletion in India. Nature 460(7258):999–1002. doi:10.1038/nature08238

Shannon SR, Payne AJ, Bartholomew ID, Van Den Broeke MR, Edwards TL, Sole AJ, Van De Wal RSW, Zwinger T (2013) Enhanced basal lubrication and the contribution of the Greenland ice sheet to future sea-level rise. doi:10.1073/pnas.1212647110/-/DCSupplemental.www.pnas.org/cgi/doi/10.1073/pnas.1212647110

Shepherd A et al (2012) A reconciled estimate of ice-sheet mass balance. Science 338(6111):1183–1189. doi:10.1126/science.1228102

Simpson MJR, Breili K, Kierulf HP (2014) Estimates of twenty-first century sea-level changes for Norway. Clim Dyn 42(5–6):1405–1424. doi:10.1007/s00382-013-1900-z

Slangen ABA, van de Wal RSW (2011) An assessment of uncertainties in using volume-area modelling for computing the twenty-first century glacier contribution to sea-level change. Cryosphere 5:673–686. doi:10.5194/tc-5-673-2011

Slangen ABA, Katsman CA, van de Wal RSW, Vermeersen LLA, Riva REM (2012) Towards regional projections of twenty-first century sea-level change based on IPCC SRES scenarios. Clim Dyn 38(5–6):1191–1209. doi:10.1007/s00382-011-1057-6

Slangen ABA, Carson M, Katsman CA, van de Wal RSW, Koehl A, Vermeersen LLA, Stammer D, Van De Wal RSW (2014) Projecting twenty-first century regional sea-level changes. Clim Change 124:317–332. doi:10.1007/s10584-014-1080-9

Slangen ABA, Church JA, Zhang X, Monselesan D (2015) The sea-level response to external forcings in CMIP5 climate models. J Clim 28(21):8521–8539. doi:10.1175/JCLI-D-15-0376.1

Taylor KE, Stouffer RJ, Meehl GA (2012a) An overview of CMIP5 and the experiment design. Bull Am Meteorol Soc 93(4):485–498. doi:10.1175/BAMS-D-11-00094.1

Taylor RG et al (2012b) Ground water and climate change. Nat Clim Change 3(4):322–329. doi:10.1038/nclimate1744

Vaughan DG et al (2013) Observations: cryosphere. In: Climate Change 2013: the physical science basis. contribution of working Group I to the fifth assessment report of the intergovernmental panel on Climate Change, pp 317–382. doi:10.1017/CBO9781107415324.012

Vizcaino M, Mikolajewicz U, Jungclaus J, Schurgers G (2010) Climate modification by future ice sheet changes and consequences for ice sheet mass balance. Clim Dyn 34:301–324

Wada Y, Bierkens MFP (2014) Sustainability of global water use: past reconstruction and future projections. Environ Res Lett 9(10):104003. doi:10.1088/1748-9326/9/10/104003

Wada Y, van Beek LPH, Weiland FCS, Chao BF, Wu Y-H, Bierkens MFP (2012) Past and future contribution of global groundwater depletion to sea-level rise. Geophys Res Lett. doi:10.1029/2012GL051230

Wada Y, Lo M-H, Yeh PJ-F, Reager JT, Famiglietti JS, Wu R-J, Tseng Y-H (2016) Fate of water pumped from underground and contributions to sea-level rise Nat. Clim Change. doi:10.1038/NCLIMATE3001

Weertman J (1974) Stability of the junction of an ice sheet and an ice shelf. J Glaciol 13:3–11

Winkelmann R, Levermann A, Martin MA, Frieler K (2012) Increased future ice discharge from Antarctica owing to higher snowfall. Nature 492(7428):239–242. doi:10.1038/nature11616

Winkelmann R, Levermann A, Ridgwell A, Caldeira K (2015) Combustion of available fossil fuel resources sufficient to eliminate the Antarctic Ice Sheet. Science 1(September):1–5

Wisser D, Frolking S, Hagen S, Bierkens MFP (2013) Beyond peak reservoir storage? A global estimate of declining water storage capacity in large reservoirs. Water Resour Res 49(9):5732–5739. doi:10.1002/wrcr.20452

Wong PP, Losada IJ, Gattuso J-P, Hinkel J, Khattabi A, McInnes K, Saito Y, Sallenger A (2013) Chapter 5: coastal systems and low-lying areas. In: IPCC WGII AR5, pp 1–85

Yin J (2012) Century to multi-century sea level rise projections from CMIP5 models. Geophys Res Lett. doi:10.1029/2012GL052947

Yin J, Griffies SM, Stouffer RJ (2010) Spatial variability of sea-level rise in 21st century projections. J Clim 23:4585–4607. doi:10.1175/2010JCLI3533.1

Zuo Z, Oerlemans J (1997) Contribution of glacier melt to sea-level rise since AD 1865: a regionally differentiated calculation. Clim Dyn 13:835–845